Lecture Notes in Computer Science 13968

Founding Editors

Gerhard Goos
Juris Hartmanis

The series Lecture Notes in Computer Science (LNCS), including its subseries Lecture Notes in Artificial Intelligence (LNAI) and Lecture Notes in Bioinformatics (LNBI), has established itself as a medium for the publication of new developments in computer science and information technology research, teaching, and education.

LNCS enjoys close cooperation with the computer science R & D community, the series counts many renowned academics among its volume editors and paper authors, and collaborates with prestigious societies. Its mission is to serve this international community by providing an invaluable service, mainly focused on the publication of conference and workshop proceedings and postproceedings. LNCS commenced publication in 1973.

Ying Tan · Yuhui Shi · Wenjian Luo
Editors

Advances in Swarm Intelligence

14th International Conference, ICSI 2023
Shenzhen, China, July 14–18, 2023
Proceedings, Part I

 Springer

Editors
Ying Tan 🆔
Peking University
Beijing, China

Yuhui Shi
Southern University of Science
and Technology
Shenzhen, China

Wenjian Luo
Harbin Institute of Technology
Shenzhen, China

ISSN 0302-9743   ISSN 1611-3349 (electronic)
Lecture Notes in Computer Science
ISBN 978-3-031-36621-5   ISBN 978-3-031-36622-2 (eBook)
https://doi.org/10.1007/978-3-031-36622-2

This Springer imprint is published by the registered company Springer Nature Switzerland AG
The registered company address is: Gewerbestrasse 11, 6330 Cham, Switzerland

Preface

This book and its companion volume, LNCS vols. 13968 and 13969, constitute the proceedings of the Fourteenth International Conference on Swarm Intelligence (ICSI 2023) held during July 14–18, 2023 in Shenzhen, China, both onsite and online.

The theme of ICSI 2023 was "Serving Life with Swarm Intelligence." ICSI 2023 provided an excellent opportunity for academics and practitioners to present and discuss the latest scientific results and methods, innovative ideas, and advances in theories, technologies, and applications in swarm intelligence. The technical program covered a number of aspects of swarm intelligence and its related areas. ICSI 2023 was the fourteenth international gathering for academics and researchers working on most aspects of swarm intelligence, following successful events in Xi'an (ICSI 2022) virtually, Qingdao (ICSI 2021), Serbia (ICSI 2020) virtually, Chiang Mai (ICSI 2019), Shanghai (ICSI 2018), Fukuoka (ICSI 2017), Bali (ICSI 2016), Beijing (ICSI-CCI 2015), Hefei (ICSI 2014), Harbin (ICSI 2013), Shenzhen (ICSI 2012), Chongqing (ICSI 2011), and Beijing (ICSI 2010), which provided a high-level academic forum for participants to disseminate their new research findings and discuss emerging areas of research. The conference also created a stimulating environment for participants to interact and exchange information on future challenges and opportunities in the field of swarm intelligence research.

Due to the continuing global COVID-19 pandemic, ICSI 2023 provided both online and offline presentations. On one hand, ICSI 2023 was normally held in Shenzhen, China. On the other hand, the ICSI 2023 technical team provided the ability for the authors of accepted papers who had restrictions on overseas travel to present their work through an interactive online platform or video replay. The presentations by accepted authors were made available to all registered attendees onsite and online.

The host city of ICSI 2023 was Shenzhen, China, which, located on the southern coast of China, adjacent to Hong Kong, is an international modern city in Guangdong Province. It has developed rapidly due to Reform and Opening-up. It is currently the third largest city in mainland China in terms of economic aggregate. Shenzhen is an important high-tech R&D and manufacturing base in southern China, and is often referred to as the "Silicon Valley of China". It is the most suitable dynamic city for talents at home and abroad to expand their business.

ICSI 2023 received a total of 170 submissions and invited submissions from about 452 authors in 15 countries and regions (Armenia, Brazil, Canada, Chile, China, France, India, Japan, Pakistan, Russia, South Africa, Taiwan(China), Thailand, UK, and USA) across 5 continents (Asia, Europe, North America, South America, and Africa). Each submission was reviewed by at least 2 reviewers, and on average 2.8 reviewers. Based on rigorous reviews by the program committee members and reviewers, 81 high-quality papers were selected for publication in this proceedings volume with an acceptance rate of 47.65%. The papers are organized into 12 cohesive sections covering major topics of swarm intelligence research and its development and applications along with a competition session entitled "Competition on Single-Objective Bounded Optimization Problems (ICSI-OC'2023)".

On behalf of the Organizing Committee of ICSI 2023, we would like to express our sincere thanks to the International Association of Swarm and Evolutionary Intelligence (IASEI) (iasei.org), which is the premier international scholarly society devoted to advancing the theories, algorithms, real-world applications, and developments of swarm intelligence and evolutionary intelligence. We would also like to thank Harbin Institute of Technology Shenzhen, Peking University, and Southern University of Science and Technology for their co-sponsorships, and Computational Intelligence Laboratory of Peking University and IEEE Beijing Chapter for their technical co-sponsorships, Nanjing Kanbo iHealth Academy for its technical and financial co-sponsorship, as well as to our supporters of International Neural Network Society, World Federation on SoftComputing, MDPI's journal Entropy Beijing Xinghui Hi-Tech Co., and Springer Nature.

We would also like to thank the members of the Advisory Committee for their guidance, the members of the International Program Committee and additional reviewers for reviewing the papers, and the members of the Publication Committee for checking the accepted papers in a short period of time. We are particularly grateful to the proceedings publisher Springer for publishing the proceedings in the prestigious series of Lecture Notes in Computer Science. Moreover, we wish to express our heartfelt appreciation to the plenary speakers, session chairs, and student helpers. In addition, there are still many more colleagues, associates, friends, and supporters who helped us in immeasurable ways; we express our sincere gratitude to them all. Last but not least, we would like to thank all the speakers, authors, and participants for their great contributions that made ICSI 2023 successful and all the hard work worthwhile.

May 2023

Ying Tan
Yuhui Shi
Wenjian Luo

Organization

Honorary Co-chairs

Russell C. Eberhart — IUPUI, USA
Yan Jia — Harbin University of Technology, Shenzhen, China

General Co-chairs

Ying Tan — Peking University, China
Wenjian Luo — Harbin University of Technology, Shenzhen, China

Programme Committee Chair

Yuhui Shi — Southern University of Science and Technology, China

Advisory Committee Chairs

Gary G. Yen — Oklahoma State University, USA
Xin Yao — Southern University of Science and Technology of China, China
Yaochu Jin — Bielefeld University, Germany
Xuan Wang — Harbin University of Technology, Shenzhen, China

Technical Committee Co-chairs

Kay Chen Tan — Hong Kong Polytechnic University, China
Qingfu Zhang — City University of Hong Kong, China
Haibo He — University of Rhode Island Kingston, USA
Martin Middendorf — University of Leipzig, Germany
Xiaodong Li — RMIT University, Australia
Hideyuki Takagi — Kyushu University, Japan
Ponnuthurai Nagaratnam Suganthan — Nanyang Technological University, Singapore
Mengjie Zhang — Victoria University of Wellington, New Zealand

Nikola Kasabov	Aukland University of Technology, New Zealand
Jinliang Ding	Northeast University, China
Shengxiang Yang	De Montfort University, UK
Yew-Soon Ong	Nanyang Technological University, Singapore
Andreas Engelbrecht	University of Pretoria, South Africa
Yun Li	Shenzhen Higher Academy, Electronic Technology University, Shenzhen, China
Honggui Han	Beijing University of Technology, China
Ling Wang	Tsinghua University, China
Haibin Duan	Beihang University, China

Plenary Session Co-chairs

Zhexuan Zhu	Shenzhen University, China
Andres Iglesias	University of Cantabria, Spain
Chaoming Luo	University of Mississippi, USA

Invited Session Co-chairs

Zhihui Zhan	South China University of Technology, China
Zhun Fan	Shantou University, China
Maoguo Gong	Xidian University, China

Special Sessions Co-chairs

Ben Niu	Shenzhen University, China
Yan Pei	University of Aizu, Japan
Ming Jiang	Xiamen University, China

Tutorial Co-chairs

Jiahai Wang	Sun Yatsan University, China
Junqi Zhang	Tongji University, China
Han Huang	South China University of Technology, China

Publications Co-chairs

Swagatam Das	Indian Statistical Institute, India
Radu-Emil Precup	Politehnica University of Timisoara, Romania

Publicity Co-chairs

Fernando Buarque	Universidade of Pernambuco, Brazil
Eugene Semenkin	Siberian Aerospace University, Russia
Jing Liu	Xidian University, Guangzhou Institute of Technology, China
Hongwei Mo	Harbin Engineering University, China
Liangjun Ke	Xi'an Jiaotong University, China
Shenli Xie	South China University of Technology, China
Qiuzhen	Shenzhen University, China
Mario F. Pavone	University of Catania, Italy

Finance and Registration Chairs

Andreas Janecek	University of Vienna, Austria
Suicheng Gu	Google Corporation, USA

Local Arrangement Chairs

Qing Liao	Harbin University of Technology, Shenzhen, China
Shuhan Qi	Harbin University of Technology, Shenzhen, China
Zhaoguo Wang	Harbin University of Technology, Shenzhen, China

Conference Secretariat

Yifan Liu	Peking University, China

Program Committee

Ashik Ahmed	Islamic University of Technology, Bangladesh
Shakhnaz Akhmedova	Robert Koch Institute, Germany
Abdelmalek Amine	Tahar Moulay University of Saida, Algeria
Sz Apotecas	.
Sabri Arik	Istanbul University, Turkey
Nebojsa Bacanin	Singidunum University, Serbia

Carmelo J. A. Bastos Filho	University of Pernambuco, Brazil
Heder Bernardino	Universidade Federal de Juiz de fora, Brazil
Sandeep Bhongade	Shri G.S. Institute of Technology and Science, India
Josu Ceberio	University of the Basque Country, Spain
Gang Chen	Victoria University of Wellington, New Zealand
Junfeng Chen	Hohai University, China
Yiqiang Chen	Institute of Computing Technology, Chinese Academy of Sciences, China
Long Cheng	Institute of Automation, China
Shi Cheng	Shaanxi Normal University, China
Prithviraj Dasgupta	U. S. Naval Research Laboratory, USA
Rituparna Datta	Korea Advanced Institute of Science and Technology, South Korea
Kusum Deep	IIT Roorkee, India
Mingcong Deng	Tokyo University of Agriculture and Technology, Japan
Yu Dengxiu	University of Macau, China
Khaldoon Dhou	Texas A&M University Central Texas, USA
Bei Dong	Shaanxi Normal University, China
Haibin Duan	Beijing University of Aeronautics and Astronautics, China
Xuting Duan	Beihang University, China
Kyle Erwin	University of Pretoria, South Africa
Qinqin Fan	Shanghai Maritime University, China
Wei Fang	Jiangnan University, China
Philippe Fournier-Viger	Shenzhen University, China
Hongyuan Gao	Harbin Engineering University, China
Shangce Gao	University of Toyama, Japan
Maoguo Gong	.
Jia Guo	Hubei University of Economics, China
Ping Guo	Beijing Normal University
Weian Guo	.
Zhigao Guo	Queen Mary University of London, UK
Guosheng Hao	Jiangsu Normal University, China
Ilya Hodashinsky	TUSUR, Russia
Wenjing Hong	Southern University of Science and Technology, China
Ziyu Jia	Beijing Jiaotong University, China
Changan Jiang	Osaka Institute of Technology, Japan
Mingyan Jiang	Shandong University, China
Colin Johnson	University of Nottingham, UK

Liangjun Ke	Xi'an Jiaotong Univ, China
Waqas Haider Khan	Kohsar University Murree, Pakistan
Feifei Kou	.
Jakub Kudela	Brno University of Technology, Czech Republic
Germano Lambert-Torres	PS Solutions, Brazil
Tingjun Lei	Mississippi State University, USA
Xj Lei	Northwest A&F University, China
Bin Li	University of Science and Technology of China, China
Lixiang Li	Beijing University of Posts and Telecommunications, China
Xiaobo Li	Lishui University, China
Xuelong Li	Chinese Academy of Sciences, China
Ya Li	Southwest University, China
Jing Liang	Zhengzhou University, China
Peng Lin	Capital University of Economics and Business, China
Xin Lin	Nanjing University of Information Science and Technology, China
Jianhua Liu	Fujian University of Technology, China
Jing Liu	University of New South Wales, Australia
Ju Liu	Shandong University, China
Qunfeng Liu	Dongguan University of Technology, China
Wenlian Lu	Fudan University, China
Chaomin Luo	Mississippi State University, USA
Dingsheng Luo	Peking University, China
Wenjian Luo	Harbin Institute of Technology, China
Lianbo Ma	Northeastern University, China
Chengying Mao	School Jiangxi University of Finance and Economics, China
Michalis Mavrovouniotis	University of Cyprus, Cyprus
Yi Mei	Victoria University of Wellington, New Zealand
Carsten Mueller	Baden-Wuerttemberg Cooperative State University, Germany
Sreeja N. K.	PSG College of Technology, India
Qingjian Ni	Southeast University, China
Ben Niu	Shenzhen University, China
Lie Meng Pang	Southern University of Science and Technology, China
Endre Pap	Singidunum University, Serbia
Mario Pavone	University of Catania, Italy
Yan Pei	University of Aizu, Japan
Thomas Potok	ORNL, USA

Mukesh Prasad	University of Technology Sydney, USA
Radu-Emil Precup	Politehnica University Timisoara, Romania
Robert Reynolds	Wayne State University, USA
Yuji Sato	Hosei University, Japan
Carlos Segura	Centro de Investigación en Matemáticas, A.C. (CIMAT), Mexico
Timothy Sellers	Mississippi State University, USA
Kevin Seppi	Brigham Young University, USA
Ke Shang	Southern University of Science and Technology, China
Zhongzhi Shi	Institute of Computing Technology Chinese Academy of Sciences, China
Suwin Sleesongsom	KMITL, Thailand
Joao Soares	GECAD/ISEP, Portugal
Wei Song	North China University of Technology, China
Jianyong Sun	University of Essex, UK
Meng Sun	Peking University, China
Yifei Sun	Shaanxi Normal University, China
Ying Tan	Peking University, China
Qirong Tang	Tongji University, China
Anastasiia Timofeeva	Novosibirsk State Technical University, Russia
Eva Tuba	University of Belgrade, Serbia
Mladen Veinović	Singidunum University, Serbia
Gai-Ge Wang	Ocean University of China, China
Guoyin Wang	Chongqing University of Posts and Telecommunications, China
Hong Wang	Shenzhen University, China
Zhen Wang	Northwestern Polytechnical University, China
Wenjie Yi	University of Nottingham, UK
Ka-Chun Wong	City University of Hong Kong, China
Ning Xiong	Mälardalen University, Sweden
Benlian Xu	.
Rui Xu	Hohai University, China
Yu Xue	Nanjing University of Information Science & Technology, China
Wu Yali	Xi'an University of Technology, China
Yingjie Yang	De Montfort University, UK
Guo Yi-Nan	China University of Mining and Technology, China
Peng-Yeng Yin	National Chi Nan University, Taiwan
Jun Yu	Niigata University, Japan
Ling Yu	Jinan University, China

Zhengfei Yu	National University of Defense Technology, China
Zhi-Hui Zhan	South China University of Technology, China
Fangfang Zhang	Victoria University of Wellington, New Zealand
Jie Zhang	Newcastle University, UK
Jiwei Zhang	Beijing University of Posts and Telecommunications, China
Junqi Zhang	Tongji University, China
Lifeng Zhang	Renmin University of China, China
Tao Zhang	Tianjin University, China
Xiangyin Zhang	Beijing University of Technology, China
Xiaoming Zhang	Anhui Agriculture University, Hefei
Xingyi Zhang	.
Zili Zhang	Deakin University, Australia
Xinchao Zhao	Beijing University of Posts and Telecommunications, China
Zheng PKU	.
Yujun Zheng	Zhejiang University of Technology, China
Mengchu Zhou	New Jersey Institute of Technology, USA
Yun Zhou	National University of Defense Technology, China
Tao Zhu	University of South China, China
Mingyin Zou	National University of Defense Technology, China

Additional Reviewers

Cai, Gaocheng
Chen, Wei
Cheng, Dongdong
Elbakri, Idris
Jia, Zhan Xiao
Jiang, Yi
Khan, Muhammad Saqib Nawaz

Li, Xinxin
Liu, Jianhua
Qiu, Haiyun
Wang, Yixin
Yang, Ming
Zivkovic, Miodrag

Contents – Part I

Particle Swarm Optimization Algorithms

Genetic Algorithms

Optimization Computing Algorithms

Neural Network Search and Large-Scale Optimization

Multi-objective Optimization

Contents – Part II

Machine Learning

Data Mining

Routing and Scheduling Problems

Stock Prediction and Portfolio Optimization

ICSI-Optimization Competition

Swarm Intelligence Computing

Swarm Intelligence Algorithms and Applications: An Experimental Survey

Anasse Bari[✉], Robin Zhao, Jahnavi Swetha Pothineni, and Deepti Saravanan

Computer Science Department, Courant Institute of Mathematical Sciences, New York University, New York, NY 10012, USA
{abari,bz1037,jp5867,ds6812}@nyu.edu

Abstract. Swarm Intelligence draws inspiration from the collective intelligent behavior of animals such as birds, fish, and bees. It refers to the collective behavior of decentralized, self-organized systems composed of many simple agents that interact with each other and their environment. As the Swarm Intelligence (SI) field expands, more SI meta-heuristics are incorporated into developing new Artificial Intelligence (AI) tools. This study presents a comprehensive survey of state-of-the-art Swarm Intelligence algorithms and their applications. Seven SI algorithms are described – Artificial Bee Colony Optimization, Bat-inspired algorithm, Flock by Leader algorithm, Grey Wolf Optimizer, Flower Pollination Algorithm, Whale Optimization, and Moth Flame Optimization, with their applications in three diverse fields: Scheduling Problem, Data Science and Robotics, and Route and Layout Optimization. The performance of the algorithms is compared experimentally, and discussions about future directions in Swarm Intelligence research are outlined.

Keywords: Swarm Intelligence · Artificial Intelligence · Artificial bee colony optimization · Whale Optimization · Moth Flame Optimization · Flower Pollination Algorithm · Bat Algorithm · Grey Wolf Optimization · Bird Flocking · Flock by Leader · Swarm Intelligence Applications

1 Introduction

When it comes to inspiration for solutions to computer science problems, nature is the perfect place to look. Natural phenomena, from bird flocks searching for food to interactions in an ecosystem, demonstrate that nature can always find optimal, yet simple, solutions to extremely complex problems on a global scale. This analogy is relevant to the computer science ideology, which seeks solutions that offer fast execution and simple implementation. As a result of this relationship, the field of Swarm Intelligence emerged, and its bio-inspired algorithms are a heuristics approach that models nature's behaviors for solving problems like optimization.

For the past decade, much research has concentrated on Swarm Intelligence. "Nature-inspired Algorithms for Optimization" by Yang [1] provides a thorough overview of the state-of-the-art optimization techniques that have been inspired by natural processes and

© The Author(s), under exclusive license to Springer Nature Switzerland AG 2023
Y. Tan et al. (Eds.): ICSI 2023, LNCS 13968, pp. 3–17, 2023.
https://doi.org/10.1007/978-3-031-36622-2_1

phenomena. A Survey of Bio-inspired Optimization Algorithms by Binitha and Sathya [2] presents a comprehensive survey of various bio-inspired optimization algorithms, including their concepts, underlying principles, and applications in different domains. "A Review of Applications of Bio-Inspired Algorithms in Engineering" by Kar (2016) presents a review of bio-inspired computing algorithms and their applications in various domains [3].

The bio-inspired algorithms covered in this study are organized by the chronological order in which they were proposed. The algorithms covered are Artificial Bee Colony Optimization, Bat-inspired algorithm, Flock by Leader, Grey Wolf Optimizer, Flower Pollination Algorithm, Whale Optimization, and Moth Flame Optimization.

The organization of the paper is as follows: Sect. 2 explains, in detail, the above-mentioned Swarm Intelligence algorithms. Section 3 provides an experimental comparison. Section 4 discusses the algorithms' applications in diverse fields. Lastly, the discussion and conclusion explore the evolution and impact these algorithms have on solving many complex optimization problems in the future.

2   Swarm Intelligence Algorithms

2.1   Artificial Bee Colony Optimization (ABC)

Proposed by Karaboga et al. [4], artificial bee colony optimization is a swarm-based metaheuristic algorithm, inspired by the foraging behavior of honey bees. The objective of the problem is to find such rich food sources. Following the initialization phase, the algorithm iterates over the Employed Bees Phase, Onlooker Bees Phase, and Scout Bees Phase until the maximum number of iterations or maximum CPU time is achieved.

Initialization Phase: The algorithm generates a randomly distributed initial population of an 'x' number of solutions. Each solution x_i is a D-dimensional vector where D is the number of optimization parameters.

Employed Bees Phase: Now the iteration begins. Initially, the bees randomly select food sources and share their nectar information with the hive at the dance area. The employed bees go to the food areas from their personal experience and then choose new food sources via visual information in their neighborhood using Eq. (1).

$$v_{ij} = x_{ij} + \phi_{ij}(x_{ij} - x_{kj}) \tag{1}$$

where k ∈ {1, 2,..., n}, is chosen randomly with n = number of employed bees, and j ∈ {1, 2,..., D} randomly chosen indices, $\phi_{ij} \in [-1,1]$ and is a random number. As the search approaches an optimal solution, the step length adaptively reduces. The employed bees report their new nectar information to the onlookers at the dance area.

Onlooker Bees Phase: The probability of the food source being chosen by the onlookers is directly proportional to the nectar amount of the food source. The process repeats where the onlookers find new food sources based on personal and visual information in their neighborhood. In every iteration, the number of employed and onlooker bees are equal, which is equal to the number of solutions in the population.

Scout Bees Phase: When the nectar of a food source is abandoned using greedy selection, a scout bee chooses a new food source at random to replace the poor solution. In every iteration, at most one scout bee goes outside searching for a new food source.

The performance of ABC was compared against Particle Swarm Optimisation (PSO) and Differential Evolution (DE) in the original paper for 13 test functions over 30 independent runs. ABC achieved competitive results, finding a global minimum for seven test functions and close to global optima for five test functions. It was also found to be relatively insensitive to parameter change and was able to handle a wide range of constrained optimization problems.

2.2 Bat-Inspired Algorithm

Bat Algorithm is another metaheuristic algorithm introduced by Xin-She Yang based on the echolocation of microbats [5]. Naturally, a microbat emits loud sound pulses and listens to their echoes from surrounding objects. These pulses are sent at an increased rate when closer to the prey, but the sound gets quieter when they get very close.

In the Bat Algorithm, all bats b_i are initialized randomly at random positions x_i, velocity v_i and frequency f_i with a fixed frequency range $[f_{min}, f_{max}]$ and varying emission rate $r_i \in (0,1]$ and loudness A_i. Evaluate the fitness F_i of each bat b_i as $x*$ with fitness $F*$.

After every iteration t, the new positions, and velocities of the bats are given by Eqs. (2), (3) and (4).

$$f_i = f_{min} + (f_{max} - f_{min})\beta \tag{2}$$

$$v_i^{t+1} = v_i^t + (x_i^t - x*)f_i \tag{3}$$

$$x_i^{t+1} = x_i^t + v_i^{t+1} \tag{4}$$

where $\beta \in [0, 1]$ is a random vector drawn from uniform distribution and $x*$ is the current best optimal position. The fitness of the new solutions is evaluated and the best optimal solution $x*$ is updated based on the new fitness values.

Once the best solution is selected, the x_{old} of all the bats is updated using Eq. (5).

$$x_{new} = x_{old} + \epsilon A^{(t)} \tag{5}$$

where ϵ is $[-1, 1]$ and $A^{(t)}$ is the average loudness of all the bats at this iteration. In addition to updating the positions and velocities, BA uses frequency tuning and parameter control to strike a balance between exploration and exploitation.

The loudness A_i and pulse rate r_i of each bat is regulated after every iteration using Eqs. (6) and (7).

$$A_i^{t+1} = \alpha A_i^t \tag{6}$$

$$r_i^{t+1} = r_i^0[1 - exp(-\gamma t)], \tag{7}$$

where $0 < \alpha < 1$ and $\gamma > 0$ are constants. This is continued until the maximum number of iterations or desired level of fitness is reached. The optimal solution is then returned.

Bat Algorithm was tested on benchmark functions such as Rosenbrock's Bat Algorithm in the original paper and produced competitive results with that of PSO and Genetic Algorithm in terms of convergence speed and quality of solutions. The Bat Algorithm was also found to be relatively insensitive to parameter change.

2.3 Flock by Leader: Bird Flocking Algorithm

In 2012, Bari et al. further elevated the Flock Clustering model [6] by introducing flock leaders into the algorithm [7]. Inspired by pigeon flocks, Flock by Leader (FBL) algorithm minimized the moves of agents for data clustering, and is parameter free, while the original model required predefined maximum neighboring distance and number of iterations. The components of the Swarm Clustering Framework are Swarm Metric Space, Swarm Virtual Space, Agents Position Graph, and Feature Similarity Graph.

A Swarm Metric Space M_S is defined as (X, ρ), with $X \in R^d$, $X\{x_1, x_2, ..., x_d\}$, and distance $\rho : X \times X \to R^d$. The distance ρ satisfies Reflexivity, Symmetry, and Triangle inequality. An instance of distance is given by Eq. (8).

$$d(X_i^t, X_b^t) = \sqrt{\sum_{d=1}^{D} (x_{d,i} - x_{d,j})^2} \tag{8}$$

where D is the total number of dimensions of a swarm agent and $x_{d,i}$ is the dth element in agent i's position vector.

The Swarm Virtual Space V_s is a set $X \in R^d$, $X\{x_1, x_2, ..., x_d\}$ defined in the Euclidean 2-dimensional space, where n swarm agents are deployed at random. It is used to visualize clusters in 2D space.

The Agent Position Graph G_p is defined as $G_p = \{V, E_p\}$, where V is the set of vertices from 1 to n, and E is the set of edges between two vertices. G_p maintains the position of each agent throughout the algorithm, and the distance between two agents is stored in A_p, the adjacent matrix of G_p. Therefore A_p has a size of $|v| \times |v|$ such that

$$a_{i,j} = ||p_i - p_j||_2 \tag{9}$$

The Feature Similarity Graph $G_s\{V, E_f\}$ stores the similarity value between agents calculated using distance formula specified in the Swarm Metric Space. Therefore, the Swarm Clustering Framework is defined as (M_s, V_s, G_p, G_s).

Using the same set of flocking rules, Bird Flocking by Leader Algorithm implements dynamic maximum distance between agents. First, the leaders of boids and their maximum distances are found by neighborhood and reverse neighborhood analysis. Then for each leader, its followers within its maximum distance start moving according to separation, cohesion, and alignment. Each follower is marked as visited, and the algorithm ends when there are no more unvisited agents.

To find the leaders and their maximum distances in the swarm, neighborhood and reverse neighborhood analysis is done. Let $X \in R^d$, $X\{x_1, x_2, ..., x_d\}$ to be the swarm population, $\rho(x_i, x_j)$ to be the distance between agent i and agent j The k-neighborhood of x_i in iteration t $(kNB_t(x_i))$ is defined as all agents within a circle with x_i being the center and radius d^{Li}_{max} such that

$$d^{Li}_{max} = max_{o \in Knn(x_i)} \rho(x_i, o) \tag{10}$$

where $Knn(x_i)$ is the set of k nearest neighbors of x_i.

The reverse neighborhood of x_i, denoted as $(Dr - kNB_t(x_i))$, is the set of data points whose kNB contains x_i at iteration t. In order to determine the density of swarm agents, Agent Role Factor $ARF_t(x_i)$ is introduced where

$$ARF_t(x_i) = \frac{|DR - kNB_t(x_i)|}{|DR - kNB_t(x_i)| + |DkNB_t(x_i)|} \tag{11}$$

The larger the $|DR - kNB_t(x_i)|$, the larger the $ARF_t(x_i)$ and the denser x_i situated in the swarm. In particular, if $ARF_t(x_i) \geq 0.5$, x_i is a flock leader. If $ARF_t(x_i)$ is close to 0, then x_i is an outlier.

Experiments were conducted using two datasets, one with 100 news articles from cyberspace and one being the benchmark iris dataset. The Flock by Leader Algorithm achieves 98.66% reduction in the number of iterations, a 7.5%, 16.5% increase in accuracy in terms of F-measure than other Flocking algorithms and K-means algorithms respectively.

2.4 Grey Wolf Optimization

Grey wolf optimizer (GWO), inspired by the leadership and hunting mechanism of grey wolves, is a new metaheuristic algorithm proposed by Mirjalili et al. [8]. There are three levels of ranking in a wolf pack – alpha(α), beta(β), and omega(δ). The alpha is the leader, the beta aids the alpha in various pack activities, and the omega is like the scapegoat. The rest of the pack that does not fall under any of the above ranks is the delta.

The algorithm performs initialization of the grey wolf population $X_i(i = 1, ..., n)$ where n is the size of the population. The goal of the algorithm is to find the best solution (alpha). The wolves encircle the prey while hunting, which is mathematically expressed at iteration t as Eqs. (12) and (13).

$$D = |CX^t_p - X^t| \tag{12}$$

$$X^{t+1} = X^t_p - AD \tag{13}$$

where X^t_p is the position of the prey at iteration t, X^t is the position vector of the wolf at iteration t and A, C values are given by Eqs. (14) and (15)

$$A = 2ar_1 - a \tag{14}$$

$$C = 2r_2 \tag{15}$$

with r_1, r_2 as random numbers in $[0,1]$ and the components of a linearly decreased from 2 to 0 over the cycles. The values of a, A and C are initialized before the iteration begins. The fitness of each search agent is then calculated.

The iteration now begins. For every search agent, the position of the current search agent is updated using Eqs. (16), (17), (18) and (19).

$$X^{t+1} = (X_1 + X_2 + X_3)/3 \tag{16}$$

$$X_1 = X_\alpha - A_1(C_1 X_\alpha - X^t) \tag{17}$$

$$X_2 = X_\beta - A_2(C_2 X_\beta - X^t) \tag{18}$$

$$X_3 = X_\delta - A_3(C_3 X_\delta - X^t) \tag{19}$$

Then, update the values of the vectors a, A and C. Repeat this till the maximum number of iterations is attained. The final value X_α is the best solution obtained.

In the original paper, the algorithm was tested on 29 benchmark minimization functions and gave competitive results compared to PSO, DE, Gravitational Search Algorithm (GSA), and Fast Evolutionary Programming (FEP). It was particularly effective in solving problems with multiple local optima and high-dimensional search spaces.

2.5 Flower Pollination Algorithm

Flower Pollination Algorithm is a nature-inspired metaheuristic algorithm proposed by Yang in 2012 that mimics the behavior of pollination of flowering plants [9]. Pollination in plants occurs in two forms – abiotic and biotic. In abiotics, the pollen is transferred by a pollinator like insects and animals. In biotic, they do not need any pollinators. Instead, the wind and diffusion act as a medium to transfer pollen. Pollination can be achieved either by self-pollination or cross-pollination.

The Flower Pollination Algorithm is governed by the following rules:

Rule 1: Biotic and cross-pollination by pollinators obeying Levy flight is considered a global pollination process.

Rule 2: Abiotic and self-pollination are considered to be local pollination.

Rule 3: Flower constancy can also be treated as reproduction probability which is proportional to the similarity of the flowers.

Rule 4: Local or global pollination can be managed by selecting switch probability which lies between 0 and 1.

To optimize the objective function, a population of n flowers is initialized with random solutions and a switch probability p. Then, the best solution from the initial population is selected as g^*. For every iteration, every x_i of the flowers is updated as follows:

1. If the randomly selected probability value is less than p, global pollination is chosen. Since the insects can fly over long distances, global pollination ensures the survival

of the fittest. At every iteration $t + 1$, the updation the x_i with global pollination is given by Eq. (20).

$$x_i^{t+1} = x_i^t + L(x_i^t - g^*) \tag{20}$$

where L is a step vector that follows Levy distribution.

2. If the randomly selected probability value is greater than p, then we opt for local pollination. At any iteration $t + 1$, the updation of x_i with local pollination is given by Eq. (21).

$$x_i^{t+1} = x_i^t + \epsilon(x_j^t - x_k^t) \tag{21}$$

where ϵ is drawn from the uniform distribution [0,1] and j, k are random numbers from all the solutions.

The population is updated if they are better than the existing ones. The current best solution g* is updated accordingly in every iteration. This entire process is repeated till the maximum number of iterations is reached.

The performance of the Flower Pollination Algorithm (FPA) was compared with PSO and Genetic Algorithm in the original paper in terms of the number of iterations. 100 independent runs were carried out for a population size of 25. FPA outperformed both algorithms with an exponential convergence rate. The algorithm was also successfully applied to the pressure vessel design problems.

2.6 Whale Optimisation Algorithm

The Whale Optimization Algorithm (WOA) is a metaheuristic optimization algorithm that was first proposed by Seyedali Mirjalili in 2016 [10]. It is inspired by the hunting techniques and social behavior of humpback whales. The algorithm begins by randomly initializing a population of candidate solutions, "whales". These are iteratively updated using Exploration, Exploitation, and Encircling.

Exploration (*Bubble-net attacking method*) is used to search for new candidate solutions by randomly moving some whales in the search space. The position of each whale is updated using the spiral Eq. (22).

$$\vec{X}_{t+1} = \vec{D}.e^{bl}.cos(2\Pi l) + \vec{X}^*_t \tag{22}$$

where D is the distance of the ith whale to the prey (best solution obtained so far), b is constant of definition of the logarithmic spiral shape and l is a random number in [-1, 1], imitating the helix-shaped movement of humpback whales and shrinking encircling mechanism.

Exploitation (*Search for prey*) focuses the search around the best candidate solution found so far to improve its quality where the position is updated using Eq. (23).

$$\vec{X}_{new} = \vec{X}_{rand} - \vec{A}.\vec{D} \tag{23}$$

where X_{rand} is a randomly chosen whale from the population, A is a coefficient that decreases linearly from 2 to 0 and D is the distance between the current position of the whale and X_{rand}.

Encircling forces other whales to converge towards the best candidate solution found so far. The position is updated using Eqs. (24) and (25).

$$\vec{D} = |\vec{C}.\vec{X^*}_t - \vec{X}_t|$$ (24)

$$\vec{X}_{t+1} = \vec{X^*}_t - \vec{A}.\vec{D}$$ (25)

where t = current iteration, X_t^* is the position vector of the best candidate solution, A and C represent coefficient vectors and D is the distance between the current position of the whale and X_t^*. A and C are computed as in Eqs. (26) and (27).

$$\vec{A} = 2.\vec{a}.\vec{r} - \vec{a}$$ (26)

$$\vec{C} = 2.\vec{r}$$ (27)

where a is a vector that decreases linearly from 2 to 0 and r is a random vector in [0,1].

The performance of Whale Optimization Algorithm (WOA) was tested on benchmark functions and compared with PSO, GSA, DE, and FEP in the original paper. The WOA showed high exploration ability, local optima avoidance, and convergence speed.

2.7 Moth Flame Optimisation Algorithm

Moth Flame optimization is a population-based meta-heuristic algorithm proposed by Mirajalili [11]. They navigate in the night using moonlight to travel in a straight line. But with the artificial lights, the moths fly in spiral paths. In this algorithm, both moths and flames act as solutions. Moths are actual search agents that go around the flame to explore the search area and the flames are the best positions of moths obtained so far.

The MFO algorithm starts by randomly initializing the position of moths M moving in d-dimensional space. The fitness function of each moth is evaluated using Eq. (28).

$$M(i,j) = (ub(i) - lb(i)) * rand() + lb(i)$$ (28)

where i represents the ith moth and j represents the current dimension of the space, ub and lb are the upper and lower bounds of the ith variable. The MFO algorithm also randomly initializes the flames F_j.

During every iteration, the positions of flames are updated and sorted based on the fitness values of moths. The moths then update their positions based on their corresponding flames using Eqs. (29) and (30).

$$M_i = S(M_i, F_j)$$ (29)

$$S(M_i, F_j) = D_i e^{bt} cos(2\pi t) + F_j$$ (30)

where M_i being the ith moth, F_j being the jth flame, S being the spiral function, D_i i being the distance of the ith moth from the jth flame, b being a constant for defining the shape of the logarithmic spiral, and t being a random number in [−1, 1].

This allows the moths to explore in a higher space, which potentially leads to better solutions. But updating the moths' positions around n flames may decrease the chance of exploitation around the best flame. Hence, the number of flames decreases after every iteration as in Eq. (31).

$$Flame\ nums = round(N - I * \frac{N - 1}{T}) \tag{31}$$

where N is the maximum number of flames, I is the current iteration number and T indicates the maximum iteration numbers. The flame with maximum fitness after the maximum number of iterations is reached or the convergence criteria are met is the best optimal solution found.

The Moth Flame Optimization (MFO) Algorithm was tested on 19 benchmarked functions and the performance was compared with other algorithms such as PSO and GA in terms of balancing exploration and exploitation in the original paper. MFO Algorithm produced competitive results with high exploitation ability. The algorithm was also successfully applied to constraint optimization problems.

Table 1 below summarizes the parameters and properties of the seven algorithms discussed.

Table 1. SI Algorithms - Parameters and Properties

SI Algorithm	Parameters	Properties
Artificial Bee Colony Algorithm	Colony Size, Number of employed, Onlooker and scout bees, Limit parameter, Neighborhood size, Maximum number of iterations	Balances exploration and exploitation, Solution evaluation using an objective function, Parameter Tuning, Convergence overtime
Bat Algorithm	Population size, Loudness, Pulse rate, Minimum and maximum frequency, Alpha (influence of best bat on movement of other bats), Gamma (Rate of loudness reduction overtime), Maximum number of iterations	Balances exploration and exploitation, Echolocation behavior using Frequency and Loudness, Solution evaluation using an objective function, Parameter Tuning, Convergence overtime
Bird Flocking: Flock by Leader Algorithm	Number of birds, Minimum Distance	Adaptive, Self-organization, Emergent behavior, Decentralized decision making
Grey Wolf Optimizer	Population size, Alpha (best wolf), Beta (second-best wolf), Delta (third-best wolf), Initial search range, Convergence criterion, Search and Randomness, Maximum number of iterations	Mimics social and hunting behavior of wolves, Parameter Tuning, Convergence overtime

(continued)

Table 1. (*continued*)

SI Algorithm	Parameters	Properties
Flower Pollination Algorithm	Population size, Pollination rate, Levy exponent, Initial step size, Scaling factor, Convergence criterion, Maximum number of iterations	Mimics the pollination process of flowers, Balances exploration and exploitation, Adaptive Search Strategy, Parameter Tuning, Convergence overtime
Whale Optimization Algorithm	Population size, A (Search agent's encircling behavior), C (balance the exploration and exploitation ability), B (improve exploitation ability), I (improve exploration ability), Maximum number of iterations	Combines exploration and exploitation, Applied in continuous search space, Global Optimization, Fast convergence
Moth Flame Algorithm	Population size, Light intensity, Attraction and Mutation coefficients, Convergence criterion, Maximum number of iterations	Balances exploration and exploitation, Mimics moths behavior of attraction to light, Stochasticity, Global Optimization, Fast convergence

3   Experimental Analysis

This section provides a comparison of the performance of the seven algorithms in identifying the optimal hyperparameters for a random forest classifier for the MNIST data. The MNIST database is a popular dataset of handwritten digits that is widely used for training image processing systems. It contains 60,000 examples for training and 10,000 examples for testing. The dataset is a subset of a larger set available from the National Institute of Standards and Technology (NIST). The digits in the dataset have been standardized in size and centered in a fixed-size image.

The following range of values are considered for each of the five hyperparameters of the Random Forest algorithm:

1. n_estimators (number of decision trees in the forest) was optimized over the range of 10 to 100.
2. max_depth (maximum depth of each decision tree) was optimized over the range of 5 to 50.
3. min_samples_split (minimum number of samples required to split an internal node) was optimized over the range of 2 to 11.
4. min_samples_leaf (minimum number of samples required to be at a leaf node) was optimized over the range 1 to 11.
5. max_features (the maximum number of features to consider when looking for the best split) was optimized over the range of 1 to 64.

We evaluated the seven algorithms for 100 iterations to find the optimal parameters for the Random Forest Classification on the above-mentioned dataset. Table 2 provides the optimal hyperparameters identified by the seven algorithms and the accuracy attained by the random forest classifier with the corresponding hyperparameters.

Table 2. Random Forest Classification Parameters Estimation using SI Algorithms

	ABC	Bat	BFL	GWO	FPA	WOA	MFO
n_estimators	75	51	59	87	40	20	27
max_depth	48	17	25	49	29	9	18
min_samples_split	4	9	11	3	2	3	9
min_samples_leaf	1	8	1	1	2	1	3
max_features	23	29	21	13	14	5	28
ACCURACY	98.3	95.8	96.9	97.7	**98.6**	98.3	98.0

As can be seen from Table 2, Flower Pollination Algorithm (FPA) performed the best with an accuracy of 98.6%. ABC and WOA performed the second best with an accuracy difference of 0.3.

4   Applications of Swarm Intelligence Algorithms

In this section, three major fields of applications are discussed in Table 3 where each of the Swarm Intelligence algorithms was employed for optimization purposes.

Table 3. SI Algorithms and their applications

Swarm Intelligence Algorithm	Scheduling and Assignment Problem	Data Science, ML, and Robotics	Route, Network and Layout Optimization
Artificial Bee Colony Algorithm (ABC)	**Task Scheduling** [12] Hybrid of ABC and PSO was used to maximize resource utilization and maintain load balance in cloud computing	**Image Contrast Enhancement** [13] ABC was used to replace grey levels of the input image with new enhanced values	**Vehicle Routing** [14] An improved ABC algorithm to solve the Vehicle Routing Problem (VRP) using combinatorial optimization
Bat Algorithm (BA)	**Quadratic assignment problem** [15] Discrete BA with 2-exchange neighborhood and a modified uniform crossover mechanism were used for quadratic assignment problems	**Satellite Image Enhancement** [16] Histogram equalized image was contrast enhanced where control parameters are optimized using BA	**Energy Consumption in Wireless Sensor Networks** [17] BA was used to select the optimal sensor and route to reduce energy consumption

(continued)

Table 3. (*continued*)

Swarm Intelligence Algorithm	Scheduling and Assignment Problem	Data Science, ML, and Robotics	Route, Network and Layout Optimization
Bird Flocking: Flock by Leader (FBL)	**Load Balancing** [18] Bird swarm optimization was applied for load balancing in cloud computing	**Diabetes Disease Classification** [19] Flock Optimization was used to tune hyperparameters to classify diseases	**Traveling Salesman Problem** [20] Traveling salesman problem was solved by an improved multi flock optimization
Grey Wolf Optimizer (GWO)	**Scheduling Workflow of Applications** [21] Distributed GWO was used to map dependent workflow tasks for low total execution cost	**Robot Path Planning** [22] GWO was used for robot path planning to avoid 3 different circular obstacles	**Aerial Vehicle Path Planning** [23] GWO was used to find optimal 2-D path for unmanned combat aerial vehicles
Flower Pollination Algorithm (FPA)	**Task scheduling in cloud computing environment** [24] Exploration Enhanced FPA was proposed in the cloud model to allocate tasks to the virtual machines with reduced makespan	**Multi-level Image thresholding** [25] Local pollination with Euclidean distance ratio and global pollination with random position vector for image segmentation	**Urban transit routing** [26] Improved population generative method and FPA with average travel time and the number of transfers as the optimization objective
Whale Optimization Algorithm (WOA)	**Job shop scheduling** [27] Hybrid WOA enhanced with Levy flight and differential evolution was proposed to solve the job shop scheduling problem	**Multi-robot space exploration** [28] Integration of deterministic coordinated multi robot exploration and Frequency Modified WOA	**Traffic-aware routing in urban VANET** [29] Multi-objective auto-regressive WOA for traffic-aware routing in urban VANET
Moth Flame Algorithm (MFO)	**Flow shop scheduling** [30] Improvised MFO was developed to solve green reentrant hybrid flow shop scheduling that minimizes the maximum completion time and reduces the comprehensive impact of resources and the environment	**Feature selection for medical diagnosis** [31] Enhanced MFO integrating Levy flight operators into its structure with transfer functions was developed as a search strategy within the wrapper feature selection	**Fault resilient routing: underwater wireless sensors** [32] A fault resilient routing based on MFO with a novel fitness function was proposed to transfer packets towards base stations through autonomous underwater vehicles

5 Discussion

Nature-inspired algorithms can effectively solve complex optimization problems that are difficult, or impossible, to solve with traditional methods. These algorithms do not require a large amount of problem-specific knowledge, can be implemented relatively easily, and can provide the ability to balance exploration and exploitation during the optimization process.

However, they may suffer from slow convergence rates and sensitivity to parameter selection. Due to their heuristic nature, these algorithms often lack theoretical guarantees of their performance or convergence, but they have been shown to be highly effective in solving complex optimization problems. It is important to have a comprehensive understanding of the algorithms, as well as the problem statement, in order to determine which algorithm is most suitable given the constraints. Hybrid versions of these algorithms are currently being created that aim to overcome some of these weaknesses, given an objective and application area.

6 Conclusion

Swarm Intelligence models the behavior of nature to work with different kinds of data and machine learning problems from diverse fields. The capability to apply these algorithms in various cross-disciplinary fields makes Swarm Intelligence a promising area of research in the field of Artificial Intelligence. These algorithms offer a powerful and flexible approach to problem-solving, and their ability to handle non-linear and non-differentiable objective functions makes them particularly useful in practical applications where other optimization techniques may be limited. This review provided a comprehensive overview of seven Swarm Intelligence algorithms developed over the years which took inspiration from various animals and birds around us. Several potential application areas for these algorithms were also discussed and appreciated. All algorithms discussed (Artificial Bee Colony Optimization, Bat-inspired algorithm, Flock by Leader, Grey Wolf Optimizer, Flower Pollination Algorithm, Whale Optimization, and Moth Flame Optimization) were well-explored and applied for scheduling, machine learning, robotics, and route optimization problems, providing effective results and future directions. Currently, several ongoing projects are focused on creating a hybrid of these algorithms that could be applied in more diverse and complex domains.

Acknowledgements. This study was partially supported by the 2023 Courant Institute Suzanne McIntosh Research Fellowship. Special thanks to Marie Bogdanoff for her valuable insights. Marie is a member of the Predictive Analytics and AI Research Lab at New York University, led by Prof. Anasse Bari.

References

1. Yang, X.S.: Nature-inspired algorithms for optimization. In: Springer Science & Business Media (2010)
2. Binitha, S., Sathya, S.S.: A survey of bio-inspired optimization algorithms. In: Singh, P.K., Wierzchoń, S.T., Tanwar, S., Ganzha, M., Rodrigues, J.J..P.C.: Proceedings of the Second International Conference on Intelligent Computing and Communication, LNNS, pp. 633–640. Springer (2018). https://doi.org/10.1007/978-981-16-0733-2
3. Kar, A.K.: Bio-inspired computing – a review of algorithms and scope of applications. Expert Syst. Appl. **59**, 20–32 (2016). https://doi.org/10.1016/j.eswa.2016.04.018

4. Karaboga, D., Basturk, B.: A powerful and efficient algorithm for numerical function optimization: artificial bee colony (ABC) algorithm. J. Glob. Optim. **39**, 459–471 (2007). https://doi.org/10.1007/s10898-007-9149-x

5. Yang, X.-S.: A new metaheuristic bat-inspired algorithm. In: González, J.R., Pelta, D.A., Cruz, C., Terrazas, G., Krasnogor, N. (eds.) Nature Inspired Cooperative Strategies for Optimization (NICSO 2010), pp. 65–74. Springer Berlin Heidelberg, Berlin, Heidelberg (2010). https://doi.org/10.1007/978-3-642-12538-6_6

6. Reynolds, C.W.: Flocks, herds, and schools: a distributed behavioral model, in computer graphics, In: SIGGRAPH 1987 Conference Proceedings, vol. 21, Issue 4, pp. 25–34 (1987)

7. Bellaachia, A., Bari, A.: Flock by leader: a novel machine learning biologically inspired clustering algorithm. In: Tan, Y., Shi, Y., Ji, Z. (eds.) ICSI 2012. LNCS, vol. 7332, pp. 117–126. Springer, Heidelberg (2012). https://doi.org/10.1007/978-3-642-31020-1_15

8. Mirjalili, S., Mirjalili, S.M., Lewis, A.: Grey wolf optimizer. Adv. Eng. Softw. **69**, 46–61 (2014)

9. Zipser, D.: Distributed control of complex arm movements. In: Prescott, T.J., Lepora, N.F., Mura, A., Verschure, P.F.M.J. (eds.) Living Machines 2012. LNCS (LNAI), vol. 7375, pp. 309–320. Springer, Heidelberg (2012). https://doi.org/10.1007/978-3-642-31525-1_27

10. Mirjalili, S., Gandomi, A.H., Mirjalili, S.M.: Whale optimization algorithm. Adv. Eng. Softw. **95**, 51–67 (2016). https://doi.org/10.1016/j.advengsoft.2016.01.008

11. Mirjalili, S.: Moth-flame optimization algorithm: a novel nature-inspired heuristic paradigm. Knowl. Based Syst. **89**, 228–249 (2015). ISSN 0950-7051

12. Maheswari, P., Edwin, B., Thanka, R.: A hybrid algorithm for efficient task scheduling in cloud computing environment. Int. J. Reason. Intell. Syst. **11**, 134 (2019)

13. Draa, A., Bouaziz, A.: An artificial bee colony algorithm for image contrast enhancement. Swarm Evol. Comput. **16**, 69–84 (2014)

14. Yao, B., Yan, Q., Zhang, M., Yang, Y.: Improved artificial bee colony algorithm for vehicle routing problem with time windows. PLoS One **12**(9), e0181275 (2017). https://doi.org/10.1371/journal.pone.0181275. PMID: 28961252; PMCID: PMC5621664

15. Riffi, M.E., Saji, Y., Barkatou, M.: Incorporating a modified uniform crossover and 2-exchange neighborhood mechanism in a discrete bat algorithm to solve the quadratic assignment problem. Egyptian Inform. J. **18**(3), 221–232 (2017)

16. Asokan, A., Popescu, D.E., Anitha, J., Hemanth, D.J.: Bat algorithm based non-linear contrast stretching for satellite image enhancement. Geosciences **10**(2), 78 (2020). https://doi.org/10.3390/geosciences10020078

17. Sangaiah, K., Sadeghilalimi, M., Hosseinabadi, A.A.R., Zhang, W.: Energy consumption in point-coverage wireless sensor networks via bat algorithm. IEEE Access **7**, 180258–180269 (2019). https://doi.org/10.1109/ACCESS.2019.2952644

18. Mishra, K., Majhi, S.K.: A binary bird swarm optimization based load balancing algorithm for cloud computing environment. Open Comput. Sci. **11**(1), 146–160 (2021). https://doi.org/10.1515/comp-2020-0215

19. Balasubramaniyan, D., Husin, N.A., Mustapha, N., Sharef, N.M., Aris, T.N.M.: Flock optimization induced deep learning for improved diabetes disease classification. Expert Syst. (2023). https://doi.org/10.1111/exsy.13305

20. Tongur, V., Ülker, E.: PSO-based improved multi-flocks migrating birds optimization (IMFMBO) algorithm for solution of discrete problems. Soft. Comput. **23**(14), 5469–5484 (2018). https://doi.org/10.1007/s00500-018-3199-5

21. Abed-alguni, H., Alawad, N.A.: Distributed grey wolf optimizer for scheduling of workflow applications in cloud environments. Appl. Soft Comput. **102**, 107113 (2021)

22. Doğan, L., Yüzgeç, U.: Robot path planning using gray wolf optimizer (2018)

23. Zhang, S., Zhou, Y., Li, Z., Pan, W.: Grey wolf optimizer for unmanned combat aerial vehicle path planning. Adv. Eng. Softw. **99**, 121–136 (2016). https://doi.org/10.1016/j.advengsoft.2016.05.015

24. Bezdan, T., Zivkovic, M., Antonijevic, M., Zivkovic, T., Bacanin, N.: Enhanced flower pollination algorithm for task scheduling in cloud computing environment. In: Joshi, A., Khosravy, M., Gupta, N. (eds.) Machine Learning for Predictive Analysis. LNNS, vol. 141, pp. 163–171. Springer, Singapore (2021). https://doi.org/10.1007/978-981-15-7106-0_16

25. Shen, L., Fan, C., Huang, X.: Multi-level image thresholding using modified flower pollination algorithm. IEEE Access **6**, 30508–30519 (2018). https://doi.org/10.1109/ACCESS.2018.2837062

26. Fan, L., Chen, H., Gao, Y.: An improved flower pollination algorithm to the urban transit routing problem. Soft. Comput. **24**(7), 5043–5052 (2019). https://doi.org/10.1007/s00500-019-04253-3

27. Liu, M., Yao, X., Li, Y.: Hybrid whale optimization algorithm enhanced with Lévy flight and differential evolution for job shop scheduling problems. Appl. Soft Comput. **87**, 105954 (2020)

28. Gul, F., Mir, I., Rahiman, W., Islam, T.U.: Novel implementation of multi-robot space exploration utilizing coordinated multi-robot exploration and frequency modified whale optimization algorithm. IEEE Access **9**, 22774–22787 (2021). https://doi.org/10.1109/ACCESS.2021.3055852

29. Rewadkar, D., Doye, D.: Multi-objective auto-regressive whale optimisation for traffic-aware routing in urban VANET. IET Inform. Secur. **12**(4), 293–304 (2018). https://doi.org/10.1049/iet-ifs.2018.0002

30. Xu, F., et al.: Research on green reentrant hybrid flow shop scheduling problem based on improved moth-flame optimization algorithm. In: Processes, vol. 10, no. 12 (2022)

31. Abu Khurmaa, R., Aljarah, I., Sharieh, A.: An intelligent feature selection approach based on moth flame optimization for medical diagnosis. Neural Comput. Appl. **33**(12), 7165–7204 (2020). https://doi.org/10.1007/s00521-020-05483-5

32. Kumari, S., Mishra, P.K., Anand, V.: Fault resilient routing based on moth flame optimization scheme for underwater wireless sensor networks. Wireless Netw. **26**(2), 1417–1431 (2019). https://doi.org/10.1007/s11276-019-02209-x

Overview of Complex Intelligent System Reliability Technology

Zhaoyang Zeng, Wensheng Peng, and Jiao Li[(✉)]

CPAE, Beijing, China
zengzy003@avic.com, lijiao2016@sina.com

Abstract. With the wide application of intelligent technology, model theft of face and fingerprint recognition may lead to information leakage, and even in some extreme cases, such as data poisoning and confrontational attacks may lead to loss of life. In this paper, the SMART architecture is introduced to explore the quantification of intelligent system training set, adversarial attack, model accuracy and uncertainty from three dimensions: operating environment, data, and model. The coupling of intelligent module and system leads to the uncertainty of complex and dynamic structure, which leads to procedural faults. This paper focuses on exploring the variable content of improved reliability and puts forward the power point of reliability analysis and verification method according to the fault characteristics of intelligent system. It also provides ideas for in-depth research on the reliability of intelligent systems.

Keywords: Intelligent system · Failure mechanism · Reliability parameters · Reliability model · Experimental validation

1 Introduction

Intelligent technology [1] (unmanned, mobile computing, intelligent endurance, image recognition), gradually widely used in unmanned vehicles, drones, unmanned boats and other industries, the failure of artificial intelligence systems may lead to economic losses, and even in some extreme cases will lead to loss of life. As of November 2022, the AI Incident Database (AIID) showed that more than 2,000 AI accidents have occurred in industrial applications, 29 of which involved death or injury, demonstrating that reliability issues can lead to serious losses. Musk pointed out that "if AI is not treated carefully, it may become the terminator of humanity in the future." Musk has expressed his concerns about intelligent systems on public social platforms more than once, and the reliability of intelligent systems has attracted more and more attention.

Shiwen Zhong [2] et al. point out that unexplained artificial intelligence is an important source factor of unreliability. At present, the research on explainable artificial intelligence is not mature enough, and there is no unified theory and standard, especially the remote sensing is still in the infancy field.

Bo Zhang [3] others pointed out that the data-driven property and opacity of AI algorithms have led to new unreliability problems, such as uninterpretability, and easy to be deceived, which greatly limits the application of AI technology.

Y. Tan et al. (Eds.): ICSI 2023, LNCS 13968, pp. 18–29, 2023.
https://doi.org/10.1007/978-3-031-36622-2_2

Although several work has been done on the reliability of AI, more emphasis is placed on the perspective of software design, and there is no systematic and comprehensive perspective to propose parameter systems and assessment methods. Focusing on data, model (algorithm) and operating environment, this paper introduces "SMART" [4] framework (System feature, Mechanism of fault, Arguments of Reliability, Reliability Modeling and Test Verification), systematically analyzes the reliability analysis and verification work of intelligent systems.

2   Intelligent System Characteristics and Risk Analysis

2.1   Definition and Characteristics of Intelligent System

Wang [5] et al. argue that an intelligent system is an advanced system for implementing complex cognitive abilities in machines, from reflexive, imperative, and adaptive intelligence aggregation to autonomous and cognitive intelligence. Intelligent systems are often composed of perception modules, decision-making modules, and control modules [6].

Stranieri [7] et al. argue that an intelligent system is a system that simulates, automates, and enhances certain human intelligent behaviors. Expert systems and knowledge-based systems are examples of intelligent systems.

Köse U [8] et al. argue that an intelligent system is a system based on methods or techniques in the field of artificial intelligence to perform more accurate and efficient operations to solve related problems.

The Subcommittee on Artificial Intelligence (ISO/IEC JTC 1/SC 42) [9] defines the AI system as an engineering system that generates outputs content, such as predictions, recommendations, or decisions for a given set of human-defined goals (Fig. 1).

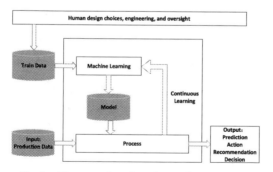

Fig. 1. AI system function view defined by SC42

Engineering systems can develop [10] a model to represent data, knowledge, processes by using various techniques and methods related to artificial intelligence, which can be used to perform tasks such as classification, regression, ranking, clustering, and dimensionality reduction.

2.2 Composition of Intelligent System

By physical composition [11]: hardware system and software system.

In addition to the typical hardware in the product, hardware used for AI computing [12] can include central processing units (CPUs), application-specific integrated circuits (ASICs), graphics processing units (GPUs), tensor processing units (TPUs), and intelligent processing units (IPUs). There are also various types of cameras, sensors, and devices that collect images, sounds, and other data formats fed into AI systems, as well as network infrastructure.

The core of the software system [13] includes machine learning/deep learning (ML/DL)-based algorithms and other rule-based algorithms. These algorithms include image recognition and speech recognition, CV, NLP, and classification, as well as data collection, processing, and other common software functions.

2.3 Life Cycle Risk Analysis of Intelligent Systems

According to ISO/IEC 23053:2022 (E), intelligent systems mainly include seven stages [14]: concept, design and development, validation and verification, deployment, operational supervision, revaluation and scrapping, including three processes of data set, model creation and evaluation and operation.

The main work contents and existing risks [15] at each stage mainly include:

Step 1 The concept stage, according to the goal, set the task, which is the basis for algorithm selection;

Step 2 Data poisoning [16] and model theft are also used to achieve the attack goal, on the one hand, deliberately influencing the training data to manipulate the results of the predictive model; On the other hand, by copying the internal functions of the model to "steal" the model, resulting in a breach of security and privacy.

Step 3 Validation and verification, including model evaluation and systematic evaluation.

Step 4 Deployment, once the ML model has been trained and evaluated, it needs to be deployed to the target platform in order to use it to make predictions on production data.

Step 5 Operational supervision, antagonistic attacks [17] involve providing slightly perturbed input data to a valid model, such as providing a slightly modified traffic sign to an autonomous vehicle system, in an attempt to trick it into misclassifying the input data.

Step 6 Revaluation, because factors such as data drift [18] (changes in the statistical characteristics of production data), conceptual drift (changes in decision boundaries), or poor generalization [19] can affect prediction accuracy, and may require continuous learning to make new evaluations of the optimized model.

Step 7 Scrap.

3 Analysis of Failure Mechanism of Intelligent System

Taking the intel UAV cluster intelligent inspection [12] as an example, the fault mechanism of the intelligent system is analyzed. In the process of intelligent inspection of UAV clusters, when the identification and perception of the current environment has a failure, it is necessary to re-plan the patrol path, extract the feature data through the recognition perception, generate the optimal path, carry out the fault patrol work, and complete the original path and find the fault path according to the specified time [13].

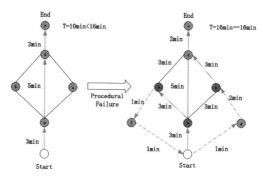

Fig. 2. Procedural failure of UAV cluster inspection system

As shown Fig. 2, assuming that the UAV inspection time cannot exceed 16min, the patrol path a → d cannot reach point E at the specified time, so the UAV cluster triggers the protection mechanism and decides to ignore the unfinished inspection path, or other members replace the work and arrive on time At point e, this phenomenon causes the drone to fail to inspect all nodes and edges on the original path as planned, resulting in a procedural failure.

- **Perception and identifies failures**

The obstacle data information in the environment is extracted, and then the extracted data is clustered. The extracted data information coverage is not comprehensive, the key feature data information is missing, and the data features of different obstacles are confused will cause the inconsistent between complexity of environmental obstacles and the actual inspection environment, which leads to the perception and recognition error of the intelligent system, resulting in the obstacle avoidance failure of the inspection UAV.

- **Learning mistakes**

Due to the improper setting of parameters such as the learning rate of the AI system, the interference of extreme environments in the learning process, and the exist of incorrect input of information sample data, the independent learning of the intelligent system will be wrong and resulting in the failure of obstacle avoidance of the inspection drone.

- **Incorrect prediction results**

Due to incomplete coverage of target training/test data samples and errors in data input, the prediction results of the intelligent system are incorrect. Cause one: Prediction failure is usually caused by distribution shift, which usually means that the operating environment is different from the training set environment, and model deviation causes model prediction to be inaccurate and resulting in failure; Cause two, if the data input to the algorithm is noisy, contaminated, or comes from a faulty sensor, it may also fail.

- **Incorrect AI algorithm decisions**

  Improper feature classification of input data, lack of environmental factors, obstacle discrimination rules and other constraints, resulting in decision-making problems.

- **The action could not be performed**

  Failure of actuators, transmission mechanisms, etc.

  In summary, the failure factors of intelligent systems mainly include operating environment, data and models (Fig. 3).

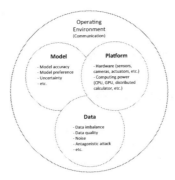

Fig. 3. Intelligent System Failure Factors

4   Intelligent System Reliability Characterization Parameters

4.1   Intelligent System Reliability Definition

Intelligent system reliability [14] is defined as its ability to consistently and correctly deliver the required predictions, recommendations, and decisions during its operational phase. The intelligent system ability can be measured by reliability, and the reliability of an intelligent system is shown in the formula:

$$R = 1 - \frac{T_{off}}{T} \tag{1}$$

Among them, it refers to the length of time that the intelligent system T_{off} stops running due to the failure of the intelligent system within the specified time, and its faults include the process faults caused by the operating environment stability, data quality, model reliability, etc. of the intelligent system, and T refers to the total operation time of the intelligent system.

4.2 Overall Framework of Reliability Parameters of Intelligent Systems

The reliability parameter system of intelligent system is mainly composed of 1 first-level index (i.e. comprehensive index), 3 s-level indicators, 10 third-level indicators, and 19 fourth-level indicators, as shown in Fig. 5 for details.

4.3 Intelligent System Reliability Parameters and Measurement Model

Operating Environment Stability

Operational stability [15] refers to the ability to avoid system failure caused by the operating environment, which can be divided into platform reliability and environmental reliability according to the type of failure. Among them, platform reliability includes hardware reliability and computing power reliability, while the main content of environmental reliability is communication reliability.

Data Quality

Data includes data [16] samples from the training, validation and testing phases, as well as production data in the use phase, data quality refers to the degree to which the data is correct for the specified task and user target running process and producing results, including data sample quality and (online) data quality.

Model Reliability

Model reliability [17] is also known as algorithm reliability, model reliability refers to the ability to avoid AI model failure in intelligent systems and cause intelligent system function failure, usually described by generalization, generalization is determined by the ability of learning algorithms, the adequacy of data and the difficulty of the learning task itself, is the key to model reliability (Fig. 4).

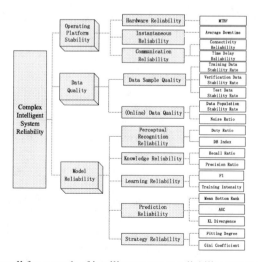

Fig. 4. Overview framework of intelligent system reliability parameter system

Model reliability is greatly affected by training intensity and data, and in-depth analysis is required, this paper describes the model reliability from the perspective of model function implementation, and from the perspective of perception recognition, learning, prediction and decision-making.

5 Intelligent System Reliability Modeling and Evaluation Method

5.1 Reliability Modeling Method Adaptability Research and Analysis

After investigation and analysis, the existing reliability modeling methods are mainly carried out from two aspects: system structure and state space, and their characteristics are as follows:

(1) Reliability modeling methods for system structures

The reliability modeling methods of system structures such as RBD, FTA, and BDD are mainly aimed at traditional series-parallel systems, based on system components or events, and it is difficult to describe such models for complex relationships such as time series, dependence, and environment.

(2) State-space reliability modeling method

For state-space modeling methods such as Markov and Petri Net, there is a spatial state explosion, and most current dynamic model applications of continuous-time Markov (CTMC) have "failure of elements follows an exponential distribution greatly limits its application in intelligent system reliability modeling.

However, intelligent systems have image perception, autonomous planning and decision-making need to go through multiple feedback calculations, high structural coupling, dynamic evolution leads to changeable states, for intelligent systems with autonomy, intelligence, and collaboration, it is inevitable that there will be problems such as inconsistent modeling scale, complex internal logic of the model, and state explosion. Therefore, a new approach to reliability modeling analysis is imperative.

5.2 Overview of Causal Analysis Methods

In the 80s of the 20th century, Pearl, Spirtes, Glymour, Scheines and other pioneers of causality studied the causality inference method of non-temporal observation data from an academic point of view. The proposed causal network is explained by combining causal relationships on the basis of Bayesian networks.

In 2017, the journal Science organized a column discussing causality and prediction of machine learning systems.

In June 2020, Turing Award winner and Bayesian founder Judea Pearl emphasized the "causal revolution" It is a revolution that changes data science, which involves mechanism policy, explainability, mechanism generalization, etc., and is a breakthrough in strong artificial intelligence, in data source interpretation, intervention formulation, multi-domain learning, etc., therefore, machine learning and causal analysis research has attracted more and more attention.

5.3   Overview of Reliability Modeling Methods for Intelligent Systems Based on Causal Analysis

Mathematical Expression of Causal Models

Taking the object, fault coupling relationship and dynamic process as the three elements, an intelligent system evaluation model is established to evaluate the operation of the intelligent system within the specified time and output the reliability of the system.

The model can be formally expressed as:

$$N = \{\mathbb{G}, \mathbb{EC}, R\} \tag{2}$$

where: G is an intelligent system object t;EC is an external perturbation of an intelligent system, which represents evolutionary conditions; R represents the Evolutionary rules.

Causal Model Transformation Methods

Intelligent system objects are divided into two parts, software system and hardware system, to describe intelligent systems.

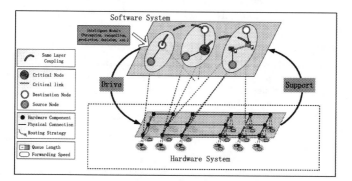

Fig. 5. Intelligent Systems Model

Intelligent system is abstracted into a system topology composed of sensors, actuators, routers and other hardware systems and interconnected links, represented by a weighted undirected diagram G (V,E), where:

$$V = \{S1, S2, \ldots, Sn\} \tag{3}$$

V represents the set of nodes, and Si represents a node. Depending on the device type and status, several properties are defined on the node.

$$E = \{l1, l2, \ldots, ln\} \tag{4}$$

E represents the link set, and li represents the link between two devices. For each edge, several attributes are defined based on its type and status: link type, link distance, channel conditions, link rate, and so on.

- External perturbations

According to the above analysis of intelligent system objects, combined with the external factors of intelligent system operation (environment, human factors, user needs, etc.), analyze the factors that induce the state change of intelligent system.

1) Changes in the state of the components of the intelligent system: equipment failure, state failure or delay change caused by factors such as occlusion, high temperature, electromagnetism or equipment movement; Status failures, base station failures, edge computing node failures and so on.;
2) Changes brought about by the status of sample data: irregular standard data sets, data set interference points, noise points, and outlier ratios, large sample data set interference and so on.
3) The algorithm design is insufficient to identify sample data with large interference, algorithm crash, algorithm response timeout, as well as identification and decision-making failures caused by reduced algorithm accuracy and functional faults in intelligent systems.
4) Changes in system status due to new technologies and frameworks.
5) Changes in system state due to functional characteristics

- Effect evolutionary rules

According to these conditions, the protocols and configuration and deployment mechanisms between different intelligent system devices are combined. Following aspects are mainly considered:

1) AI recognizes causal changes caused by faults and recovery of functions and hardware system components.
2) When multiple processes are deployed at the same time, the causal changes brought about by deployment strategies such as priority assignment.
3) Causal changes triggered by quantitative changes in processes and their path generation and ending.

- Temporal variation characteristics of cause and effect

The temporal process describes the process by which the causal relationship between factors changes dynamically over time.

Specifically, it refers to the direction and strength of the causal arrow between the intelligent system rules of the intelligent system and the AI intelligent perception, recognition, decision-making and other factors. According to the above analysis of external disturbance, the factors affecting the cause and effect of the system include AI intelligent module, equipment failure of the system, deployment of system process and configuration information (Fig. 6).

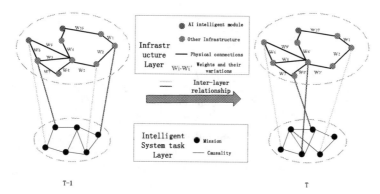

Fig. 6. Causality over time

Reliability Evaluation Methods

The running time of abnormal operation caused by system stop due to the above factors is used to calculate the reliability.

The intelligent system reliability algorithm process and steps are shown in the following Fig. 8.

Input: the topology $G(V, E)$ of the intelligent system, the collection App of intelligent system tasks, the sequence S of external interference conditions of the intelligent system, and the relevant parameters of causality rules, such as failure detection rate;

Output: task reliability R.

6   Intelligent System Reliability Test Verification Method

6.1   Intelligent System Reliability Test Evaluation Requirements

The intelligent system reliability test and evaluation environment shall meet the minimum environmental requirements required by the target intelligent system. The test profile should be consistent with the stress type, stress level, etc., and the reliability test requirements of the intelligent system should not be less than:

- Hardware and system software must not be modified during the test;
- The evaluation process should ensure that failure mechanisms do not occur.

Under the condition of ensuring the above requirements, the processing time and processing cost are minimized as much as possible.

6.2   AT Adaptability Analysis

In traditional reliability, the basic idea of AT is to carry out reliability tests under high utilization rate, temperature, vibration and other accelerated stresses, and extrapolate the reliability level under normal stress levels according to the acceleration situation to quickly verify product reliability Provides an effective method.

In section 5.3.12 of ISO/IEC TR 24372 states "Machine learning models should use the same test environment to use accelerated situations during training or testing.

Each machine learning method under test should be optimized to take advantage of acceleration where available and appropriate."

Traditional stress type settings are usually temperature, vibration, electrical stress, etc., but these stresses are not suitable for AI systems. Different from traditional equipment, the core intelligent module (AI model) of intelligent systems is related to the frequency of use and interference environment, such as increasing the amount of input data, failure Injection, interference with the environment and so on.

In summary, the idea of AT can be applied to AI systems, requiring additional modeling work to collect AT data for reliability prediction. The key step is to model the acceleration factor and link reliability performance to normal usage conditions.

7   Conclusion and Outlook

Different from traditional reliability, intelligent system reliability has its own characteristics, through the analysis of intelligent system SMART framework, the following preliminary exploration is mainly carried out:

1. Intelligent feature introduction, intelligent system has the characteristics of strong data dependence, model generalization, and learning adaptability;
2. Uncertainties such as complex state and dynamic state trigger the process failure mechanism of identification, prediction and decision-making;
3. Build a comprehensive intelligent system reliability parameter system around data, models and operating environments;
4. Compared with traditional methods such as structure and state space, causal analysis theory has the ability to describe the reliability of intelligent systems;
5. Accelerated stress and feasibility analysis of accelerated test.

In terms of system characteristics and fault mechanisms, it is still necessary to collect information in practical application scenarios for continuous iterative improvement. Further discussion on the comprehensiveness, correctness and necessity of the parameter system is still needed; The maturity of the causal model needs to be improved, and it is necessary to pay attention to the comprehensiveness of fault logic relationships and external disturbances under multi-scenario applications. In addition, how to carry out test verification, test profile and acceleration factor determination still need to be further determined, and the reliability of intelligent systems still needs to be studied in depth.

References

1. Chen, R., Yang, C., Han, S., Zheng, J.: Dynamic path planning of USV with towed safety boundary in complex ocean environment. In: Proceeding of the 33rd China Conference on Control and Decision Making, pp. 444–449. IEEE (2021)
2. Xiao, J., Lu, J., Li, X.: Davies Bouldin Index based hierarchical initialization K-means. Intell. Data Anal. **21**(6), 1327–1338 (2017)
3. Zhou, Z.: Machine Learning, p. 01. Tsinghua University Press, Beijing (2016)
4. Mnih, V., Kavukcuoglu, K., Silver, D., et al.: Human-level control through deep reinforcement learning. Nature **518**(7540), 529–533 (2015)

5. Zhang, Y.: Research on knowledge graph representation learning based on entity attribute information. Jilin Univ. (2022). https://doi.org/10.27162/d.cnki.gjlin.2022.004382
6. Van Erven, T., Harremos, P.: Rényi divergence and Kullback-Leibler divergence. IEEE Trans. Inf. Theory **60**(7), 3797–3820 (2014)
7. Liu, X., Li, Y., Zhang, J., et al.: Self-adaptive dynamic obstacle avoidance and path planning for USV under complex maritime environment. IEEE Access (2019)
8. Zhang, B.: Cymbals Artificial intelligence enters the post-deep learning era. Chin. J. Itell. Sci. Technol. **1**(01), 4–6 (2019)
9. Cui, T., Li, S.: Research on the principle of failure analysis of artificial intelligence system. J. Intell. Syst. **16**(4), 785–791 (2021)
10. Shi, W., Zhang, M.: Artificial intelligence for remote sensing target reliability recognition: overall framework design, current analysis and prospect. J. Surv. Map. **50**(8), 1049–1058 (2021). https://doi.org/10.11947/j.AGCS.2021.20210095
11. Yang, Z., Yang, C., Chen, F., et al.: Parameter estimation of machining center reliability model based on PSO algorithm and SVR model. J. Jilin Univ. Eng. Sci. **3**, 8 (2015)
12. Zou, Q., Liu, Y., He, M., et al.: Online evaluation method for CPS system reliability based on machine learning. Comput. Eng. Appl. **50**(10), 128–130 (2014)
13. Information Technology Artificial Intelligence Quality Elements and Test Methods of Machine Learning Models and Systems: T/CESA 1036-2019 (20 19)
14. Dong, R., Zhang, H., Liu, W.: Quality elements and testing methods of machine learning systems. Electron. Test **2021**(9), 92–94, 103 (2021). https://doi.org/10.3969/j.issn.1000-8519.2021.09.037
15. Wang, Y.X., Hou, M., Plataniotis, K.N., et al.: Towards a theoretical framework of autonomous systems underpinned by intelligence and systems sciences. IEEE-CAA J. Autom. Sinica **8**(1), 52–63 (2021)
16. Zhu, X., Wang, H., You, H.: Review of research on testing of autonomous driving intelligent systems. J. Softw. **32**(7), 2056–2077 (2020)
17. Mankad, K.B.: An intelligent process development using fusion of genetic algorithm with fuzzy logic. Artificial Intelligence: Concepts, Methodologies, Tools, and Applications, pp. 245–281. IGI Global (2017)
18. Stranieri, A., Sun, Z.: Only can AI understand me?: Big data analytics, decision making, and reasoning. Intelligent Analytics With Advanced Multi-Industry Applications pp. 46–66. IGI Global (2021)
19. Köse, U., Arslan, A.: Chaotic systems and their recent implementations on improving intelligent systems. Handbook of Research on Novel Soft Computing Intelligent Algorithms: Theory and Practical

The Swarm Hierarchical Control System

E. V. Larkin[1], T. A. Akimenko[1(✉)], and A. V. Bogomolov[2]

[1] Tula State University, Tula 300012, Russia
tantan72@mail.ru
[2] St. Petersburg Federal Research Center of the Russian Academy of Sciences, St. Petersburg 199178, Russia

Abstract. A hierarchical multi-controller system designed to manage the swarm of unmanned vehicles, moving on the terrestrial surface, is investigated. The flowchart of the hierarchical control system is worked out, in which the system distributed controller is performed at the strategic (the organizing controller) and functional-logical (vehicle onboard controllers) levels. It is shown, that mathematical model of the onboard controller consists of two parts: part, describing the actual semantics of embedded control algorithm, and part, describing interpretation of control algorithm by Von Neumann type controller, unfolding in real physical time. The formula for time delay estimation, when embedded algorithm is the linear-cyclic one, is obtained. A time of delay is included into the matrix equation describing the semantics of data processing, which together with the description of the vehicle with caterpillar propellers movement on terrestrial vehicle forms a complete model of the swarm unit. Description of the organizing controller operation is obtained when realization steps of algorithm if accessing the milestone, including formulae for time estimation of accessing the milestone by the only unit and by the whole swarm, as far as formula for evaluation of time which only unit spend waiting the whole swarm. Theoretical results are confirmed by the five-units-swarm operation modeling.

Keywords: Swarm · controller · hierarchical level · control algorithm · delay · time interval · waiting time

1 Introduction

Effectiveness in solving problem by autonomous unmanned vehicles, operating on terrestrial surface, increases significantly, when they are gathered in swarms [1–4]. Managing by swarm is a more difficult problem, than control of a single swarm unit, due to necessity of synchronization of units operation. For example, when the problem, solving by swarm, is to reach some milestone together, as organized group [5], then there is necessary periodically align both velocities, and positions of all units on the terrestrial surface. The complexity of problems performed implies the complexity of control system structure, which is expediently divided into two hierarchical levels; strategic and functional-logical, each of which solves its own range of tasks [6, 7]. Strategic level is represented by organizing controller, whose function is to divide the general problem, solving by swarm as a whole, onto local tasks, solving by units, to transmit tasks onto functional-logical level, and to receive confirmation of tasks completion. Unit onboard controllers, forming functional-logical level, are switched onto feedbacks, which control current state of unmanned vehicles.

Effectiveness of swarm units controllable operation is determined, on the one hand, by control algorithm, embedded into controller, and on the other hand, by Von Neumann type controller performance, as a physical device, which interpret control algorithm operator-by-operator, and spend on the interpretation a time. In turn, computer «thinking» time causes both data skews, and time delays in feedback contours [8–10]. Thus, the digital controller, in addition to the algorithmic implementation of the control law [11], contributes to delays in the control process, which, in turn, affects the quality characteristics of the swarm unit control system [12, 13]. To simulate time delays, one can use the apparatus of the semi-Markov process [14–16], since each control algorithm processes random data generated at the outputs of the sensors and includes decision-making operators at branching points.

On the strategy level swarm units onboard computers compete with each other for execution of their tasks, which affect on the solving of the global problem [17]. Thus, prediction of unit behavior during competition is necessary for a design of the effective swarm managing algorithm, embedded into the organizing controller.

Methods of hierarchical control system design, which take into account real properties of Von Neumann type controller, are not sufficiently developed in engineering practice, which explains the urgency and relevance of the study.

2   Configuration of the Hierarchical Swarm Control System

Structure of swarm under managing of the two level hierarchical control system, is shown on the Fig. 1.

The control system strategy level is performed by the organizing controller, whose destination is to generate the global aim of swarm functioning, to divide the global problem onto local tasks, solving by swarm units, and to transmit tasks to each performer. When swarm includes N unmanned vehicles, $U_1, \ldots, U_n, \ldots, U_N$, moving on a terrestrial surface, aim may be formulated as gathering to the milestone at the predetermined time.

So, organizing controller, receiving information from $U_1, \ldots, U_N$ about their current Cartesian co-ordinates $(\mathbf{z}_1, \ldots, \mathbf{z}_n, \ldots, \mathbf{z}_N)$ should compute for n-th unit the aim vector $\mathbf{f}_n = (f_{n,v}, f_{n,\psi})$, where $f_{n,v}$ is the desired longitudinal velocity of U_n; $f_{n,\psi}$ is its desired movement azimuth angle. In the controller of unmanned vehicle with caterpillar propellers vector $\mathbf{f}_n$ is transformed to action data vector $\mathbf{u}_n = (u_{n,l}, u_{n,r})$ of signals, applied to n-th left and right actuators accordingly. Under affection of $u_{n,l}$ and $u_{n,r}$ sum the vehicle longitudinal velocity is formed.

Their difference gives the rapidity of azimuth angle change. The unit U_n current state $\mathbf{x}_n = (x_{n,v}, x_{n,\psi})$, where $x_{n,v}$ and $x_{n,\psi}$ are current longitudinal velocity and azimuth angle correspondingly, is measured with the sensor subsystem, which forms output data vector $\mathbf{y}_n = (y_{n,v}, y_{n,\psi})$, where $y_{n,v}$ and $y_{n,\psi}$ are results of measurement of current longitudinal velocity and azimuth angle. Onboard controller transforms $y_{n,v}$ and $y_{n,\psi}$ to digital codes, computes current Cartesian coordinates of n-th unit, and transmits them to organizing controller, which forms regular local tasks for swarm units.

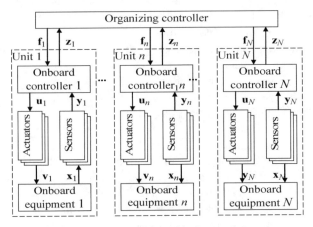

Fig. 1. Flowchart of hierarchical control system

3   Operation of Control System as (N+1)-Parallel Semi-markov Process

Effectiveness of swarm managing is determined, on one hand by control algorithms semantics, and on the other hand, by properties of Von Neumann type controllers as physical devices, generating time intervals between transactions when algorithms interpretation. Due to data, processed in controllers, are random, control algorithm has number of branching point and computational complexity of different branches may be different, time intervals between transactions are random too. So the model of hierarchical control system may be performed as the $(N + 1)$-parallel semi-Markov process [17]

$$\tilde{\mu} = \{\mu_0(t), \ \mu_1(t), \ ..., \ \mu_n(t), \ ..., \ \mu_N(t)\} \tag{1}$$

where $\mu_0(t)$ describes the organizing controller operation; $\mu_n(t)$, $1 \leq n \leq N$, describes the operation of n-th controller; t is the time;

$$\mu_n = \{A_n, \ \mathbf{h}_n(t)\}. \tag{2}$$

In (2) $A_n = \{a_{1(n)}, \ ..., \ a_{j(n)}, \ ..., \ a_{J(n)}\}$ are states of semi-Markov process $\mu_n(t)$, $\mathbf{h}_n(t) = \mathbf{g}_n(t) \otimes \mathbf{p}_n = [h_{j(n),l(n)}]$ is the $J(n) \times J(n)$ semi-Markov matrix; $\mathbf{g}_n(t) = [g_{j(n),l(n)}(t)]$ is the $J(n) \times J(n)$ matrix of pure time densities; $\mathbf{p}_n = [p_{j(n),l(n)}]$ is the $J(n) \times J(n)$ stochastic matrix. In the context of problem being solved, only time intervals between transactions are significant, so $J(0) = 2N$, due to organizing controller quests N onboard controllers when input data from them and when output data to them; $J(n) = 4$, $1 \leq n \leq N$, due to every onboard controller inputs data from sensors and organizing controller, and output data to actuators and to organizing controller back. The control algorithm may have arbitrary high structural complexity, while the principle of data acquisition precedence before function computation itself must be strictly followed. So, semi-Markov process (2) may be reduced to process with the structure, shown on the Fig. 2a, where $a_{1(n)} \div a_{3(n)}$ are operators of reading data from peripherals to the n-th,

$1 \le n \le N$, onboard controller (two sensors and organizing controller); $a_{4(n)} \div a_{6(n)}$ are operators of writing data to peripherals of n-th, $1 \le n \le N$, onboard controller (two actuators and organizing controller); DP - data processing operator.

Transactions flow is shown on the Fig. 2b, where arrows with round tip indicate data inputs, while arrows with square tip indicate data outputs. Similar structure may be obtained from 0-th algorithm, describing organizing controller operation.

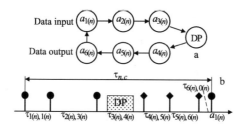

Fig. 2. Reduced structure of control algorithm

With use methods, described in [10, 18], for every neighboring pair of transactions time delays $\tau_{\xi(n),\zeta(n)}$, where $\xi(n), \xi(n) \in \{1(n), ..., 6(n)\}$, for $1 \le n \le N$, and $\xi(0), \xi(0) \in \{1(0), ..., N(0), N(0) + 1, ..., N(0) + N(0)\}$, may be determined. Common control program runtime τ may be determined as follows:

$$\tilde{\tau}_n = \begin{cases} \sum_{\xi(n)=1(n)}^{6(n)} \tau_{\xi(n),\xi(n)+1}, & \text{when } 1 \le n \le N; \\ \sum_{\xi(0)=1(0)}^{N(0)+N(0)} \tau_{\xi(0),\xi(0)+1}, & \text{when } n = 0. \end{cases} \tag{3}$$

In addition to $\tilde{\tau}_n$ dispersions $D_{\xi(n),\zeta(n)}$ of time delays should be evaluated too, for example with use the method, described in [10, 18]. In the case of linear-cyclic algorithm, shown on Fig. 2, common dispersion is as follows

$$\tilde{D}_n = \begin{cases} \sum_{\xi(n)=1(n)}^{6(n)} D_{\xi(n),\xi(n)+1}, & \text{when } 1 \le n \le N; \\ \sum_{\xi(0)=1(0)}^{N(0)+N(0)} D_{\xi(0),\xi(0)+1}, & \text{when } n = 0; \end{cases} \tag{4}$$

In (3) and (4) $7(n) = 1(n)$; $N(0) + N(0) + 1 = 1(0)$.

Time intervals between transactions may be estimated according either expectation (3), or either using «the three sigma rule» [19]:

$$\tau_{n,} = \begin{cases} \tilde{\tau}_n, & \text{when mean value;} \\ \tilde{\tau}_n + 3\sqrt{\tilde{D}_n}, & \text{when maximal value.} \end{cases} \tag{5}$$

Time interval (3) between similar transactions should meet the Nyquist conditions of sampling theorem [20].

Pure delay of controller response on signals obtained from peripherals may be estimated as half of $\tau_{n,c}$, i.e.

$$\tau_n = \tau_{n,c}/2. \tag{6}$$

In common, semantics of control algorithm, embedded into organizing controller, supposes computation of the vector function

$$\widetilde{\boldsymbol{\varphi}}_0 = \left(\widetilde{\mathbf{f}}_1\left[\mathbf{u}_1(t), ..., \mathbf{u}_n(t), ..., \mathbf{u}_N(t), t\right] ... \widetilde{\mathbf{f}}_n\left[\mathbf{u}_1(t),\right.\right. \tag{7}$$
$$\left.\left. \mathbf{u}_n(t), ..., \mathbf{u}_N(t), t\right] ... \widetilde{\mathbf{f}}_N\left[\mathbf{u}_1(t), ..., \mathbf{u}_n(t), ..., \mathbf{u}_N(t), t\right]\right),$$

where $\widetilde{f}_1, ..., \widetilde{f}_n \ \widetilde{f}_N$ are data, which are formed in the controller memory until they are downloaded onto proper peripheral via the interface.

After downloading the vector function takes the form

$$\boldsymbol{\varphi}_0 = \left(\mathbf{f}_1\left[\mathbf{u}_1(t-\tau_0), ..., \mathbf{u}_n(t-\tau_0), ..., \mathbf{u}_N(t-\tau_0), t-\tau_0\right] ... \mathbf{f}_n\left[\!\left[\mathbf{u}_1(t-\tau_0), ,,. \mathbf{u}_n(t-\tau_0),\right.\right.\right.$$
$$\left.\left.\left. ..., \mathbf{u}_N(t-\tau_0), t-\tau_0\right] ... \mathbf{f}_N\left[\mathbf{u}_1(t-\tau_0), ..., \mathbf{u}_n(t-\tau_0), ..., \mathbf{u}_N(t-\tau_0), t-\tau_0\right]\right), \tag{8}$$

where τ_0 is the delay time.

Similarly common semantics of control algorithms, embedded into onboard controllers, suppose computation of the vector function.

$$\widetilde{\boldsymbol{\varphi}}_n = \left(\widetilde{u}_n \ \left[\mathbf{f}_n(t), \mathbf{y}_n(t), t\right] \ \widetilde{z}_n\left[\mathbf{f}_n(t), \mathbf{y}_n(t), t\right]\right), \ 1 \leq n \leq N, \tag{9}$$

where $\widetilde{u}_n, \widetilde{z}_n$ are data, which are formed in the n-th onboard controller memory until they are downloaded onto n-th peripherals via interface.

After downloading vector function takes the form

$$\boldsymbol{\varphi}_n = \left(u_n \ \left[\mathbf{f}_n(t-\tau_n), \mathbf{y}_n(t-\tau_n), t-\tau_n\right], \ z_n\left[\mathbf{f}_n(t-\tau_n), \mathbf{y}_n(t-\tau_n), t-\tau_n\right]\right), \tag{10}$$

where τ_n is the delay time in the n-th controller, $1 \leq n \leq N$.

For simplification of simulation it is more conveniently use instead the description of vectors $\mathbf{f}_n(t), \mathbf{u}_n(t), \mathbf{v}_n(t), \mathbf{x}_n(t), \mathbf{y}_n(t), \mathbf{z}_n(t)$ in time domain their Laplace transforms [21] as follows: $\mathbf{F}_n(s) = L[\mathbf{f}_n(t)]; \ \mathbf{X}_n(s) = L[\mathbf{x}_n(t)]; \ \mathbf{U}_n(s) = L[\mathbf{u}_n(t)]; \mathbf{V}_n(s) = L[\mathbf{v}_n(t)]; \ \mathbf{X}_n(s) = L[\mathbf{x}_n(t)]; \ \mathbf{Y}_n(s) = L[\mathbf{y}_n(t)], \mathbf{Z}_n(s) = L[\mathbf{z}_n(t)]$, where $L[...]$ is the direct Laplace transform operation, s is the Laplace variable (differentiation operator). With use the Laplace transforms of signal vectors delays in controllers may be described as

$$\mathbf{F}(s) = L[\mathbf{f}(t-\tau_0)] = \widetilde{\mathbf{F}}(s) \cdot \exp(-\tau_0 s);$$
$$\mathbf{U}(s) = L[\mathbf{u}_n(t-\tau_n)] = \mathbf{U}(s) \cdot \exp(-\tau_n s). \tag{11}$$

Expressions (7)–(11) completely describe controllers operation, as the physical device, that processes information that generate action data based on the information received.

4   The Swarm Unit Description

Structure of swarm unit, as the object under control, is shown on the Fig. 3. In the unit vector control signal $\left[U_{n,l}(s),\ U_{n,r}(s)\right]$ is sent to left and right actuators, which through transmission are linked with left and right caterpillar propellers. The middle velocity of caterpillars gives the unit longitudinal speed $Y_{n,v}$, while integrated difference of velocities gives the movement azimuth angle $Y_{n,\psi}(s)$.

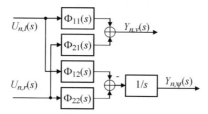

Fig. 3. Object under control

The matrix equation, describing the object under control, is as follows:

$$\begin{bmatrix} Y_{n,v}(s) \\ Y_{n,\psi}(s) \end{bmatrix} = \begin{bmatrix} 1 & 0 \\ 0 & 1/s \end{bmatrix} \begin{bmatrix} \Phi_{11}(s) & \Phi_{21}(s) \\ -\Phi_{12}(s) & \Phi_{22}(s) \end{bmatrix} \begin{bmatrix} U_{n,l}(s) \\ U_{n,r}(s) \end{bmatrix}, \tag{12}$$

where $\Phi_{11}(s), \Phi_{12}(s)\ \Phi_{21}(s),\ \Phi_{22}(s)$ are transfer functions, which determine the n-th swarm unit dynamics; $1/s$ is the transfer function of the physical integrator, which is the vehicle trunk, driven by caterpillar propellers.

When sensors are practically inertia-less, the performance of the n-th swarm unit is determined by acceleration characteristic of loaded actuators only. So, for symmetrical vehicle $\Phi_{11}(s) = \Phi_{21}(s)\frac{k_1}{Ts+1}$; $\Phi_{21}(s) = \Phi_{22}(s)\frac{k_2}{Ts+1}$, where T is the time constant of acceleration characteristics.

5   The Organizing Controller Operation

One of possible task, which may solve the organizing controller, is the managing by all swarm units reaching the milestone simultaneously. For managing by n-th unit step-by-step cyclic algorithm is realized. On every step unit desired velocity $f_{n,v}$ is set up by controller. The place in the formation is set by direction of the desired azimuth angle $f_{n,\psi}$. Pre-determined distance ϑ_n till the milestone with the desired velocity $f_{n,v}$ n-th unit would cover in the time $\theta_n = \vartheta_n/f_{n,v}$, while random factors, such as deviation of current longitudinal velocity $x_{n,v}$ and azimuth angle $x_{n,\psi}$ due to complexity of terrestrial relief, presence of obstacles to movement and n-th unit control system operation exactness, make the time under consideration the random one. According to the central limit theorem [18, 22] a time of reaching the milestone is normally-bounded distributed:

$$g_n(t) = \frac{\eta\left(t - \theta_{n,\min}\right)\frac{1}{\sqrt{2\pi}\sigma_n}\exp\left[-\frac{(t-\theta_n)^2}{2\sigma_n^2}\right]}{\int\limits_{\theta_{n,\min}}^{\infty}\frac{1}{\sqrt{2\pi}\sigma_n}\exp\left[-\frac{(t-\theta_n)^2}{2\sigma_n^2}\right]}, \tag{13}$$

where σ_n is the standard deviation of milestone reaching time by n-th unit distance; $\eta(t - \theta_{n,\min})$ is the shifted Heaviside identity function.

Restriction of the definition domain in (12) is linked with limitation of n-th unit velocity, namely $\theta_{n,\min} = \vartheta_n/f_{n,v,\max}$, when azimuth angle is directed strongly on the milestone.

Time of reaching the milestone by the whole swarm is distributed as follows [18]:

$$g_w(t) = \frac{d \prod_{n=1}^{N} G_n(t)}{dt}, \tag{14}$$

where $G_n(t) = \int_{-\infty}^{t} g_n(\tilde{t})d\tilde{t}$.

Backlog in getting the milestone of the swarm, as a whole, from any unit may be evaluated as [16]:

$$g_{n \to w}(t) = \frac{\eta(t) \int_0^{\infty} g_n(\xi)g_w(\xi + t)d\xi}{\int_0^{\infty} G_n(t)dG_w(t)}, \tag{15}$$

where $\eta(t)$ is the Heaviside identity function; $G_w(t) = \prod_{n=1}^{N} G_n(t)$.

Mean value of backlog is as follows: $T_{n \to w} = \int_0^{\infty} t \cdot g_{n \to w}(t)dt$ So, organizing controller managing algorithm should be designed to minimize $T_{n \to w}$.

6   The Example

To confirm main theoretical results, discussed above, let's consider hierarchical control system of homogeneous five units swarm. In the structure shown on the Fig. 3, and description (12) the time constant of acceleration characteristics is equal to $T = 0, 1$ s Transmission ratio of longitudinal movement channel is $k_1 = 1, 5$. Transmission ratio of azimuth angle maneuver channel is $k_2 = 8$. All sensors, in comparison with actuators and object under control itself, are considered as inertia-less links. Onboard controller computes the error between desired controllable parameters $\mathbf{f}_n$ and results of measurement of current state of unit $\mathbf{y}_n$ and forms the action signal $\mathbf{u}_n$, transmitted to actuators. Unit response on standard control action $[f_{n,v}(t), f_{n,\psi}(t)] = [\eta(t), \eta(t)]$ is shown on the Fig. 4.

Graphs, shown on the Fig. 4a, display a change of the longitudinal velocity depending on the delay, born by onboard controllers. Graphs, shown on the Fig. 4b, display a change of the azimuth angle depending on the delay, born by the onboard controller. Curves 1 were obtained, when the delay is equal to zero; curves 2 were obtained, when the delay estimated without «the three sigma rule», and is equal to 0,02 s; curves 3 were obtained, with taking into account «the three sigma rule», and is equal to 0,03 s. As one can see

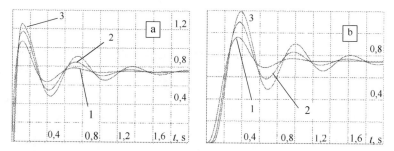

Fig. 4. Reaction of the swarm unit on the Heaviside function

from plots, decreasing of the onboard controller performance leads to an overshoot and the transient time increasing.

Figure 5 shows time characteristics of getting the milestone by swarm when realizing step-by-step algorithm. During every step organizing controller predetermines the distance, which n-th unit should overcome during 0,07 s.

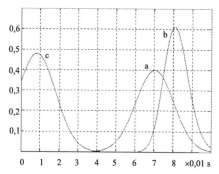

Fig. 5. Time distribution of getting the milestone by n-th unit (a), by swarm as a whole (b) and the backlog in getting the milestone by the swarm respectively to n-th unit

Curve 5a shows the time distribution density of overcoming the distance by sole swarm unit. Curve 5b shows the time distribution density of overcoming the distance by five units swarm. As one can see from the curves, the swarm covers the distance more slowly, then one swarm unit. This fact creates the problem a time overhead by «competition winner» which is the greater, the more units in the swarm. Curve 5c shows the time distribution density of waiting by sole unit when whole swarm overcomes the distance. As one can see from plots, waiting time depend on overcoming the distance by sole unit density dispersion. So managing algorithm, embedded into the organizing controller, should operate in such a way, as to increase the consistency of units movement at all movement stages.

7 Conclusion

As a result, a simple engineering method for investigation of swarm two-level hierarchical control system is proposed. Semi-Markov processes fundamental theory put on the forefront of method. The approach proposed allows take into account time characteristics of Von Neumann type controllers, which are basis of control system hardware. It is shown, that distributed controller hardware hierarchical configuration permits to unload organizing controller from dispatching functions and reduce the managing by swarm to sequence of simple tasks of achieving the milestone from a current state.

Further investigations in the domain may be directed to working out the model of interaction between organizing computer and controllers in time domain, based on the concurrency theory. The question of optimal division the distance onto stages for increasing the swarm effectiveness, also needs to be resolved.

Acknowledgments. The research was carried out within the grant № 22-26-00808 of Russian Scientific Foundation.

References

1. Tzafestas, S.G.: Introduction to Mobile Robot Control, 750 p. Elsevier (2014)
2. Godwin, M.F., Spry, S.C., Hedrick, J.K.: A Distributed System for Collaboration and Control of UAV Groups: Experiments and Analysis, p. 224. Center for the Collaborative Control of Unmanned Vehicles University of California, Berkeley (2007)
3. Larkin, E.V., Privalov, A.N., Bogomolov, A.V.: Discrete approach to simulating synchronized relay races. Autom. Document. Math. Linguist. **54**(1), 43–51 (2020). https://doi.org/10.3103/S0005105520010082
4. Yu Chun, Y., Ming, H.W., Xiao, S.H.: Vehicle swarm motion coordination through independent local-reactive agents. In: Wu, Y. (ed.) Advanced Materials Research. vol. 108–111, pp. 619–624 (2010)
5. Larkin, E., Bogomolov, A., Antonov, M.: Modeling of increased rigidity of industrial manipulator. In: Ronzhin, A., Rigoll, G., Meshcheryakov, R. (eds.) ICR 2020. LNCS (LNAI), vol. 12336, pp. 170–178. Springer, Cham (2020). https://doi.org/10.1007/978-3-030-60337-3_17
6. Aström, J., Wittenmark, B.: Computer Controlled Systems: Theory and Design, 557 p. Tsinghua University Press. Prentice Hall (2002)
7. Landau, I.D., Zito, G.: Digital Control Systems, Design, Identification and Implementation, 484 p. Springer (2006)
8. Arnold, K.A.: Timing analysis in embedded systems. In: Ganssler, J., Arnold , K., et al.: Embedded hardware. 01803, pp. 239–272. Elsevier Inc, USA (2008)
9. Larkin, E.V., Akimenko, T.A., Bogomolov, A.V.: Modeling the reliability of the onboard equipment of a mobile robot. Izvestiya of Saratov University. New Series. Series: Mathematics. Mechanics. Informatics, vol. 21, no. 3, pp. 390–399 (2021)
10. Larkin, E.V., Nguyen, V.S., Privalov, A.N.: Simulation of digital control systems by nonlinear objects. Lect. Notes Data Eng. Commun. Technol. **124**, 711–721 (2022)
11. Monson, H.: Statistical Digital Signal Processing and Modeling, 624 p. Willey, N.Y. (2009)
12. Zhang, X.M., Min, W.U., Yong, H.E.: Delay dependent robust control for linear systems with multiple time-varying delays and uncertainties. Control Decision **19**(5), 496–500 (2004)

13. Li, D., Chen, G.: Impulses-induced p-exponential input-to-state stability for a class of stochastic delayed partial differential equations. Int. J. Control **92**(8), 1805–1814 (2019)
14. Janssen, J., Manca, R.: Applied Semi-Markov processes, p. 310. Springer, US (2006)
15. Larkin, E.V., Nguyen, V.S., Privalov, A.N., Dobrovolsky, N.N.: Relay races along a pair of selectable routes. Bulletin of the South Ural State University. Series: Mathematical Modelling, Programming and Computer Software, vol. 11, no 1, pp. 15–26 (2018)
16. Larkin, E.V., Nguyen, V.S., Privalov, A.N., Dobrovolsky, N.N.: Discrete model of paired relay-race. Bulletin of the South Ural State University. Series: Mathematical Modelling, Programming and Computer Software, vol. 11, no. 3, pp. 72–84 (2018)
17. Girault, C., Valk, R.: Petri Net technology for systems Engineering. A Guide to Modelling, Verification and Applications, 621 p. Springe (2001)
18. Kobayashi, H., Mark, B.L., Turin, W.: Probability, Random Processes, and Statistical Analysis. Cambridge University Press (2012). https://doi.org/10.1017/CBO9780511977770
19. Pukelsheim, F.: The three sigma rule. Am. Statist. **48**(2), 88–91 (1994). https://doi.org/10.1080/00031305.1994.10476030
20. Yeh, Y.-C., Chu, Y., Chiou, C.W.: Improving the sampling resolution of periodic signals by using controlled sampling interval method. Comput. Electric. Eng. **40**(4), 1064–1071 (2014)
21. Bogomolov, A.V., Bychkov, E.V., Kotlovanov, KYu.: Stochastic mathematical model of internal waves. Bull. South Ural State Univ. Ser. Math. Model. Program. Comput. Softw. **13**(2), 33–42 (2020). https://doi.org/10.14529/mmp200203
22. Bárány, I., Van, V.: Central limit theorems for Gaussian polytopes. Ann. Probabil. **35**(4),(2007). https://doi.org/10.1214/009117906000000791

Swarm Intelligence Optimization Algorithms

Multi-Modal Honey Bee Foraging Optimizer for Weapon Target Assignment Problem

Beena S. Lawrence[(⊠)] and Chandra S. S. Vinod

Machine Intelligence Research Laboratory, Department of Computer Science,
University of Kerala, Thiruvananthapuram, India
{beena,vinod}@keralauniversity.ac.in

Abstract. Weapon Target Assignment(WTA) problem, a classical combinatorial optimization problem, has been a consistent threat while considering the field of Air Defence or Military Operations research and the threat is continually increasing as technological advances increase and decision-making is becoming extremely intricate. As the problem falls into the category of NP-complete, heuristic algorithms compete over traditional methods and give near optimal solutions in less time. The solution to this problem will give insight into various real-life applications such as Media Allocation, Product Positioning and Market Research. In this paper, we proposed the energy-efficient Multi-Modal Honey Bee Foraging optimizer to solve the computationally complex WTA problem. The results show the efficiency of the proposed method in solving the problem with 100% convergence and provide statistically better results in all 12 different problem scenarios with low to high dimensions.

Keywords: Combinatorial search · NP-Complete · Multi-Modal Honey Bee Foraging Optimizer · Weapon Target Assignment · Military Operations Research

1 Introduction

Research on effective allocation of Air Defence Resources attracted great attention in recent years due to the advancements in the area of Modelling and Solution techniques. While considering the field of Automating the Military Command and Control (C2), The Weapon Target Assignment [4] becomes a pivotal decision making problem which is highly complex and its scenarios are always uncertain, stochastic in nature. An astute solution to the problem is always demanding in the modern war field which is equipped with sophisticated weapon genres like Remotely-Piloted Aircrafts (RPA).

In optimization and operations research, the WTA problem can be generally defined as the optimal allocation of various weapons of various types to a set of targets. In literature WTA problem can be viewed in two different perspectives - Defensive and Offensive. In Defensive perspective the problem can be defined

Y. Tan et al. (Eds.): ICSI 2023, LNCS 13968, pp. 43–54, 2023.
https://doi.org/10.1007/978-3-031-36622-2_4

as finding an ideal cost effective energy efficient assignment of weapons of various genres to incoming targets or missiles conducive to minimize the targets' probability of destroying the protected assets. In the Offensive perspective, the problem is to optimally assign various weapons to maximize the total damage of the enemy protected assets and targets. Both the above mentioned problems are under the category of asset based WTA, where the value of threatened assets are employed directly. If the goal is to minimize the total threat of targets and the value of threatened assets are not considered then they come under the category of target based WTA.

Research on WTA can be done in two different versions to resolve it by considering various aspects like time, complexity and finiteness. They are Static WTA(SWTA) and Dynamic WTA(DWTA). In Static WTA, all problem attributes are fixed and the weapons are launched at targets at once. This single stage allocation of weapons allow the damage assessment only after the mission is completed and no subsequent engagements are possible in SWTA. Thus it can be viewed as assigning weapons for temporary defense purposes. The solution to SWTA provides information about how many weapons are launched to each target as a whole. In contrary, DWTA have multiple stages for weapon allocation and time is a dimension considered. The first stage of DWTA is the same as the SWTA and for each subsequent stage the surviving targets are aimed and assigned weapons to completely destroy them by assessing the damage after each engagement in no time. DWTA solutions informs about how many weapons to shoot for the initial engagement and how many to reserve for subsequent engagements in order to optimize the use of weapons and destruction of targets. Variants of DWTA include - Multi-stage DWTA and Shoot-look-shoot DWTA. In Multi-stage DWTA, the allocation of weapons in different stages is computed based on some probability distribution. Shoot-Look-Shoot DWTA enables the defense to assess the damage at each stage and its solution informs the defense about how many weapons to be reserved for the next level to re-engage any missile leakage. Thus the complexity is more in the case of DWTA.

The decision making problem of WTA can be proved as NP-Complete [7]. According to computational complexity theory, it is a good sign if a problem can be solved fast, in a manner that is roughly linear, or at the very least, a small polynomial function of the input size. Yet, NP completeness is a sort of proof that there are numerous significant real-world issues that cannot be addressed rapidly. The difficulty of the challenges is categorised in computational complexity theory. Some of the common and useful classifications include P, NP, PSPACE, EXPTIME, and Undecidable. The challenge of applying complexity theory to specific issues is addressed by the NP completeness theory, which focuses on the distinction between P and NP classes. It is defined more explicitly in terms of "reduction," which can be viewed as a manner.

In different ways one can approach NP complete problems based on the impact of those problems in real scenarios. Either go for finding an exact solution or solve it approximately. If an exact solution is really in need then develop an exponential time algorithm without worrying about the complexity. But if quick solutions are the priority then settle down for near optimal solutions instead of exact solutions. Here comes the importance and applicability of Heuristic, Meta-heuristic and Hyper-heuristic techniques. All these techniques provide quite effective and fast strategies for finding solutions to complex optimization problems. Nature Inspired Optimization techniques are the kind of techniques which draw great attention in recent research, in the field of complex optimization problems, which helps in finding quick and near optimal solutions.

Nature inspired optimization algorithms are the algorithms inspired from optimization processes that occur in nature. They are kind of general purpose "black box" optimization algorithms that exploit only a limited knowledge regarding the problem and thus well fit for real world complex scenarios with uncertainties which always lack the whole physics of the problem. Evolutionary algorithms, Swarm Intelligence based algorithms, algorithms inspired from physics and chemistry and other natural behavior based algorithms are from the field of nature inspired computing techniques [16] and are effectively applied for finding the solutions of many complex applications. [1,6,8,15,17]are some of the Nature inspired algorithms used to solve the WTA problem.In this paper, we tried to solve the more complex NP complete WTA problem using Multi Modal Honey Bee Foraging Optimizer [11] which mimics the honey bees' foraging in multiple bee colonies for multiple commodities.

2   Mathematical Modelling of WTA

Flood presented the WTA problem to the field of operations research for the first time in 1957, while the mathematical formulation was contributed to the literature by Manne in 1958 [9]. There are different formulations in literature for both SWTA and DWTA. For better understanding of the problem some of the formulations are explained below.

2.1   Formulations on SWTA

Manne introduced the problem formulation with the following assumptions as there are w_i weapon types to defend against j targets. Each weapon has a probability of destroying the target as p_{ij}, target has a destructive value of v_j and number of weapons of type i assigned to target j is taken as x_{ij}. Thus the general mathematical definition for the basic SWTA problem as follows:

$$min \sum_{j=1}^{n} v_j \prod_{i=1}^{m} (1 - p_{ij})^{x_{ij}}$$

$$st \sum_{j=1}^{n} x_{ij} \leq w_i, \quad for\ i = 1,2,3, ...m \tag{1}$$

$$x_{ij} \in \mathbb{Z}_+, \quad for\ i = 1,2,3, ...m,\ j = 1,2,3,, ...n$$

Here the goal is to minimize expected survival value of target while maximizing the expected damage such that the number of weapons of each kind allocated must not exceed the number of weapons that are available and the assignments should be integer. Due to the complexity, many simpler formulations of the above mentioned non linear objective function are used in the literature [5] to make the problem easily optimizable but their solutions reduces the real applicability of the problem and they do not guaranteed to be optimal for the original problem formulated in (1).

Another variant of SWTA formulation is defined by Shang [13] in which the importance of the protected assets is taken into consideration. This makes the problem more complex but in real scenario the inclusion of the value of protected assets makes it more relevant in the field of Missile defense. The formulation can be defined as:

$$min \sum_{k=1}^{K} a_k \prod_{j=1}^{n_k} [\gamma_{jk}(1 - p_{ij}x_{ij})]$$

$$st \sum_{k=1}^{K} n_k = n,$$

$$\sum_{j=1}^{n} x_{ij} = 1, \quad for \ i = 1, 2, 3, ...m$$

$$x_{ij} \in \{0, 1\} \quad for \ i = 1, 2, 3, ...m, \ j = 1, 2, 3, ...n$$

(2)

where a_k is the value of asset k having values from 1 to K and γ_{jk} is the probability of destroying the asset k by target j.

2.2   Formulations on DWTA

The variants of DWTA also have different formulations in the literature. For DWTA, the problem becomes more complex and more realistic by considering parameters in every aspect of the air defense system. As the complexity increases, finding solutions turns out to be more expensive in terms of computations which in turn makes the research more challenging. From literature, some of the formulations which consider most of the dynamic scenarios of the problem are given in this paper for better understanding of the mathematical enigma.

Burr et all [2] put forth a model that aims to minimise projected damage to assets that are safeguarded while increasing damage to targets using minimum number of weapons as

$$\min_{d(k,j)} \sum_{k=1}^{K} \sum_{j=1}^{a(k)} d(k, j)$$

$$st \ V(d, a) \leq \frac{V}{n} \sum_{k=1}^{K} a(k), \quad for \ all \ a$$

(3)

where $a(k)$ is the attack strategy which identifies the number of targets aimed at asset k, $d(k, j)$ is the defense strategy which defines how many weapons to fire at the target j directed at asset k, V is the total value of all protected assets and n is the number of targets. $V(d, a)$, the estimated destruction to the guarded asset with a targets, and deployment strategy d where a_k is the value of asset k, can be calculated as

$$V(d, a) = \sum_{k=1}^{K} a_k \left(1 - \prod_{k=1}^{a(k)} (1 - q^{d(k,j)}) \right) \tag{4}$$

In this multi stage DWTA, it is assumed that the targets are arrived sequentially with sufficient time between arrivals but for each stage the number of targets aimed at the same asset is not known for the defender which is considered only as a probability distribution so that the defender forced to assign weapons in advance to potentially arriving targets.

Murphey [10] proposed another formulation of multi stage DWTA by assuming that the partial assignments of weapons are required at different stages where, for each stage, a subset of n targets is known and the remainder are not known with certainty. It is also assumed that there is a cost associated with weapon assignment $c(t)$, based on the two facts - missed reacquisition and attrition. Thus the deterministic formulation for the problem defined is as follows

$$min \sum_{t=1}^{T} c(t) \sum_{j=1}^{n(t)} V_j q_j^{x_j(t)}$$

$$st \sum_{t=1}^{T} \sum_{j=1}^{n(t)} x_j(t) = m, \tag{5}$$

$$V_j \in \mathbb{V} \in \mathbb{R}_+^n \quad for \ j = 1, 2, 3 ...... n(T)$$

$$x^t \in \mathbb{Z}_+^{n(T)} \quad for \ t = 1, 2, 3, ... T$$

The WTA problem becomes dynamic when partial information is available initially, and more information are revealed as the computation progresses. The sub-problem in each stage is the replica of static WTA with time as a dimension. The real scenario of WTA is always dynamic and is computationally more complex to solve optimally. In general, in the case of multi-stage dynamic WTA, the decision-making is easier in later stages due to the reduction of targets and available weapons.

3   Multi-Modal Honey Bee Foraging Optimization Algorithm

In accordance with their hive's needs, bees forage for water, pollen, honey and resins. This algorithm takes inspiration from a different epitome of foraging behavior performed by honey bees of multiple colonies. A mathematical framework is being developed to simulate the distributed computational method used by bees from numerous colonies who are employed to forage various resources while using the least amount of energy possible to maximise profit. Thus the optimizer provides a solution to optimization problems in complex real scenarios with multi dimensional parameters.

The forager bees in a bee colony are responsible for the collection of different commodities from multiple sources based on the requirement of their hives. A group of foraging bees known as scouts explore the search area to find fresh and quality sources of different products, and they use waggle dance to communicate their discoveries to the other bees in the hive. So, employed bees analyse the profitability of each source based on the information provided by the scout bees, and products are optimally collected and delivered to the hive to maximise profitability at each hive. Scouts and employed bees perform their tasks incrementally until the needs of the hive are best met. Three crucial components make up the computational model of the algorithm: bee colonies, sources of various supplies, and possible pathways. Scout bees transmitted a variety of information, including the calibre of the food source, the distance from the hive, and the source's orientation with relation to the colony. If Q_{ijk} is the quality of the j_{th} source of the k_{th} commodity as estimated by the scout from the i_{th} colony and d_{ijk} is the distance between the food source and beehive for the k_{th} commodity, then the energy $E_{ijk}(t)$ is measured [12], with ρ as the proportionality constant used to quantify the energy, as follows:

$$E_{ijk}(t) = \rho * \delta_{ijk} \tag{6}$$

The profitability P_{ijk} is then estimated based on the energy E_{ijk} and the quality Q_{ijk} with μ as an adjustment parameter and N_{ijk} keeps track of the number of scouts visiting a source of a particular commodity.

$$P_{ijk}(t) = \mu * (Q_i(t)) / (E_{ijk}(t) * N_{ijk}(t)) \tag{7}$$

Until the objective function is optimised, this process is carried out concurrently in each of the hives. The reticent bees then observe various scout bees' waggle dances to determine the most advantageous sources of each commodity to meet the needs of the hive. They visit the source with profitability higher than a particular threshold T_{ijk} which is fixed based on the requirements at each colony. The total number of forager bees accessible divides proportionally to visit different nodes to meet the diverse needs of the hive when more than one commodity is required at a location. This division is based on the desirability, λ_{ijk} of each source.

$$\lambda_{ijk}(t) = \frac{P_{ijk}(t)}{\sum_{i=1}^{n} \sum_{j=1}^{m} P_{ijk(t)}} \tag{8}$$

The number of foragers, $F_{ijk}(t)$, visiting a source node $S_{jk}(t)$ of the $k_{th}(t)$ commodity is based on $\lambda_{ijk}(t)$, where $X_i(t)$ is the total number of foragers available at a specific colony D_i.

$$F_{ijk}(t) = X_i(t) * \lambda_{ijk}(t) \tag{9}$$

Honey bee foraging is thus described as a multi-modal, multi-dimensional optimization problem with an objective function TP_i designed to optimise overall profitability at each bee colony.

$$MaximizeTP_i = \sum_{j=1}^{m} \sum_{k=1}^{p} P_{ijk} * F_{ijk} \qquad (10)$$

The WTA problem can be resolved using this algorithm more effectively and its mapping is done as follows.

4 WTA Problem Model Using Multi-modal Honey Bee Foraging Optimizer

Figure 1 shows how the WTA problem is mapped into the HBF optimizer. This computational model consists of parameters such as multiple weapon types, multiple targets and multiple potential assignments.

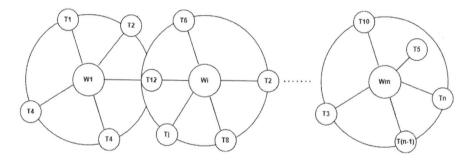

Fig. 1. Mapping of the proposed WTA into the HBF optimizer

There are m weapon types with each weapon type having w_i weapons and n targets. There should be a potential path between weapons and targets if i^{th} weapon is assigned to j^{th} target. The graphical representation keeps changing in the dynamic scenario of the problem as computation proceeds. More target nodes will be added when new targets come. An existing target node can be removed when it is completely destroyed. The same thing can happen at the weapon side also. If the number of weapons of a type is completely used then that node will be removed. Thus the nodes and the potential assignments keep on changing at different stages of the assignments. From the figure it is clear that each weapon platform acts as bee hives and each target acts as a food source. The scouts from each weapon platform visit targets and collect information about them. They communicate the gathered information such as the destructive value of targets V_j, threatened asset by the target and value of the asset with the reticent bees. From the information passed by the scouts, reticent bees compute the efficacy, ε_{ij} of assigning a weapon type i to target j to protect asset k using:

$$\varepsilon_{ij} = \rho * c_{ij} \tag{11}$$

A target can be destroyed by many weapon types. Therefore assigning a weapon type to a target decreases as the number $(N_{ijk}(t))$ of weapon types to that particular target increases. Now the profitability, P_{ij} of assigning w_i to T_j can be calculated using,

$$P_{ij} = \mu * \frac{V_j}{\varepsilon_{ij} * N_{ij}} \tag{12}$$

The forager bees at a weapon node, choose worthwhile targets with an Efficacy above a particular threshold and thus an optimized assignment can be made. Only targets with E_{ijk} threshold are taken by the reticent bees. Based on the availability of the weapon types and the value of protected assets, the threshold is fixed. Now the objective function for the WTA problem can be defined as

$$Min \sum_{j=1}^{n} \sum_{i=1}^{m} (1 - P_{ij}) x_{ij}$$

$$st \sum_{j=1}^{n} x_{ij} \leq w_i, \quad for\ i = 1, ...m \tag{13}$$

$$x_{ij} \in \mathbb{Z}_+, \quad for\ i = 1, ...m,\ j = 1, ...n$$

Thus in this work the WTA is proposed as a scenario of combinatorial optimiza-

---

Algorithm 1. At Target node

1: For each weapon type w_i from 1 to n in parallel
2: For each asset a_k from 1 to K
3:      Visit the target nodes in the vicinity
4:      Asses the number of bees from other weapon types visiting Target j as N_{ijk}.
5:      Estimate the destructive value of the target as V_j
6: Determine the cost required to destroy the target j by w_i as c_{ij}
7: Estimate the efficacy of assigning w_i to target j as ε_{ij}

---

tion problem which is intended to optimally assign available weapons to incoming targets so as to minimize the target's probability of destroying a protected asset by maximising the probability of destruction of targets. The pseudocode of the proposed HBF algorithm consists of three procedures as given below. Algorithm 1 details the steps performed at each Target node Algorithm 2 details the steps performed at each Weapon node Algorithm 3 outlines the performance of the honeybee foraging in DWTA.

Algorithm 2. At Weapon node

1: In each target j from 1 to n in parallel until the targets are destroyed completely.
2: For each asset a_k from 1 to K
3:     Scout bees estimate the profitability,P_{ij} of different weapon types.
4:     Profitability is transferred to the foragers.
5:     Reticent bees at w_i chooses one from among the targets based on the probability
6: Discard the target j when it is completely destroyed.

Algorithm 3. At Assignment phase

1: For each weapon type w_i and target j in parallel until the targets are destroyed optimally repeat steps 2-5.
2: Scouts randomly search of target.
3: Scout from w_i measure the profitability of target j using (12)
4: Profitability is transferred to forager bees w_i to determine probability of a target using (8)
5: Discard the targets those are completely destroyed.

---

5   Results and Discussion

We explain the experimental results of the proposed method on various problem instances. Table 1 shows the details regarding the 12 different problem scenarios from D-I to D-XII used for conducting simulations.

Table 1. Problem scenarios

Scenario	Weapon	Target	Scenario	Weapon	Target
I	5	5	VII	60	60
II	10	10	VIII	70	70
III	20	20	IX	80	80
IV	30	30	X	90	90
V	40	40	XI	100	100
VI	50	50	XII	200	200

In each set of data the number of weapon types, the number of targets and their assignments are different such that the problem dimension could be explored from small to high.

The dataset includes from data of Weapon=5 and Target=5 with 25 possible assignments(D-I) to data of Weapon=200 and Target=200 with 40000 possible assignments(D-XII). The targets are assigned values at random from a uniform distribution in the range of 25 to 100. For weapon-target assignments, random numbers from a uniform distribution are used to create the probabilities of destroying targets in the range of 0.60–0.90. [14].

Table 2. Optimal Assignments for 10 iterations

WTA	1	2	3	4	5	6	7	8	9	10	Average
D-I	48.364	48.364	48.364	48.364	48.364	48.364	48.364	48.368	48.364	48.364	**48.364**
D-II	119.71	110.03	115.59	121.65	111.35	115.95	114.69	119.71	**109.20**	113.35	**115.123**
D-III	216.70	200.44	**195.57**	207.12	198.27	208.42	207.02	209.20	211.45	202.30	**205.649**
D-IV	414.29	**382.32**	390.42	397.79	410.92	419.29	422.42	397.81	416.88	415.68	**406.782**
D-V	531.52	542.22	557.53	**530.92**	566.31	548.11	561.15	533.26	546.50	534.09	**545.141**
D-VI	648.97	652.91	626.28	**621.24**	662.92	648.23	666.58	670.01	659.31	647.79	**650.424**
D-VII	**772.45**	802.55	779.45	787.39	831.97	817.20	787.69	800.74	803.98	818.96	**800.238**
D-VIII	940.68	983.60	975.42	999.24	985.87	**929.68**	998.78	973.19	984.97	958.19	**972.962**
D-IX	1072.00	1065.26	**1032.75**	1060.71	1044.33	1075.78	1077.50	1071.04	1062.24	1065.26	**1062.687**
D-X	1219.35	1200.79	1211.37	1212.24	**1154.16**	1225.31	1232.43	1214.71	1218.98	1201.93	**1209.127**
D-XI	1416.18	1420.77	1414.51	1424.55	1427.77	1429.80	1399.72	**1394.54**	1411.70	1406.53	**1414.607**
D-XII	2805.84	2818.52	2843.38	2846.06	2848.13	2852.27	**2768.37**	2789.41	2821.86	2838.05	**2823.189**

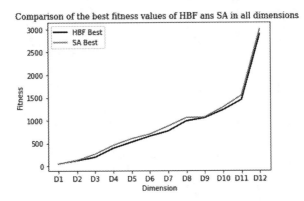

Fig. 2. Best Fitness values with HBF and SA algorithms in all problem scenarios

From the results it shows that our method give statistically best results for the objective function in less time compared to other algorithms in the literature. The performance of the algorithm in the above mentioned problem scenarios in terms of it's best fitness values based on the objective function is given in Table 2 .The best 10 CPU runs and their averaged results are highlighted in the table.

From the literature it is noted that the Simmulated Annealing(SA) algorithm shows the appealing results in the WTA problem and therefore the proposed HBF is compared with the SA algorithm and the results are visualized in Fig. 2.

From the figure it is clear that the proposed method gives the best results in some dimensions and in other cases it shows similar results with SA. But in all cases HBF algorithm converge to the optimal values in less time than SA. For better visualization of the results the best fitness values in some dimensions from D-I to D-XII using HBF optimizer is shown in Fig. 3

Research in this area, particularly in Dynamic WTA [3], can't advance a certain point due to lack of data availability and this could be addressed in the future works.

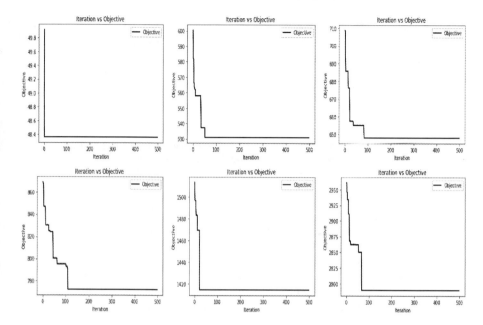

Fig. 3. The fitness curves for dimensions of WTA (D-I, D-V, D-VI, D-VII, D-XI and D-XII) using HBF

6   Conclusions

This work uses the multi-modal foraging behaviour of honey bee swarms to solve the classic combinatorial, NP-Complete, Weapon Target assignment problem. Results show that the algorithm works well in various problem scenarios with promising values in linear time. The literature indicates that the honey bee foraging optimizer is an appealing alternative for solving WTA problems in less CPU time which helps in the automation of Command and Control, media selection and product positioning in market research.

References

1. Anand, H., Saritha, R., Vinod Chandra, S.: Many objective optimization algorithm based on mixed species particle flocking. J. Ambient Intell. Hum. Comput., 1–17 (2022)
2. Burr, S.A., Falk, J.E., Karr, A.F.: Integer prim-read solutions to a class of target defense problems. Oper. Res. **33**(4), 726–745 (1985)
3. Cai, H., Liu, J., Chen, Y., Wang, H.: Survey of the research on dynamic weapon-target assignment problem. J. Syst. Eng. Electr. **17**(3), 559–565 (2006)
4. Flood, M.: Target-assignment model. In: Proceedings of the Princeton University Conference on Linear Programming, Princeton (NJ) (1957)
5. Kline, A., Ahner, D., Hill, R.: The weapon-target assignment problem. Comput. Oper. Res. **105**, 226–236 (2019)

6. Lee, Z.J., Lee, C.Y., Su, S.F.: An immunity-based ant colony optimization algorithm for solving weapon-target assignment problem. Appl. Soft Comput. **2**(1), 39–47 (2002)
7. Lloyd, S.P., Witsenhausen, H.S.: Weapons allocation is np-complete. In: 1986 Summer Computer Simulation Conference, pp. 1054–1058 (1986)
8. Luo, W., Lü, J., Liu, K., Chen, L.: Learning-based policy optimization for adversarial missile-target assignment. IEEE Trans. Syst. Man Cybern.: Syst. **52**(7), 4426–4437 (2021)
9. Manne, A.S.: A target-assignment problem. Oper. Res. **6**(3), 346–351 (1958)
10. Murphey, R.A.: An approximate algorithm for a weapon target assignment stochastic program. Approximation Complex. Numer. Optim.: Continuous Discrete Prob. 406–421 (2000)
11. Saritha, R., Chandra, S.V.: Multimodal foraging by honey bees toward optimizing profits at multiple colonies. IEEE Intell. Syst. **34**(1), 14–22 (2018)
12. Saritha, R., Vinod Chandra, S.: Multi dimensional honey bee foraging algorithm based on optimal energy consumption. J. Inst. Eng. (India): Ser. B **98**, 527–531 (2017)
13. Shang, G., Zaiyue, Z., Xiaoru, Z., Cungen, C.: Immune genetic algorithm for weapon-target assignment problem. In: Workshop on Intelligent Information Technology Application (IITA 2007), pp. 145–148. IEEE (2007)
14. Sonuc, E., Baha, S., Bayir, S.: A parallel simulated annealing algorithm for weapon-target assignment problem. Int. J. Adv. Comput. Sci. Appl. **8**(4) (2017)
15. SS, V.C.: Smell detection agent based optimization algorithm. J. Inst. Eng. (India): Ser. B **97**, 431–436 (2016)
16. Ss, V.C., Hs, A.: Nature inspired meta heuristic algorithms for optimization problems. Computing **104**(2), 251–269 (2022)
17. Yanxia, W., Longjun, Q., Zhi, G., Lifeng, M.: Weapon target assignment problem satisfying expected damage probabilities based on ant colony algorithm. J. Syst. Eng. Electr. **19**(5), 939–944 (2008)

A Dual-Strategy Contest Harmony Search Algorithm Based on Population State for the System Reliability-Redundancy Allocation Problems

Siqi Liang[1], Chang He[1], and Haibin Ouyang[1,2(✉)]

[1] School of Mechanical and Electric Engineering, Guangzhou University, Guangzhou 510006, China
Oyhb1987@163.com

[2] School of Computer Science and Engineering, South China University of Technology, Guangzhou, China

Abstract. At present, many improved algorithms have effectively improved the ability to converge to the optimal value. However, there are still difficulties: how to effectively balance the deep development ability of the algorithm in a small range and the extensive exploration ability in the global search field in the iterative process. For this reason, this topic first proposes a population state evaluation indicator, which continuously detects the state and diversity of the population with iteration. On this basis, the dual-strategy competition mechanism is adopted. Through the population state evaluation index, the strategy can select a strategy that is more conducive to rapid convergence according to the changes of the population during the iteration process. Through the advantage competition of the two strategies, the development ability of the algorithm has been steadily and persistently improved. However, there is still the risk of falling into the local optimal dilemma. Therefore, the change of population diversity is considered in the evaluation index of population status in this study. In order to further increase the activity of the global search range, the computational individuals used in the proposed environmental assessment index of population diversity are randomly selected. In general, in the greedy comparison of the two strategies, this study still ensures the consideration and influence of diversity, and effectively balances the ability of algorithm exploration and development.

Keywords: Harmony Search Algorithm · Dual-strategy contest mechanism · Population state evaluation · Diversity evaluation · System reliability

1 Introduction

The development of society and the advancement of technology have led to a significant increase in the sensitivity and complexity of engineering problems. At the same time, during the operation process, equipment aging, operating environment changes and other reasons will affect the reliability of the system. Reliability refers to the probability that the system runs smoothly in a certain time and environment [1]. The operation efficiency of the system can be improved by optimizing the design of the system reliability. The importance of system reliability optimization design in engineering practice is reflected in the accurate grasp of risks and costs. As a classical NP-hard problem, reliability redundancy allocation (RRAP) and component reliability design and redundancy allocation belong to the optimization method of system reliability [2].

In order to cope with increasingly complex reliability problems, Huang et al. transformed RRAP into an unconstrained optimization problem, simplifying the difficulty of the solution [3]. Li et al. proposed a more suitable reliability for multi-source, multi-sink complex modern network system, called gRRAP, and designed PIPSO algorithm by measuring the individual uniformity of the population according to the distance between the individuals to solve it effectively[4]. Caserta and Voß simplified the difficulty of solving by designing RRAP as a multiple-choice backpack problem [5]. Muhuri and Nath designed RRAP as a new two-layer optimization model and used a two-layer evolutionary algorithm based on quadratic approximation to obtain more efficient results[6]. Yeh et al. considered cost constraints, established a two-objective RRAP model, and solved it effectively by the proposed algorithm SSO [7].

An effective solution can be obtained by simplifying the RRAP model. Moreover, after in-depth research by researchers, the application of the solution method will also affect the quality of the solution. When traditional mathematical-based techniques are adopted to solve RRAP, their own shortcomings, such as the limitation of problem size and high computational cost, make it difficult to calculate the global optimal value in a short time. In order to obtain the results of RRAP more efficiently, researchers employ swarm intelligence optimization algorithm.

In recent years, swarm intelligence optimization algorithms inspired by natural phenomena have shown better optimization performance. Among them, HS algorithm has attracted wide attention due to its advantages of less parameters, no initialization of decision variables, and simple structure. At present, in the study of HS improvement, the most commonly used improvement methods are adjusting parameters, hybrid algorithm improvement and designing new parameters [8]. AIHS based on parameter correction proposed by Boor et al. mainly includes two stages. Firstly, dynamic HMCR and bw are designed to adjust HM, and then new HM is obtained according to the information of old HM [9]. In terms of hybrid algorithm, Shaikh combines the advantages of SA and proposes HS-SA, which significantly improves the global search ability of the algorithm and avoids the premature situation [10]. In the study of introducing parameters, Zhu et al. successfully improved the development performance of the algorithm by using the convergence coefficient, and improved the exploration ability of the algorithm through the nonlinear convergence region [11].

Many studies have shown that the improved HS has good ability to solve reliability problems. Ouyang et al. used different methods to simply improve HS, obtained four improved HS algorithms, and improved the solution ability of the algorithm [12]. Based on HS and combined with the advantages of PSO, Zou et al. obtained the EGHS algorithm which can effectively solve the reliability problem [13]. In order to deal with large-scale reliability problems, Ouyang et al. proposed a dimension-based strategy selection strategy based on the global situation, and combined with the improved bw fine-tuning strategy, proposed the AHS algorithm [14].

However, the disadvantage of the existing HS algorithm is that it is difficult to solve large-scale problems efficiently, and there is still much room for improvement in the face of complex RRAP. Therefore, this paper proposes a dual-strategy competition harmony search (DCHS-PS) algorithm based on population state. Specifically, according to the designed comprehensive population evaluation index, the most suitable strategy for population development in the current environment is obtained. The comprehensive population evaluation index includes the population comprehensive state index and the diversity environment evaluation index. The comprehensive state and environmental changes of the population in the iterative process can be obtained by weighted combination.

The content of this research is distributed as follows: The second section mainly introduces several related algorithms. The third section introduces in detail the DCHS-PS proposed in this paper. The fourth section presents and analyzes the experimental results. The last section is the concluding remarks.

2 Harmony Search Algorithm (HS)

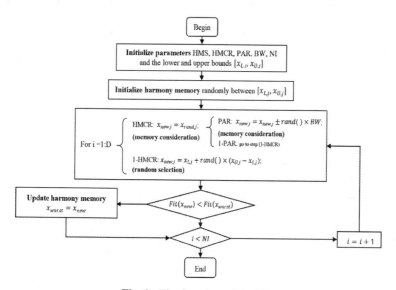

Fig. 1. The flowchart of the HS.

The harmony search algorithm is an intelligent optimization algorithm that simulates the orchestra's continuous search for the best harmony during the music playing process[15]. After initializing the parameters and harmony memory, the algorithm selects the new harmony generation method according to the HMCR probability, and then selects whether to jitter the individual according to PAR probability. When the new individual has a better fitness value than the worst individual in the population, the worst individual is replaced. The population is continuously updated until the maximum number of updates is reached. Specifically, the specific flow chart of generating new individuals after initialization is shown as follow (see Fig. 1).

3 Our Algorithm

The classical HS algorithm has attracted much attention for its simple structure, high efficiency and high robustness. However, in complex large-scale optimization problems, the convergence ability of HS is still worrying. Therefore, this study has developed a dual-strategy contest harmony search algorithm based on population state (DCHS-PS), in which the development trend of the population under different strategies is obtained through the population state evaluation indicator and the population diversity environment indicator, and then the strategy operator is selected.

In this section, we will give a detailed explanation of the population state evaluation indicator, the population diversity environment indicator, and the improved HS algorithm based on the dual-strategy contest mechanism. At the end, we will give the flow chart of the proposed algorithm and its detailed introduction.

3.1 Strategy Feedback Indicator Based on Past-Current-Future

Different strategy operators will lead to different directions of the population, and different population states will also affect the performance of the strategy operator. In order to explore the specific situation of population state change under the guidance of this strategy, this paper uses the strategy feedback indicator proposed by Han [16]. Specifically, according to the population's past experience, current status and future predictions, evaluation indicators can be obtained in different strategy feedback.

Past experience: In order to judge the impact of the current strategy on the state of the population, the past experience is very meaningful. Each decision of the strategy is contested by the population state evaluation indicator. Therefore, the comparison of the number of times of the strategy used in the history and the total number of times can directly reflect the popularity of the strategy in the historical environment, that is, the probability of having a positive guiding effect on the population.

$$past_{str}(gn) = time_{str}/gn \tag{1}$$

where gn represents the current iteration number, $past_{str}(gn)$ represents the past experience under the adopt of the strategy str at the gn iteration, $time_{str}$ refers to the cumulative number of times to use strategy str.

Current state: The most effective way to obtain the current state of a population at the current time is to compare the fitness values of individuals. By comparing the fitness value of the newly generated harmony with the fitness value generated in the previous iteration, the direct impact of the strategy on the population under the current number of iterations can be directly obtained.

$$current_{str}(gn) = 1 - min[Fit(x_{current}), Fit(x_{new})]/Fit(x_{current}) \tag{2}$$

Here $Fit(x_{current})$ represents the fitness value of $x_{current}$, and $Fit(x_{new})$ represents the fitness value of x_{new}.

Future projection: In addition to past experience and current state, it is also necessary to combine the macro prediction of the future to grasp the comprehensive state of the population more accurately. The upper confidence limit (UCB) algorithm is a reinforcement learning algorithm, which can quickly explore the environment through its own prior knowledge and experience. The future forecast obtained by using UCB is as follows:

$$future_{str}(gn) = \theta \left[\frac{use_{str}^{indi}(gn)}{win_{str}^{indi}(gn)} + sqrt\left(\frac{Co * log(gn)}{use_{str}^{indi}(gn)} \right) \right]$$

$$+ (1 - \theta) \left[\frac{use_{str}^{pop}(gn)}{win_{str}^{pop}(gn)} + sqrt\left(\frac{Co * log(NP * gn)}{use_{str}^{pop}(gn)} \right) \right] \quad (3)$$

Here $use_{str}^{indi}(gn)$ and $use_{str}^{pop}(gn)$ represents the cumulative number of times individual and population use the str strategy, and $win_{str}^{indi}(gn)$ and $win_{str}^{pop}(gn)$ represents the cumulative number of times individual has used the str strategy and succeeded. NP is the population size. θ and Co are the balance coefficients, which is 0.5 and 0.7 respectively.

The final Strategy feedback indicator is summed by the weight system as follows:

$$Feedback_{str}(gn) = \alpha * past_{str}(gn) + \beta * current_{str}(gn) + \gamma * future_{str}(gn) \quad (4)$$

Here, α, β and γ are weight coefficients, which are 0.1, 0.4 and 0.5 respectively.

3.2 Population Diversity Evaluation Indicator Based on Random Individuals

Diversity is an important indicator of population convergence. At present, many studies have calculated the distance between all individuals or specific individuals through Euclidean distance, Markov distance, etc., so as to obtain the distribution of individual population. Considering the waste of computing resources and in order to further increase diversity, this paper reflects the diversity of population through random individual selection. The specific formula is as follows:

$$Diversity(gn) = distance(x_{rand1}^{current}, x_{rand2}^{current})/distance(x_{rand1}^{initial}, x_{rand2}^{initial}) \quad (5)$$

where $x_{rand1}^{current}$ and $x_{rand2}^{current}$ represent two randomly selected individuals in the population under the current number of iterations gn. What's more, $x_{rand1}^{initial}$ and $x_{rand2}^{initial}$ refer to two individuals randomly selected in the initial population, both of which are used regularly during the iteration process.

3.3 A Dual-Strategy Contest Harmony Search Algorithm Based on Population State

As the iterative process progresses, the population state has been changing, and the demand for search capabilities is also fluctuating. If the search ability is not adjusted with iteration, the population will fall into local optimum. At present, most of the improved algorithms use a single strategy to search the optimal solution. In particular, in order to balance the exploration and development of algorithms, researchers choose strategies or parameters that can be adaptively adjusted. However, when the algorithm enhances one ability, other performance will also be affected. In other words, it is difficult to perfectly balance exploration and development capabilities through the efforts of a single strategy itself. Based on this, we propose a dual strategy competition harmony search algorithm based on population state.

Specifically, according to the strategy feedback index and the population diversity evaluation index, the population status evaluation index of different strategies is obtained. Then, through the two-strategy competition mechanism, the strategy more suitable for the rapid development of the population is selected. The formula of population status evaluation index is as follows:

$$State_{str}(gn) = a * Feedback_{str}(gn) + b * Diversity(gn) \qquad (6)$$

Here a and b are weight coefficients, which are 0.7 and 0.3 respectively.

Then, the strategy is selected by the dual-strategy contest mechanism. Because the local search ability of NGHS algorithm [17] is very strong, and the nonlinear convergence region of AHS-HCM algorithm[11] also significantly improves the global search ability of the algorithm, the DCHS-PS uses these two strategies to generate new harmony (see Algorithm.1).

Algorithm.1   The flow chat of DCHS-PS algorithm

1:Initialize the parameter and set $Status_1$=0, $Status_2 = 0$;
2:Initialize the harmony memory (HM) and compute the fitness.
3:**for**   gn < NI   **do**
4:      **if** $State_1 < State_2$ **do**   str = 1;   /Select NGHS strategy.
5:      **elseif** $State_1 > State_2$ **do**   str = 2;   /Select AHS-HCM strategy.
6:      **else do** str = randi(2);   /Random select strategy.
7 :   **end if**
8:      **if**   str = 1   **do**   Algorithm1;   /Improvise new harmony x_{new}.
9:      **elseif** str =2   **do**   Algorithm2;
10:     **end if**
11:     Calculate the fitness of new harmony x_{new} ;
12:     Update the HM;
13:     Calculate strategy feedback indicator according to Eq. (4) ;
15:     Calculate population diversity evaluation indicator according to Eq. (5) ;
16:     Calculate population state evaluation indicator $Status_{str}$ according to Eq. (6) ;
17:     gn = gn+1;
18: **end for**

---

4   Experimental Results and Discussions

The experiments of this paper are run on Intel(R) Core(TM) i5-8250U CPU dual core processor @ 1.60GHz 1.80 GHz. In addition, the algorithms used in the experiment are implemented by Matlab 2018a software in Microsoft Windows 10 Professional environment.

4.1 Reliability Optimization Problems and Algorithm Configurations

The DCHS-PS algorithm proposed in this study is verified on eight classic RRAPs: P1:complex (bridge) system, P2:series system, P3:series-parallel system, P4:overspeed protection system, P5:large-scale system, P6:convex quadratic reliability problem,P7: imperfect fault detecting the switching, P8:mixed series–parallel system. Ouyang et al. has introduced these eight reliability systems in detail, the specific parameters and models can be understood from the paper [20]. To verify the performance of the DCHS-PS, it is compared with HS and its four variants, including IGHS, NGHS, IMGHSA, AHS-HCM. The specific parameters of DCHS-PS and comparison algorithms are shown as follow (see Table 1). The maximum number of iterations and the set population size are 50000 and 10, respectively. In addition, each optimization algorithm operates 30 times independently.

Table 1. Parameter setting.

Method	Parameters	Method	Parameters
DCHS-PS	HMS $= 5$, pm $= 0.005$	**NGHS** [17]	HMS $= 5$, pm $= 0.005$
HS [15]	HMS $= 5$, HMCR $= 0.9$, PAR $= 0.33$, BW $= 0.5 \times 0.0001 \times$ (xmax-xmin)	**AHS-HCM** [11]	HMS $= 5$, HMCR $= 0.99$, PAR $= 1$
IGHS [18]	HMS $= 5$, HMCR $= 0.99$, PARmax $= 0.99$, PARmin $= 0.01$, BWmax $=$ (xmax-xmin)/20, Bwmin $= 0.0001$	**IMGHSA** [19]	HMS $= 5$, HMCR $= 0.9$, PAR $= 0.3$, BW $= 0.01$, pm $= 0.005$, u1 $= 0.7$, u2 $= 0.3$

4.2 Experimental Results and Analysis

In order to verify the performance of DCHS-PS, this study independently ran 30 times in 8 reliability problems, with 50000 times of evaluation each time. Tables 2, 3 and 4 shows the optimal value (Best), the worst value (Worst), the mean value (Mean), the standard deviation (SD) value and the average running time (AT) of the 30 results, and Fig. 2 shows the convergence of the operation. In addition, this section will also carry out rigorous analysis.

It can be seen from Tables 2, 3 and 4 that the performance of DCHS-PS algorithm is obviously better than that of HS algorithm and its variants. From the results of the mean, it can be seen that DCHS-PS has the best performance in the P2–7 problem, although it is slightly inferior to the NGHS algorithm in p1 and p8. In terms of the optimal value, the DCHS-PS algorithm is superior to or equal to other algorithms in the P1 and P4 problems. Among the five dimensions of P5, DCHS-PS performs best in the two dimensions of D $= 38$ and D $= 40$. In addition, in the reliability problem of P6-P8,

DCHS-PS converges to the optimal result as other HS and their variants. In terms of the worst value, DCHS-PS is superior to or equal to other algorithms except for P1, P5 (D = 40), and P8.

Figure 2 shows the convergence of DCHS-PS and five comparison algorithms on eight reliability problems. Obviously, DCHS-PS can converge to the optimal value in all problems except P8. In terms of process, DCHS-PS reaches the inflection point earliest among the problems except P1 and P5 (D = 36, D = 50). It can be seen from P1, P2 and P6 that the algorithm reaches the optimal result through stepwise increase after reaching the inflection point.

Table 2. Comparison of the results for the system reliability problems (P1–4).

Problemms	Algorithms	Best	Worst	Mean	SD	AT
P1	**DCHS-PS**	**0.99988949**	0.999738788	0.999849287	3.741E-05	1.931769997
	HS	0.999885075	0.998945771	0.999690577	0.000227181	1.072285307
	IGHS	0.9998894	0.999348635	0.999758639	0.000147273	5.264756617
	NGHS	0.999889227	**0.999827733**	**0.999873177**	**1.980E-05**	**1.056769733**
	IMGHSA	0.999889393	0.989006223	0.998832149	0.002182989	1.061563037
	AHS-HCM	0.996394716	0.83552052	0.966630908	0.036083532	1.13733571
P2	**DCHS-PS**	0.93163048	**0.911877156**	**0.924861552**	**0.005104738**	1.420239457
	HS	0.929374983	0.826528559	0.910607813	0.0211494	1.083150923
	IGHS	**0.931673846**	0.799847026	0.902034791	0.038490374	5.029522173
	NGHS	0.931658771	0.909808484	0.920859672	0.007394873	**1.05294231**
	IMGHSA	0.931667023	0.664559589	0.848234485	0.074130852	1.13016803
	AHS-HCM	0.766166877	0.026202952	0.204794686	0.155624036	1.410771753
P3	**DCHS-PS**	0.999976526	**0.999924134**	**0.999957048**	**1.543E-05**	1.892944803
	HS	0.999975887	0.999092517	0.999890416	0.000166036	**1.09499925**
	IGHS	**0.999976646**	0.999822008	0.999945348	3.332E-05	5.341472037
	NGHS	0.999976584	0.99992236	0.999949393	1.876E-05	1.18606839
	IMGHSA	**0.999976601**	0.987167058	0.999101404	0.00256718	1.34703023
	AHS-HCM	0.999518597	0.937485487	0.987943264	0.013161873	1.248711827
P4	**DCHS-PS**	**0.999954629**	**0.999906764**	**0.999939274**	**1.337E-05**	1.633397857
	HS	0.999951741	0.999526589	0.999908404	8.052E-05	1.382458513
	IGHS	**0.99995466**	0.999175767	0.999839547	0.000169272	4.600951413
	NGHS	**0.999954672**	0.999867838	0.99992735	2.460E-05	**1.081482313**
	IMGHSA	0.999946125	0.997142082	0.999728244	0.000540966	1.122629623
	AHS-HCM	0.999911243	0.834648667	0.956623061	0.048854106	1.13254599

Table 3. Comparison of results for the large-scale system reliability problem (P5).

Dim	Algorithms	Best	Worst	Mean	SD	AT
36	**DCHS-PS**	0.518966303	**0.496166659**	**0.508999944**	0.005836658	2.549979837
	HS	0.518966303	0.490826183	0.504515427	0.006327197	1.487402037
	IGHS	0.519975965	0.484501848	0.502938053	0.008111166	15.53685977
	NGHS	**0.573811698**	0.469099299	0.502711238	0.023454988	1.129058507
	IMGHSA	0.503848438	0.458655655	0.480715959	0.011969109	**1.094909667**
	AHS-HCM	0.519471134	0.505317384	0.510604972	**0.002504992**	1.174596833
38	**DCHS-PS**	**0.510988596**	**0.499477024**	**0.508862662**	**0.003933751**	2.271332847
	HS	0.506561846	0.477684816	0.489932729	0.008172906	1.129643407
	IGHS	0.50655806	0.466550422	0.489220103	0.011068092	14.38584762
	NGHS	**0.510988596**	0.49563213	0.507613268	0.004497043	**1.04790997**
	IMGHSA	0.496587089	0.441076795	0.472412089	0.013197301	1.09168656
	AHS-HCM	0.510492491	0.49562749	0.500425957	0.004428592	1.160549403
40	**DCHS-PS**	**0.503292493**	0.491000899	**0.502736905**	0.002249843	3.515532153
	HS	0.501348306	0.460402282	0.485169746	0.011383023	2.02353336
	IGHS	0.497959892	0.46180145	0.482969437	0.009602077	33.09714022
	NGHS	**0.503292493**	**0.499413553**	0.50238634	**0.001213293**	2.668196543
	IMGHSA	0.490278615	0.446159569	0.467098002	0.011164677	**2.011598677**
	AHS-HCM	0.503292493	0.490053935	0.498333919	0.004929802	2.142113737
42	**DCHS-PS**	0.476631086	**0.47204722**	**0.475681129**	**0.001436022**	2.59469941
	HS	0.470672325	0.430957057	0.455072947	0.010235012	**1.19909695**
	IGHS	0.47204722	0.434318784	0.455111595	0.010748988	17.87614158
	NGHS	**0.47689093**	0.467042174	0.475154809	0.002128428	1.266196207
	IMGHSA	0.455819714	0.422628942	0.437125438	0.009633479	1.33336796
	AHS-HCM	0.476631086	0.459182164	0.469731748	0.005655439	1.470191003
50	**DCHS-PS**	0.40499557	**0.395871112**	**0.400521026**	**0.0017909**	2.478005053
	HS	0.390885517	0.358336247	0.375817485	0.009254477	1.290760733
	IGHS	0.390547212	0.360489097	0.373670682	0.008333426	26.96447769
	NGHS	**0.405781969**	0.394342617	0.39964992	0.003219312	**1.125923727**
	IMGHSA	0.391301914	0.352711411	0.366236411	0.008448105	1.173674223
	AHS-HCM	0.40460237	0.383230773	0.394764306	0.004948462	1.286520627

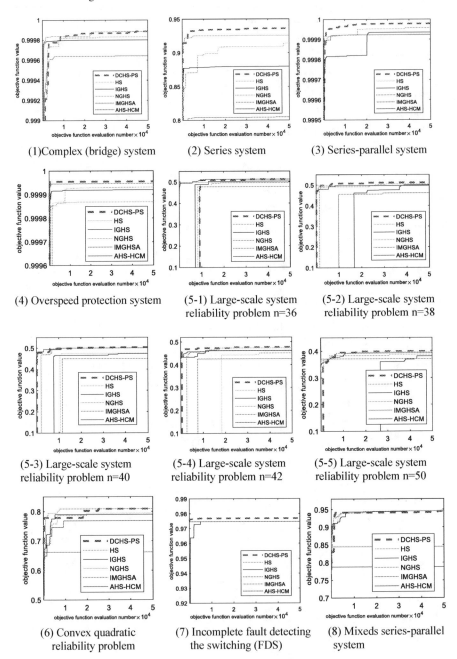

Fig. 2. Optimization curves for the 8 reliability systems.

Overall, DCHS-PS has better performance than HS and its variants. The reason for starring is that the randomness of HS is too strong, resulting in weak local development ability. IGHS algorithm relies on the optimal individual in the population to converge, which increases the risk of falling into local optimum. The NGHS algorithm considers the value of the optimal value and the worst value in the population, but ignores other information. The IMGHSA algorithm considers a variety of information to guide convergence. However, it does not adaptively change according to iteration, and there is a risk of falling into local optimum. The AHS-HCM algorithm uses multiple random individuals to affect the convergence direction, which has a certain blindness. In general, a single strategy is used to guide the algorithm to converge and escape the trap of local optimum. Therefore, DHS-PS adopts a two-strategy competition mechanism, and adjusts the use of strategies according to the change of population state in the iterative process, and obtains better performance.

Table 4. Comparison of the results for the system reliability problems (P6–8).

Problems	Algorithms	Best	Worst	Mean	SD	AT
P6	**DCHS-PS**	**0.80884419**	**0.80884419**	**0.80884419**	**4.516E-16**	1.840035713
	HS	**0.80884419**	0.790126302	0.803959298	0.008257845	**1.074279117**
	IGHS	**0.80884419**	0.790126302	0.803852753	0.008418865	4.948427937
	NGHS	**0.80884419**	0.801970591	0.80861507	0.001254942	1.13403228
	IMGHSA	0.772267977	0.531025966	0.674345351	0.056433843	1.089696147
	AHS-HCM	0.793322659	0.710565677	0.757435994	0.020029903	1.16098746
P7	**DCHS-PS**	**0.974565216**	**0.974565216**	**0.974565216**	**4.516E-16**	0.885785727
	HS	**0.974565216**	**0.974565216**	**0.974565216**	**4.516E-16**	**0.40381468**
	IGHS	**0.974565216**	**0.974565216**	**0.974565216**	**4.516E-16**	1.988072033
	NGHS	**0.974565216**	0.972758812	0.97444479	0.000458299	0.409127053
	IMGHSA	**0.974565216**	0.90054607	0.96624382	0.016232073	0.4321373
	AHS-HCM	**0.974565216**	0.961511963	0.971269411	0.004840797	0.47373094
P8	**DCHS-PS**	**0.945613357**	0.942024984	0.94502027	0.000769725	1.226668817
	HS	**0.945613357**	0.94329638	0.94493318	0.000596379	0.633056487
	IGHS	**0.945613357**	0.940998073	0.944662128	0.001293266	5.969324493
	NGHS	**0.945613357**	**0.944748485**	**0.945382725**	**0.000388999**	0.567154587
	IMGHSA	0.935198506	0.629161374	0.842698267	0.071828656	0.58825435
	AHS-HCM	0.788244997	0.267229612	0.534756508	0.200886679	**0.566109077**

5   Conclusions

In order to help the algorithm, achieve a good balance between the deep development in a small range and the overall search in a large scale, this study proposes a dual-strategy contest harmony search algorithm based on the population state. Specifically, it is to make use of the complementary advantages of the two algorithms to achieve a significant improvement in the optimization performance of the algorithm. First of all, a population state evaluation index is proposed to obtain the impact of different strategies on the population through the strategy feedback parameters, and further consider the diversity. According to the population status, the strategy contest mechanism is used to determine the strategy that is more conducive to the optimization, so as to ensure that the algorithm is in long-term and efficient development. In order to help the population to break the deadlock of local optimization, the diversity of the algorithm is effectively improved by adopting the dynamic population diversity environmental evaluation index within a flexible range. Finally, it is concluded that the dual-strategy contest mechanism adopted in this study can combine the advantages of the two strategies to achieve $1 + 1 > 2$, which has a very broad application prospect. The consideration of the population status evaluation index further ensures the reliability and practicability of the algorithm.

References

1. Yeh, W.C., Hsieh, T.J.: Solving reliability redundancy allocation problems using an artificial bee colony algorithm. Comput. Oper. Res. **38**(11), 1465–1473 (2011)
2. Kundu, T., Deepmala, Jain P.K.: A hybrid salp swarm algorithm based on TLBO for reliability redundancy allocation problems. Appl. Intell. **52**(11), 12630–12667 (2022)
3. Huang, X., Coolen, F., et al.: A heuristic survival signature based approach for reliability-redundancy allocation. Reliab. Eng. Syst. Saf. **185**, 511–517 (2019)
4. Li, S., Chi, X., Yu, B.: An improved particle swarm optimization algorithm for the reliability–redundancy allocation problem with global reliability. Reliab. Eng. Syst. Saf. **225**, 108604 (2022)
5. Caserta, M., Voß, S.: An exact algorithm for the reliability redundancy allocation problem. Eur J Oper Res **244**(1), 110–116 (2015)
6. Muhuri, P.K., Nath, R.: A novel evolutionary algorithmic solution approach for bilevel reliability-redundancy allocation problem. Reliab. Eng. Syst. Saf. **191**, 106531 (2019)
7. Yeh, W.: Simplified swarm optimization for bi-objection active reliability redundancy allocation problems. Appl. Soft Comput. **106**, 107321 (2020)
8. Qin, F., Zain, A.M., Zhou, K.Q.: Harmony search algorithm and related variants: a systematic review. Swarm Evol. Comput. **74**, 101126 (2022)
9. Loor, A., Bidgoli, M., Hamid, M.: Optimization and buckling of rupture building beams reinforced by steel fibers on the basis of adaptive improved harmony search-harmonic differential quadrature methods. Case Stud. Construct. Mater. **15**, e00647 (2021)
10. Shaikh, T.A., AliAn, R.: An intelligent healthcare system for optimized breast cancer diagnosis using harmony search and simulated annealing (HS-SA) algorithm. Inform. Med. Unlock. **21**, 100408 (2020)
11. Zhu, Q., Tang, X.: An ameliorated harmony search algorithm with hybrid convergence mechanism. IEEE Access **9**, 9262–9276 (2021)
12. Ouyang, H., Wu, W., Zhang, C., Li, S., Zou, D., Liu, G.: Improved harmony search with general iteration models for engineering design optimization problems. Soft. Comput. **23**(20), 10225–10260 (2018). https://doi.org/10.1007/s00500-018-3579-x
13. Zou, D.X., Gao, L.Q., Li, S., et al.: An effective global harmony search algorithm for reliability problems. Expert Syst. Appl. **38**(4), 642–4648 (2011)
14. Ouyang, H.-B., Gao, L.-Q., Li, S.: Amended harmony search algorithm with perturbation strategy for large-scale system reliability problems. Appl. Intell. **48**(11), 3863–3888 (2018). https://doi.org/10.1007/s10489-018-1175-5
15. Geem, Z.W., Kim, J.H., Loganathan, G.V.: A new heuristic optimization algorithm: harmony search. Simulation **76**(2), 60–68 (2001)

16. Han, Z., Chen, M., Shao, S., et al.: Improved artificial bee colony algorithm-based path planning of unmanned autonomous helicopter using multi-strategy evolutionary learning. Aerosp. Sci. Technol. **122**, 107374 (2022)
17. Zou, D., Gao, L., Wu, J., Li, S.: Novel global harmony search algorithm for unconstrained problems. Neurocomputing **73**(16–18), 3308–3318 (2010)
18. El-Abd, M.: An improved global-best harmony search algorithm. Appl. Math. Comput. **222**, 94–106 (2013)
19. Gholami, J., Ghany, K.K.A., Zawbaa, H.M.: A novel global harmony search algorithm for solving numerical optimizations. Soft. Comput. **25**(4), 2837–2849 (2020). https://doi.org/10.1007/s00500-020-05341-5
20. Ouyang, H., Gao, L., Li, S., et al.: Improved novel global harmony search with a new relaxation method for reliability optimization problems. Inf. Sci. **305**, 14–55 (2015)

An Improved Fire Hawks Optimizer for Function Optimization

Adnan Ashraf[1]([⊠]), Aliza Anwaar[2], Waqas Haider Bangyal[3], Rabia Shakir[4], Najeeb Ur Rehman[5], and Zhao Qingjie[1]

[1] Beijing Institute of Technology (BIT), Beijing 100081, China
adnannashraf@gmail.com
[2] University of Sialkot (USKT), Sialkot 51310, Pakistan
[3] Kohsar University, Murree 47150, Pakistan
[4] Department of Computer Science, FUUAST, Islamabad, Pakistan
[5] University of Gujrat, Gujrat 50700, Pakistan

Abstract. Fire hawk Optimizer (FHO) is a relatively new intake in the family of evolutionary algorithms for a distinct type of optimization problem. Initialization of the population plays a significant role in solving classical optimization issues. Incorporating quasi-random sequences such as the sobol, halton, and torus sequences, this study demonstrates novel ways for swarm initiation. The outcomes of our proposed techniques display outstanding performance as compared with the traditional FHO. The exhaustive experimental results conclude that the proposed algorithm remarkably superior to the standard approach. Additionally, the outcomes produced from our proposed work exhibits anticipation that how immensely the proposed approach highly influences the value of cost function, convergence rate, and diversity.

Keywords: Fire Hawk Optimizer · Swarm Intelligence · FHO · SO-FHO · H-FHO · TO-FHO · Quasi-Random Sequence

1 Introduction

The field of computer science has made some fascinating strides recently, and one of these is evolutionary computing (EC). Algorithm development, implementation, and analysis guided by the principles of natural selection as articulated by Darwin amounts to this [1]. The Evolutionary Computation (EC) family of algorithms is a branch of AI that seeks global optimization solutions by drawing parallels between biological evolution and AI (AI). EC is a method of optimization that takes cues from nature. In technical terms, EC is a category of problem solvers that utilizes a meta-heuristic or stochastic optimization based on a population-based trial and error strategy [2].

Meta-heuristic optimization approaches start with a population or swarm [3–5]. These computations affect diversity and combination in setup selection [6]. Meta-heuristic methods that identify and account for mixing have been improved by experts [7–9]. Since computer science and technology have rapidly developed different software

Y. Tan et al. (Eds.): ICSI 2023, LNCS 13968, pp. 68–79, 2023.
https://doi.org/10.1007/978-3-031-36622-2_6

programs and high-speed parallel processors, engineering and economics experts have focused on optimization. Calculus-based optimization algorithms cannot find global optimum solutions, its main drawback. Gradient-based algorithms can't solve complex optimization issues since they require differentiable objective functions. Thus, meta-heuristic algorithms have been recommended as a practical, accurate strategy for max-imising a range of variables [10]. This is the first metaheuristic algorithm inspired by fire hawk hunting methods. Discrete optimization methods considered this. Birds' coor-dinated behaviours inspired the programme. Craig W. Reynolds' research expands our flocking model for birds. Iztok Lebar Bajec and Frank Heppner study bird flocking behaviour using massive human social interaction. Fire Hawk Optimizer, an impor-tant contribution to optimization issue research, depends on research into these topics [11–14].

Fire Hawk optimizater intrigues, the whistler's kite, black kite, and brown falcon inspired the FHO metaheuristic algorithm, according to the author. These birds use fire to catch their prey. The optimization method evaluated 233 mathematical functions with dimensions from 2 to 100 and ran 100 optimization runs. Benchmarking used each of the 10 metaheuristic algorithms, and normal statistical methods were used to evaluate FHO's performance. We evaluated FHO's computing cost and complexity research, constraint problem competitions, and real-world optimization competitions while rating it [10].

Fire hawks' methods for lighting fires and catching prey are modelled by the FHO metaheuristic algorithm. Fire hawks and their prey are located using a pool of potential replies (X). Random initialization inserts these vectors into (1) search space.

$$
X = \begin{bmatrix} X_1 \\ X_2 \\ \vdots \\ X_i \\ \vdots \\ X_N \end{bmatrix} \begin{bmatrix} X_1^1 X_1^2 \cdots X_1^j \cdots X_1^d \\ X_2^1 X_2^2 \cdots X_2^j \cdots X_2^d \\ \vdots\vdots\ddots\vdots \\ X_i^1 X_i^2 \cdots X_i^j \cdots X_i^d \\ \vdots\vdots\ddots\vdots \\ X_N^1 X_N^2 \cdots X_N^j \cdots X_N^d \end{bmatrix}, \begin{cases} i = 1, 2, \ldots, N. \\ j = 1, 2, \ldots, d. \end{cases} \tag{1}
$$

The Fire Hawks start fires with blazing sticks from the main fire. The dropping of a burning stick on the territory of the attacking bird scares away the prey. As can be seen in (2), these two actions can be combined in FHO's main search loop.

$$
FH_l^{new} = FH_l + (r_1 \times GB - r_2 \times FH_{Near}), l = 1, 2, \ldots, n, \tag{2}
$$

The next step of the algorithm takes into account the importance of animal behavior, specifically how mobile prey are within each Fire Hawk's territory, to the location update process. When a Fire Hawk drops a burning stick, its prey either hides, runs away, or accidentally runs into the bird of prey. To take these factors into account during the position update process, the following equation might be used:

$$
PR_q^{new} = PR_q + (r_3 \times FH_l - r_4 \times SP_l), \begin{cases} l = 1, 2, \ldots, n. \\ q = 1, 2, \ldots, r. \end{cases} \tag{3}
$$

Targets of ambushes may wander into the territory of other Fire Hawks or look for a place to hide outside of the Fire Hawks' territory in the hope of finding some relief. You can account for these changes when changing a position using (4).

$$PR_q^{new} = PR_q + (r_5 \times FH_{Alter} - r_6 \times SP), \begin{cases} l = 1, 2, \ldots, n. \\ q = 1, 2, \ldots, r. \end{cases} \quad (4)$$

The following mathematical presentations of Safe Place (SPl) and Safe Place (SP) are based on the observation that, when faced with a threat, the vast majority of animals congregate in one secure location in the wild:

$$SP_l = \frac{\sum_{q=1}^{r} PR_q}{r}, \begin{cases} q = 1, 2, \ldots, r. \\ l = 1, 2, \ldots, n. \end{cases} \quad (5)$$

$$SP = \frac{\sum_{k=1}^{m} PR_k}{m}, \{k = 1, 2, \ldots, m. \quad (6)$$

PRq is the qth prey in the search space, surrounded by the lth fire hawk (FHl), PRk is the kth prey [10]. The effectiveness of four different variants of FHO algorithms was evaluated by comparing them to one another. The mean calculated from a total of 10 individual optimization runs. These results show that the proposed improved version TO-FHO with torus distribution outperformed the other FHO variants with uniform random distribution, Sobol sequence and Halton sequence.

2   Literature Review

Native Australians have traditionally used fire to control and sustain the local ecosystem. Fires whether caused intentionally or by lightning are especially dangerous to native species and terrain. Black kites, whistling kites, and brown falcons spark fires nationwide, according to study [10]. "Fire Hawks" are accused of carrying flaming sticks to start fires. Hawks readily catch rats, snakes, and other creatures that are startled by these small fires and flee.

Author of this article [15] explain the behave or, how Raptors leap headlong into fires to grab flaming wood, then carry them up to a kilometer before releasing them in dense undergrowth or a field. Sticks may come from campfires or forests. Raptors set fire to unburned regions to scare away prey from the other side of a stream, road, or firefighter's break. During the fire, one or a few raptors may exhibit this behavior. Despite being dropped, the burning sticks may miss the target or fail to ignite the plants.

Nevertheless, Computer-based evolutionary processes can handle modelling, machine learning, constraint, and optimization problems. Evolving system simulations can also examine life, business, and society phenomena. Darwinian art and design are also developing. In science, commerce, and entertainment, evolutionary computing applications will grow [1]. Based on their evolution over the previous few decades, metaheuristic algorithms can be divided into four types. Human and animal lifestyles inspire evolutionary algorithms, physics-inspired algorithms, swarm intelligence, and algorithms [16].

This article [17] concentrates on optimization meta-heuristics and population initialization. Instead of uniform distribution, this study suggests three low discrepancy sequences, WELL, Knuth, and Torus [18]. PSO, BA [19], and DE methods were tested for benchmark test issues and artificial neural network learning. Proposed variant beat uniform random numbers with low discrepancy sequence.

BA is a renowned meta-heuristic algorithm for optimization [19, 20]. BA uses echolocation via hunting like bats. BA's local search accelerates convergence. Torus stroll may improve local search. The torus walk increased diversity and convergence, improving BA. 15 benchmark problems test MCBA. Our technique beats PSO and BA. Benchmark issues and artificial neural network training assessed the suggested MCBA, BPA, Standard PSO, and Standard BA (ANN).

This paper [7, 21] analyses semi-arbitrary PSO initialization methods. A theoretical model separates the population introduction cycle into two phases, which each approach uses for restoration. The report identifies population initialization and research deficiencies but focuses on future development and improvement. This approach computes a model for the cutting edge of exploration.

PSO is a popular metaheuristic and stochastic optimization algorithm [22]. Population initialization helps meta-heuristic algorithms optimize conventionally. 15 unimodal and multimodal benchmark problems were tested with population initialization. The Paper's suggested technique influences cost function, convergence rate, and variety. QRS population initialization enhanced the PSO algorithm's convergence rate and diversity. For population initialization, TO-PSO uses the torus sequence.

Nematollahi et al. [23] GROM is a metaheuristic algorithm based on the golden ratio in plant and animal growth. Natural growth optimizes GROM. Two-phase updates use Fibonacci's golden ratio. 11 popular optimization strategies are compared. 30 trials statistically test each function. Five mechanical and one high-constraint electrical optimization challenges test the approach.

3 Methodology

3.1 Initialization

In this research work, four various population initiation methodologies have been implemented on standard FHO algorithm, which are Uniform Pseudo-Random [24], Halton Sequence [25], Sobol Sequence [26] and Torus Sequence [27]. Three proposed techniques which are based on Quasi-Random Sequences (QRS) achieve optimal results. Torus is a technique that improves upon the efficiency of regular FHO and produces low discrepancy sequences as a result.

Uniform Pseudo-Random Initialization. The pseudo-random sequence generated by a uniform distribution [24] generates random numbers and is characterized by a constant probability density function (PDF). Given in (7) as:

$$f(a) = \begin{cases} \frac{1}{x-y} & for \; x < a < y \\ 0 & for \; a < x \; or \; a > y \end{cases}, \tag{7}$$

whereas, x and y are maximum likelihood parameters. Altering the integral $f(a)da$ Over any other interval due to 0. (8) provides a probability function to find the most likely parameter.

$$l(x, y|a) = nlog(y - x), \tag{8}$$

Halton Quasi-Random Initialization. The Halton sequence is J. Halton's [25] enhanced Van Dar Corput [28] sequence. Halton sequence generates random sequences using co-prime base. Following is Haltom sequence pseudo-code:

Halton Sequence.
> //input: Initial index = *s* & base = *co-prime*
>> // output: instances = *h*
>>> Interval lower and upper limits
>>> a)   *minimum* = 0
>>> b)   *maximum* = 1
>> Foreach epoch $k_1, k_2, k_3 \dots k_n$:do
>>> a.   For each particle {p₁, p₂, p₃, ...pₙ}
>>> ▪ *max = max/coprime*
>>> ▪ *min = min + max * s mod b*
>>> ▪ *s = s/b*
>> Return h

Sobol Quasi-Random Initialization. Russian mathematician Ilya M. Sobol [26] introduced Sobol [17] sequence. Sobol sequence reconstructs coordinates. Each dimension's coordinates have linear recurrence relations. A non-negative instance has a binary expression, where s is as follows in (9):

$$s = s_1 2^0 + s_2 2^1 + s_3 2^2 + \dots + s_w 2^{w-1}, \tag{9}$$

Then, Using the Eq. (10) below, the ith instance of dimension D is generated as follows.

$$X_i^D = i_1 v_1^D + i_2 v_2^D + \dots + i_w v_w^D, \tag{10}$$

However, the binary function are signified by v_1^D followed by i^{th} direction instance and dimension D. . The direction instance is generated by using (11) and c_q is a polynomial coefficient where $i > q$.

$$V_i^D = c_1 v_{i-1}^D + c_2 v_{i-2}^D + \dots + c_q v_{q-1}^D + \left(\frac{v_{i-q}^D}{2^q} \right), \tag{11}$$

Torus Quasi-Random Initialization. A geometric term named as TORUS was coined in 1987 [27], Torus sequence mesh structure is also the core of game development. Torus-mesh facilitates left-right hand coordination. Different dimensions of torus have different forms. For example, 1D shapes are lines, 2D shapes are circles and rectangles, and 3D shapes are donuts. The above combination of Eqs. (12), (13) and (14) generates 3D shapes:

$$a(\theta, \delta) = (D + rcos\theta)cos\delta, \tag{12}$$

$$b(\theta, \delta) = (D + r\cos\theta)\sin\delta, \tag{13}$$

$$c(\theta, \delta) = r\sin\delta, \tag{14}$$

whereas, the angles of the circles are defined as θ, δ and D, and the radius of the circle is represented with r, the tube-Torus distance. The Torus sequence in (15):

$$\alpha_k = \left(f\left(k\sqrt{s_1}\right), \ldots, f\left(k\sqrt{s_d}\right) \right), \tag{15}$$

Here, i^{th} prime-numbers and fraction evaluated by $f = a - floor(a)$ are represented by s_1 and f respectively. Dimensions beyond 100,000 require manual numbering.

4 Results and Discussion

4.1 Experimental Settings of Fire Hawk Optimizer

Parameter Settings. Search space limits are determined as per the undergoing function, as a constant value population size set to 50 agents, and 5 dimensional levels (10, 20, 30, 40, and 50) and 5 iterations (1000, 2000, 3000, 4000, and 5000) are settled respectively. To get the balanced results, 10 runs of execution are conduction and their average value calculated accordingly (Table 1).

Table 1. Parameters settings for experimental setup

Parameters	Values
Search-Space	Upper and lower boundaries of function/s
Dim	(10, 20, 30, 40 and 50)
Iterations	(1000, 2000, 3000, 4000 and 5000)
Pop-Size	50
FHO Runs	10

Functions. Standard benchmark unimodal functions [29] are in given below Table 2. Which are taken from the literature and used by different researchers, across the continuous range of values to implement the evolutionary computing algorithm/s.

4.2 Initialization Techniques Results and Implementation

In Table 3, the outcomes of a comparison between four FHO initialization techniques and ten benchmark test functions are displayed. The calculated mean value is shown in the column called "Mean." The values in bold show which of the four FHO algorithms is the winner. The TO-FHO method works much better than the traditional FHO, SO-FHO, and H-FHO methods. On benchmark functions for minimization optimization, the TO-FHO sequence based on QRS outperformed alternative initialization methods in terms of enhanced convergence, speed, consistency, and longevity.

Table 2. Optimum values of benchmark objective functions

Function Names	Function Equation	Search Space	Optimum Val
F1: Bukin6	$100\sqrt{\left\|x_2 - 0.01x_1^2\right\|} + 0.01\|x_1 + 10\|$	$[-15, -5]$	0
F2: Deb1	$\frac{-1}{D}\sum_{i=1}^{D} \sin^6(5\pi x_i)$	$[-1, 1]$	-1
F3: Schwefel 2.22	$\sum_{i=1}^{D}\|x_i\| + \prod_{i=1}^{n}\|x_i\|$	$[-100, 100]$	0
F4: Trid 6	$\sum_{i=1}^{D}(x_i - 1)^2 - \sum_{i=1}^{D} x_i \; x_i - 1$	$[-36, 36]$	-50
F5: Trid 10	$\sum_{i=1}^{D}(x_i - 1)^2 - \sum_{i=1}^{D} x_i \; x_i - 1$	$[-100, 100]$	-210
F6: Styblinski_Tang	$\frac{1}{2}\sum_{i=1}^{n}(x_i^4 - 16x_i^2 + 5x_i)$	$[-5, 5]$	-78.322 -1958.3
F7: Glodenstein Price	$\left[1 + (x_1 + x_2 + 1)^2(19 - 14x_1 + 3x_1^2 - 14x_2 + 6x_1x_2 + 3x_2^2)\right] \times \left[30 + (2x_1 - 3x_2)^2(18 - 32x_1 + 12x_1^2 + 48x_2 - 36x_1x_2 + 27x_2^2)\right]$	$[-2, 2]$	3
F8: XinShe1	$\sum_{i=1}^{D} \varepsilon_i \|x_i\|^i$	$[-5, 5]$	0 -1
F9: XinShe2	$\left(\sum_{i=1}^{D} \varepsilon_i\|x_i\|\right)\exp[-\sum_{i=1}^{D} \sin(x_i^2)]$	$[-2\pi, 2\pi]$	0
F10: Ackley1	$-20e^{-0.02\sqrt{D^{-1}\sum_{i=1}^{D} x_i^2}} - e^{D-1}\sum_{i=1}^{D}\cos(2\pi x_i) + 20 + e$	$[-35, 35]$	0

4.3 Graphs of FHO Initialization Techniques

Following graphical representation from Figs. 1, 2, 3, 4, 5, 6, 7, 8, 9 and 10 has been prepared from the results of above Table 3.

Table 3. Comparative results across 10 standard test functions and four FHO versions

Functions	Iterations	Dim.	Std-FHO	SO-FHO	H-FHO	TO-FHO
			Mean	*Mean*	*Mean*	*Mean*
F1	1000	10	2.2918E+02	2.2918E+02	2.2918E+02	**1.7498E−101**
	2000	20	2.2918E+02	2.2918E+02	2.2918E+02	**7.1065E−213**
	3000	30	2.2918E+02	2.2918E+02	2.2918E+02	**0.0000E+00**
	4000	40	2.2918E+02	2.2918E+02	2.2918E+02	**0.0000E+00**
	5000	50	2.2918E+02	2.2918E+02	2.2918E+02	**0.0000E+00**
F2	1000	10	−9.5284E−01	−9.4085E−01	−9.6466E−01	**−9.7495E−01**
	2000	20	−9.7072E−01	−9.7779E−01	−9.7794E−01	**−9.8791E−01**
	3000	30	−9.8403E−01	−9.8211E−01	−9.7963E−01	**−9.8671E−01**
	4000	40	−9.8890E−01	−9.8908E−01	−9.9328E−01	**−9.9501E−01**
	5000	50	−9.9328E−01	−9.9336E−01	−9.9399E−01	**−9.9607E−01**
F3	1000	10	6.2833E−25	**3.7202E−25**	1.9700E−22	5.0727E−23
	2000	20	1.2593E−49	8.0838E−48	7.6713E−48	**9.5991E−50**
	3000	30	7.2068E−75	6.4106E−75	7.3864E−75	**6.9677E−76**
	4000	40	1.8456E−99	5.9702E−100	9.7802E−101	**5.8469E−101**
	5000	50	3.9864E−125	1.4698E−122	1.4497E−126	**8.9452E−127**
F4	1000	10	−2.0447E+02	−2.0522E+02	−2.4609E+03	**−2.5466E+03**
	2000	20	−7.4799E+02	−7.5834E+02	−2.5064E+03	**−2.5794E+03**
	3000	30	−1.2553E+03	−1.0524E+03	−2.5066E+03	**−2.5938E+03**
	4000	40	−1.7578E+03	−1.7186E+03	−2.4816E+03	**−2.5375E+03**
	5000	50	−2.3952E+03	−2.3789E+03	−2.4281E+03	**−2.5477E+03**
F5	1000	10	−2.0388E+02	−2.0411E+02	−3.2440E+03	**−4.6700E+03**
	2000	20	−7.6250E+02	−6.1824E+02	−5.1202E+03	**−5.9809E+03**
	3000	30	−1.5396E+03	−1.0360E+03	−5.1410E+03	**−6.0472E+03**
	4000	40	−2.0374E+03	−3.1353E+03	−5.2147E+03	**−6.0314E+03**
	5000	50	−4.3074E+03	−4.8064E+03	−4.9751E+03	**−5.7401E+03**
F6	1000	10	−3.6887E+02	−3.6350E+02	−1.6023E+03	**−1.6882E+03**
	2000	20	−7.2825E+02	−7.3019E+02	−1.6427E+03	**−1.7748E+03**
	3000	30	−1.0382E+03	−1.0987E+03	−1.7588E+03	**−1.7887E+03**
	4000	40	−1.4215E+03	−1.3767E+03	−1.7347E+03	**−1.7972E+03**
	5000	50	−1.8290E+03	−1.4094E+03	−1.7079E+03	**−1.8477E+03**
F7	1000	10	3.0005E+00	3.0013E+00	3.0007E+00	**3.0005E+00**

(*continued*)

Table 3. (*continued*)

Functions	Iterations	Dim.	Std-FHO	SO-FHO	H-FHO	TO-FHO
			Mean	*Mean*	*Mean*	*Mean*
	2000	20	3.0002E+00	3.0007E+00	3.0004E+00	**3.0001E+00**
	3000	30	3.0002E+00	3.0003E+00	3.0002E+00	**3.0002E+00**
	4000	40	3.0002E+00	3.0002E+00	3.0003E+00	**3.0002E+00**
	5000	50	3.0002E+00	**3.0001E+00**	3.0002E+00	3.0002E+00
F8	1000	10	5.9517E−113	2.4453E−111	1.1719E−109	**3.8257E−115**
	2000	20	1.3488E−216	5.9554E−218	1.2823E−219	**4.2165E−221**
	3000	30	0.0000E+00	0.0000E+00	0.0000E+00	**0.0000E+00**
	4000	40	0.0000E+00	0.0000E+00	0.0000E+00	**0.0000E+00**
	5000	50	0.0000E+00	0.0000E+00	0.0000E+00	**0.0000E+00**
F9	1000	10	6.0172E−04	6.0754E−04	1.2102E−20	**1.2080E−20**
	2000	20	5.1649E−08	5.1677E−08	1.2075E−20	**1.2075E−20**
	3000	30	3.5136E−12	3.5129E−12	1.2075E−20	**1.2076E−20**
	4000	40	2.1272E−16	2.1273E−16	1.2075E−20	**1.2075E−20**
	5000	50	1.2075E−20	1.2074E−20	1.2075E−20	**1.2074E−20**
F10	1000	10	8.9249E−08	8.9249E−08	8.8818E−16	**8.8818E−16**
	2000	20	8.9249E−08	8.9249E−08	8.8818E−16	**8.8818E−16**
	3000	30	8.8818E−16	8.8818E−16	8.8818E−16	**8.8818E−16**
	4000	40	8.8818E−16	8.8818E−16	8.8818E−16	**8.8818E−16**
	5000	50	8.8818E−16	8.8818E−16	8.8818E−16	**8.8818E−16**

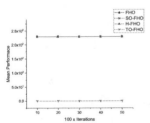

Fig. 1. Performance curve for Bukin6 function has been shown.

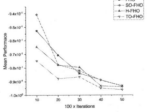

Fig. 2. Performance curve for Deb1 function has been shown.

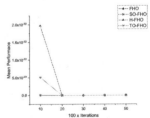

Fig. 3. Performance curve for Schwefel 2.22 function has been shown.

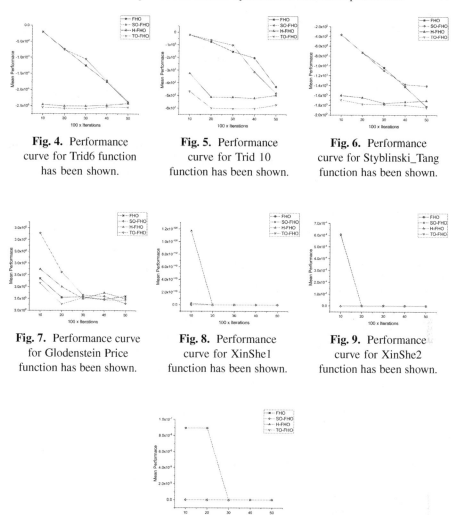

Fig. 4. Performance curve for Trid6 function has been shown.

Fig. 5. Performance curve for Trid 10 function has been shown.

Fig. 6. Performance curve for Styblinski_Tang function has been shown.

Fig. 7. Performance curve for Glodenstein Price function has been shown.

Fig. 8. Performance curve for XinShe1 function has been shown.

Fig. 9. Performance curve for XinShe2 function has been shown.

Fig. 10. Performance curve for Ackley1 function has been shown.

5 Conclusion

This study proposes a new FHO population initialization method. Ten different benchmark test functions, both unimodal and multimodal, are used in the experiment to ensure accuracy. The proposed technique improves convergence speed, search ability, and swarm diversity. The technique also emphasizes that if no candidate solutions are known, the best candidate solution can be chosen first. The results show that TO-FHO for initialization outperform standard FHO and other variants such SO-FHO with Sobol sequence and H-FHO with Halton sequence. This research's primary goal is universal but relevant to different stochastic as well as metaheuristic algorithms that shape our future on various fields.

References

1. Eiben, A.E., Schoenauer, M.J.: Evolutionary computing. Inf. Process. Lett. **82**(1), 1–6 (2002)
2. Zhou, A., et al.: Multiobjective evolutionary algorithms: a survey of the state of the art. Swarm Evol. Comput. **1**(1), 32–49 (2011)
3. Blum, C., Li, X.: Swarm intelligence in optimization. In: Blum, C., Merkle, D. (eds.) Swarm Intelligence. NCS, pp. 43–85. Springer, Heidelberg (2008). https://doi.org/10.1007/978-3-540-74089-6_2
4. Bangyal, W.H., et al.: An improved bat algorithm based on novel initialization technique for global optimization problem. Int. J. Adv. Comput. Sci. Appl. **9**(7), 158–166 (2018)
5. Bangyal, W.H., et al.: An analysis of initialization techniques of particle swarm optimization algorithm for global optimization. In: 2021 International Conference on Innovative Computing (ICIC). IEEE (2021)
6. Poli, R., Kennedy, J., Blackwell, T.J.: Particle swarm optimization. Swarm Intell. **1**(1), 33–57 (2007)
7. Pervaiz, S., et al.: Comparative research directions of population initialization techniques using PSO algorithm. Intell. Autom. Soft Comput. **32**(3), 1427–1444 (2022)
8. Bangyal, W.H., et al.: Constructing domain ontology for Alzheimer Disease using deep learning based approach. Electronics **11**(12), 1890 (2022)
9. Bangyal, W., Ahmad, J., Abbas, Q.: Recognition of off-line isolated handwritten character using counter propagation network. Int. J. Eng. Technol. **5**(2), 227 (2013)
10. Azizi, M., Talatahari, S., Gandomi, A.H.J.: Fire Hawk Optimizer: a novel metaheuristic algorithm. Artif. Intell. Rev. **56**, 1–77 (2022)
11. Reynolds, C.W.: Flocks, herds and schools: a distributed behavioral model, vol. 21. ACM (1987)
12. Pervaiz, S., et al.: A systematic literature review on particle swarm optimization techniques for medical diseases detection. Comput. Math. Methods Med. **2021**, 10 (2021)
13. Bangyal, W., Ahmad, J., Abbas, Q.: Analysis of learning rate using CPN algorithm for hand written character recognition application. Int. J. Eng. Technol. **5**(2), 187 (2013)
14. Ul Hassan, N., et al.: Improved opposition-based particle swarm optimization algorithm for global optimization. Symmetry **13**(12), 2280 (2021)
15. Bonta, M., et al.: Intentional fire-spreading by "Firehawk" raptors in Northern Australia. J. Ethnobiol. **37**(4), 700–718 (2017)
16. Bangyal, W.H., et al.: A new initialization approach in particle swarm optimization for global optimization problems, vol. 2021 (2021)
17. Ashraf, A., et al.: Studying the impact of initialization for population-based algorithms with low-discrepancy sequences. Appl. Sci. **11**(17), 8190 (2021)
18. Ashraf, A., et al.: Training of artificial neural network using new initialization approach of particle swarm optimization for data classification. In: 2020 International Conference on Emerging Trends in Smart Technologies (ICETST). IEEE (2020)
19. Bangyal, W.H., et al.: New modified controlled bat algorithm for numerical optimization problem. Comput. Mater. Continua **70**(2), 2241–2259 (2022)
20. Abbas, Q., Bangyal, W.H., Ahmad, J.: The impact of training iterations on ANN applications using BPNN algorithm. Int. J. Future Comput. Commun. **2**(6), 567 (2013)
21. Bangyal, W.H., et al.: Comparative analysis of low discrepancy sequence-based initialization approaches using population-based algorithms for solving the global optimization problems. Appl. Sci. **11**(16), 7591 (2021)
22. Ashraf, A., et al.: Particle swarm optimization with new initializing technique to solve global optimization problems. Intell. Autom. Soft Comput. **31**(1), 191–206 (2022)

23. Nematollahi, A.F., Rahiminejad, A., Vahidi, B.: A novel meta-heuristic optimization method based on golden ratio in nature. Soft. Comput. **24**(2), 1117–1151 (2020)
24. Uy, N.Q., et al.: Initialising PSO with randomised low-discrepancy sequences: the comparative results. In: 2007 IEEE Congress on Evolutionary Computation. IEEE (2007)
25. Wang, X., Hickernell, F.J.: Randomized halton sequences. Math. Comput. Model. **32**(7–8), 887–899 (2000)
26. Pant, M., et al.: Particle swarm optimization using Sobol mutation. In: 2008 First International Conference on Emerging Trends in Engineering and Technology. IEEE (2008)
27. Nikulin, V.V., Shafarevich, I.R.: Geometries and Groups. Springer, Heidelberg (2012). https://doi.org/10.1007/978-3-642-61570-2
28. Van der Corput, J.: Verteilungsfunktionen: Mitteilg 7. NV Noord-Hollandsche Uitgevers Maatschappij (1936)
29. Jamil, M., Yang, X.-S.: A literature survey of benchmark functions for global optimization problems (2013)

Hybrid Artificial Ecosystem-Based Optimization Algorithm Solves Photovoltaic Models Parameter Identification Problem

Xiaoqiu Yu[✉]

College of Computer and Communication Engineering, Zhengzhou University of Light Industry, Zhengzhou, China
rachelby@163.com

Abstract. Photovoltaic (PV) models have both nonlinear and multimodal characteristics, and traditional algorithms are prone to fall into local optimum and insufficient recognition accuracy in identifying their parameters. This paper proposes a hybrid artificial ecosystem-based optimization (HAEO) algorithm for fast, accurate, and reliable identification of PV model parameters under different environmental conditions. In HAEO, a Nelder-Mead polyhedral search strategy is used to update the poorer individuals. In addition, a chaotic local search strategy is introduced to enhance the ability of populations to jump out of the local optimum by using the ergodic and non-repetitive nature of chaos. To verify the effectiveness of the proposed algorithm, it is used for the parameter identification of thin-film ST40 PV modules under different irradiation intensities and temperatures conditions. Simulation results show that the HAEO algorithm is more competitive than the existing algorithms in terms of accuracy and reliability. Therefore, HAEO can be considered an effective PV model parameter identification method.

Keywords: PV model · Parameter identification · Artificial ecosystem-based optimization

1 Introduction

Photovoltaic (PV) systems can convert solar radiation into electrical energy. They are successfully used in many fields [1], such as lighting and heating, because of their simple structure and easy installation and maintenance. Photovoltaic models can effectively describe the nonlinear behavior of PV systems, and the relevant parameters mainly influence their accuracy. These parameters are crucial for the performance evaluation and quality control of PV models and have an essential role in PV systems' maximum power point tracking. Therefore, it is critical to identify the unknown parameters of PV models accurately and quickly.

Due to the multimodal and nonlinear nature of PV models, the traditional analytical methods [2] are prone to fall into local optima and have low search accuracy when performing parameter identification. In contrast, metaheuristic algorithms such as differential evolution [3] and particle swarm optimization [4] have achieved remarkable research

© The Author(s), under exclusive license to Springer Nature Switzerland AG 2023
Y. Tan et al. (Eds.): ICSI 2023, LNCS 13968, pp. 80–91, 2023.
https://doi.org/10.1007/978-3-031-36622-2_7

results in the field of PV parameter identification due to their advantages of not requiring an explicit formulation of the problem and being good at handling multimodal and non-linear problems. In [3], a self-adaptive ensemble-based differential evolution algorithm is proposed to balance diversity and convergence by combining three different variational strategies and an adaptive scheme. To address the drawback of premature convergence of the PSO algorithm, the literature [4] proposed an enhanced leader particle swarm optimization algorithm by introducing an adaptive variational strategy and a five-stage continuous variational strategy. In addition, novel intelligent optimization algorithms, continuously applied in PV model parameter identification, show great potential. Yu et al. [5] proposed a multiple learning backtracking search algorithm, optimizing each subpopulation by introducing a multivariate learning strategy and a chaotic local search strategy.

Much effort has been made in the existing literature on the PV model parameter identification problem, but there are still drawbacks. Most studies only consider parameter identification under normal environmental conditions, e.g., the literature [3–5] can only perform experiments when temperature and irradiation intensity is maintained at a certain level, ignoring the performance under extreme irradiation intensity or temperature conditions. However, severe weather is something that PV systems will inevitably encounter during operation. Therefore, the problem of identifying PV model parameters in extreme environments deserves further investigation.

In order to effectively identify unknown parameters of PV models under different irradiation intensities and temperatures, a hybrid artificial ecosystem-based optimization (HAEO) algorithm is proposed in this paper for the PV module parameter identification problem. Artificial ecosystem-based optimizer (AEO) [6] has been used in many application areas [7] as it has fewer parameters, a simple structure, and easy to implement than metaheuristic algorithms. However, the AEO algorithm has poor exploration capabilities, so it can easily fall into a local optimum. Once it falls into a local optimum, the randomness of the Levy flight alone is not enough to produce a good detachment, and it is impossible to search other regions of the problem space by more iterations. Therefore, we propose a Nelder-Mead polyhedral search strategy to update the worse individuals through four strategies: reflection, expansion, contraction, and compression, using reflection operation to expand the search space of individuals, expansion operation to move individuals away from the better ones to prevent them from falling into the local optimum, and compression operation to lead individuals close to the best ones to improve the convergence speed. In addition, a chaotic local search strategy is also introduced to enhance its ability to jump out of the local optimum by exploiting the randomness and non-repetitiveness of chaos. We conduct simulation experiments using thin-film ST40 of PV modules with actual data obtained from the manufacturer's datasheet to compare the proposed algorithm with five established algorithms and check the accuracy and stability of the algorithm under different temperature and irradiation intensity conditions with extreme environmental changes. The experimental results show that HAEO possesses higher accuracy and stability than the proposed parameter identification methods.

2 PV Model

2.1 PV Module Model

As shown in Fig. 1, a single-diode PV panel consists of a photo generated current, N_s series-connected single diodes, N_p sets of parallel-connected single diodes, a series resistor, and a shunt resistor. The output current of the single diode PV module model is represented as follows.

$$I/N_p = I_{ph} - I_{sd}\left[\exp(\frac{q(V_t/N_s + I_tR_s/N_p)}{zkT}) - 1\right] - \frac{(V_t/N_s + I_tR_s/N_p)}{R_{sh}} \quad (1)$$

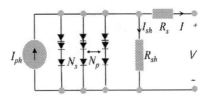

Fig. 1. Equivalent circuit of photovoltaic module model.

2.2 Objective Function Definition

The primary objectives of identifying PV model parameters are improving parameter accuracy and minimizing the discrepancy between measured and predicted values [5]. To achieve this, the root mean square error (RMSE) is used as the objective function for the PV model, expressed as follows:

$$RMSE(X) = \sqrt{\frac{1}{N}\sum_{k=1}^{N} f_k(V, I, X)^2} \quad (2)$$

where $\sum_{k=1}^{N} f_k(V, I, X)$ is the error function of the measured and predicted values of the PV model, N is the size of the experimental data, and X is the parameter to be estimated by the model.

3   Hybrid Artificial Ecosystem-Based Optimization Algorithm

In this paper, we propose a hybrid artificial ecosystem-based optimization algorithm by adding Nelder-Mead based polyhedral optimization search strategy and chaos local search strategy to the basic artificial ecosystem optimization algorithm.

3.1   Basic Artificial Ecosystem-Based Optimization

Artificial ecosystem-based optimization algorithm is proposed based on ecosystem energy flow [6]. Ecosystems contain producers, consumers, and decomposers. In the production phase, the producer update equation is as follows.

$$X_1(t + 1) = (1 - \alpha)X_{best}(t) + \alpha X_{rand}(t) \tag{3}$$

where X_1 is the worst individual, X_{best} is the worst individual, $\alpha = (1 - t/T)r_1$ is the linear weight coefficient, r_1 denotes a random number in the range [0, 1], T and t are the maximum and current number of iterations, respectively, and X_{rand} is a randomly generated individual in the decision space.

In the consumption stage, consumers X_i are divided into three types, herbivores, carnivores and omnivores, updated according to Eqs. (4), (5) and (6), respectively.

$$X_i(t + 1) = X_i(t) + K(X_i(t) - X_1(t)) \tag{4}$$

where $i = [2, ..., NP]$, NP denotes the population size, $K = \frac{v_1}{2|v_2|}$ is the consumption factor based on the Lévy flight improvement, v_1 and v_2 are random numbers obeying standard normal distribution.

$$X_i(t + 1) = X_i(t) + K \cdot (X_i(t) - X_v(t)) \tag{5}$$

where X_v is a random consumer with a higher energy level and v denotes a random number between $[2, i - 1]$.

$$X_i(t + 1) = X_i(t) + K(r_2(X_i(t) - X_1(t)) + (1 - r_2)(X_i(t) - X_v(t)) \tag{6}$$

where r_2 denotes a random number in the range [0, 1].

In the decomposition phase, each organism is broken down by decomposers in the ecosystem, and the renewal equation is shown in Eq. (7).

$$X_i(t + 1) = X_{best}(t) + D(eX_n(t) - h \cdot X_i(t)) \tag{7}$$

where $D = 3u$ is the decomposition coefficient, u is a random number that obeys the standard normal distribution, $e = r_3 \cdot randi([1, 2]) - 1$, $i = [1, ..., NP]$, and r_3 is a random number between [0, 1].

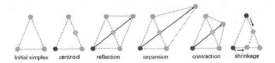

Fig. 2. Simplex manipulation diagram for 2D problems.

3.2 Nelder-Mead Based Polyhedral Optimization Search Strategy

This paper proposes a Nelder-Mead [7] based polyhedral optimization technique that may increase the HAEO algorithm's local search capacity and convergence time. The strategy uses five steps: sorting, reflection, expansion, contraction, and compression, and is coupled with a greedy strategy to find the better point to replace the worst point. Figure 2 illustrates the Nelder-Mead operation for a two-dimensional optimization problem.

step 1. The fitness values of each individual are calculated and ranked, and the two best points X_a, X_b and the worst point X_z are selected, corresponding to the fitness values $f(X_a), f(X_b), f(X_z)$. The central position is X_c for Eq. (8).

$$X_c = (X_a + X_b)/2 \tag{8}$$

step 2. The reflection operation is performed on the worst X_z points, X_d is the reflection point, and α is the reflection coefficient taking the value of 1.

$$X_d = X_c + \alpha(X_c - X_z) \tag{9}$$

step 3. If $f(X_d) < f(X_a)$, the reflection is in the correct direction, and the expansion operation is performed next. X_e is the point after performing the expansion operation, i.e., the expansion point. λ is the expansion factor, usually $\lambda = 2$. If $f(X_e) < f(X_a)$, X_z the point is replaced by X_e; otherwise, X_z is replaced by X_d, which is the closest.

$$X_e = X_c + \lambda(X_d - X_c) \tag{10}$$

step 4. If $f(X_d) > f(X_a)$, the reflection is in the wrong direction, and the compression operation is performed next. X_f is the point after the compression operation, i.e., the compressed point and β is the expansion factor taking the value of 0.5. If $f(X_f) < f(X_z)$, the point X_z will be replaced by X_f.

$$X_f = X_c + \beta(X_z - X_c) \tag{11}$$

step 5. If $f(X_z) > f(X_d) > f(X_a)$, the next inward contraction operation is performed. X_g is the point after the contraction operation, i.e., the contraction point. If $f(X_g) < f(X_z)$, then point X_z will be replaced by X_g; conversely, point X_z will be replaced by X_d.

3.3 Chaos Local Search Strategy

The chaos local search (CLS) [8] strategy exploits the ergodic and non-repetitive nature of chaos, which allows for population diversification and, thus, the opportunity to explore a vast search space. We use the CLS strategy only for the locally optimal individuals in the population and explore only around the best solution, which enhances its exploitation and reduces the algorithm execution time. Local optimal solution updated using Eq. (12).

$$newX_{best}(t + 1) = X_{best}(t) + (\omega^{k+1} - 0.5) \cdot (X_{r1}(t) - X_{r2}(t)) \tag{12}$$

$$\omega^{k+1} = \begin{cases} \frac{\omega^k}{P}, 0 \leq \omega^k \leq P \\ \frac{\omega^k - P}{0.5 - P}, P \leq \omega^k \leq 0.5 \\ \frac{1 - P - \omega^k}{0.5 - P}, 0.5 \leq \omega^k \leq 1 - P \\ \frac{1 - \omega^k}{P}, 1 - P \leq \omega^k \leq 1 \end{cases} \tag{13}$$

where $X_{r1}(t)$ and $X_{r2}(t)$ are two solutions chosen randomly, and the parameter P is 0.4 according to the literature [8]. ω^{k+1} is the chaos coefficient generated by the chaos mapping with an initial value of a random number between [0, 1].

3.4 The Concrete Flow of Hybrid Artificial Ecosystem Optimization Algorithm

Algorithm 1 is the pseudo-code for the hybrid artificial ecosystem-based optimization algorithm proposed in this paper. Firstly, the artificial ecological population is initialized, the fitness values of the individuals in the population are calculated, and the individuals in the population are arranged in descending order according to the fitness values. The producers in the population are updated in the producer phase using Eq. (3). Then, enter the consumer stage, where each individual in the population is randomly classified as herbivore, omnivore, and carnivore according to the same probability and updated according to Eqs. (4)–(6), respectively. After entering the decomposition phase, the fitness values of individuals in the population are calculated to find the best individual, and all consumers are decomposed according to Eq. (7). Finally, Nelder-Mead based polyhedral optimization search strategy and chaotic local search strategy are executed. If the algorithm stopping condition is satisfied, the optimal global solution is output; otherwise, it enters the step 2 update loop until the termination condition is satisfied.

Algorithm 1 HAEO

Input: *MaxFes* (maximum number of function evaluations), *FES* (current function evaluation times), *NP* (Population size), *D* (individual dimensions)

Output: Optimal solution X_{best}

1. Initializing ecological populations $POP = \{X_1, X_2, \cdots, X_{NP}\}$;

2. $FES = 0$;

3. **While** $FES < MaxFes$ **do**

4.   Calculate the fitness value of individuals in the population, $FES = FES + NP$;

5.   Arrange in descending order according to fitness value;

6.   Update the producer's position according to Eq.(3) ;// Production stage

7.   **For** $i = 2$ **to** NP // Consumption stage

8.   **If** $rand < \dfrac{1}{3}$ // Herbivores

9.   Update herbivores according to Eq.(4);

10.   **Else if** $\dfrac{1}{3} \leq rand \leq \dfrac{2}{3}$ // Carnivores

11.   Update carnivores according to Eq.(5);

12.   **Else** // Omnivores

13.   Update omnivores according to Eq.(6);

14.   **End if**

15. **End for**

16. Calculation of fitness values of individuals in the population to find the best individual, $FES = FES + NP$;

17. Update each individual in the population using Eq.(7); // decomposition stage

18. Perform Nelder-Mead based polyhedral optimization search strategy to the worst individual, $FES = 1 + FES$;

19. Perform chaotic local search strategy to update local optimal individual, $FES = 1 + FES$;

20. **End while**

21. **Return** the global optimal solution X_{best}.

---

4   Experimental Analysis

4.1   Experimental Settings

We used thin film ST40 PV modules with real data from the manufacturer's datasheet [9] to test the accuracy of HAEO algorithm. The experimental data used were extracted directly from the current-voltage curves for five different irradiance levels (1000 w/m², 800 w/m², 600 w/m², 400 w/m², and 200 w/m²) and different temperature levels (25 °C, 40 °C, 55 °C and 70 °C) given in the manufacturer's datasheet. Four advanced algorithms were used to compare with the proposed HAEO, including IJAYA [9], PGJAYA [10], MLBSA [5], STLBO [11] and AEO [6], and their relevant parameters were set in line with the original paper. The maximum number of evaluations was set to 50,000, and each algorithm was run 30 times on its own to show that the comparison was fair. The simulation experiments were performed using a computer with an Intel Core(R) Core (TM) i5 CPU @ 2.8 GHz and 8 GB RAM, and all algorithms were written in MATLAB R2018b software.

4.2   Results of Thin Film ST40 at Different Irradiation Levels

The results of 30 independent runs of the experiment were statistically examined to demonstrate the superiority of the suggested approach, and the findings are shown in Table 1. From Table 1, it can be seen that HAEO achieves the minimum RMSE value both at high and low irradiance intensity. Although the PGJAYA algorithm achieves the same minimum value as HAEO at 200 w/m² and 600 w/m², HAEO outperforms PGJAYA in terms of other metrics. Especially in terms of standard deviation, HAEO is far ahead of the other algorithms. It indicates the HAEO's superior robustness.

In order to confirm the correctness of the acquired parameters, we also plotted the *I_V* characteristic curves of the ST40 at 25 °C and various irradiance intensities. As seen in Fig. 3, the predicted data fit the actual current and voltage data very well, thus verifying the accuracy of HAEO in the ST40 PV module simulation experiments.

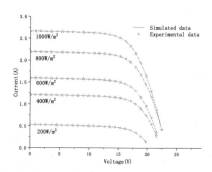

Fig. 3. Comparisons of *I-V* characteristics different irradiance levels at 25 °C.

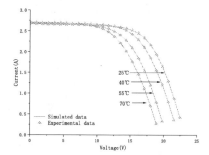

Fig. 4. Comparisons of *I-V* characteristics different temperature levels at G = 1000 w/m²

Table 1. At various irradiance levels and T = 25 °C, comparison of statistical results.

G	Algorithm	RMSE			
		Max	Min	Mean	Std
200 (W/m^2)	IJAYA	5.4454E−04	5.0746E−04	5.0746E−04	2.4167E−05
	PGJAYA	4.7725E−04	**4.7720E−04**	4.7721E−04	1.4355E−08
	MLBSA	4.7965E−04	4.7965E−04	4.7965E−04	4.1922E−06
	STLBO	4.9232E−04	4.7723E−04	4.7879E−04	4.7562E−06
	AEO	5.1066E−04	4.9813E−04	5.0244E−04	1.0438E−05
	HAEO	**4.7720E−04**	4.7720E−04	**4.7720E−04**	**5.4471E−17**
400 (W/m^2)	IJAYA	1.8392E−03	1.3299E−03	1.3299E−03	4.4001E−04
	PGJAYA	6.5425E−04	6.3321E−04	6.3321E−04	7.4019E−06
	MLBSA	9.8761E−04	6.8119E−04	6.8119E−04	1.0880E−04
	STLBO	6.4355E−04	6.3078E−04	6.3468E−04	4.3095E−06
	AEO	7.4713E−04	7.1300E−04	7.2453E−04	3.5591E−05
	HAEO	**6.3072E−04**	**6.3072E−04**	**6.3072E−04**	**5.9304E−17**
600 (W/m^2)	IJAYA	3.2574E−03	9.5342E−04	1.6749E−03	7.5009E−04
	PGJAYA	6.7845E−04	**6.7404E−04**	6.7467E−04	1.3887E−06
	MLBSA	1.5383E−03	6.9483E−04	8.6889E−04	2.6031E−04
	STLBO	7.2792E−04	6.7604E−04	6.9103E−04	2.0675E−05
	AEO	1.3051E−03	7.3591E−04	8.6973E−04	2.3784E−04
	HAEO	**6.7404E−04**	6.7404E−04	**6.7404E−04**	**9.5038E−17**
800 (W/m^2)	IJAYA	5.4637E−03	3.9243E−03	3.9243E−03	1.0357E−03
	PGJAYA	8.3347E−04	7.8619E−04	7.8619E−04	1.9928E−05
	MLBSA	1.8046E−03	1.2194E−03	1.2194E−03	3.2604E−04
	STLBO	8.5954E−04	7.7519E−04	8.0070E−04	2.7862E−05
	AEO	1.9289E−03	8.7774E−04	1.2808E−03	4.0774E−04
	HAEO	**7.7391E−04**	**7.7391E−04**	**7.7391E−04**	**2.4932E−15**
1000 (W/m^2)	IJAYA	6.6664E−03	2.2409E−03	3.8218E−03	1.4061E−03
	PGJAYA	7.4652E−04	7.3512E−04	7.3792E−04	4.1893E−06
	MLBSA	1.7135E−03	8.0924E−04	1.1343E−03	3.3154E−04
	STLBO	8.2506E−04	7.3414E−04	7.7796E−04	3.9768E−05
	AEO	1.8865E−03	7.4210E−04	1.2127E−03	4.2340E−04
	HAEO	**7.3410E−04**	**7.3410E−04**	**7.3410E−04**	**1.7165E−15**

Table 2. At various temperature levels and G = 1000 w/m^2, comparison of statistical results.

T	Algorithm	RMSE			
		Max	Min	Mean	Std
20 °C	IJAYA	5.4596E−04	4.7789E−04	5.1624E−04	2.1275E−05
	PGJAYA	4.7728E−04	4.7722E−04	4.7723E−04	1.7929E−08
	MLBSA	4.8027E−04	4.7722E−04	4.7826E−04	1.1435E−06
	STLBO	4.7854E−04	4.7721E−04	4.7747E−04	4.2363E−07
	AEO	3.6927E−03	7.8962E−04	1.3842E−03	8.8628E−04
	HAEO	**4.7720E−04**	**4.7720E−04**	**4.7720E−04**	**7.6383E−17**
40 °C	IJAYA	1.9294E−03	7.1619E−04	1.1237E−03	4.3281E−04
	PGJAYA	6.3076E−04	6.3073E−04	6.3074E−04	9.1781E−09
	MLBSA	1.1842E−03	6.3074E−04	7.2078E−04	1.7010E−04
	STLBO	6.4165E−04	6.3074E−04	6.3309E−04	3.3488E−06
	AEO	1.8531E−03	1.3214E−03	1.4118E−03	1.6667E−04
	HAEO	**6.3072E−04**	**6.3072E−04**	**6.3072E−04**	**4.8863E−16**
55 °C	IJAYA	3.9595E−03	8.3775E−04	1.9915E−03	9.1769E−04
	PGJAYA	7.9164E−04	**6.7404E−04**	6.8588E−04	3.7161E−05
	MLBSA	1.1415E−03	7.0776E−04	8.5141E−04	1.3705E−04
	STLBO	6.9070E−04	6.7495E−04	6.8158E−04	5.8511E−06
	AEO	1.8477E−03	1.8233E−03	1.8280E−03	9.0978E−06
	HAEO	**6.7404E−04**	**6.7404E−04**	**6.7404E−04**	**3.2771E−16**
70 °C	IJAYA	7.4297E−03	1.4925E−03	3.6130E−03	1.7787E−03
	PGJAYA	7.8883E−04	7.7392E−04	7.7686E−04	5.0036E−06
	MLBSA	1.6832E−03	9.0195E−04	1.1836E−03	2.4405E−04
	STLBO	9.6599E−04	7.8267E−04	8.3689E−04	6.1190E−05
	AEO	7.8023E−04	7.7772E−04	7.7798E−04	7.9106E−07
	HAEO	**7.7391E−04**	**7.7391E−04**	**7.7391E−04**	**4.5821E−16**

4.3 Results for Thin Film ST40 at Different Temperature Levels

Table 2 gives the statistical results at different temperatures for a fixed irradiation intensity of 1000 w/m^2. In terms of minimum values, PGJAYA achieves the same good RMSE as HAEO at 55 °C, followed by STLBO, MLBSA, and IJAYA; in terms of maximum and mean values, HAEO provides the best RMSE values; in terms of standard deviation, HAEO has the s irradianceest standard deviation of RMSE values at different temperature conditions, which indicates that HAEO possesses better stability.

To further high irradiance the precision of the acquired parameters, we drew the I_V characteristic curves of the ST40 at 1000 w/m^2 and various temperature levels. As clearly seen in Fig. 4, the predicted data fit the actual current and voltage data very well, thus verifying the accuracy of HAEO in the ST40 PV module simulation experiments.

4.4 Experimental Comparison with the Basic AEO Algorithm

The last two rows of Tables 1 and 2 show the experimental results of the basic AEO and the improved algorithm under different conditions. It can be seen that adopting the Nelder-Mead based polyhedral optimization search strategy and the chaotic local optimum strategy makes the algorithm significantly more capable of exploring and jumping out of the local optimum in the face of complex nonlinear multimodal problems, and the accuracy and stability are also improved considerably.

5   Conclusion

This paper proposes an hybrid artificial ecosystem algorithm combining a Nelder-Mead based polyhedral search strategy and a chaotic local search strategy to determine the unknown parameters of PV modules under various weather conditions. The effectiveness of the proposed algorithm applied to the PV model location parameter identification problem is demonstrated through simulation experiments on thin-film ST40 models with different irradiation intensities and different temperatures, and the results show that HAEO has advantages over four other advanced algorithms. In future work, we plan to use HAEO to solve more complex nonlinear multimodal problems, such as rolling bearing design problems, vehicle side impact problems, etc.

References

1. Pillai, D.S., Rajasekar, N.: Metaheuristic algorithms for PV parameter identification: a comprehensive review with an application to threshold setting for fault detection in PV systems. Renew. Sustain. Energy Rev. **82**(3), 3503–3525 (2018)
2. Silva, E.A., Bradaschia, F., Cavalcanti, M.C., et al.: Parameter estimation method to improve the accuracy of photovoltaic electrical model. IEEE J. Photovoltaics **6**(1), 278–285 (2015)
3. Liang, J., Qiao, K., Yu, K., et al.: Parameters estimation of solar photovoltaic models via a self-adaptive ensemble-based differential evolution. Sol. Energy **207**(3), 336–346 (2020)
4. Jordehi, A.R.: Enhanced leader particle swarm optimisation (ELPSO): a new algorithm for optimal scheduling of home appliances in demand response programs. Artif. Intell. Rev. **53**(3), 2043–2073 (2020)
5. Yu, K., Liang, J.J., Qu, B.Y., et al.: Multiple learning backtracking search algorithm for estimating parameters of photovoltaic models. Appl. Energy **226**(15), 408–422 (2018)
6. Zhao, W., Wang, L., Zhang, Z.: Artificial ecosystem-based optimization: a novel nature-inspired meta-heuristic algorithm. Neural Comput. Appl. **32**(13), 9383–9425 (2020)
7. Izci, D., Hekimoğlu, B., Ekinci, S.: A new artificial ecosystem-based optimization integrated with Nelder-Mead method for PID controller design of buck converter. Alexandria Eng. J. **61**(3), 2030–2044 (2022)
8. Barshandeh, S., Piri, F., Sangani, S.R.: HMPA: an innovative hybrid multi-population algorithm based on artificial ecosystem-based and Harris Hawks optimization algorithms for engineering problems. Eng. Comput. **38**(2), 1581–1625 (2022)

9. Yu, K., Liang, J.J., Qu, B.Y., et al.: Parameters identification of photovoltaic models using an improved JAYA optimization algorithm. Energy Convers. Manag. **150**(15), 742–753 (2017)
10. Yu, K., Qu, B., Yue, C., et al.: A performance-guided JAYA algorithm for parameters identification of photovoltaic cell and module. Appl. Energy **237**(1), 241–257 (2019)
11. Niu, Q., Zhang, H., Li, K.: An improved TLBO with elite strategy for parameters identification of PEM fuel cell and solar cell models. Int. J. Hydrogen Energy **39**(8), 3837–3854 (2014)

Improved Ant Colony Optimization Based Binary Search for Point Pattern Matching for Efficient Image Matching

N. K. Sreeja[✉] and N. K. Sreelaja

PSG College of Technology, Coimbatore, India
sreeja.n.krishnan@gmail.com

Abstract. Point Pattern Matching (PPM) is an approach to establish correspondence between two related patterns by pairing up of points. PPM is widely used in the field of computer vision and pattern recognition. The existing approaches for PPM has either high computational complexity or the search space is large. To overcome this drawback, an Improved Ant Colony Optimization based Binary Search for Point Pattern Matching (IACOBSPPM) has been proposed. The algorithm chooses a query image point value from the query image point pattern and finds the reduced search space in the stored image point pattern based on the length of the point value. The query image point value is searched only in the reduced search space. When a match is found, the next query image point value of the same length is searched only from the matching position of the previous query image point value, further reducing the search space. The computational complexity of the proposal is less compared to the existing approaches for point pattern matching. Experimental results prove the efficiency of IACOBSPPM algorithm.

Keywords: Point Pattern matching · Ant colony optimization · Image matching

1 Introduction

Point Pattern Matching (PPM) is a significant problem in the field of pattern recognition, computational geometry and computer vision. It aims to find the correspondence between the points in two images of a same scene [2]. An exact match between the point sets is termed complete match. Otherwise, it is termed incomplete or partial match. Applications of PPM include model building for navigation in complex environments, image registration, image recognition etc. Methods based on relaxation, alignment, fuzzy relaxation nearest neighbor, probabilistic nearest neighbor and rigid transformations have been proposed for PPM. However, the computational complexity of these algorithms range between $O(n^2)$ to $O(n^6)$ [9]. Sreeja and Sankar proposed an Ant Colony Optimization based Binary Search for PPM (ACOBSPPM) [7]. Although the time complexity of ACOBSPPM was less than conventional binary search, it utilized the entire search space to find a match for a query image point value. To overcome the drawbacks in the existing approaches and to reduce the search space of ACOBSPPM algorithm, an Improved Ant

© The Author(s), under exclusive license to Springer Nature Switzerland AG 2023
Y. Tan et al. (Eds.): ICSI 2023, LNCS 13968, pp. 92–105, 2023.
https://doi.org/10.1007/978-3-031-36622-2_8

Colony Optimization based Binary Search for PPM (IACOBSPPM) has been proposed. The edge map of the stored images and query image are obtained and converted to a point pattern. IACOBSPPM chooses a group of ant agents equivalent to the number of stored image point patterns. Each ant agent finds the length of every query image point value and finds the reduced search space in the stored image point pattern based on the length. Therefore, a query image point value of length L is searched only among the stored image point values of the same length, reducing the search space. This is repeated for all point values in the query image point pattern and the number of matches is found. If the percentage of matches is at least 0.05, the stored image corresponding to the matching stored image point pattern is considered as the matching image.

The rest of the paper is organized as follows. Section 2 describes the related work, Sect. 3 briefs point pattern matching, Sect. 4 gives an overview of Ant colony optimization, Sect. 5 describes the improved ant colony optimization based binary search algorithm for point pattern matching, Sect. 6 discusses a case study, Sect. 7 discusses the experimental results and the conclusions are drawn in Sect. 8.

2 Related Work

Wayman et al. [10] proposed an approach based on point pattern matching for minutiae matching. Chui and Rangarajan [3] proposed an approach for medical image registration using point pattern matching. Chen et al. proposed a method based on rigid transformations for PPM. However, the complexity of the algorithm is $O(m^3n^2\log mn)$ where m and n denotes the number of points in the model and scene respectively. Van et al. proposed a randomized algorithm called Fast Expected Time (FET) algorithm for PPM. The complexity of the algorithm is $O(n(\log m)^{3/2})$ where m and n denotes the number of points in the model and scene respectively. However, the method fails if the model has regularly spaced points [3]. Aiger and Kedem [1] proposed an input sensitive algorithm for PPM. The complexity of the proposal was $O(n \log n + km \log n)$ where m and n denotes the number of points in the model and scene respectively. However, input sensitive algorithm is not suitable for large number of points. Sreeja and Sankar proposed an Ant Colony Optimization based binary search for PPM [7]. Although, the time complexity of ACOBSPPM was less than conventional binary search, it uses the entire search space for matching a point value in the query image point pattern. Sreelaja and Sreeja [8] proposed an ant colony optimization based approach for binary search to categorize the search space and perform binary search in the reduced search space, reducing the computational complexity of search operation. Inspired by the idea of search space reduction, an improved ant colony optimization based binary search for point pattern matching is proposed. The computational complexity of the proposal is less compared to other point pattern matching approaches. Experiments prove the efficiency of IACOBSPPM for image matching.

3 Point Pattern Matching

Point pattern of an image is set of integers representing row and column positions corresponding to the edges of the image. Figure 1(a) and (b) shows an image and the edge map generated from the image respectively. The edge map is represented as a Boolean matrix shown in Fig. 1(c). A value '1' in the matrix represents the presence of an edge and a value '0' denotes absence of an edge. Since, the presence of an edge is significant, each edge in the matrix (corresponding to a value '1') may be represented as a point value by concatenating its row and column position in the matrix. For example, the point value for the edge highlighted in Fig. 1(c) is 14 as the row and column position of its occurrence are 1 and 4 respectively. Similarly, the point value for the remaining edges in the matrix may be computed. Therefore, the point pattern for the image in Fig. 1(a) is {14, 23, 25, 32, 36, 41, 42, 43, 44, 45, 46, 47}. Each value in the point pattern is termed as a point value.

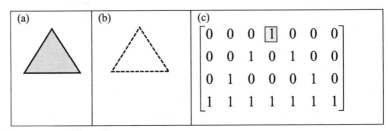

Fig. 1. (a). A sample image **(b)** Edge map generated for the image in (a) and **(c)** Edge matrix corresponding to (b)

4 Ant Colony Optimization

Ant Colony Optimization (ACO), inspired by the foraging behavior of ants, communicate with the help of a chemical substance called pheromone that helps in finding the shortest path to the food source. A colony of artificial ants that denote a set of computationally concurrent and asynchronous ant agents moves through states of the problem that corresponds to partial solutions [4]. They move by applying a stochastic local decision policy based on trails and attractiveness. As they move, they construct solutions incrementally. They evaluate the solutions generated and modifies the trail value on the components in its solution. This directs the search of future ants. ACO algorithm includes two other factors called pheromone evaporation and daemon action. Pheromone evaporation may be related to forgetfulness and daemon action is used to perform centralized actions that cannot be performed by a single ant agent such as updating the global information [6].

5   Improved ACO Based Binary Search for Point Pattern Matching

This section discusses the proposed algorithm and its computational complexity.

5.1   Image Matching Using IACOBSPPM Algorithm

Consider a database with n stored images. Every stored image is represented as a point pattern as discussed in Sect. 3. Let Q be the query image for which a matching stored image has to be found using IACOBSPPM algorithm. The point pattern of the query image is obtained. Let the point pattern of the query image Q be I. The point values in the point pattern of each stored image and the query image is sorted. IACOBSPPM algorithm chooses n ant agents where n denotes the number of stored image point patterns. Each ant agent is assigned a distinct stored image point pattern. Every ant agent matches the query image point pattern with the stored image point pattern assigned to it. The ant agents maintain a global value to store the maximum number of matches between the stored and query image point values and a global set to store the corresponding stored image point pattern. Initially, global set $= \{\Phi\}$ and global value $= 0$. Each ant agent has a tabulist denoting memory. Consider that the ith ant agent is assigned the stored image point pattern St_i. The tabulist of the ith ant agent contains three lists denoted by C_i, P_i and S_i. C_i contains the number of the point values of each length in St_i. If the stored image is of size $u \times v$, then C_i contains $j = length(u\|v)$ entries as in Eq. (1) where $\|$ denotes concatenation.

$$C_i = \begin{bmatrix} count_1 \ count_2 \ count_3 \ldots \ count_j \end{bmatrix} \quad 1 \leq i \leq n \tag{1}$$

where $count_k$ denotes the number of point values of length k and $1 \leq k \leq j$

P_i contains the end position of the point values of a particular length in the stored image point pattern St_i and is computed from C_i as in Eqs. (2) and (3).

$$P_i = \begin{bmatrix} ep_1 \ ep_2 \ ep_3 \ldots \ ep_j \end{bmatrix} \quad 1 \leq i \leq n \tag{2}$$

$$ep_k = \begin{cases} 0 & if \ count_k = 0 \\ \sum_{t=1}^{k} count_t & otherwise \end{cases} \quad 1 \leq k \leq j \tag{3}$$

S_i contains the starting position of the point values of a particular length in the stored image point pattern St_i and is computed as in Eqs. (4) and (5).

$$S_i = \begin{bmatrix} sp_1 \ sp_2 \ sp_3 \ldots \ sp_j \end{bmatrix} \quad 1 \leq i \leq n \tag{4}$$

$$sp_k = ep_k - count_k + 1 \quad \forall \ count_k \neq 0 \quad 1 \leq k \leq j \tag{5}$$

The tabulist of the ith ant agent also contains three variables namely low, high and match. Consider that the ith stored image point pattern St_i contain 'm' point values and the query image point pattern I contain 'r' point values as in Eqs. (6) and (7) respectively.

$$St_i = \{sq_1 \ sq_2 \ sq_3 \ldots \ sq_m\} \quad where \ sq_1 < sq_2 < sq_3 < \ldots < sq_m \tag{6}$$

$$I = \{ip_1\ ip_2\ ip_3 \ldots ip_r\} \qquad where \ \ ip_1 < ip_2 < ip_3 < \ldots < ip_r \qquad (7)$$

Each point value in I is matched with the point values in St_j. The ith ant agent chooses the first point value ip_1 from I and computes its length. Let *len* be the length of ip_1. Since, the length of ip_1 is *len*, the ant agent searches only among the point values of length *len* in St_j. Therefore, the initial values of low, high and match are sp_{len}, ep_{len} and 0 respectively. Let the point values in St_j corresponding to positions sp_{len} and ep_{len} be sq_i and sq_j respectively. Therefore, the reduced search space in St_j to find a match for ip_1 is $[sq_i, sq_{i+1}, \ldots, sq_j]$. The ant agent deposits pheromone by computing the value of mid as in Eq. (8).

$$mid = \left\lceil \frac{(low + high)}{2} \right\rceil \qquad (8)$$

If $sq_{mid} < ip_1$ then the energy value of the ant agent is 1. The position denoted by the pheromone deposition is incremented by 1 and stored in the variable low in the tabulist as in Eq. (9).

$$low = mid + 1 \qquad (9)$$

If $sq_{mid} > ip_1$, then the energy value of the ant agent is -1. The position denoted by the pheromone deposition is decremented by 1 and stored in the variable high in the tabulist as in Eq. (10).

$$high = mid - 1 \qquad (10)$$

If low $\leq$ high, it indicates that the search is not complete. The pheromone evaporates and the ant agent moves to the next trail. The ant agent continues to deposit pheromone to find a match for the point value ip_1 in the stored image point pattern St_i by computing the value of mid.

If $sq_{mid} = ip_1$, then the energy value of the ant agent is 0. The position denoted by the pheromone deposition is incremented by 1 and stored in sp_{len} as in Eq. (11). This indicates that a match has occurred and therefore match is incremented by 1 as in Eq. (12).

$$Sp_{len} = mid + 1 \qquad (11)$$

$$match = match + 1 \qquad (12)$$

When low $>$ high, the ant agent stops the matching process and chooses the next point value ip_2 from the query image point pattern to find a match in the stored image point pattern St_j. The process continues until all point values in I are compared with the point values in St_j. When the ant agent completes the process, the value in the variable match in the tabulist denotes the number of point values in the query image point pattern I matching with the point values in the stored image point pattern St_j. The ant agent computes the match percentage by dividing the value in match with the total number of point values in the query image point pattern as in Eq. (13).

$$match\_percentage = \frac{match}{r} \tag{13}$$

If match_percentage is greater than a threshold of 0.05, the ant agent compares the match value in its tabulist with the global value. If the global value is less than match, the ant agent updates the global value with the match value and the global set with the corresponding stored image point pattern. A threshold of 0.05 ensures that IACOBSPPM retrieves stored images with at least 5% of its point values matching with the point values of the query image. Each ant agent repeats the process of matching the query image point pattern with the stored image point pattern assigned to it. If the global value is greater than or equal to 0.95, remaining ant agents stops the matching process. Finally, the global set contains the matching stored image point pattern for the query image point pattern I. The stored image corresponding to the point pattern in the global set is the matching image.

5.2 Computational Complexity

Consider that the number of point values in the stored image and query image point pattern are m and r respectively. Assume that the majority of the point values in the stored image point pattern are of length '*len*'. Let C_{len} be the number of point values of length '*len*'.

Case 1 - No Point Value in the Query Image Matches with the Point Values of the Stored Image: The time complexity to find a match for each point value of a particular length in the query image point pattern is $O(log\ C_{len})$. Since there are r point values in the query image point pattern, the worst case time complexity of IACOBSPPM is $O(r\ log\ C_{len})$ where $C_{len} < m$.

Case 2 - Every Point Value in the Query Image Point Pattern Matches with the Point Values in the Stored Image Point Pattern: The number of comparisons required to match the first point value of a particular length in the stored image point pattern is $O(log\ C_{len})$. Similarly, the number of comparisons required for the second point value of a particular length is $O(log\ (C_{len} - 1))$. Therefore, the best case time complexity of IACOBSPPM algorithm is computed as in Eq. (14).

$$log\ C_{len} + log(C_{len} - 1) + log(C_{len} - 2) + \ldots\ log(C_{len} - C_{len}) = log(C_{len}!) \tag{14}$$

6 Case Study

Consider three stored images with point patterns as given in Table 1. Since there are three stored image point patterns, three ant agents are selected. Let ant agents A1, A2 and A3 be assigned stored image point patterns St_1, St_2 and St_3 respectively.

The working of A1 has been discussed below. Since St_1 contains four point values of length 2, three point values of length 3 and two point values of length 4, the count list

Table 1. Stored image point pattern.

Stored image	Stored image point pattern
St_1	{11, 12, 13, 14, 290, 300, 310, 4130, 4150}
St_2	{35, 45, 47, 62, 92}
St_3	{11, 12, 19, 291, 300, 310, 4130, 4150}

in the tabulist of ant agent A1 contains values $C_1 = \{Count_1 = 0, Count_2 = 4, Count_3 = 3, Count_4 = 2\}$. The end position of the point values is computed from C_1 as $P_1 = \{ep_1 = 0, ep_2 = 4, ep_3 = 7, ep_4 = 9\}$ and stored in the tabulist of A1. The starting position of the point values is computed from C_1 and P_1 as $S_1 = \{sp_1 = 0, sp_2 = 1, sp_3 = 5, sp_4 = 8\}$ and stored in A1's tabulist. Let the query image point pattern be $I = \{12, 14, 310\}$. A1 chooses the first point value from I and finds its length. Since the length of 12 is 2, A1 chooses sp_2 and ep_2. Since $sp_2 = 1$ and $ep_2 = 4$, A1 searches among the point values between position 1 and 4 in the stored image point pattern St_1. Therefore, the reduced search space to find a match for the first point value 12 is {11, 12, 13, 14}. Since $sp_2 = 1$ and $ep_2 = 4$, the value of low and high are 1 and 4 respectively. Since low $\leq$ high, A1 deposits pheromone by computing mid = ceil((1 + 4)/2) = 3. Since the stored image point value in position 3 is 13 which is greater than the query image point value 12, the energy value of A1 is -1 and A1 updates the value of high to (mid $-$ 1) which is equal to 2. The pheromone deposition of A1 evaporates and it moves to the next trail. Since low $\leq$ high, A1 deposits pheromone and computes mid = ceil((1 + 2)/2) = 2. The query image point value 12 is compared with the stored image point value at position 2. Since the stored image point value at position 2 is 12, a match is found. The energy value of the ant agent is 0 and the variable match in the tabulist of A1 is incremented by 1. Since a match was found at position mid = 2, A1 updates the value of sp_2 in its tabulist to mid + 1 = 3 and moves to the next trail. A1 chooses the next query image point value 14 and finds its length. Since the length of 14 is 2, A1 chooses sp_2 and ep_2. As $sp_2 = 3$ and $ep_2 = 4$, the values of low and high are 3 and 4 respectively. Since low $\leq$ high, A1 finds a match for the query image point value 14 in the search space {13, 14}. A1 deposits pheromone by computing mid = ceil((3 + 4)/2) = 4. Since the stored image point value in position 4 is 14, a match is found. The energy value of the ant agent A1 is 0 and variable match is incremented and updated to 2. A1 moves to the next trail and chooses the next query image point value 310 and computes its length. Since the length of the query point value is 3, A1 chooses sp_3 and ep_3 from its tabulist. Since $sp_3 = 5$ and $ep_3 = 7$, low and high values in the tabulist of A1 are 5 and 7 respectively. The search space to find a match for the query image point value 310 is {290, 300, 310}. A1 deposits pheromone by computing mid = ceil((5 + 7)/2) = 6. A1 compares the stored image point value at position 6 with the query image point value 310. Since the stored image point value at position 6 is 300 which is less than the query image point value, the energy value of the ant agent is $+$ 1 and the low value in the tabulist of A1 is updated as (low = mid + 1) which is equal to 7. A1 moves to the next trail and deposits pheromone by computing mid = ceil ((7 + 7)/2) = 7. Since the stored image point value at position 7 is 310, a match is found. The energy value of the ant agent is 0 and match is incremented by 1 and updated to 3. As all point values in the query image point pattern have been compared, A1 computes the match percentage. Since the match percentage is equal to 1 which is greater than the threshold 0.5, A1 compares match value with the global value. Since match value is greater than the global value, global value is updated to 1 and the global set is updated with the stored image point pattern St_1. Since the global value is greater than 0.95, ant agents A2 and A3 stops matching I with the stored image point patterns assigned to it. Table 2 shows A1 finding a match for the query image point pattern I. Similarly, ant agents A2 and A3 matches the query image point pattern I with St_2 and St_3 respectively. Table 3 shows A2 and A3 matching I with the stored image point pattern assigned to it. It may be observed from Table 3 that since the match percentage of the ant agents A2 and A3 are 0 and 0.66 respectively, the global value is not updated. Finally, the global value contains value 1 and the global set contains the matching stored image point pattern St_1.

Table 2. Working of ant agent A1 to find a match for the query image point pattern I.

Trail	Query image point value	Reduced Search space	Pheromone deposition	Compared stored image point value	Energy	Match
0	-	-	-	-	-	-
1	12	{11, 12, 13, 14}	3	13	-1	-
2	12		2	12	0	1
3	14	{13, 14}	4	14	0	2
4	310	{290, 300, 310}	6	300	1	
5	310		7	310	0	3

Table 3. Ant agents matching the query image point pattern I with the corresponding stored image point pattern.

Stored image point pattern	Assigned ant agent	Match	Match percentage	Global value	Global set
St_1	A1	3	1	1	St_1
St_2	A2	0	0		
St_3	A3	2	0.66		

7   Experimental Results

Efficiency of IACOBSPPM algorithm in matching query images with (i) distinct stored images and (ii) confusing patterns in an indoor environment was studied. All experiments were performed in a system configuration of Intel Pentium CPU B960 @ 2.20 GHz and 1954 Mbytes RAM.

Case 1: Performance of IACOBSPPM in Matching Distinct Images
The performance of IACOBSPPM was studied using five real images [1]. Figure 2 shows the real images and the corresponding edge map obtained. IACOBSPPM was used to match the query images in Fig. 3 with the stored images in Fig. 2. It may be noted from Fig. 3 that query images Q1, Q2 and Q3 was matched with the stored images Image 1, Image 2 and Image 3 respectively.

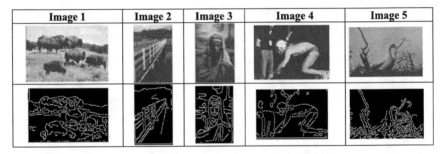

Fig. 2. Stored real images (Aiger and Kedem, 2010) and the corresponding edges

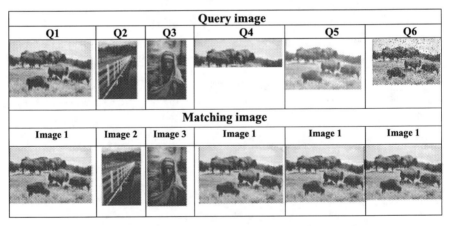

Fig. 3. Query images and the corresponding matching stored image

Query images Q4, Q5 and Q6 shows partial, brightened and noisy versions of the stored image 'Image 1'. It may be noted from Fig. 3 that query images Q4, Q5 and Q6 was matched with stored image 'Image 1', proving the efficiency of IACOBSPPM algorithm in matching images that are identical to the stored images. Table 4 shows the number of point values of the query image matching with the stored images.

Table 4. Number of points values of the query image matching with the stored images.

Stored image	Query image					
	Q1	Q2	Q3	Q4	Q5	Q6
Image 1	338	6	11	22	35	71
Image 2	6	152	54	1	6	32
Image 3	11	54	171	5	12	29
Image 4	12	1	10	6	22	32
Image 5	12	9	2	2	21	44

Case 2: Performance of IACOBSPPM in Matching Images in a Confusing Environment

Indoor environments contain self-repetitive structures and confusing patterns that makes image matching challenging. Such man-made structures look similar at the first sight, but on closer inspection it may be found that they contain subtle details that distinguish them. IACOBSPPM algorithm is efficient in locating individuals in confusing environments in times of emergency. Figure 4 shows nine stored images in an indoor environment and Fig. 5 shows query images Q7 and Q8 and the corresponding matching stored image. It may be noted that Q7 contains the image of a person and Q8 has some modifications to the original image. It may be observed from Fig. 4 that IACOBSPPM was able to match Q7 and Q8 to the most similar stored image proving the robustness of IACOBSPPM algorithm.

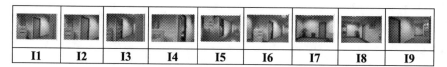

| I1 | I2 | I3 | I4 | I5 | I6 | I7 | I8 | I9 |

Fig. 4. Stored images [5] in an indoor environment

| Q7 | Matching image - I8 |

| Q8 | Matching image -I2 |

Fig. 5. Query images matched using IACOBSPPM

7.1 Comparison of IACOBSPPM Algorithm with the Existing Approaches

Comparison of Number of Searches. The number of searches taken by IACOBSPPM to find a match for the query images in Figs. 3 and 5 is compared with ACOBSPPM and binary search. It may be noted from Table 5 that the number of searches taken by IACOBSPPM is less than ACOBSPPM and binary search.

Table 5. Comparison of number of searches

Query	Number of searches		
	Binary search	ACOBSPPM	IACOBSPPM
Q1	2540	2210	**1659**
Q2	969	824	**606**
Q3	1121	957	**721**
Q4	1270	1097	**944**
Q5	7958	5892	**4986**
Q6	18184	17820	**11511**
Q7	3855	3296	**3076**
Q8	6558	5453	**4852**

Comparison of Computational Complexity. The computational complexity of IACOBSPPM algorithm is compared with the existing approaches for point pattern matching like FET, AIS and ACOBSPPM algorithm. Table 6 shows the worst case computational complexity of these approaches with m and n denoting the number of point values in the model and scene. It may be observed from Table 6 that the time complexity of IACOBSPPM algorithm is less compared to the existing approaches for point pattern matching.

Table 6. Comparison of computational complexity

Fast Expected Time	Approximate input sensitive algorithm	ACOBSPPM	IACOBSPPM
$O(n (\log m)^{3/2})$	$O(n \log n + km \log n)$	$O(m \log n)$	$O(m \log C_{len})$ where $C_{len} < n$

Comparison of Matching Time. The matching time of IACOBSPPM algorithm is compared with the matching time of FET, AIS and ACOBSPPM algorithm. 'm' point values were chosen at random from point pattern of Image 1 in Fig. 2 to form a model. Similarly, 'n' point values were chosen at random from the point pattern to form a scene such that $n = 2m$. The time taken to match m point values in the model with n point values in the scene was obtained for each image and the average time was computed. Table 7 shows the average time taken to match the point values in the model with the point values in the scene.

Table 7. Average matching time of the existing algorithms

m	Time (seconds)		
	$n = 2m$		
	FET	ACOBSPPM	IACOBSPPM
50	0.02	0.00405	**0.0009**
100	0.051	0.011	**0.001**
200	0.13	0.006	**0.002**

The computation time of IACOBSPPM algorithm is compared with that of AIS and ACOBSPPM algorithm for five real images shown in Fig. 2. The experiment was performed by choosing 8, 16, 32, 64, 128 and 256 point values from Image 1 in Fig. 2 and compared with the point patterns of other real images. Table 8 shows the average computation time taken by AIS, ACOBSPPM and IACOBSPPM for the real images. It may be observed from Tables 7 and 8 that the average matching time of IACOBSPPM is less compared to FET, AIS and ACOBSPPM algorithms.

Table 8. Average computation time for real images

Image	Algorithm	Average computation time					
		8	16	32	64	128	256
Image 1	AIS	0.29	0.1	0.08	0.1	0.17	0.2
	ACOBSPPM	**0.001**	0.001	0.02	0.02	0.02	0.03
	IACOBSPPM	**0.001**	**0.0001**	**0.002**	**0.0001**	**0.001**	**0.002**
Image 2	AIS	0.25	0.15	0.1	0.1	0.13	0.15
	ACOBSPPM	0.02	0.02	0.01	0.01	0.01	0.03
	IACOBSPPM	**0.001**	**0.0001**	**0.0001**	**0.0001**	**0.002**	**0.003**
Image 3	AIS	0.1	0.05	0.05	0.09	0.2	0.23
	ACOBSPPM	0.03	0.03	0.02	0.02	0.03	0.03
	IACOBSPPM	**0.002**	**0.001**	**0.002**	**0.0001**	**0.001**	**0.002**
Image 4	AIS	0.11	0.05	0.06	0.09	0.18	0.2
	ACOBSPPM	0.01	0.02	0.01	0.03	0.03	0.05
	IACOBSPPM	**0.0001**	**0.002**	**0.001**	**0.002**	**0.0001**	**0.002**
Image 5	AIS	1.3	0.1	0.09	0.09	0.2	0.25
	ACOBSPPM	**0.0001**	**0.0001**	0.02	**0.001**	0.02	0.01
	IACOBSPPM	**0.0001**	**0.0001**	**0.0001**	**0.001**	**0.001**	**0.002**

8 Conclusion

This paper proposes an improved ant colony optimization based binary search algorithm to match the query image point pattern with the stored image point patterns. Each point value in the query image point pattern is searched in a reduced search space, reducing the computational complexity. IACOBSPPM is efficient in matching partial and noisy images as well as matching confusing images in indoor environment. The computational complexity of IACOBSPPM is less compared to the existing methods for point pattern matching like FET, AIS and ACOBSPPM algorithm.

References

1. Aiger, D., Kedem, K.: Approximate input sensitive algorithms for point pattern matching. Pattern Recogn. **43**, 153–159 (2010)
2. Li, B., Meng, Q., Holstein, H.: Point pattern matching and applications-a review. In: SMC 2003 Conference Proceedings. 2003 IEEE International Conference on Systems, Man and Cybernetics. Conference Theme - System Security and Assurance (Cat. No.03CH37483), vol. 1, pp. 729–736 (2003)
3. Chui, H., Rangarajan, A.: A new point matching algorithm for non-rigid registration. Comput. Vis. Image Underst. **89**, 114–141 (2003)
4. Dorigo, M., Stutzle, T.: Ant Colony Optimization, pp. 37–38. Prentice Hall of India Private Limited, New Delhi (2005)

5. Kang, H., Efros, A.A., Hebert, M., Kanade, T.: Image matching in large scale indoor environment. Carnegie Mellon University: School of Computer Science (2009)
6. Maniezzo, V., Gambardella, L.M., de Luigi, F.: Ant colony optimization (2004). http://www. idsia.ch/~luca/aco2004.pdf
7. Sreeja, N.K., Sankar, A.: Ant colony optimization based binary search for efficient point pattern matching in images. Eur. J. Oper. Res. **246**(1), 154–169 (2015)
8. Sreelaja, N.K., Sreeja, N.K.: An ant colony optimization based approach for binary search. In: Tan, Y., Shi, Y. (eds.) ICSI 2021. LNCS, vol. 12689, pp. 311–321. Springer, Cham (2021). https://doi.org/10.1007/978-3-030-78743-1_28
9. Van Wamelen, P.B., Li, Z., Iyengar, S.S.: A fast expected time algorithm for the 2-D point pattern matching problem. Pattern Recogn. **37**, 1699–1711 (2004)
10. Wayman, J.L., Jain, A.K., Maltoni, D., Maio, D.: Biometric Systems: Technology, Design and Performance Evaluation. Springer London (2005). https://doi.org/10.1007/b138151

Hierarchical Heterogeneous Ant Colony Optimization Based Approach to Optimize File Similarity Searches Using ssDeep

N. K. Sreelaja[(✉)] and N. K. Sreeja

PSG College of Technology, Coimbatore, India
sreelajank@gmail.com

Abstract. Identifying files similar to a particular file helps forensic investigators to identify malwares. The computational complexity of the existing approaches in literature for identifying similar files using ssDeep signatures is high. Brian Wallace had proposed an approach to optimize ssDeep comparisons. However, the drawback is that the substrings of the incoming chunks are checked for a match with the substrings of chunks in the reference list before an edit distance method is applied. Thus to further optimize the search space, a Hierarchical Heterogeneous Ant Colony Optimization based approach to detect similarity in ssDeep signatures (HHACOS) algorithm is proposed. The substrings of the chunks and double chunks of the incoming ssDeep message digest is compared with the substrings of the chunks and the double chunks of the ssDeep digests in the reference list. An ant agent identifies the search space and the number of substrings of the chunks and double chunks of the message digest in the reference list matching with the incoming ssDeep digest is found and the similarity between the files is computed. It is shown that HHACOS algorithm scales well compared to the existing approaches in terms of computational complexity. Also, the accuracy of detecting file similarity is efficient.

Keywords: Hierarchical Heterogeneous Ant Colony Optimization · ssDeep signature · File Similarity

1 Introduction

Identifying files similar to a particular file helps forensic investigators to identify malwares. Files can be compared using message digests. NIST [3] has defined approximate matching functions as a "promising technology designed to identify similarities between two digital artifacts. It is used to find objects that resemble each other or find objects that are contained in another object". Broder [2] defines resemblance when two objects resemble each other, while containment is when one object is contained inside another. Matching of objects byte wise is considered more efficient because the matching is based on the sequence of bytes. Also, the structure of data is not considered. Hence such methods are termed similarity hashing or fuzzy hashing. ssDeep, a Content Triggered Piecewise Hashing (CTPH) detects similarities at the file level. To find similarities between two digests, ssDeep uses an edit distance algorithm [4] and a score in the range of 0–100 denotes the similarities. However, there exists certain scaling issues with ssDeep. In order to compare one ssDeep hash with 100 ssDeep hashes, ssDeep compare function must be called 100 times. Brian Wallace [1] had proposed an approach to optimize ssDeep comparisons. However, the drawback is that the computational complexity is high since the matching strings are found before applying edit distance. Finding similarities between files is efficient when correct optimization methods are employed [6]. Moia et al., [9] have stated that the existing methods in literature doesn't have low computational complexity and results in more false positives.

Thus to further optimize the search space, a Hierarchical Heterogeneous Ant Colony Optimization [7] based approach to detect similarity in ssDeep signatures (HHACOS) algorithm is proposed. In this approach, an Ant Manager sends the incoming message digest of the file to the ant agent at level 2. The ant agent at Level 2 has the message digests in the reference lists which are to be compared with the incoming digest. The ant agent at level 2 sends the 7 character substring of the chunks and double chunks of each message digest in the reference list to the working agent. Similarly, the count and the last position of the substrings beginning with a character for the chunks and double chunks of each message digest in the reference list are also sent to the working ant agent. The working ant agent identifies the search space in the substring of the message digest in the reference list and compares the substrings of the incoming message digest in the search space. Thus the count of character substrings in the search space matching with the incoming substrings is computed for both chunk and double chunk. The total count gives the file similarity score which is returned to the Ant Manager. Thus the ant manager identifies the files similar to an incoming file.

It is shown that HHACOS algorithm scales well compared to the existing approaches in terms of computational complexity. Also, the accuracy of detecting file similarity is efficient. The rest of the paper is organized as follows. Section 2 describes the related work. Section 3 describes ssDeep digests. Section 4 describes the system model. Section 5 describes HHACOS algorithm. Section 6 describes case study. The experimental results are discussed in Sect. 7. The conclusions are presented in Sect. 8.

2   Related Work

Naive Brute Force approach is proposed for searching similar ssDeep digests where the target digest is compared with all digests in the reference list. However, the time complexity is O(r) for a single query where r is the number of digests stored in the reference list. For larger data sets, the search can take days or weeks using common hardware [9]. Jianzhong et al. [5] have proposed iCTPH, in which the similar message digests are mapped into near clusters using iDistance technique. In the comparison phase, the message digest of the incoming object is computed. A query interval is generated by iCTPH for every cluster related to the queried item and a comparison is performed with all digests of the clusters which corresponds to that interval and the count of similar digests is retrieved. The drawback of this approach is that there are certain unnecessary lookups for a digest which would result in an increase time complexity equal to that of a brute force. Also, ssDeep tool used to lookup digests is computationally heavy at scale.

Winter et al., [10] have proposed a Fast Forensic Similarity search (F2S2) approach for similarity search. However, the drawback is that F2S2 approach uses ssDeep tool. Brian Wallace have proposed an optimizing approach for ssDeep using integer DB optimization. However, the drawback is that similar 7 character substrings are identified before applying the edit distance which increases the computational complexity. Sreelaja and Sreeja [8] proposed an ACO based approach for binary search to categorize the search space. Inspired by the idea of search space reduction, an Hierarchical Heterogeneous ACO based approach for similarity detection in ssDeep signatures is proposed.

3   ssDeep

ssDeep [11] is a fuzzy hashing algorithm which employs a similarity digest in order to determine whether the hashes that represent two files have similarities. Modification in a single byte in a file results in a more similar ssDeep hashes between the original and modified file. The hash of each block of input data is computed in ssDeep. Using ssDeep comparison to find the similarity between files, replaces the need of comparing every byte in each file. To compare two files, the digests of two files are chosen and ssDeep uses edit distance algorithm [4]. According to edit distance algorithm, the count of minimum number of operations needed to transform one string to another is found. Thus strings of equal size must be taken for comparison. Comparing two files using ssDeep, results in a score ranging from 0 to 100. A value of zero indicates the two files are not similar and a score of 100 indicates higher similarity. For instance, two similar malware files are chosen and if a configuration is statically inserted into a stub sample, the similarity between the two files will be high.

To compute the ssDeep hash of a file, the file is split into chunks and a hash function is run on each of these chunks. The final hash result is constructed as the result of each hash. Hence, the hash value of a file is changed slightly if a few bytes of a file is modified. The ssDeep hash is formatted as chunksize:chunk:double_chunk. chunksize is an integer describing the chunks size of the ssDeep hash. Every character in the chunk denotes a part of the file of length chunksize. The data used for chunk is used to compute double_chunk with chunksize * 2.

4   System Model

The system is modeled using a Hierarchical Heterogenous Ant Colony Optimization based approach. The system has an Ant Manager, an Ant agent at Level 2 and working ant agents. In this model, the number of working ant agents is determined based on the count of the chunk sizes of the message digests in the reference list as shown in Eq. (1).

$$\text{Number of Working ant agents} = \text{count of chunksizes} \qquad (1)$$

The Ant manager passes the ssDeep digest of the incoming file to the ant agent at level 2. The Ant manager has a list to store the similarity score value and the compared ssDeep digest. The ant manager identifies the similar files based on the similarity score value.

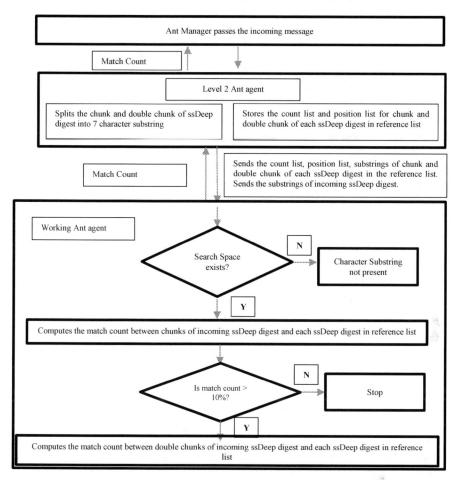

Fig. 1. System Model

The ant agent at level 2 has the ssDeep digests in the reference list. These digests are compared with the incoming ssDeep digest. The ant agent at level 2, will separate the chunks of each message digest and a sliding window is moved over the digest to read 7 character substrings from the chunk. These 7 character substrings are sorted in ascending order. The substrings are categorized based on the first character. The first character can be either digits 0 to 9, alphabets a-z, '+' or special characters. The ant agent at level 2 stores the count of the 7 character substrings beginning with each of these first characters for each chunk of the message digest. Also, the position of last occurrence of the 7 character substrings beginning with each of these first characters are stored for each chunk. Similarly, the double chunk of each message digest is split into 7 character substring and stored in the sorted order. The ant agent at level 2 stores the count of the 7 character substrings beginning with each of these first characters for each double chunk of the message digest. Also, the position of last occurrence of the 7 character substrings beginning with each of these first characters are stored for each double chunk. The ant agent at level 2 splits the chunks and double chunk of the incoming ssDeep digest into 7 character substring and sends it to the working ant agent corresponding to the chunk size. The Ant agent at level 2 has the count of the message digests in the reference list. The ant agent at level 2 deposits pheromone and passes the substrings, the count value, position value of the chunks and double chunks of the first message digest in the reference list to working ant agent.

The working ant agent deposits pheromone and identifies the search space to match the 7 character strings of the incoming ssDeep digest with the character strings of the chunk of the message digest received from the ant agent at level 2 and counts the number of character strings of the received ssDeep digest matching with the incoming character string. If the match count value is greater than 10%, the pheromone deposition evaporates and the working ant agent repeats the same process for double chunk of the incoming ssDeep digest with the double chunk of the ssDeep digest received from the ant agent at level 2. The working ant agent sums up the match count value and sends to the ant agent at level 2. If the match count value received from the working ant agent is greater than 10%, the ant agent at level 2 sends it to the ant manager. The ant manager stores this match count value in similarity count along with the matching ssDeep digest. The ant agent at level 2 increments the count value. The pheromone deposition of the ant agent evaporates and ant agent at level 2 chooses the next ssDeep digest from the reference list and the process is repeated until the count value is equal to the total number of ssDeep digests in the reference list for the specific chunk size. The ant manager identifies the files similar to the incoming file based on the similarity count. Figure 1 shows the model of the system.

5 Hierarchical Heterogeneous Ant Colony Optimization Based Approach to Detect Similarity in ssDeep Signatures

The ant manager sends the incoming ssDeep digest to the ant agent at level 2 for comparison with the ssDeep digests in the reference list. The ant agent at level 2 decides the number of working ant agents based on the chunk size of the ssDeep digests in the reference lists. The number of working ant agents is equal to the count of the chunk size since, ssDeep messages are compared based on the similar chunk sizes. Consider an ant agent at level 2 having 'm' ssDeep digests in the reference list. The ant agent at level 2 split the chunks and double chunks of the ssDeep digests into 7 character substrings. Let $SC_{rc} = [SC_{rc1}, SC_{rc2}....SC_{rcn1}]$ be a set of character substrings for the chunks in the ssDeep digest in ascending order where r ranges between 1 and m. Let $SC_{rdc} = [SC_{rd1}, SC_{rd2}....SC_{rdn2}]$ be a set of character substrings for the double chunks in the ssDeep digest in ascending order where r ranges between 1 and m. The ant agent at level 2 splits the incoming ssDeep digest into 7 character substrings. Let $IC_c = [IC_{c1}, IC_{c2}....IC_{cn3}]$ and $IC_{dc} = [IC_{dc1}, IC_{dc2}....IC_{dcn4}]$ be a set of character sub strings of the chunk and double chunk respectively for an incoming ssDeep digest. The ant agent at level 2 has a tabu-list denoting its memory. The tabu-list of the ant agent at Level 2 has a count list which stores the count of the substring of the chunks of each message digest beginning with a special character or alphabet or digit as shown in Eq. (2). The position of last occurrence of the character substrings beginning with a character are stored as shown in Eq. (3).

$$C_{rc} = \{C_{rc/}, C_{rc+}, C_{rc0}, C_{rc1}, C_{rc2}, C_{rc3}, C_{rc4}, C_{rc5}, C_{rc6}, C_{rc7}, C_{rc8}, C_{rc9}, C_{rca-A}, C_{rcb-B}, C_{rcc-C},$$
$$C_{rcd-D}, C_{rce-E}, C_{rcf-F}, C_{rcg-G}, C_{rch-H}, C_{rci-I}, C_{rcj-J}, C_{rck-K}, C_{rcl-L}, C_{rcm-M}, C_{rcn-N}, C_{rco-O}, C_{rcp-P},$$
$$C_{rcq-Q}, C_{rcr-R}, C_{rcs-S}, C_{rct-T}, C_{rcu-U}, C_{rcv-V}, C_{rcw-W}, C_{rcx-X}, C_{rcy-Y}, C_{rcz-Z}\} \tag{2}$$

where $C_{rc}(i)$ denotes the count of character substrings of the chunk of the ssDeep digest 'r' beginning with the special character or alphabet or digit 'i'. The position of last occurrence of the character substring (P_{rci}) beginning with the special character or alphabet or digit 'i' is found using Eq. (3).

$$P_{rci} = \begin{cases} \sum_{j=1}^{i} C_{rcj} & if\ C_{rcj} \neq 0 \\ 0 & if\ C_{rcj} = 0 \end{cases} \tag{3}$$

The position list $P_{rc} = [P_{rc/}, P_{rc+}, P_{rc0}, P_{rc1}, P_{rc2}, P_{rc3}, P_{rc4}, P_{rc5}, P_{rc6}, P_{rc7}, P_{rc8}, P_{rc9}, P_{rca-A}, P_{rcb-B}, P_{rcc-C}, P_{rcd-D}, P_{rce-E}, P_{rcf-F}, P_{rcg-G}, P_{rch-H}, P_{rci-I}, P_{rcj-J}, P_{rck-K}, P_{rcl-L}, P_{rcm-M}, P_{rcn-N}, P_{rco-O}, P_{rcp-P}, P_{rcq-Q}, P_{rcr-R}, P_{rcs-S}, P_{rct-T}, P_{rcu-U}, P_{rcv-V}, P_{rcw-W}, P_{rcx-X}, P_{rcy-Y}, P_{rcz-Z}]$ is stored in the tabu-list of the ant agent at level 2. Similarly, the count list and position list for the double chunks of each ssDeep digests are shown in Eqs. (4) and (5) respectively.

$$C_{rdc} = [C_{rdc/}, C_{rdc+}, C_{rdc0}, C_{rdc1}, C_{rdc2}, C_{rdc3}, C_{rdc4}, C_{rdc5}, C_{rdc6}, C_{rdc7}, C_{rdc8}, C_{rdc9}, C_{rdca-A},$$
$$C_{rdcb-B}, C_{rdcc-C}, C_{rdcd-D}, C_{rdce-E}, C_{rdcf-F}, C_{rdcg-G}, C_{rdch-H}, C_{rdci-I}, C_{rdcj-J}, C_{rdck-K}, C_{rdcl-L},$$
$$C_{rdcm-M}, C_{rdcn-N}, C_{rdco-O}, C_{rdcp-P}, C_{rdcq-Q}, C_{rdcr-R}, C_{rdcs-S}, C_{rdct-T}, C_{rdcu-U}, C_{rdcv-V}, C_{rdcw-W}, C_{rdcx-X},$$
$$C_{rdcy-Y}, C_{rdcz-Z}] \tag{4}$$

$$P_{rdc} = [P_{rdc/}, P_{rdc+}, P_{rdc0}, P_{rdc1}, P_{rdc2}, P_{rdc3}, P_{rdc4}, P_{rdc5}, P_{rdc6}, P_{rdc7}, P_{rdc8}, P_{rdc9}, P_{rdca-A}, P_{rdcb-B},$$
$$P_{rdcc-C}, P_{rdcd-D}, P_{rdce-E}, P_{rdcf-F}, P_{rdcg-G}, P_{rdch-H}, P_{rdci-I}, P_{rdcj-J}, P_{rdck-K}, P_{rdcl-L}, P_{rdcm-M}, P_{rdcn-N},$$
$$P_{rdco-O}, P_{rdcp-P}, P_{rdcq-Q}, P_{rdcr-R}, P_{rdcs-S}, P_{rdct-T}, P_{rdcu-U}, P_{rdcv-V}, P_{rdcw-W}, P_{rdcx-X}, P_{rdcy-Y}, P_{rdcz-Z}] \tag{5}$$

The ant agent at level 2 has a list of ssDeep digests in the reference list. The ant agent at level 2 deposits pheromone and passes IC_c, SC_{1c}, C_{1c}, P_{1c}, IC_{dc}, SC_{1dc}, C_{1dc}, P_{1dc} to the working ant agent. The working ant agent has a start position sp and an end position ep and a count variable. Initially, the values of sp and ep are zero. The working ant agent chooses the first character 'i' in IC_{c1}. To identify the search space, the working ant agent deposits pheromone by choosing the position value P_{1ci}. If P_{1ci} is zero, the search space is not found and the pheromone deposition evaporates and the working ant agent chooses IC_{c2}. If P_{1ci} is greater than zero, the count value C_{1ci} is chosen from the count list. The position value P_{1ci} is stored in ep. The count value C_{1ci} is subtracted from the position value P_{1ci} and the subtracted value $(P_{1ci} - C_{1ci}) + 1$ is stored in sp of the working ant agent as shown in Eq. (6).

$$sp = (P_{1ci} - C_{1ci}) + 1 \tag{6}$$

The range [sp, ep] denotes the search space and the working ant agent compares IC_{c1} with the substrings in the search space. If there are more than 2 characters in the pattern of the compared substrings, then the match count is incremented by 1. Thus the match count denotes the number of substrings in the search space matching with IC_{c1}. The pheromone deposition evaporates and the working ant agent chooses IC_{c2}, increments the value of count by 1 and moves to the next trail. The process is repeated until the count value is equal to the total number of substrings in the incoming chunk. If the match count is greater than 10%, the working ant agent repeats the same procedure for double chunk of the incoming ssDeep digest. The working ant agent computes the sum of the match count for both chunk and double chunk and computes the match count percentage and sends to the ant agent at level 2. If the match count percentage is greater than 10% the ant agent at level 2 sends the match count along with the matching ssDeep digest to the ant manager. The pheromone deposition of the ant agent at level 2 now evaporates and passes SC_{2c}, C_{2c}, P_{2c}, SC_{2dc}, C_{2dc}, P_{2dc} to the working ant agent and the process is repeated. The ant agent at level 2 increments the count value by 1. The process is stopped when the count value is equal to the count of the ssDeep digests in the reference list. The ant manager has the list of ssDeep digests along with the match count which denotes the similarity scores. Thus the ssDeep digests having higher score will represent the more similar files with the incoming file.

6 Case Study

A case study is discussed to find the similarity between the ssDeep digest of incoming file against the ssDeep digests of files stored as shown in Tables 1 and 2 respectively. HHACOS algorithm is invoked to find the similarity between files. The ant agent at level 2 has only one ssDeep digest in the reference list. The ant agent at level 2 splits the chunk of ssDeep digest in the reference list into seven character substring (SC_{1c}) in sorted order as shown in Table 3. Thus there are 55 character substrings for the chunk. The tabu list of the ant agent at level 2 has the count list (C_{1c}) and position list (P_{1c}) for the chunk of ssDeep digest in the reference list as shown in Tables 4 and 5 respectively. Since there is only 1 chunk size, there is only one working ant agent. The ant manager sends the ssDeep message digest of the incoming file to the ant agent at level 2. The ant agent at level 2 splits the chunk and the double chunk of the incoming file into character substrings denoted as IC_c and IC_{dc} respectively. The substrings of IC_c is shown in Table 6. The ant agent at level 2 passes IC_c, SC_{1c}, C_{1c}, P_{1c}, IC_{dc}, SC_{1dc}, C_{1dc}, P_{1dc} to the working ant agent.

Table 1. ssDeep digests of Incoming file

File Name	ssDeep digests
Incoming File	3:yionv//thPl0l/TtA4RthwkBDsTBZtz9MAYHAamWQ5/hdmojpuavgzW9pPSt0+Ga: 6v/lhPuTtA4nDspzmAYgah2hdfjpV4zd

The working ant agent chooses the first substring yionv// from the chunk of the incoming ssDeep digest and finds the search space in SC_{1c} shown in Table 3. The working ant agent chooses the first character 'y' from the substring and deposits pheromone by selecting the value 53 corresponding to P_{1cyY}. Thus, the ending position ep in the tabu list has the value 53. The working ant agent chooses the corresponding count value C_{1cyY} as 2. Thus the starting position sp is computed as $(ep - sp) + 1 = 52$. Thus the search space lies between 52 and 53.Thus the substrings in the positions 52 and 53 in

Table 2. ssDeep digests in reference list

File Name	ssDeep digests
File1	3:yionv//thPl0l/TtA4RthwkBDsTBZtuAkxGAYHWG4zGCkQ2FTEFaMHUb+R/jp: 6v/lhPuTtA4nDspdnAY2G2GCh2FTMaMp

Table 3. Character substrings for the chunk of the ssDeep digest in reference list in sorted order

	Position		Position		Position
//thPl0	1	G4zGCkQ	19	RthwkBD	37
/thPl0l	2	GAYHWG4	20	sTBZtuA	38
/TtA4Rt	3	GCkQ2FT	21	tA4Rthw	39
0l/TtA4	4	hPl0l/T	22	TBZtuAk	40
2FTEFaM	5	HUb+R/j	23	TEFaMHU	41
4RthwkB	6	HWG4zGC	24	thPl0l/	42
4zGCkQ2	7	hwkBDsT	25	thwkBDs	43
A4Rthwk	8	ionv//t	26	TtA4Rth	44
AkxGAYH	9	kBDsTBZ	27	tuAkxGA	45
aMHUb+R	10	kQ2FTEF	28	uAkxGAY	46
AYHWG4z	11	kxGAYHW	29	Ub+R/jp	47
BDsTBZt	12	l/TtA4R	30	v//thPl	48
BZtuAkx	13	l0l/TtA	31	WG4zGCk	49
CkQ2FTE	14	MHUb+R/	32	wkBDsTB	50
DsTBZtu	15	nv//thP	33	xGAYHWG	51
EFaMHUb	16	onv//th	34	YHWG4zG	52
FaMHUb+	17	Pl0l/Tt	35	yionv//	53
FTEFaMH	18	Q2FTEFa	36	zGCkQ2F	54
				ZtuAkxG	55

SC_{1c} are compared with the string yionv// to check for more than 2 patterns matching with the string. Thus it is found that all seven characters in yionv// matches with the string at 53 position. The pheromone deposition of the working ant agent evaporates and moves to the next trail and chooses IC_{c2} and increments the count value by 1. The procedure continues until the count value is equal to strings in IC_c. Thus it is found that the characters in 40 substrings in SC_{1c} matches with IC_c substrings as shown in Table 6. Since, the match count percentage is 72, the working ant agent repeats the same procedure for the double chunk IC_{dc}, SC_{1dc}, C_{1dc}, P_{1dc} and the match count is found to be 20. Thus the working ant agent computes the sum of match count as 60. The working

Table 4. Count list in ant agent's tabu list

$C_{1c/}$	C_{1c+}	C_{1c0}	C_{1c1}	C_{1c2}	C_{1c3}	C_{1c4}	C_{1c5}	C_{1c6}	C_{1c7}	C_{1c8}	C_{1c9}	C_{1caA}
3	0	1	0	1	0	2	0	0	0	0	0	4
$C_{1c\,bB}$	$C_{1c\,cC}$	C_{1cdD}	$C_{1c\,eE}$	$C_{1c\,fF}$	$C_{1c\,gG}$	$C_{1c\,hH}$	$C_{1c\,iI}$	$C_{1c\,jJ}$	$C_{1c\,kK}$	$C_{1c\,lL}$	$C_{1c\,mM}$	$C_{1c\,nN}$
2	1	1	1	2	3	4	1	0	3	2	1	1
$C_{1c\,oO}$	$C_{1c\,pP}$	$C_{1c\,qQ}$	$C_{1c\,rR}$	$C_{1c\,sS}$	$C_{1c\,tT}$	$C_{1c\,uU}$	$C_{1c\,vV}$	$C_{1c\,wW}$	$C_{1c\,xX}$	$C_{1c\,yY}$	$C_{1c\,zZ}$	
1	1	1	1	1	7	2	1	2	1	2	2	

Table 5. Position list in ant agent's tabu list

$P_{1c/}$	P_{1c+}	P_{1c0}	P_{1c1}	P_{1c2}	P_{1c3}	P_{1c4}	P_{1c5}	P_{1c6}	P_{1c7}	P_{1c8}	P_{1c9}	P_{1caA}
3	0	4	0	5	0	7	0	0	0	0	0	11
P_{1cbB}	P_{1ccC}	P_{1cdD}	$P_{1c\,eE}$	$P_{1c\,fF}$	$P_{1c\,gG}$	$P_{1c\,hH}$	$P_{1c\,iI}$	$P_{1c\,jJ}$	$P_{1c\,kK}$	$P_{1c\,lL}$	$P_{1c\,mM}$	$P_{1c\,nN}$
13	14	15	16	18	21	25	26	0	29	31	32	33
$P_{1c\,oO}$	$P_{1c\,pP}$	$P_{1c\,qQ}$	$P_{1c\,rR}$	$P_{1c\,sS}$	P_{1ctT}	P_{1cuU}	P_{1cvV}	P_{1cwW}	P_{1cxX}	P_{1cyY}	$P_{1c\,zZ}$	
34	35	36	37	38	45	47	48	50	51	53	55	

ant agent computes the match count percentage as $(60/81) * 100 = 74\%$ and sends to the ant agent at level 2. The ant agent at level 2 sends the match count and match count percentage along with the ssDeep digest of file 1. Since there are no more digests in the reference list, the ant agent at level 2 stops the process and the ant manager finds file 1 and file 2 have a high similarity since the similarity score is 74%.

7   Experimental Results

A similarity score is computed between the incoming SSDeep digest 3: yionv//thPl0l/TtA4RthwkBDsTBZtuAkxGAYHWG4zGCkQ2FTEFaMHUb+R/jp: 6v/lhPuTtA4nDspdnAY2G2GCh2FTMaMp and the ssDeep digests [12] shown in Table 7 invoking HHACOS algorithm. The same digest is compared with 41 SSDeep digests [12] and the time performance is shown in Fig. 2. It is inferred from Table 7 that the digests of the files from positions 1 to 5 are similar and the digests of the file in positions 6 to 8 are dissimilar. The efficiency of HHACOS algorithm in finding similarity between files is shown in Table 7 by comparing with the benchmark ssDeep tool.

7.1   Comparison of HHACOS and Integer DB Optimization for ssDeep

Brian Wallace [1] has proposed an approach for optimizing ssDeep. Integer DB method is used in which seven character string is used. This is compared against the seven character string stored. According to this approach, the strings which match with the strings of the ssDeep digests is found before the edit distance method is used which increases the computational complexity.

In HHACOS algorithm, the search space is identified and the substrings are compared which reduces the computational complexity.

Table 6. Character substrings for the chunk of the incoming digest

	Count of substring matching with the compared digest		Count of substring matching with the compared digest		Count of substring matching with the compared digest
'yionv//'	1	'hwkBDsT'	2	'Q5/hdmo'	0
'ionv//t'	1	'wkBDsTB'	1	'5/hdmoj'	0
'onv//th'	1	'kBDsTBZ'	2	'/hdmojp'	0
'nv//thP'	1	'BDsTBZt'	1	'hdmojpu'	0
'v//thPl'	1	'DsTBZtz'	1	'dmojpua'	0
'//thPl0'	2	'sTBZtz9'	1	'mojpuav'	0
'/thPl0l'	1	'TBZtz9M'	1	'ojpuavg'	0
'thPl0l/'	2	'BZtz9MA'	1	'jpuavgz'	0
'hPl0l/T'	2	'Ztz9MAY'	1	'puavgzW'	0
'Pl0l/Tt'	1	'tz9MAYH'	0	'uavgzW9'	0
'l0l/TtA'	1	'z9MAYHA'	0	'avgzW9p'	0
'0l/TtA4'	1	'9MAYHAa'	0	'vgzW9pP'	0
'l/TtA4R'	1	'MAYHAam'	0	'gzW9pPS'	0
'/TtA4Rt'	2	'AYHAamW'	1	'zW9pPSt'	0
'TtA4Rth'	1	'YHAamWQ'	1	'W9pPSt0'	0
'tA4Rthw'	1	'HAamWQ5'	0	'9pPSt0+'	0
'A4Rthwk'	1	'AamWQ5/'	1	'pPSt0+G'	0
'4RthwkB'	1	'amWQ5/h'	0	'PSt0+Ga'	0
'RthwkBD'	1	'mWQ5/hd'	0		
'thwkBDs'	3	'WQ5/hdm'	0		

7.2 Comparison of HHACOS with Naive Brute Force Approach for ssDeep

In Naive Brute force approach [9], every object in the target system is compared with all objects in the reference list. The best matching file is the one having the highest similarity. If there are 'n' strings in the compared digest and 'm' strings in the incoming digest, the number of comparisons is for one string in the incoming digest is o(n).

In HHACOS algorithm, the search space is identified containing the strings of the ssDeep digest. If there are 'n' strings in the compared digest and 'm' strings in the incoming digest, the number of comparisons is for one string in the incoming digest is o(s) where 's' is the count of strings in the search space s < n.

Table 7. ssDeep digests and the similarity score

	ssDeep digests	Match score in percentage using HHACOS algorithm	Match score using SSDeep tool	Similarity
1	3:yiomv/thPl0l/TtA4RthwkBDsTBZtz9MAYHAamWQ5/hdmojpuavgzW9pPSt0+Ga: 6v/lhPuTtA4nDspzmAYgah2hdfjpV4zd	74	63	Similar
2	3:yiomv/thPl38/A/40Rg9RthwkBDsTBZt5BxdSkltmfwk/weolPp: 6v/lhP6I/xGjnDsp5BZlVp	54.3	52	Similar
3	3:yiomv/thPloxtmtdllljK9Ag9RthwkBDsTBZtjjmoaIJZFlGW4/jldp:6v/lhPcmtdlllm9AgjnDspjyoaIO3jTp	51.8	54	Similar
4	3:yiomv/thPlEdtl/mRg9RthwkBDsTBZt69/j1SwdNjQGR7l1VdRVvdh9KBFHyMN: 6v/lhP8JmRgjnDspeblDdWGRM1VTVZAX	50.6	46	Similar
5	3:yiomv/thPlhyttQWl9IK9Ag9RthwkBDsTBZtw9FEu7czKtB1p: 6v/lhPettQr9AgjnDspgEaczKtjp	49.4	50	Similar
6	3:0KTXbdh39KVcXDDVXaAa2cWk0gwFJOFWLL3jYWFGG3: hvtOcMA/k0gwNjYW7	1.2	0	No Similarity
7	3:aKVh+yEilSlJlqXMLLkFlVlRDBWjUoFY91lrAlxUOHnlUES1: aK6XioJqcLwVIRNWwou91lUlxUOiES1	2.5	0	No Similarity
8	3:It1MHtEJ5EOXVn0e+U: e1MHc59ND	4.9	0	No Similarity

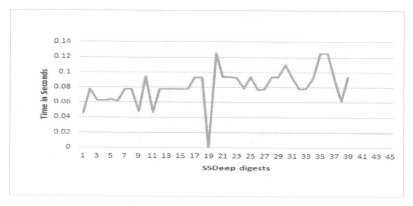

Fig. 2. Time performance of HHACOS algorithm

7.3 Comparison of HHACOS with iCTPH Distributed Hash Tables

Jianzhong et al., [5] have proposed iCTPH which uses ssDeep tool to lookup for digests. Also, ssDeep tool used to lookup digests is computationally heavy at scale. It uses an edit distance method which compares each string in the incoming message digest with the complete set of strings of the compared message digest. If there are 'n' strings in the compared digest and 'm' strings in the incoming digest, the number of comparisons is for one string in the incoming digest is o(n).

In HHACOS algorithm, the search space is identified containing the strings of the compared digest. The strings of the incoming digest is compared with the strings in the search space, which reduces the computational complexity. If there are 'n' strings in the compared digest and 'm' strings in the incoming digest, the number of comparisons for one string in the incoming digest is o(s) where 's' is the count of strings in the search space s < n.

7.4 Comparison of HHACOS with Fast Forensic Similarity Search (F2S2)

Winter et al. [10] have proposed Fast Forensic Similarity Search (F2S2) for similarity digest search. This approach uses ssDeep tool to lookup digests. An edit distance approach is used for comparing the strings in the stored digest and the incoming digest. If there are 'n' strings in the compared digest and 'm' strings in the incoming digest, the number of comparisons is for one string in the incoming digest is o(n).

In HHACOS algorithm, the search space is identified containing the strings of the compared digest. The strings of the incoming digest is compared with the strings in the search space, which reduces the computational complexity. If there are 'n' strings in the compared digest and 'm' strings in the incoming digest, the number of comparisons is for one string in the incoming digest is o(s) where 's' is the count of strings in the search space s < n.

7.5 Comparison of HHACOS with Sequential Search

The incoming ssDeep digest given in Sect. 7 is compared with eight ssDeep digests in the reference list shown in Table 7. The chunks and double chunks of ssDeep digests in Table 7 and the incoming ssDeep digest is split into substrings and a comparison is made between the substrings of the incoming digest and the substrings of the chunks and double chunks of each ssDeep digest in the reference list using HHACOS algorithm and Sequential search. Figure 3 shows a comparison between the number of searches using HHACOS and sequential search. It is shown that HHACOS algorithm outperforms sequential search.

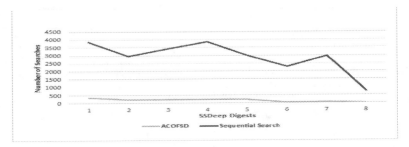

Fig. 3. Comparison of HHACOS and sequential search

8 Conclusion

A Hierarchical Heterogeneous Ant Colony Optimization based approach to detect similarity in ssDeep Signatures (HHACOS) algorithm is proposed to find the search space to find similar files. The existing approaches employs ssDeep tool to compute similarity thereby increasing the computational complexity. It is shown that HHACOS algorithm scales well compared to the existing approaches in terms of computational complexity. Also, the efficiency of HHACOS algorithm is proved by comparing with the benchmark ssDeep tool.

References

1. Wallace, B.: Optimizing ssDeep for Use at Scale. Cylance, USA (2015). Grooten, M. (ed.)
2. Broder, A.Z.: On the resemblance and containment of documents. In: Proceedings of the 1997 International Conference on Compression and Complexity of Sequences, pp. 21–29, June 1997 (1997)
3. Breitinger, F., Guttman, B., McCarrin, M., Roussev, V., White, D.: Approximate matching: definition and terminology. Natl. Inst. Stand. Technol. **800**, Article ID 168 (2014)
4. Wu, G.: String Similarity Metrics, Edit Distance. https://www.baeldung.com/cs/author/gangwu

5. Jianzhong, Z., Kai, P., Yuntao, Y., Jingdong, X.: iCTPH: an approach to publish and lookup CTPH digests in chord. In: Hsu, C.-H., Yang, L.T., Park, J.H., Yeo, S.-S. (eds.) ICA3PP 2010. LNCS, vol. 6082, pp. 244–253. Springer, Heidelberg (2010). https://doi.org/10.1007/978-3-642-13136-3_25

6. Abrahamy, J.: Intezer Community Tip: How to Optimize ssDeep Comparisons with Elastic-Search. https://www.intezer.com/blog/malware-analysis/intezer-community-tip-ssdeep-comparisons-with-elasticsearch/. Accessed 19 Sept 2017

7. Rusin, M., Zaitseva, E.: Hierarchical heterogeneous ant colony optimization. In: Proceedings of the IEEE Federated Conference on Computer Science and Information Systems, pp. 197–203 (2012)

8. Sreelaja, N.K., Sreeja, N.K.: An ant colony optimization based approach for binary search. In: Tan, Y., Shi, Y. (eds.) ICSI 2021. LNCS, vol. 12689, pp. 311–321. Springer, Cham (2021). https://doi.org/10.1007/978-3-030-78743-1_28

9. Moia, V.H.G., Henriques, M.A.A.: Similarity digest search: a survey and comparative analysis of strategies to perform known file filtering using approximate matching. Secur. Commun. Netw. **2017**, 17 (2017). Article ID 1306802

10. Winter, C., Schneider, M., Yannikos, Y.: F2S2: fast forensic similarity search through indexing piecewise hash signatures. Digit. Investig. **10**(4), 361–371 (2013)

11. http://ssdeep.sourceforge.net/ (http://ssdeep.sourceforge.net/)

12. https://www.nist.gov/itl/ssd/software-quality-group/ssdeep-datasets

Particle Swarm Optimization
Algorithms

PSO with Local Search Using Personal Best Solution for Environments with Small Number of Particles

Yuji Sato[1] , Yuma Yamashita[1], and Jia Guo[2(✉)]

[1] Hosei University, Tokyo 184-8584, Japan
yuji@hosei.ac.jp
[2] Hubei University of Economics, Wuhan 430205, China
guojia@hbue.edu.cn

Abstract. Particle swarm optimization (PSO) has been actively studied as an effective search method using populations. The basis of search algorithms using particle swarms is efficient convergence from exploiting the ability to move toward the best individual (G_{best}) within the current population. Therefore, PSO has the property of quickly converging to the optimal solution, especially for unimodal search problems. For multimodal search problems, there are also reported methods, such as Standard PSO (SPSO), which divides a population into several subgroups, and research examples in which several particles are used to improve the local search capability. However, using a part of the particles in populations for a local search increases the search time in environments with fewer particles because the force toward G_{best} of the entire population tends to be relatively weaker than that of the original PSO. To solve this problem, this paper proposes a search method in which all particles toward to G_{best} while conducting a local search using their respective past best solutions, instead of dividing some particles into local searches. Using typical benchmark functions and existing algorithms, we show that the proposed method works well even with a small number of particles.

Keywords: Swarm intelligence · Local search · Personal best solution · Optimization problems

1 Introduction

The Particle Swarm Optimization (PSO) [1] algorithm is an optimization method using populations of individuals (particle swarms) that was proposed by J. Kennedy and R. Eberhart in 1995. PSO is easy to implement, and has superior convergence and global search capabilities. For this reason, research has been actively carried out in recent years, and many excellent and improved algorithms have been proposed [2–8]. Particle swarm optimization algorithms generally use the force toward the best solution (G_{best}) in populations in the current generation to search. Therefore, they have the advantage of fast convergence for unimodal space search problems [1, 4]. However, it is difficult to adjust the parameters of the algorithms, and there is a high possibility that they will fall into a local solution in a multimodal space search problem. Therefore, algorithms have been proposed to improve the local search capability [3, 5–8]. A typical example is Standard PSO (hereafter SPSO) [3]. In addition, BBPSO [2] has been proposed as a method that solves the problem of the design parameters in PSO being difficult to adjust, and Pairwise BBPSO (PBBPSO) [6] also been proposed to improve the local search ability of BBPSO. SPSO improves the local search ability by dividing a particle swarm into several subpopulations, and it improves the local search ability of PSO by using a topology that connects the whole particle swarm. BBPSO is a particle swarm optimization algorithm that uses a Gaussian distribution search method to solve the parameter design problem. PBBPSO adds a method called pairwise strategy to BBPSO, and the algorithm of BBPSO can be performed while half of the particles perform a local search.

However, in previous research that allocates some particles in a particle group to local search, there are cases where it is not possible to set enough particles for the size of the search space due to problems such as memory size and execution time. In such cases, these methods may not provide sufficient convergence towards G_{best}. For this problem, by adjusting the ratio of the number of particles heading to G_{best} and the number of particles in charge of local search, it is possible to handle many types of problems, from unimodal space search problems to multimodal ones. However, the appropriate ratio at which to distribute the number of particles is problem-dependent and must be determined by trial and error. Also, it is difficult to determine an appropriate ratio for new problems. Here, we propose a method to add a local search function without allocating some particles of the particle swarm to local search.

Below, Sect. 2 presents an example of conventional research for adding a local search capability to PSO, and Sect. 3 proposes a local search method using the personal best solution for each particle. Section 4 presents the experimental method, experimental results, and their discussion and concludes with a summary.

2   Background and Related Works

Particle swarm optimization (PSO) is an effective population-based searching method, and many improved algorithms for PSO have already been proposed. Here, what many of them have in common is that they are searches that mainly use the power toward the leader in the group or the best individual (G_{best}) in the current generation. This feature has the advantage of showing fast convergence to the optimal solution in problems where the gradient method works effectively, such as search problems in unimodal spaces. However, depending on the values of the design parameters, the problem of searching a multimodal space may fall into local minima. BBPSO has been proposed to avoid the difficulty of adjusting multiple design parameters, but it still tends to fall into local minima in multimodal space search problems.

As an example of previous research to improve the local search ability, the fireworks algorithm [5] shows good effects but requires many particles. Here, we present a standard SPSO that divides a population into several subgroups and a pairwise BBPSO that uses half the particles in the population to improve the local search ability.

2.1 PSO and SPSO

PSO. In the original PSO, each particle has a position vector and a velocity vector to calculate the evaluation value and the particle's movement destination. After calculating the evaluation value at the current position of each particle using the position vector, the update of the velocity vector and the position vector is given, as shown in the following equation.

$$v(t + 1) = w * v(t) + c_1 * rand_1 * (P_{bestx} - x(t))$$
$$+ c_2 * rand_2 * (G_{bestx} - x(t)) \tag{1}$$

$$x(t + 1) = x(t) + v(t + 1) \tag{2}$$

where $v(t + 1)$ and $v(t)$ are the velocity vectors of each particle at generations $t + 1$ and t, respectively, and $x(t)$ is the position vector of each particle at generation t.

P_{bestx} represents the highest-evaluation position vector for each particle that has been searched so far, and G_{bestx} represents the highest-evaluation position vector among P_{bestx} for all particles in the current generation.

Also, w, c_1 and c_2 are positive constants, and $rand_1$ and $rand_2$ are random numbers in the range [0, 1]. Changing these design variables allows PSO to be widely applied to various search problems. However, these design variables have a problem in that it is difficult to decide on the appropriate parameters for real-world problems.

SPSO. Standard PSO (hereafter SPSO) is an extended algorithm of PSO. On the basis of the particle swarm topology defined by the designer, the best solution between each particle and its neighboring particles is used as L_{best} instead of G_{best} for searching. SPSO changes the motion of the entire particle group by changing the topology. For example, the original PSO can be regarded as a topology in which the whole particle swarm is connected. In this case, since all particles always share G_{best}, the convergence is high, and the solution search ability is high for unimodal functions.

On the other hand, Fig. 1 shows the ring topology, which is generally used as the topology of SPSO. In this topology, information is shared with neighboring particles. The finer the topology is subdivided, the less information the particles share. Therefore, the convergence to G_{best} is lower than that of the original PSO, and a high-precision search can be performed for multimodal functions with many local minima.

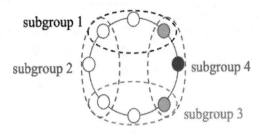

Fig. 1. Example of ring topology, which is generally used as topology of SPSO.

The formula for the velocity of SPSO is shown below. In SPSO, the inertia term χ is used to control the speed at which particles converge.

$$v(t+1) = \chi(v(t) + c_1 * rand_1 + c_2 * rand_2 * (L_{bestx} - x(t))) \tag{3}$$

$$\chi = \frac{2}{\left|2 - \varphi - \sqrt{\varphi^2 - 4\varphi}\right|} \tag{4}$$

$$\varphi = c_1 + c_2 \tag{5}$$

In SPSO, the motion of the entire particle group changes depending on the value of φ. For values of φ less than 4.0, the particle swarm repeatedly converges and diverges, so finding the optimal solution is not guaranteed. If it is greater than 4, then the particle swarm is guaranteed to converge. Therefore, SPSO generally uses a combination of $c_1 = c_2 = 2.05$.

2.2   BBPSO and Pairwise BBPSO

BBPSO. BBPSO is a method for calculating the position update formula of particles using a Gaussian distribution. It has been shown in some benchmark problems that it is effective for the problem of parameter setting for unknown problems, which is a problem in PSO [3]. In BBPSO, the mean, variance, and position of the search solution are defined by the following formulas.

$$\mu = \frac{p_{best} + g_{best}}{2}, \quad \delta = \left|p_{best} - g_{best}\right| \tag{6}$$

$$x(t+1) = \begin{cases} N(\mu, \delta) & if \ (\omega > 0.5) \\ p_{best} & else \end{cases} \tag{7}$$

where p_{best} is the best position information for a single particle so far, and g_{best} is the best position in the whole group. ω is a random number between 0 and 1, μ is the mean, and δ is the variance.

BBPSO uses Gaussian distribution (μ, δ) and iteratively updates the position information of particles. BBPSO solves the design variable problem of PSO but finds the optimal solution by g_{best}. Therefore, the iterative process tends to converge to a local solution early, which may reduce the search accuracy.

Pairwise BBPSO. In Pairwise BBPSO, pairs of two particles are randomly determined from the population. If the individual population size is N, then there are $N/2$ pairs. Then, we compare the evaluation values of the two-particle pairs and define the superior particle as the leader and the particle with the lower evaluation value as the follower. Here, the leader particle performs a global search using the same algorithm as BBPSO, and the follower particle performs a local search using a normal distribution with the leader particle. Specific search algorithms for leader and follower particles are shown in Eqs. (1) and (2) below.

Leader particles search with the following equation:

$$\mu = \frac{p_{best}(i) + g_{best}}{2}, \quad l = |p_{best}(i) - g_{best}| \tag{8}$$

$$x_t(i) = \begin{cases} N(\mu, l) \; if \; (\omega > 0.5) \\ p_{best} \quad else \end{cases} \tag{9}$$

where $p_{best} = (p_{best}(1), p_{best}(2), \ldots, p_{best}(n))$ is a matrix used for recording the best position each element has ever reached, g_{best} is the best position that all elements have ever reached, and $N(u, l)$ is a Gaussian distribution with mean u and standard deviation l. This equation is inherited from the standard BBPSO.

Conversely, follower particles aim at supporting the leader. The next position of a follower particle is randomly selected from a Gaussian distribution with a mean $(LP(i) + FP(i))/2$ and a standard division $|LP(i) - FP(i)|$. This means the follower particle becomes the evolute with the following equation:

$$x_t(i) = N\left(\frac{LP(i) + FP(i)}{2}, |LP(i) - FP(i)|\right) \tag{10}$$

where $LP(i)$ is the position of the leader particle, and $FP(i)$ is its follower.

Relatively, if the distance between G_{best} and the leader particle is larger than the distance between the leader particle and the follower particle, it means that the local search function using the follower particle is added to BBPSO. Conversely, when the distance between G_{best} and the leader particle is short, and the distance between the leader particle and the follower particle is long, the possibility of BBPSO falling into a local optimum is reduced by a global search using follower particles.

3 Local Search of PSO in Environment with Small Number of Particles

3.1 Difficulties in Designing Local Search Functions for PSO

As shown in the previous section, SPSO and Pairwise PSO are effective algorithms for improving the local search ability of PSO and BBPSO, respectively. On the other hand, SPSO can adjust the ratio of global search and local search by how many subgroups the population is divided into, but the appropriate ratio depends on the problem and needs to be adjusted by trial and error. Also, when Pairwise BBPSO cannot set enough particles in accordance with the size of the search space due to memory size and execution time, allocating half of the particles to the local search reduces the convergence power of each particle toward G_{best}. Therefore, the execution time until convergence tends to increase. In both cases, by adjusting the ratio of the number of particles heading to G_{best} and the number of particles in charge of local search, we can handle many types of problems, from unimodal space search problems to multimodal ones. However, the appropriate particle numbers to be allocated is problem-dependent and must be determined by trial and error. Also, it is difficult to determine an appropriate ratio for a new real problem. Research that dynamically determines this ratio is also conceivable. However, we here propose a method to add a local search function without allocating some particles of the particle swarm to the local search, that is, without reducing the convergence force of the entire particle swarm toward G_{best}. As a first step for that purpose, we consider how to improve the local search ability based on PSO without lowering the convergence power toward G_{best}.

3.2 Proposal of PSO with Local Search Using Personal Best Solution

We consider adding local search functionality based on the original PSO, which is the most popular algorithm. We propose a method in which all particles use the information of their past searches to perform local and peripheral searches for each individual. By doing this, we aim to improve the local search ability without allocating particles that only perform a peripheral search or local search, and without weakening the force toward G_{best} of the entire group. Specifically, the current position of each particle and the position (P_{best}) where each particle has the highest evaluation in the past searches are used to search with a Gaussian distribution defined by the following equation.

$$\mu = \frac{x(t) + p_{best}(i)}{2}, \quad \delta = |x(t) - p_{best}(i)| \tag{11}$$

$$x(t+1) = x(t) + N(\mu, \delta) \tag{12}$$

where $x(t+1)$ and $x(t)$ are the position vectors at generations t and $t+1$, respectively, $p_{best}(i)$ is the p_{best} of particle i, and μ and δ are the mean and variance of the Gaussian distribution, respectively. In the proposed method, the position update by PSO is performed in parallel with the position update by the Gaussian distribution described above, and the highly evaluated result for each particle is adopted. The pseudocode of the proposed algorithm is shown below.

Algorithm 1: Pseudo code of the proposed method

```
1  let N = the number of particles
2  for i = 1 to iterations do
3  end for
4    for j = 1 to N
5      particleA[j] = update via PSO algorithm
         (x(t + 1) = x(t) + v(t + 1))
6      particleB[j] = update via suggest algorithm
         (x(t + 1) = x(t) + N(μ, δ))
7      particle[j] = choose the better of particleA
         and particleB
8      if Gbestx change rate is 0.01% or less for 10
         consecutive times
9        stop exploration
10     end if
11   end for
```

Figure 2 shows a conceptual diagram showing the motion of particles when the proposed method is used. Each particle performs a local search using its own past best solution (P_{best}) and a global search according to the original PSO algorithm using G_{best}, and it moves to a position vector with a higher evaluation value. Therefore, the proposed algorithm searches for the optimum solution while performing local search and wide-area search at the same time.

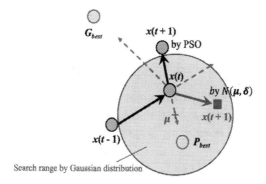

Fig. 2. Conceptual diagram showing motion of particles when proposed method is used.

4 Evaluation

4.1 Experimental Method

Four types of benchmark functions, Sphere, Ackley, Rastrigin, and Griewank, and their respective shifted functions were used for evaluation experiments. Sphere is a unimodal function used to evaluate the convergence speed toward Gbest, and Ackley, Rastrigin, and Griewank are multimodal functions with different shapes used to evaluate the local search ability. Shift functions are used to evaluate the ability to explore edge regions of the solution space. Using these benchmark problems, we will compare the search accuracy between PSO, which is based on the proposed method, SPSO, which is a representative algorithm based on PSO with a local search function added, and the proposed method. This time, to evaluate the robustness of the proposed method for the case where it is not possible to allocate a sufficient number of particles in accordance with the size of the search space, we set 5 dimensions with 1000 particles, 10 dimensions with 500 particles, and 50 dimensions with 100 particles as Experiments 1, 2, and 3, respectively. In each search, when the rate of change in the evaluation value of the best evaluation position was 0.01% or less for 10 consecutive generations, the search was considered to have converged at that point and was terminated. If the convergence condition was not satisfied, the search ended when the number of generations reached 1000. Here, the number of dimensions indicates the number of parameters possessed by the particles, and the number of generations indicates the number of times the position/speed information of the particles is updated. Details on the benchmark functions used in the experiments are shown in Table 1 below.

The optimal solutions of the benchmark functions other than the shift functions are all $x_i = 0$ $(i = 0, ..., d)$, and the optimal solutions of the shift functions are $x_i = 2$ $(i = 0, ..., d)$ for Rastrigin, and $x_i = 32$ $(i = 0, ..., d)$ for otherwise. In addition, we introduce an accuracy ratio to evaluate the improvement ratio for search accuracy quantitatively. The accuracy ratio can be expressed as follows, using PSO, SPSO, and evaluation values f_{PSO}, f_{SPSO}, and f_{GPSO} of the execution result of the proposed method.

$$\text{Accuracy ratio} = \frac{\min(f_{PSO}, f_{SPSO})}{f_{GPSO}} \tag{13}$$

The higher the accuracy ratio, the better the search accuracy of the proposed method. In addition, results with higher evaluations are used for the PSO and SPSO.

The test execution environment is summarized in Table 2. Experimental results represent the average of 20 trials. The maximum number of search generations was set to 1000, and the experiments were performed using the number of dimensions and the number of particles of the problem as variables.

4.2 Experimental Results and Discussion

Experiment 1: When Number of Particles was Sufficient. As an example of the case where the number of particles is sufficient for the size of the search space, Table 3 shows the search accuracy of PSO, SPSO, and the proposed method when the number of particles was set to 1000 for 5 dimensions.

Table 1. Benchmark function used in experiments.

Function	Equation	Bounds
Sphere	$$f_1 = \sum_{i=0}^{d} x_i$$	$(-32, 32)^d$
Griewank	$$f_2 = \frac{1}{4000} \sum_{i=1}^{d} x_i - \prod_{i=1}^{d} \cos\left(\frac{x_i}{\sqrt{i}}\right) + 1$$	$(-32, 32)^d$
Ackley	$$f_3 = -20exp\left[-\frac{1}{5}\sqrt{\frac{1}{d}\sum_{i=1}^{d} x_i}\right] - exp\left[\frac{1}{d}\sum_{i=1}^{d} \cos(2\pi x_i)\right] + 20 + e$$	$(-32, 32)^d$
Rastrigin	$$f_4 = \sum_{i=0}^{d}\left[x_i^2 - 10 * \cos(2\pi x_i) + 10\right]$$	$(-5, 5)^d$
Sphere (Shifted)	$$f_1 = \sum_{i=0}^{d} (x_i - 32)$$	$(-32, 32)^d$
Griewank (Shifted)	$$f_2 = \frac{1}{4000} \sum_{i=1}^{d} (x_i - 32) - \prod_{i=1}^{d} \cos\left(\frac{(x_i-32)}{\sqrt{i}}\right) + 1$$	$(-32, 32)^d$
Ackley (Shifted)	$$f_3 = -20exp\left[-\frac{1}{5}\sqrt{\frac{1}{d}\sum_{i=1}^{d}(x_i - 32)}\right.$$ $$\left. -exp\left[\frac{1}{d}\sum_{i=1}^{d} \cos(2\pi(x_i - 32))\right] + 20 + e\right]$$	$(-32, 32)^d$
Rastrigin (Shifted)	$$f_4 = \sum_{i=0}^{d}\left[(x_i - 2)^2 - 10 * \cos(2\pi(x_i - 2)) + 10\right]$$	$(-2, 2)^d$

Table 2. Test execution environment.

CPU	Intel Core i7-8565U 1.80 GHz
GPU	Intel UHD Graphics 620
Memory	16 GB DDR4 SDRAM
OS	Microsoft Windows 10 Pro (64 bit)
Compiler	Visual C++ 12.0

Table 3 shows that the original PSO is an effective algorithm when searching with a sufficient number of particles for the size of the search space. The highest solution accuracy was obtained for unimodal functions, with or without shift. The reason PSO showed the highest solution search accuracy, even for the Griewank function, which is a multimodal function, is considered to be that the Griewank function has a large

Table 3. Search accuracy of PSO, SPSO, and proposed method. (1000 particles, 5 dimensions)

Function	f_{PSO}	f_{SPSO}	f_{GPSO}	Accuracy ratio
Sphere	**3.73*e-25**	4.77*e-5	1.31*e-21	2.85*e-4
Griewank	**2.11*e-2**	1.53*e-1	1.99*e-1	1.06*e-1
Ackley	3.28*e-7	2.51*e-6	**3.55*e-15**	9.23*e+7
Rastrigin	1.98	2.89	2.04	0.97
Sphere (Shif)	**1.26*e-29**	6.14*e-7	7.69*e-20	1.64*e-10
Griewank (Shift)	**2.47*e-2**	1.21*e-1	1.99*e-1	1.24*e-1
Ackley (Shift)	3.05*e-1	4.27*e-1	**3.55*e-15**	8.59*e+13
Rastrigin (Shift)	5.77	4.76	4.97	0.96

number of undulations, but the height difference of each undulation is small. On the other hand, for the Ackley function with deep valleys, the proposed method showed much higher precision than PSO and SPSO. In addition, the proposed method showed a higher convergence performance than SPSO, although the search performance was slightly lower than that of PSO for unimodal functions. The proposed local search function is considered to have worked effectively. The fact that SPSO showed inferior results to PSO as a whole and the reason the Rastrigin function did not reach the optimum solution are presumed to be problems in the design variables of PSO and SPSO.

Experiment 2: When Number of Particles was Small. As an example of the case where the number of particles is small for the size of the search space, Table 4 shows the search accuracy of PSO, SPSO, and the proposed method when the number of particles was set to 500 for 10 dimensions.

Similar to Experiment 1, the proposed method showed a much higher solution search accuracy than PSO and SPSO for the Ackley function when a sufficient number of particles was not assigned in accordance with the size of the search space. In addition, the proposed method suppressed PSO and showed the highest solution search accuracy even for the Griewank function, which is a multimodal function. Surprisingly, the solution search accuracy greatly exceeded that of PSO, even for Sphere, which is a unimodal function. It is possible that the proposed method also contributed to extensive searches, but detailed investigations are necessary for the future. From Table 4, it can be considered that the proposed local search function worked effectively on the whole. SPSO showed a higher solution search accuracy than PSO for the Ackley function (with or without shift) and Griewank (with shift), which are multimodal functions. It seems that the relative effectiveness of SPSO in multimodal functions for PSO increases as the number of particles decreases relative to the size of the search space.

Experiment 3: When Number of Particles was Extremely Small. As an example of the case where the number of particles is extremely small for the size of the search space,

Table 4. Search accuracy of PSO, SPSO, and proposed method. (500 particles, 10 dimensions)

Function	f_{PSO}	f_{SPSO}	f_{GPSO}	Accuracy ratio
Sphere	3.31*e-5	8.50*e-4	**6.17*e-22**	**5.36*e+16**
Griewank	1.23*e-1	1.51*e-1	**8.66*e-2**	1.42
Ackley	4.31	4.63*e-5	**2.09*e-10**	**2.09*e+5**
Rastrigin	4.91*e+4	4.95*e+4	4.91*e+4	1.00
Sphere (Shif)	1.72*e-4	9.07*e-5	**2.61*e-20**	**3.48*e+15**
Griewank (Shift)	5.32*e-1	2.54*e-1	**1.89*e-1**	1.34
Ackley (Shift)	4.06	6.64*e-1	**3.41*e-7**	**1.95*e+6**
Rastrigin (Shift)	4.92*e+4	4.99*e+4	4.91*e+4	1.00

Table 5 shows the search accuracy of PSO, SPSO, and the proposed method when the number of particles was set to 100 for 50 dimensions.

Table 5. Search accuracy of PSO, SPSO, and proposed method. (100 particles, 50 dimensions)

Function	f_{PSO}	f_{SPSO}	f_{GPSO}	Accuracy ratio
Sphere	7.23*e+2	7.93*e+2	**4.39*e-2**	**1.65*e+4**
Griewank	1.27	1.39	**2.10*e-3**	**6.05*e+2**
Ackley	1.38*e+1	1.23*e+1	**5.31**	**2.31**
Rastrigin	8.80*e+2	8.14*e+2	5.64*e+2	1.44
Sphere (Shif)	2.24*e+4	2.36*e+5	**6.42*e+2**	**3.49*e+1**
Griewank (Shift)	6.38	5.96	**3.04*e-2**	**1.96*e+2**
Ackley (Shift)	2.09*e+1	2.13*e+1	**7.61**	**2.75**
Rastrigin (Shift)	8.71*e+2	1.00*e+4	7.12*e+2	1.22

When only an extremely small number of particles could be set for the wide search space, PSO and SPSO fell into local optima for all functions and could not reach the optimal solution. In comparison, the proposed method reached the optimal solution for the Sphere function and the Griewank function (with or without shift), and even when it fell into local minima for the other functions, the values were closer to the optimal solution than PSO and SPSO. It is thought that the proposed method has an improved ability to prevent falling into local minima by performing a local search of each particle on the basis of a Gaussian distribution.

Discussion for All Experiments and Future Work. Summarizing experiments 1 to 3, the proposed method has a great search ability using both global and local searches even under conditions where the number of particles is insufficient, where conventional PSO and SPSO tend to have difficulty converging on the optimal solution. This is because as the search progresses with a small number of particles, the parameters of the velocity vectors of the particles gradually decrease in the conventional PSO and SPSO, and convergence is judged before the optimum solution is found. In comparison, in the proposed method, movement by Gaussian distribution may change the positions of particles relatively greatly, which may increase the parameters of the velocity vectors again. In addition, since the proposed method uses PSO as the prototype algorithm, it is possible to show a relatively high search accuracy even for Sphere, which is a unimodal function, and Griewank, which has properties globally close to a unimodal function.

However, the proposed method has a problem because the number of comparisons of evaluation values is greater than in the conventional algorithm when updating the particle positions, and therefore, it requires more computation time. Shortening the execution time with more effective calculation methods, such as parallel distribution, is desired for practical use. Also, in the future, it is necessary to conduct evaluation experiments for a wider variety of problems.

5 Conclusion

In this paper, instead of dividing some particles into local searches, we proposed a particle swarm optimization method with the function of performing local searches using the personal best solutions of each particle. In particular, we implemented a local search function using a Gaussian distribution, and we compared the search accuracy with existing algorithms using four types of benchmark functions, Sphere, Griewank, Ackley, and Rastrigin, and their respective shifted functions for evaluation. The results show that the proposed method improves the local search ability without degrading the global search ability, and it works effectively even in environments with few particles. However, it has a problem in that it requires more calculation time, so it is desired to shorten the execution time with a more effective calculation method. In addition, further evaluation experiments are required for a wider variety of problems.

Acknowledgment. This work was supported by JSPS KAKENHI Grant Numbers JP19K12162, JP22K12185, and the Education Department Scientific Research Program Project of Hubei Province of China (Q20222208).

References

1. Kennedy, K., Eberhart, R.: Particle swarm optimization. In: IEEE International Conference on Neural Networks, vol. 4, pp. 1942–1948. IEEE, Perth, WA, Australia (1995)

2. Kennedy, J.: Bare bones particle swarms. In: Proceedings of the 2003 IEEE Swarm Intelligence Symposium (SIS 2003), pp. 80–87. IEEE, Indianapolis, IN, USA (2003)
3. Bratton, D., Kennedy, J.: Defining a standard for particle swarm optimization. In: IEEE Swarm Intelligence Symposium 2007, pp. 120–127. IEEE, Honolulu, HI, USA (2007)
4. Bastos Filho, C.J.A., de Lima Neto, F.B., da Cunha Carneiro Lins, A.J., Nascimento, A.I.S., Lima, M.P.: Fish school search. Nat.-Inspired Algorithms Optim. **2009**, 261–277 (2009)
5. Tan, Y., Zhu, Y.: Fireworks algorithm for optimization. In: Tan, Y., Shi, Y., Tan, K.C. (eds.) ICSI 2010. LNCS, vol. 6145, pp. 355–364. Springer, Heidelberg (2010). https://doi.org/10.1007/978-3-642-13495-1_44
6. Guo, J., Sato, Y.: A pair-wise bare bones particle swarm optimization algorithm. In: Proceedings of the 2017 IEEE/ACIS 16th International Conference on Computer and Information Science (ICIS), pp. 353–358. IEEE, Wuhan, China (2017)
7. Wang, J., Xie, Y., Xie, S., Chen, X.: Cooperative particle swarm optimizer with depth first search strategy for global optimization of multimodal functions. Appl. Intell. **52**, 10161–10180 (2022)
8. Li, T., Shi, J., Deng, W., Hu, Z.: Pyramid particle swarm optimization with novel strategies of competition and cooperation. Appl. Soft Comput. **121**, 108731 (2022)

Hybrid Algorithm Based on Phasor Particle Swarm Optimization and Bacterial Foraging Optimization

Xiaole Liu[1], Chenhan Wu[1], Peilin Chen[2(✉)], and Yongjin Wang[3]

[1] College of International Education, Shenzhen 518048, China
[2] Whittle School and Studios-Shenzhen Campus, Shenzhen 518067, China
2210135009@email.szu.edu.cn
[3] Greater Bay Area International Institute for Innovation, Shezhen University,
Shenzhen 518060, China

Abstract. In response to the issues of premature convergence and instability of the phasor particle swarm optimization (PPSO) for solving function optimization problems, a new hybrid algorithm called bacteria PPSO (BPPSO) was proposed which combines the chemotaxis operation of the bacterial foraging optimization (BFO) algorithm with PPSO. In BPPSO, all individuals undergo tumbling and swimming strategies when the chemotaxis condition is met. New coefficients are introduced to update the positions of particles in BPPSO, achieving complementary advantages of BFO and PPSO. Finally, BPPSO is validated using eight benchmark functions, demonstrating its fast convergence speed, high computational accuracy, and good stability, making it a powerful global optimization algorithm.

Keywords: Phasor particle swarm optimization · Bacteria foraging optimization · Hybrid algorithm · Chemotaxis

1 Introduction

Particle swarm optimization (PSO) [1] is an optimization algorithm that belongs to swarm intelligence, and it was initially motivated by the foraging behavior of birds. The algorithm uses a simplified model of swarm intelligence based on the observation of group and animal activity behavior. PSO achieves optimal solutions by creating an evolutionary process from disorder to order in the problem-solving space, through information sharing among individuals in the swarm. Over time, many researchers have been interested in enhancing its optimization performance.

There are numerous research studies aimed at enhancing the optimization performance of PSO: Mojtaba [2] combined the phasor theory in mathematics with PSO, and proposed a phasor particle swarm optimization (PPSO) based on phase angle (θ) to model particle control parameters. J.J.Liang [3] proposed a comprehensive learning PSO (CLPSO) that employs a novel learning strategy where the historical best position of other particles serves as an example from which any particle can learn. Each dimension of each particle can learn from different examples. S. Mahdi Homayouni [4] mixed PSO with simulated annealing (SA) to solve the problem that PSO tends to fall into local optimal. Claudiu [5] added particle filtering to PSO, which can control the size of the search area to improve accuracy, but it also leads to the problem of converging too quickly. Liu [6] has proposed an algorithm for incorporating R2 [7] indicator and decomposition-based archive pruning strategy into particle swarm optimizer for addressing the balance between convergence and diversity in high-dimensional objective space multi-objective optimization problems. In response to large-scale complex problems, Li [8] proposed a multi-objective particle swarm optimization algorithm based on enhanced selection and to increase the ability of exploration and exploitation, this algorithm designed an enhanced selection strategy to update the personal optimal particles, and used objective function weighting to update the global optimal particle adaptively. In the algorithm mentioned above, PPSO is widely regarded as one of the most popular algorithms. The phase angle (θ) of each particle in the PPSO is a one-dimensional variable, so each particle is formed by a modular vector with an angle (θ). This operation turns PPSO into an adaptive, balanced, and nonparametric meta-heuristic algorithm.

Like other swarm intelligence optimization algorithms, PPSO also has its shortcomings. Compared to PSO, although PPSO improves search accuracy and avoids premature convergence to some extent, it still has limitations in complex high-dimensional function optimization problems. To improve the search accuracy of PPSO in solving complex high-dimensional function optimization problems, in this paper, we take a new perspective and consider that the chemotaxis operation of bacterial foraging optimization (BFO) [9] can help particles jump out of the minimum point and gain the ability to continuously search for local optima. Passino proposed BFO by simulating the foraging behavior of E. coli in human body in 2002. BFO is a search technology based on bacterial population, which realizes optimization through competition and cooperation between bacterial populations. At the same time, BFO is a global random search algorithm, which is simple, fast convergence speed, and does not need to optimize the gradient information of the object in the optimization process. It simulates the bacterial population in three steps: chemotaxis, reproduction and elimination dispersal. Therefore, we propose a hybrid algorithm based on PPSO and BFO. The new algorithm which named bacterial phasor particle swarm optimization (BPPSO) not only improves the accuracy, but also ensures that the algorithm does not converge too fast. In addition, we also made some small improvements in the hybrid algorithm, such as updating and adjusting the parameters.

The rest of this paper is discussed in the following order. In Sect. 2, we introduce improved hybrid algorithm. The experimental results of four algorithms on test set are compared in Sect. 4. Finally, we draw the conclusion and give future work in Sect. 5.

2 Traditional Algorithm

2.1 Bacterial Foraging Optimization

BFO, as a global random search algorithm, does not need to optimize the gradient information of the object in the optimization process, and has the advantages of simple and fast convergence. It mainly simulates the three processes of chemotaxis, reproduction and elimination dispersal.

Chemotaxis. The movement of bacteria to the area favorable to its environment is called chemotaxis, in which a trend operation includes tumbling motion and swimming motion. The movement of bacteria in any direction is called tumbling motion. The movement of the bacteria along the previous step is called swimming motion. In general, bacteria tumble more frequently in bad environments (toxic areas) and swim more in good environments (food is abundant). The entire life cycle of E. coli consists of alternating between two basic movements: swimming and tumbling, in order to find food and avoid toxins.

Suppose the population size of bacteria is S, the location of bacteria marks a candidate solution to the problem, and the information of bacteria i is marked by the D-dimension vector:

$$\theta^i = [\theta_1^i, \theta_2^i, \ldots, \theta_D^i] \tag{1}$$

$i = 1, 2, \ldots, S$, $\theta i(j,k,l)$ represents the position of bacteria i after the jth convertitive operation, the kth replication operation, and the lth migration operation. Bacteria i updates its position after each step of the convergent operation through formula (2).

$$\theta^i(j+1, k, l) = \theta^i(j, k, l) + C(i)\Phi(j) \tag{2}$$

where $C(i) > 0$ represents the step size of forward walking, and $\Phi(j)$ represents the unit direction vector randomly selected after tumbling. If the fitness at $\theta i(j + 1,k,l)$ is better than that at $\theta i(j,k,l)$, keep Φ unchanged and continue swimming in this direction until the optimal fitness position is found or the set number of convertivities is reached. Otherwise, a new Φ is generated for the next tumble.

Reproduction. The law of biological evolution is survival of the fittest. After a period of food hunting, some of the bacteria that are weak at finding food are naturally weeded out, and the remaining bacteria reproduce in order to maintain population size. This phenomenon is called reproduction in the BFO.

In BFO, the population size of the algorithm remains unchanged after the replication operation. Suppose the number of bacteria eliminated is $Sr = S/2$, first of all, the ranking is based on the position of bacteria, and then the Sr bacteria in the back are eliminated, and the remaining Sr bacteria are self-replicated to generate a new individual exactly the same as themselves, that is, the generated new individual has the same position as the original one, or has the same foraging ability.

Elimination Dispersal. The local environment in which individual bacteria live may undergo sudden changes, such as a sudden increase in temperature, or gradual changes, such as the depletion of food. This can lead to either collective death or migration of the bacterial population living in this local region. This phenomenon is referred to as elimination dispersal in BFO.

Elimination dispersal occurs with a certain probability. Given the probability of P, if a bacterial individual in the population meets the probability of migration occurrence, the bacterial individual will perish, and a new individual will be generated randomly at any position in the solution space. The new individual and the extinct individual may have different positions, that is, different foraging abilities. The new individual randomly generated by migration operation may be closer to the global optimal solution, which is more conducive to the directional operation to jump out of the local optimal solution and find the global optimal solution.

2.2 Phasor Particle Swarm Optimization

PPSO is an optimization algorithm that combines phasor theory in mathematics with PSO. This algorithm models particle control parameters by utilizing phase angle (θ). The assumption made in PPSO is that the phase angle (θ) of each particle is a one-dimensional variable. A modular vector with an angle (θ) is used to construct each particle. By adding the phase angle (θ), PSO becomes an adaptive, balanced and nonparametric meta-heuristic algorithm.

Original Particle Swarm Optimization.   PSO is an optimization algorithm that simulates the behavior of flocks of birds searching for food. In this scenario, there is only one piece of food in the area and none of the birds know where it is located. However, they are aware of their proximity to the food. The most effective strategy for finding the food is for the birds to search the area around the bird that is currently closest to the food. During the search process, the birds share information with each other about their current positions, allowing them to cooperatively judge whether or not they have found the optimal solution. This sharing of information also allows the optimal solution to be transmitted to the entire flock, ultimately leading to the flock gathering around the food source.

In PSO, each particle has an adaptation value determined by an optimized function and a speed that determines their flying direction and distance. The particles follow the current optimal particle in the search space to find the optimal solution. The PSO starts by initializing a group of random particles, which represent potential solutions. Through iteration, the optimal solution is found by updating each particle and tracking two extreme values. The first extreme value is the optimal solution found by the particle itself, known as the individual personal best. The other extreme value is the optimal solution found by the entire population, called the global best.

Specifically, a massless particle is designed in a PSO to simulate a bird in a flock, and the particle has only two properties: velocity and position. In PSO, position represent the direction of movement while speed represents the rate of movement. Birds are abstracted as particles (points) without mass or volume and extend into D-dimensional space. Each particle's position in D-dimensional space is represented as a vector $Xi = (x1,x2,...,xd)$ and its flight speed is expressed as a vector $Vi = (v1,v2,...,vd)$. Every particle in the population has a fitness value determined by the objective function, and it knows the personal best (*pbest*) and current position *Xi* it has found so far. This represents the particle's own flight experience. Additionally, each particle knows the global best (*gbest*) of all particles in the entire population so far, which represents the collective experience of its peers. The particles determine their next move by using a combination of their own experience and the best experiences of their peers. The formula for updating the particle's speed and position is:

$$v_i^d = wv_i^d + c1r1\left(pbest_i^d - x_i^d\right)$$

$$+c2r2\left(gbest_i^d - x_i^d\right) \tag{3}$$

$$x_i^d = x_i^d + v_i^d \tag{4}$$

where v_{id} and x_{id} are the velocity and position of particle i respectively; w is the inertia weight. In the standard PSO, w decreases linearly with the number of iterations. $c1$ and $c2$ are learning factors; $r1$ and $r2$ are uniformly distributed random numbers on [0,1].

Phasor Particle Swarm Optimization. The PPSO is an improvement on the PSO. PPSO is different from traditional PSO in particle swarm initialization. N particles $\overrightarrow{X} = |Xi| < \theta i (i = 1 : N)$ are randomly generated in the D-dimensional space of problem with their own phase angle θi through uniform distribution $\theta_{id}^{Iter=1}$ and with initial velocity limit $v_{max,id}^{Iter=1}$.

The ith phasor angle $\theta_{id}^{Iter=1}$ of the first iteration satisfies the uniform distribution of $[0,2\Pi]$:

$$\theta_{id}^{Iter=1} = U(0, 2\Pi) \tag{5}$$

The formula for calculating the maximum velocity v and the maximum position of the ith particle in the first iteration is:

$$v_{max,id}^{Iter=1} = 0.5 \times (x_{max,d} - x_{min,d}) \tag{6}$$

$$x_{max,id}^{Iter=1} = x_{min,d} + rand_{id}0.5 \times (x_{max,d} - x_{min,d}) \tag{7}$$

where $x_{max,d}$ and $x_{min,d}$ are the maximum and minimum positions of the *ith* particle respectively. PPSO is also different from PSO in speed and position update. The speed update formula of PPSO is as follows:

$$v_{id}^{Iter} = \left|cos\theta_i^{Iter}\right|^{2sin\theta_i^{Iter}} \times \left(pbest_{id}^{Iter} - x_{id}^{Iter}\right)+$$

$$\left|sin\theta_i^{Iter}\right|^{2cos\theta_i^{Iter}} \times \left(gbest_{id}^{Iter} - x_{id}^{Iter}\right) \tag{8}$$

We then use the following formula to determine the final velocity:

$$v_{id}^{Iter} = \min(\max\left(v_{id}^{Iter}, -v_{max,id}^{Iter}\right), v_{max,id}^{Iter}) \tag{9}$$

The position update formula of PPSO is as follows:

$$\overrightarrow{X}_{id}^{Iter+1} = \overrightarrow{X}_{id}^{Iter} + v_{id}^{Iter} \tag{10}$$

We then use the following formula to determine the final position:

$$x_{id}^{Iter+1} = \min(\max\left(x_{id}^{Iter+1}, x_{min,d}\right), x_{max,d}) \tag{11}$$

where, $pbest_i^{Iter}$ and $gbest_i^{Iter}$ are consistent with $pbest_i^{Iter}$ and $gbest_i^{Iter}$ in PSO respectively.

Then update θ_i and $v_{max,id}^{Iter}$, and the update formula is:

$$\theta_{id}^{Iter+1} = \theta_{id}^{Iter} + 2\Pi|cos\theta_i^{Iter} + sin\theta_i^{Iter}| \tag{12}$$

$$v_{max.id}^{Iter+1} = \left|cos\theta_i^{Iter}\right|^2 \times (x_{max,d} - x_{min,d}) \tag{13}$$

3   BPPSO Algorithm

3.1   Convergence Strategy

Each particle initialization in the PPSO is randomly generated. And the velocity and position updates also add phasor angle (θ). Compared with PSO, PPSO becomes an adaptive, balanced and non-parametric meta-heuristic algorithm, which can find better solutions due to larger search space. However, it still has the deficiency of fast convergence in the early stage. This is because at the beginning of the iteration of PPSO, due to the better diversity of population initialization, the algorithm can quickly converge to the local extreme value. When the individuals of the population gradually move to the optimal position of the population, we can see that the diversity of the population gradually decreases, and the value has little change or is in a stagnant state, that is, it is in the local optimal. But local optimality is not global optimality.

The chemotactic behavior of BFO involves bacteria congregating in areas rich in food, and it includes two modes of movement: tumbling and swimming. The tumbling mode involves the bacteria moving a certain distance in any direction, while the swimming mode determines whether the bacteria should continue in the current direction by evaluating whether the fitness function value of the bacteria improves after tumbling. The bacteria move along their previous step in a swimming motion. This behavior enables bacteria to continuously optimize their local environment.

Therefore, the PPSO improvement strategy proposed in this paper is to add the chemotactic behavior in BFO to the PPSO, which avoids the shortcoming of too fast convergence in the early stage through flipping and swimming modes, while preserving the diversity of the initial PPSO. In addition, in order to continue to consolidate PPSO's advantages in search space, we also added the trigonometric function of phasor angle θ in formula (14) to the position update of PPSO. Compared with PPSO, the new algorithm makes particles have more advantages in position update, so as to obtain better solutions.

$$x_{max.id}^{Iter+1} = cos\theta_i^{Iter} \times x_{max.id}^{Iter} + v_{max.id}^{Iter+1} \tag{14}$$

3.2 Algorithm Process

We would show the process of the new algorithm process using the mixed strategy as follows:

Step 1. Initializes related parameters in the BPPSO.
Step 2. Then, the control variables are input, the solution space is established, and the values are taken randomly under the premise that the upper and lower limits are not exceeded, and n D-dimensional particles are formed. The number of control variables is the dimension m of the particle. Calculate related parameters according to Eqs. (5)–(7).
Step 3. The fitness value of each particle is calculated according to the fitness function.
Step 4. At this time, we give a random number, and judge whether the random number meets the conditions of chemotactic behavior. If it meets the conditions of chemotactic behavior, we will tumble or swim the particle. The specific formula is as follows:

$$\theta^i(j+1, k, l) = \theta^i(j, k, l) + C(i)\frac{\Delta(i)}{\sqrt{\Delta^T(i)\Delta(i)}} \tag{15}$$

Step 5. Formula (8)–(9) was adopted for speed updating and formula (10)–(11) for position updating.
Step 6. The fitness value of the function was recalculated.
Step 7. The fitness value of the updated particle is compared with the local extreme value $pbest_{id}^{Iter}$ and the optimized global extreme value $gbest_{id}^{Iter}$. If any fitness value is better, $pbest_{id}^{Iter}$ and $gbest_{id}^{Iter}$ are updated.
Step 8. Formula (12) was used to update vector angle and formula (13) was used to update velocity maximum.
Step 9. Repeat steps (4)–(8) until the maximum number of iterations is met.

4 Experiments and Discussions

In this paper, we use the new algorithm on eight test functions. The experimental results of the new algorithm are compared with PSO, PPSO, CLPSO. We will look at the performance of four optimization algorithms in different test functions. All the algorithms presented in this paper were coded in MATLAB language and run on an Intel Core i7 processor with 1.8GHz CPU speed.

4.1 Experimental Setting

In this paper, the detailed parameter settings for algorithms are listed in the Table 1.

4.2 Experimental Results

We set the population size N as 40, dimension D as 30, iteration times as 2000. The average performance of four algorithms on eight test functions after 20 runs is as follows:

We saved the solutions of each round of experiments. When the number of iterations was 2000, the average value of 20 experiments was taken to draw the following figure:

Table 1. Parameter settings for four algorithms

Algorithm	n	D	max interation	c1*r1	c2*r2	w
PSO	40	30	2000	2*rand[0,1]	2*rand[0,1]	0.9;0.4
CLPSO	40	30	2000	-	-	0.5ki/2000
PPSO	40	30	2000	$\left\|\cos\theta_i^{Iter}\right\|^{2\sin\theta_i^{Iter}}$	$\left\|\sin\theta_i^{Iter}\right\|^{2\cos\theta_i^{Iter}}$	-
BPPSO	40	30	2000	$\left\|\cos\theta_i^{Iter}\right\|^{2\sin\theta_i^{Iter}}$	$\left\|\sin\theta_i^{Iter}\right\|^{2\cos\theta_i^{Iter}}$	-

Table 2. The result of running the algorithm on test function Ackley

	BPPSO	CLPSO	PPSO	PSO
Max	**1.56e-11**	2.17E-04	3.90E-04	3.90E-04
Min	**0**	6.49E-05	3.87E-06	3.87E-06
Mean	**9.18e-13**	1.29E-04	1.73E-04	1.73E-04
Std	**3.50e-12**	3.46E-05	1.08E-04	1.08E-04

Table 3. The result of running the algorithm on test function Apline

	BPPSO	CLPSO	PPSO	PSO
Max	**1.25e-11**	5.78E-04	1.34E-02	6.70E-08
Min	**5.45e-24**	1.91E-04	3.42E-05	3.19E-10
Mean	**7.77e-13**	3.58E-04	1.57E-03	1.05E-08
Std	**2.81e-12**	9.92E-05	3.15E-03	1.74E-08

Table 4. The result of running the algorithm on test function Dixo_price

	BPPSO	CLPSO	PPSO	PSO
Max	**2.37E-01**	1.48E+01	8.20E-01	1.74E+00
Min	**1.65E-01**	1.51E+00	6.67E-01	6.67E-01
Mean	**1.77E-01**	7.02E+00	6.91E-01	7.92E-01
Std	**1.49E-02**	3.62E+00	4.02E-02	2.51E-01

Table 5. The result of running the algorithm on test function Griewank

	BPPSO	CLPSO	PPSO	PSO
Max	**0**	1.25e-05	9.90E-03	4.18E-02
Min	**0**	1.06e-06	2.70E-07	4.33E-15
Mean	**0**	5.50e-06	2.49E-03	7.50E-03
Std	**0**	3.85e-06	3.93E-03	1.25E-02

Table 6. The result of running the algorithm on test function Rastrigin

	BPPSO	CLPSO	PPSO	PSO
Max	**0**	9.70e-07	1.78e-06	5.17E+01
Min	**0**	2.03e-07	2.16e-09	1.49E+01
Mean	**0**	5.93e-07	3.96e-07	3.58E+01
Std	**0**	2.30e-07	4.35e-07	9.75E+00

Table 7. The result of running the algorithm on test function Rosenbrock

	BPPSO	CLPSO	PPSO	PSO
Max	**1.24e-05**	2.68E+01	2.39E+01	8.00E+01
Min	**2.58e-10**	1.86E+01	2.49E-04	1.67E+01
Mean	**1.30e-06**	2.44E+01	2.19E+01	4.01E+01
Std	**2.88e-06**	2.12E+00	5.21E+00	2.50E+01

Table 8. The result of running the algorithm on test function Schwefel

		BPPSO	CLPSO	PPSO	PSO
Max		1.14E+04	1.07E+04	1.13E+04	**7.54E+03**
Min		1.07E+04	1.07E+04	1.07E+04	**5.72E+03**
Mean		1.07E+04	1.07E+04	1.08E+04	**6.65E+03**
Std		1.69E+02	**1.43E-07**	1.92E+02	5.71E+02

Table 9. The result of running the algorithm on test function Step

	BPPSO	CLPSO	PPSO	PSO
Max	5.77e-05	2.63e-07	1.65e-05	**3.50e-12**
Min	3.02e-09	3.24e-08	1.15e-06	**6.13e-16**
Mean	6.78e-06	1.04e-07	5.45e-06	**2.73e-13**
Std	1.30e-05	5.66e-08	3.96e-06	**7.76e-13**

4.3 Discussions

In this section, four algorithms are compared to analyze the performance of each algorithm. This experiment adopts eight different test functions, and the number of decision
variables we set is 30.

We compare the running results of four algorithms on eight test functions. It can be
seen from Tables 2, 3, 4, 5, 6, 7, 8 and 9 that the maximum value, minimum value, mean
value and variance of BPPSO on the six test functions are all smaller than those of the
other three algorithms. However, the performance on the test functions Schwefel and
Step is not as good as PSO and CLPSO. But BPPSO's performance on most of the test
functions is more reliable.

Through the analysis of Fig. 1, we can see that the search accuracy of BPPSO
on the six test functions is several orders of magnitude higher than that of the other
three algorithms, which fully reflects that BPPSO has more advantages in the accuracy
of seeking the optimal solution. Moreover, BPPSO converges after 2000 iterations on
test function 1, test function 2, test function 3, test function 4, test function 6 and test
function 8. Compared with BPPSO, the other three algorithms are very unstable in terms
of convergence, especially PSO and CLPSO, which basically do not converge after 2000
iterations.

To sum up, by comprehensively comparing the convergence curves of the maximum
value, minimum value, mean value, variance and 2000 iterations of the operation results
of BPPSO and the other three algorithms, the search accuracy of BPPSO is higher, that
is, the solutions found is better, the convergence speed is faster, and the local optimal
solution can be avoided as far as possible.

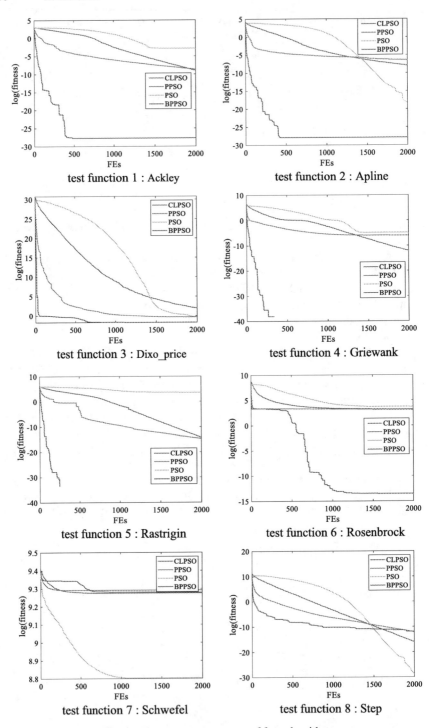

Fig. 1. The convergence curve of four algorithms

5   Conclusion and Future Work

In this paper, chemotactic behavior of BFO is added to PPSO, and the proposed new algorithm and CLPSO, PPSO,PSO are used to solve eight test functions. The results show that the maximum value, minimum value, mean value and variance of BPPSO on the six test functions are smaller than the other three algorithms, and the search accuracy of the seven test functions is several orders of magnitude higher than the other three algorithms, which fully reflects that BPPSO has more advantages in the accuracy of searching the optimal solution. All these show that the algorithm has the characteristics of high accuracy, fast convergence and good stability, and is an effective and reliable optimization algorithm. Further studies can obtain better solutions by mixing PPSO and other algorithms and how to further optimize BPPO model to solve high-dimensional complex problems.

References

1. J. Kennedy., R. Eberhart.: Particle swarm optimization. In: Proceedings of International Conference on Neural Networks, Perth, WA, Australia, pp. 1942–1948 (1995)
2. Ghasemi, M., et al.: Phasor particle swarm optimization: a simple and efficient variant of PSO. Soft Comput. **23**, 9701–9718 (2019)
3. Liang, J.J., Qin, A.K., Suganthan, P.N., Baskar S.: Comprehensive learning particle swarm optimizer for global optimization of multimodal functions. IEEE Trans. Evol. Comput. **10**(3), 281–295 (2006)
4. Fontes, D.B.M.M., Mahdi Homayouni, S., Gonçalves, J.F.: A hybrid particle swarm optimization and simulated annealing algorithm for the job shop scheduling problem with transport resources. Eur. J. Oper. Res. **306**(3), 1140–1157 (2023)
5. Pozna, C., Precup, R.-E., Horváth, E., Petriu, E.M.: Hybrid particle filter–particle swarm optimization algorithm and application to fuzzy controlled servo systems. IEEE Trans. Fuzzy Syst. **30**(10), 4286–4297 (2022)
6. Liu, J., Li, F., Kong, X., Huang, P.: Handling many-objective optimisation problems with R2 indicator and decomposition-based particle swarm optimiser. Int. J. Syst. Sci. **50**(2), 320–336 (2018)
7. Raquel, H.G., Coello Coello, C.A.: Improved metaheuristic based on the R2 indicator for many-objective optimization. In: Proceedings of the 2015 Annual Conference on Genetic and Evolutionary Computation, pp. 679–686. Association for Computing Machinery, New York (2015)
8. Li, X., Li, X.-L., Wang, K., Li Y.: A multi-objective particle swarm optimization algorithm based on enhanced selection. IEEE Access **7**, 168091–168103 (2019)
9. Passino, K.M.: Biomimicry of bacterial foraging for distributed optimization and control. IEEE Control Syst. Mag. **22**(3), 52–67 (2002)

Hybrid Algorithm Based on Phasor Particle Swarm Optimization and Firefly Algorithm

Peilin Chen[1], Chenhan Wu[2], Xiaole Liu[2(✉)], and Yongjin Wang[3]

[1] Whittle School and Studios-Shenzhen Campus, Shenzhen 518067, China
[2] College of International Education, Shenzhen 518048, China
2210135009@email.szu.edu.cn
[3] College of Management, Shenzhen University, Shenzhen 518060, China

Abstract. This paper proposes a new hybrid algorithm (HBPPSO) based on the Phasor Particle Swarm Optimization (PPSO) and the Firefly Algorithm (FA) to improve particle velocity and position updates. Initially, a trigonometric updating approach is presented for nonlinearly adjusting particle velocity. The particle position is then controlled using an exponential updating technique. Following that, six groups of experiments are performed to validate the optimization performance of the proposed HFPPSO method. The phasor particle swarm optimization algorithm and particle swarm optimization are chosen as the comparison algorithms. Finally, the experimental findings show that HBPPSO unifies the strengths of managing the learning parameters in both PPSO and FA algorithms while alleviating the weakness of premature convergence, improving searching accuracy and avoiding slipping into local optimum.

Keywords: Hybrid Algorithm · Firefly Algorithm · Phasor Particle Swarm Optimization · Trigonometric Update · Exponential Update

1 Introduction

Particle swarm optimization (PSO), which was inspired by the swarm behavior of bird flocks, was proposed by Kenny and Eberhart [1, 2] originally. Since its inception, numerous academics have a soar interest in applying PSO to solve various complicated optimization problems and proposing several techniques to improve PSO's efficiency in the case of trapping in the local optimum. Parameter modification and hybrid techniques are two common enhancement approaches of PSO.

A large amount of research [3–5] has been done to enhance the optimization performance of PSO based on parameter modification methods. Among them, Phasor Particle Swarm Optimization (PPSO) [8] is one of the most popular methods and is inspired by phasor theory in mathematics. In PPSO, the authors used phase-angle-control methods to achieve the nonlinear velocity updating process, which transformed the PSO algorithm into a self-adaptive (trigonometric), balanced, nonparametric meta-heuristic method and contributed to the faster convergence and higher searching accuracy compared with the basic PSO algorithm.

© The Author(s), under exclusive license to Springer Nature Switzerland AG 2023
Y. Tan et al. (Eds.): ICSI 2023, LNCS 13968, pp. 148–157, 2023.
https://doi.org/10.1007/978-3-031-36622-2_12

Hybrid techniques are also common in the research area of PSO enhancement [6–8]. For instance, the Hybrid Firefly and Particle Swarm Optimization Algorithm (HFPSO) merged the strengths of PSO and the Firefly Algorithm (FA) in the PSO hybrid algorithms [8], including the exponential based position updating method of FA and the concise structure of PSO, improving the convergence speed and the searching accuracy.

Inspired by the above insights, this paper conducts the research work on a hybrid algorithm based on phasor particle swarm optimization and firefly algorithm (HFPPSO), so as to further enhance the optimization performance of PPSO by learning from the position updating approach of the firefly algorithm.

This paper is organized as follows: in Sect. 2, the introduction of the phasor particle swarm optimization and firefly algorithm is given. In Sect. 3, the updating methods of the velocity and position in the proposed hybrid algorithm is demonstrated. In Sect. 4, comparison experiments and related results are given and analyzed to evaluate the performance of HFPPSO. Finally, Sect. 5 concludes the whole paper.

2 Basic Algorithms

2.1 The Phasor Particle Swarm Optimization

The phasor particle swarm optimization is an efficient form of particle swarm optimization that focuses on the phase-angle-based velocity update control mechanism to increase the convergence rate and ultimate solution quality across multiple dimensions. These objectives are attained by picking appropriate and proficient functions for PSO control parameters based on periodic trigonometric functions like and, the values of which change on a regular basis within a defined range.

First, the authors used to adjust the phase-angle in a nonlinear way. The calculation of the phase angle is given in formula (1), and θ_i^{Iter} means the phase angle of particle i in iteration $Iter$.

$$\theta_i^{Iter+1} = \theta_i^{Iter} + \left| \cos\left(\theta_i^{Iter}\right) + \sin\left(\theta_i^{Iter}\right) \right| \times (2\pi) \qquad (1)$$

Then, the update of the velocity and position of each particle will be conducted following formula (1).

2.2 The Firefly Algorithm

The Firefly Algorithm (FA) is a bio-inspired metaheuristic optimization technique that simulates the flashing behavior of fireflies at midnight [10]. Most fireflies emit brief, repetitive flashes that are often unique to a particular species. The purpose of this bioluminescent mechanism, which produces the flashing light, is still debated by researchers. Fireflies use their chemical light to communicate, hunt, and warn their predators.

There are three rules used in the construction of FA. The first is that all fireflies are unisex, meaning that any firefly can be attracted to a brighter one. The second rule regulates brightness of a firefly should be adjusted according to the position it arrives. The final rule is that attraction is directly proportional to brightness but decreases with distance. A firefly will migrate towards the brighter one and move randomly if there is no brighter one in sight.

To calculate the light intensity at a given distance from a light source, the inverse square law is used. This means that there is an inverse relationship between light intensity and distance; light intensity decreases and weakens as distance increases. Since most fireflies can be seen from hundreds of meters away due to these factors, the light intensity at a distance from the light source can be calculated using the formula below,

$$I(r) = \frac{l_s}{r^2} \tag{2}$$

where r means a specific distance and l represents the light source.

Light is absorbed with a constant coefficient of light absorption $\gamma \in [0, \infty)$ in an environment. Therefore, we can form the equation in Gaussian by the formula below,

$$B(r) = B_0 \rceil^{-\gamma r^2}, \tag{3}$$

where $B(r)$ and B_0 are the attractiveness of a firefly at r distance and the attractiveness where $r = 0$, respectively.

Assuming that a, b are two fireflies and their positions are $X_a(x_a, y_a)$ and $X_b(x_b, y_b)$. Distance r_{ab} between the two fireflies is calculated based on Euclidean by the formula below.

$$r_{ab} = \|X_a - X_b\| = \sqrt{(x_a - x_b)^2 - (y_a - y_b)^2} \tag{4}$$

Therefore, the new position of $a(X_a)$ which is the less bright firefly, and its movement toward more attractive one, the firefly b is calculated by the formula below.

$$X_a = X_a + B_0 e^{-\gamma r_{ab}^2}(X_a - X_b) + \alpha \epsilon_a \tag{5}$$

In the equation above, ϵ_a is vector of random variables and the randomization parameter $\alpha \epsilon [0, 1]$.

In summary, the Firefly Algorithm is a novel population-based nature-inspired algorithm that is capable of optimizing objectives by simulating the flashing behavior of fireflies. The algorithm utilizes the inverse square law to calculate light intensity at a distance from a light source, which determines the attractiveness of fireflies towards each other.

3   The Proposed Algorithm

The No Free Lunch Theorem (NFL) [11] states that there is no one-size-fits-all optimization technique and a single optimization algorithm may perform well in certain problems but poorly in others. Phasor Particle Swarm Optimization has the problem of premature convergence: PPSO may converge prematurely to a local optimal solution, leading to the failure of finding the global optimal solution. To deal with this, the Hybrid-Firefly and Phasor Particle Swarm Optimization (HFPPSO) algorithm is designed to address the problem of premature convergence and the limitation of other optimization techniques. HFPPSO utilizes the advantage of the Firefly Algorithm (FA) to prevent the problem of trapping in local optima or premature convergence, which is achieved by incorporating velocity characteristics that prevent the problem of fast or slow velocity. Additionally, HFPPSO incorporates learning from Phasor Particle Swarm Optimization (PPSO), resulting in a more self-adaptive, balanced, and nonparametric optimizing process.

In summary, the HFPPSO algorithm is designed to overcome the optimization limitation by incorporating the strengths of FA and PPSO algorithms, including the trigonometric update of particle velocity and the exponential update of particle position.

3.1   Trigonometric Update of Particle Velocity in HFPPSO

Similar to the update method of velocity in PPSO, in this part, the core idea is the usage of functions of sin and cos in the velocity function, as shown in formulas (6),

$$V_i^{Iter} = p\left(\theta_i^{Iter}\right) \times \left(Pbest_i^{Iter} - X_i^{Iter}\right) + g\left(\theta_i^{Iter}\right) \times \left(Gbest^{Iter} - X_i^{Iter}\right) \quad (6)$$

where $Pbest_i^{Iter}$ and $Gbest^{Iter}$ represent the historical optimal position of particle i and the global optimal position in the whole particle swarm. Besides, θ_i^{Iter} represents the phase angle of particle i in iteration $Iter$. The update of the phasor angle follows formula (7).

$$\theta_i^{Iter+1} = \theta_i^{Iter} + \left|\cos\theta_i^{Iter} + \sin\theta_i^{Iter}\right| \cdot 2\pi \quad (7)$$

Based on the evaluation and test given in PPSO, the subsequent functions of $p\left(\theta_i^{Iter}\right)$ and $g(\theta_i^{Iter})$ have been calculated as formulas (8) and (9):

$$p\left(\theta_i^{Iter}\right) = \left|\cos\theta_i^{Iter}\right|^{2*\sin\theta_i^{Iter}} \quad (8)$$

$$g\left(\theta_i^{Iter}\right) = \left|\sin\theta_i^{Iter}\right|^{2*\cos\theta_i^{Iter}} \quad (9)$$

The variation of the *sin* and *cos* functions, as well as |sin| and |cos| in the range of [0, 2 π], have different values for different phase angles in a certain range, i.e., [-1, 1] and [0, 1], respectively. Using the intermittent nature of functions, HFPPSO is able to present better parameters control ability than PSO.

3.2 Exponential Update of Particle Position in HFPPSO

The update of the particle position in the proposed hybrid algorithm could be divided into two parts. The first part follows the updating way of the basic particle swarm optimization, which is called as the initial position update process, as given in formula (10),

$$X_i^{Iter+1} = X_i^{Iter} + V_i^{Iter} \tag{10}$$

where X_i^{Iter} and V_i^{Iter} represent the position and velocity of particle i in iteration *Iter*. Then, the previous position of particle i is recorded as X_i^{temp}, following formula (11).

$$X_i^{temp} = X_i^{Iter} \tag{11}$$

Then, in the second part, the update of the particle position is more related to the position updating method in the Firefly Algorithm, which focuses on the exponential update approach, as shown in formula (12),

$$X_i^{Iter+1} = B_0 \cdot B(r) \cdot X_i^{Iter} - X_i^{temp} \tag{12}$$

where $B(r)$ and B_0 are the attractiveness of a firefly at r distance and the attractiveness where $r = 0$, respectively, similar to formula (3). The usage of the exponential control factor is multiplied to further slowdown the convergence and thus increase the accuracy of both local and global search.

Finally, the flowchart of HFPPSO is demonstrated in Fig. 1.

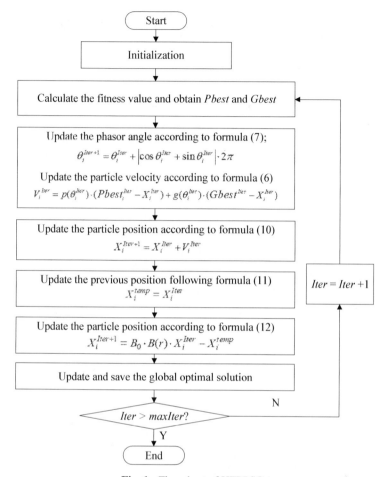

Fig. 1. Flowchart of HFPPSO.

4 Experimental Study

To validate the optimization performance of the proposed HFPPSO, this section compares HFPPSO, PPSO, and PSO on six benchmark functions and analyzes the experimental results.

4.1 Parameter Settings in HFPPSO

HFPSO, PSO algorithms are chosen as the comparison algorithms to test the performance of HFPPSO on Alpine, Ackley, Dixo_Price, Griewank, Rastrigin, and Rosenbrock benchmark functions (shown in Table 1). The experiment is carried out with the same search ranges for all three algorithms. Each of the benchmark functions runs 20 times, and the maximum iteration is set to 1500. The swarm sizes are 40, the acceleration coefficients are 2.0, the inertia weights increase from 0.9 to 0.4, and variable dimensions are 30, as shown in Table 2. The experimental equipment is configured with 11th Gen Intel (R) Core (TM) i5-1135G7 CPU and Windows 10 system.

Table 1. Search ranges for benchmark functions

Functions	Mathematical Formulas	Search Ranges
Alpine	$f(x) = \sum_{i=1}^{n} \|x_i \mathrm{Sin}x_i + 0.1x_i\|$	$[-10, 10]^n$
Ackley	$(x) = $ $-20\mathrm{Exp}\left(-0.2\sqrt{\frac{1}{n}\sum_{i=1}^{n}x_i^2}\right) - \mathrm{Exp}(\frac{1}{n}\sum_{i=1}^{n}cos2\pi x_i) + 20 + e$	$[-32, 32]^n$
Dixo_Price	$f(x) = (x_i - 1)^2 + \sum_{i=2}^{n} i(2x_i^2 - x_i1)^2$	$[-10, 10]^n$
Griewank	$f(x) = \frac{1}{4000}\sum_{i=1}^{n}x_i^2 - \prod_{i=1}^{n}\mathrm{Cos}\left(\frac{xi}{\sqrt{i}}\right) + 1$	$[-600, 600]^n$
Rastrigin	$f(x) \sum_{i=1}^{n}(x_i^2 - 10\mathrm{Cos}2\pi x_i + 10)$	$[-5.12, 5.12]^n$
Rosenbrock	$f(x) = \sum_{i=1}^{n}\left[100(x_{i+1} - x_i^2)^2 + (x_i - 1)^2\right]$	$[-2.048, 2.048]^n$

Table 2. Parameters of the three algorithms

No.	Algorithms	Parameters
1	PSO	Swarm Size = 40, Run for 20 times, acceleration coefficients $c_1 = c_2 = 2$, inertia weight = 0.9 to 0.4, iteration = 1500, Dimension = 30
2	PPSO	Swarm Size = 40, Run for 20 times, acceleration coefficients $c_1 = c_2 = 2$, iteration = 1500, Dimension = 30
3	HFPPSO	Swarm Size = 40, Run for 20 times, acceleration coefficients $c_1 = c_2 = 2$, iteration = 1500, Dimension = 30

4.2 Experimental Result and Analysis

The performance comparison among HFPPSO, PPSO, and PSO algorithms are demonstrated in Table 3, and the bold numbers represent the optimal values for each benchmark function. From Table 3, it could be seen that when dealing with Dixo_Price, Griewank, Rastrigin, and Rosenbrock benchmark functions, HFPPSO can always obtain the optimal maximum value, minimum value, mean value, and standard deviation of the solutions. And when solving Alpine and Ackley functions, HFPPSO ranks the second. Therefore, it can be concluded that HFPPSO performed better than PSO and PPSO distinctively on the majority of the scenarios except for the Alpine and Ackley functions.

Table 3. Performance comparison among HFPPSO, PPSO, and PSO algorithms

Functions		HFPPSO	PPSO	PSO
Alpine	Max	1.16E-03	1.51E-02	**3.83E-05**
	Min	1.50E-04	1.31E-04	**1.13E-07**
	Mean	4.48E-04	3.21E-03	**6.03E-06**
	Std	2.52E-04	4.50E-03	**1.09E-05**
Ackley	Max	6.56E-03	**1.47E-03**	1.50E+00
	Min	5.45E-04	**4.92E-05**	1.25E-06
	Mean	2.27E-03	**5.09E-04**	1.22E-01
	Std	2.27E-03	**3.33E-04**	3.86E-01
Dixo_price	Max	**6.80E-01**	6.67E-01	1.81E+00
	Min	**6.67E-01**	6.67E-01	6.67E-01
	Mean	**6.68E-01**	6.67E-01	7.27E-01
	Std	**2.76E-03**	6.22E-05	2.56E-01
Griewank	Max	**9.80E-04**	1.82E-01	4.66E-02
	Min	**7.27E-06**	9.29E-07	1.12E-10
	Mean	**2.35E-04**	1.02E-02	6.77E-03
	Std	**2.90E-04**	4.06E-02	1.08E-02
Rastrigin	Max	**7.78E-05**	1.99E+00	5.97E+01
	Min	**9.80E-07**	1.57E-07	1.99E+01
	Mean	**1.92E-05**	9.95E-02	3.81E+01
	Std	**2.12E-05**	4.45E-01	9.21E+00
Rosenbrock	Max	**2.82E+01**	2.50E+01	7.97E+01
	Min	**5.55E-09**	3.99E+00	1.74E+01
	Mean	**4.16E+00**	2.29E+01	2.93E+01
	Std	**1.02E+01**	4.49E+00	1.57E+01

Then, the convergence curves of HFPPSO, PPSO, and PSO algorithms on six functions are given in Fig. 2. It can be concluded that HFPPSO overcomes the premature convergence of both PSO and PPSO in solving the functions like Griewank, Rastrigin, and Rosenbrock. Furthermore, the searching accuracy is also enhanced when HFPPSO is applied. Therefore, the proposed HFPPSO dramatically alleviates the speed of convergence, and can avoid being trapped in the local optimal solution effectively.

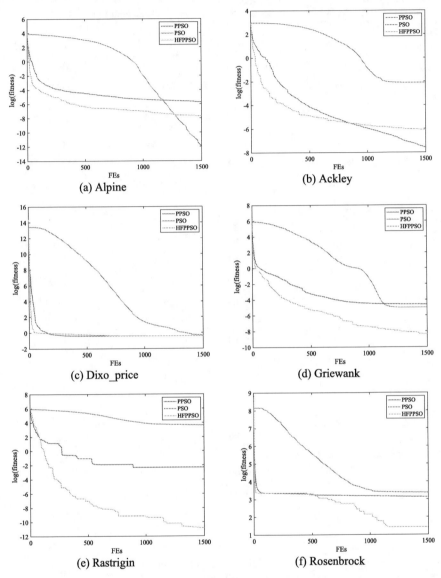

Fig. 2. Convergence curves of HFPPSO, PPSO, and PSO algorithms.

5   Conclusion

This paper is motivated by the velocity and position updating methods in the phasor particle swarm optimization and firefly algorithm respectively, and then formulates a hybrid optimization algorithm HFPPSO combing the strengths of the two algorithms. Then, six groups of experiments are given to test the performance of HFPPSO, PPSO, and PSO. The results show that HFPPSO can solve common benchmark functions effectively with good convergence and search efficiency.

In the future, HFPPSO could be used to solve practical problems as an effective optimization algorithm and more real-world applications can be studied to demonstrate its efficiency. To summarize, HFPPSO is a promising optimization algorithm, and more research could be done to investigate HFPPSO improvement methods.

References

1. Eberchart, R.C., Kennedy, J.: Particle swarm optimization. In: IEEE International Conference on Neural Networks, Perth, Australia (1995)
2. Eberchart, R.C., Kennedy, J.: A new optimizer using particle swarm theory. In: Proceedings of the 6th International Symposium on Micromachine and Human Science, Nagoya, Japan, pp. 39–43 (1995)
3. Clerc, M., Kennedy, J.: The particle swarm: explosion, stability, and convergence in multidimensional complex space. IEEE Trans. Evol. Comput. 6(1), 58–73 (2002)
4. Zhan, Z.H., Zhang, J., Yun, L., Chung, S.H.: Adaptive particle swarm optimization. IEEE Trans. Syst. Man Cybern. B Cybern. 39(6), 1362–1381 (2009)
5. Liang, J.J., Qin, A.K., Suganthan, P.N., Baskar, S.: Comprehensive learning particle swarm optimizer for global optimization of multimodal functions. IEEE Trans. Evol. Comput. 10(3), 281–295 (2006)
6. Fan, W., Luo, L., He, X.S., Yan, W.: Hybrid optimization algorithm of PSO and Cuckoo Search. International Conference on Artificial Intelligence. IEEE (2011)
7. Garg, H.: A hybrid PSO-GA algorithm for constrained optimization problems. Appl. Math. Comput. 274, 292–305 (2016)
8. Aydilek, I.B.: A hybrid firefly and particle swarm optimization algorithm for computationally expensive numerical problems. Appl. Soft Comput. 66, 232–249 (2018)
9. Ghasemi, M., Akbari, E., Rahimnejad, A., Razavi, S.E., Ghavidel, S., Li, L.: Phasor particle swarm optimization: a simple and efficient variant of PSO. Soft. Comput. 23(19), 9701–9718 (2018). https://doi.org/10.1007/s00500-018-3536-8
10. Yang, X.S.: Firefly algorithm, stochastic test functions and design optimisation. Int. J. Bio-Insp. Comput. 2(2), 78–84 (2010)
11. Wolpert, D.H., Macready, W.G.: No free lunch theorems for optimization. IEEE Trans. Evol. Comput. 1(1), 67–82 (1997)

Particle Swarm Optimizer Without Communications Among Particles

JunQi Zhang[1,2], XuRui Huang[1,2], Huan Liu[1,2], and MengChu Zhou[3(✉)]

[1] Department of Computer Science and Technology, Tongji University,
Shanghai, China
{zhangjunqi,liuhuan1912}@tongji.edu.cn, dreamlife@163.com
[2] Key Laboratory of Embedded System and Service Computing,
Ministry of Education, Shanghai, China
[3] Department of Electrical and Computer Engineering,
New Jersey Institute of Technology, Newark, NJ 07102, USA
zhou@njit.edu

Abstract. Particle Swarm Optimizer (PSO) is a kind of population-based evolutionary optimizer. Many PSO variants have been proposed and most of them require mutual communications among particles for their fitness values to find the best position, hence leading to their effective collaboration. However, some real scenes using swarm robots to perform PSO cannot provide reliable communications during their distributed search. To handle such issues, this work proposes a novel PSO variant without communications among particles, called Communication-free Particle Swarm Optimizer (CfPSO). It employs particles' detection ability instead of direct communications among them to accomplish the needed collaboration. Experimental results show that it obtains higher accurate performance than the standard PSO equipped with full communication ability, which is against human intuition.

Keywords: Communication · Communication-free · Particle Swarm Optimizer

1 Introduction

Particle swarm optimizer (PSO) is a well-known evolutionary algorithm for solving optimization problems. It is inspired by bird flocking or fish schooling [9,12] to simulate swarm behaviors of social animals. Due to its simplicity in implementation, PSO has become one of the most popular optimization techniques applied to many fields [10,11,17] since decades ago.

This work was supported by Innovation Program of Shanghai Municipal Education Commission (202101070007E00098), Shanghai Industrial Collaborative Science and Technology Innovation Project (2021-cyxt2-kj10), Shanghai Municipal Science and Technology Major Project (2021SHZDZX0100) and the Fundamental Research Funds for the Central Universities. This work was also supported in part by the National Natural Science Foundation of China (51775385, 61703279, 62073244 and 61876218), and the Shanghai Innovation Action Plan under grant no. 20511100500.

PSO contains a swarm of particles. Each particle in the swarm has a position and velocity in a D-dimensional search space. Shi *et al.* [16] have developed an inertia weight into PSO whose particles are updated as follows:

$$v_i^d = \omega v_i^d + c_1 r_1^d (p_i^d - x_i^d) + c_2 r_2^d (p_g^d - x_i^d), \tag{1}$$

$$x_i^d = x_i^d + v_i^d \tag{2}$$

where ω is an inertia weight of velocity. $d \in \{1, ..., D\}$ is the dth dimension of a search space. The swarm has N particles and $i \in \{1, ..., N\}$. v_i^d is the velocity of particle i in the dth dimension. p_i is the personal best position (*pbest*) of particle i. p_g is the global best position (*gbest*) found by the swarm. c_1 is the cognitive coefficient and c_2 the social coefficient. r_1^d and r_2^d are randomly generated in the dth dimension within $[0, 1]$. x_i^d represents the position of particle i in the dth dimension.

Bratton *et al.* [5] have proposed a standard PSO (SPSO) utilizing a local ring model. It applies the local best position (*lbest*) found by the particle's neighborhoods and a constriction factor χ [7] to the velocity update, i.e.,

$$v_i^d = \chi \left(v_i^d + c_1 r_1^d \left(p_i^d - x_i^d \right) + c_2 r_2^d \left(\tilde{p}_i^{\,d} - x_i^d \right) \right) \tag{3}$$

$$\chi = \frac{2}{\left| 2 - \varphi - \sqrt{\varphi^2 - 4\varphi} \right|}, \varphi = c_1 + c_2. \tag{4}$$

where $\tilde{p}_i$ is the *lbest* of a particle's neighborhoods.

The velocity formula (1) and (3) utilize *gbest* or *lbest* to guide particles. It indicates that particles in a swarm communicate to share their fitness values in order to find out *gbest* or *lbest*. So, the swarm requires a reliable communication to accomplish its search process.

However, relying on communications to share fitness values is practically challenging in some scenes, such as in underwater mobile sensor networks [13,14]. Source seeking is impacted by numerous factors. As PSO relies on communications, it becomes no longer feasible when facing poor or no communication scenes. Thus, a new PSO variant without communications among particles is demanded to handle such scenes. To the authors' best knowledge, this is the first such work.

In order to meet above-mentioned need, this work proposes a Communication-free Particle Swarm Optimizer (CfPSO). Particles just utilize their detection ability to sense other positions and do not need communications to share mutual fitness values. When a particle fails to update *pbest* for a pre-set number of generations, it returns to and waits at its current *pbest*, called a stop-and-wait particle. A stop-and-wait particle randomly flies by following another stop-and-wait one and a swarm center to be defined later on.

The rest of this paper is organized as follows. Section 2 reviews the related work. Section 3 develops the proposed CfPSO in detail. Section 4 carries out experiments and validates CfPSO on CEC2017's benchmark suites by comparing it with SPSO. Section 5 provides conclusions.

2   Related Work

In this section, we introduce the related work that can be divided to PSO variants with communications and source seeking algorithms without communications.

2.1   PSO Variants with Communications

Currently most PSO variants require communications. PSO variants with communications can be classified into PSOs by adjusting control parameters, topology PSOs, hybrid PSOs and multi-swarm PSOs. PSOs by adjusting control parameters try to introduce new parameters or adjust w, c_1 and c_2. The adjustment of parameters aims to improve the search efficiency and convergence speed for PSO. Topology PSOs mostly use the *gbest* or *lbest* model. Different PSO topology structures produce significant impact on PSO performance [4]. The *lbest* model allows each particle to be influenced by a smaller number of adjacent members called a particle's neighbors. Hybrid PSOs hybridize PSO with other search techniques, such as genetic algorithm (GA) [15], ant colony optimization (ACO) [8] and estimation of distribution [6]. Multi-swarm techniques can improve the convergence and diversity of the swarm. Hence, multi-swarm PSOs are suitable to resolve high-dimensional problems.

2.2   Source Seeking Algorithms Without Communications

A few non-PSO based source seeking algorithms without communications are proposed and applied to a 2D/3D search space [1,3]. These studies aim to solve unimodal and single source problems in 2 or 3 dimensions without communications. They does not require to share fitness values among swarm robots. However, these algorithms does not consider high-dimensional or multimodal scenes. Hence, they cannot be directly applied to high-dimensional or multimodal problems without much extension.

3   Communication-free Particle Swarm Optimizer

In this section, a Communication-free Particle Swarm Optimizer (CfPSO) is proposed to support PSO searching in environments without communications among particles. This work solves minimization problems.

3.1   Particle Update Rule

In communication-free environments, particles cannot obtain others' fitness values. They are equipped with detection ability, such as the radar or sonar detection, the visual observation, to sense other particles' positions. After sensing other particles' positions with its proactive detection ability, each particle calculates the swarm center by all particles' positions:

$$\bar{x}^d = \frac{1}{N} \sum_{i=1}^{N} x_i^d \qquad (5)$$

Particles follow their chosen leaders and the swarm center. Each particle randomly initializes another one as its leader. Thus, the velocity formula is:

$$v_i^d = \omega v_i^d + c_1 r_1^d (x_{l_i}^d - x_i^d) + c_2 r_2^d (\bar{x}^d - x_i^d) \tag{6}$$

where l_i is particle i's leader and $x_{l_i}^d$ is the leader's position in the dth dimension. Particles update their positions according to (2).

3.2 Stop-and-Wait Particles

A Stop-and-Wait (S&W) particle is a particle with zero velocity. Each particle maintains a failure count indicating the number of times it fails to update its *pbest*:

$$k_i = \begin{cases} 0 & \text{if } f(X_i) < f(p_i) \\ k_i + 1 & \text{otherwise} \end{cases} \tag{7}$$

where $f(\cdot)$ represents a fitness function for a minimization problem, k_i is the failure count of particle i. If particle i fails to update *pbest* for a pre-set number of generations K, it returns to its current *pbest* and becomes an S&W particle.

An S&W particle constructs an S&W set containing all other S&W ones. When the population size of the S&W set reaches a pre-set population size M, an S&W particle randomly chooses another S&W one as its leader. At the next generation, it follows its chosen leader and the swarm center:

$$i \in P \text{ if } V_i = \mathbf{0}_D \tag{8}$$

where $V_i = \mathbf{0}_D$ represents an S&W particle i with zero velocity. P represents an S&W set containing all S&W particles. The choice of an S&W particle i's leader is defined as:

$$l_i = \text{randi}(P - \{i\}) \text{ if } |P| \geq M \text{ and } i \in P \tag{9}$$

where $|P|$ is the population size of P, randi$(\cdot)$ means a function to randomly choose another S&W particle for an S&W particle i.

3.3 Overall Framework

In CfPSO, as in Algorithm 1, particles are guided by their chosen leaders and the swarm center. Then, particles become S&W particles when they fail to update their *pbest* for a pre-set number of generations. At last all particles converge to an optimal solution.

4 Experimental Settings

4.1 Environment Description

In this work, CEC2017 benchmark suites [2] are used to validate the performance of CfPSO. CEC2017 defines 30 benchmark functions. However, f_2 is deprecated as it is unstable for high dimensions [2]. Hence, the experiments are carried out on its 29 benchmark functions. All functions set the search space of each dimension as $[-100, 100]$. In the experiments, it sets the swarm size as $N = 50$ and dimensions as $D=50$. The maximum number of function evaluations is set to $10000 \times D$. The peer algorithm is SPSO [5] which requires communications. The parameters of SPSO and CfPSO are given in Table 1. Each algorithm is executed with 51 times on each function and the mean results are used for comparisons.

4.2 Parameters Analysis

Two parameters K and M need to be tuned. An orthogonal experiment is carried out by setting the parameter range as $K \in \{1, 2, 3, 4, 5\}$ and $M \in \{0.1N, 0.2N, ..., 0.9N, N\}$. This work uses the average deviation of fitness (ADF) to verify the parameter experiment's results:

$$ADF = \frac{1}{\Xi} \cdot \sum_{\xi=1}^{\Xi} \hat{f}(\xi) \tag{10}$$

where $\Xi = 29$ is the number of CEC2017 functions. $\hat{f}(\xi)$ is the normalized value of the fitness accuracy deviation with 51 repeated times on each function.

The parameter experiment's results are listed in Table 2. Figure 1 displays the ADF trend. As shown in Figs. 1(a–b), the better ADF is obtained when $M \in \{0.5N, N\}$ and $K \in \{2, 3\}$.

Algorithm 1. CfPSO

1: Randomly initialize a swarm with N particles;
2: Randomly initialize each particle's leader as another one;
3: **while** $t < T$ **do**
4:     Calculate the swarm center according to (5);
5:     **for** $i = 1 \rightarrow N$ **do**
6:         **if** Has chosen a leader **then**
7:             Update the velocity according to (6);
8:             Update the position according to (2);
9:             Evaluate a fitness $f(X_i)$;
10:            Update the failure count k_i according to (7);
11:            Update the *pbest* of p_i;
12:            **if** $k_i \geq K$ **then**
13:                $X_i = p_i$;
14:                Become an S&W particle;
15:            **end if**
16:        **end if**
17:    **end for**
18:    **for** $i = 1 \rightarrow N$ **do**
19:        **if** Is an S&W particle **then**
20:            Construct an S&W set according to (8);
21:            An S&W particle i chooses its leader according to (9);
22:        **end if**
23:    **end for**
24:    $t = t + 1$;
25: **end while**
26: **return** the best particle's position and fitness.

Table 1. Parameters in PSO Variants

Algorithm	Parameter Setting
SPSO	$\chi = 0.72984$, $c_1 = 2.05$, $c_2 = 2.05$
CfPSO	$w = 0.72984$, $c_1 = 1.496172$, $c_2 = 1.496172$

Table 2. ADF Experiments on Parameters K and M

ADF \ M K	5	10	15	20	25	30	35	40	45	50
1	0.3190	0.3178	0.3138	0.3136	0.3155	0.3098	0.3151	**0.3049**	0.3118	0.3213
2	0.2781	0.2671	0.2542	0.2645	**0.2518**	0.2683	0.2550	0.2576	0.2520	0.2726
3	0.3490	0.3222	0.2949	0.2817	0.2929	0.2947	0.2826	**0.2778**	0.2964	0.2946
4	0.4672	0.4084	0.3826	0.3259	**0.3157**	0.3172	0.3389	0.3264	0.3168	0.3306
5	0.9701	0.7653	0.6040	0.3957	0.3505	0.3404	**0.3370**	0.3490	0.3377	0.3475

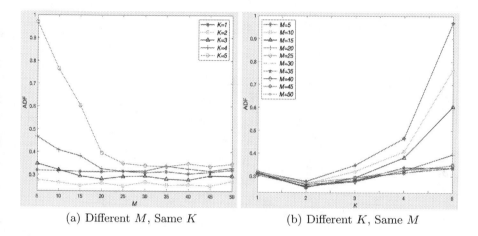

(a) Different M, Same K (b) Different K, Same M

Fig. 1. The parameter experiment for K and M on CEC2017 benchmark functions.

From the above parameter experiment, this work finally sets $K = 2$ and $M = 0.5\,N$ to run the comparison experiments.

4.3 Experimental Results

We compare CfPSO with SPSO. The results of their accuracy are listed in Table 3. In each function, the boldface indicates the better result; "Mean" denotes the mean results and "Std." denotes the standard deviations of 51 independent runs; "A.R." means the average ranking value, calculated by the sum of an algorithm's ranking values on all functions divided by 29. So, an algorithm with a less "A.R." is a better one.

With respect to the accuracy, CfPSO is significantly better than SPSO on 24 out of 29 functions but worse on 5 functions only. Overall, CfPSO's A.R. is 1.17, which is far better than SPSO.

The representative function f_1, f_9, f_{11}, f_{16}, f_{22} and f_{26} are illustrated to show the two algorithm's convergence speed. As shown in Figs. 2(a–f), CfPSO demonstrates its superior convergence accuracy and faster convergence speed than SPSO. The experiment result can be explained theoretically. During the early search stage, an appropriate M improves particles' global search ability as it increases the diversity of particles' leaders. Hence, particles are easier to escape

Table 3. PSO Accuracy Comparisons with The Peer Algorithm

	Func	SPSO	CfPSO	Func	SPSO	CfPSO	Func	SPSO	CfPSO
Mean	f_1	2.57E+03	**1.68E+03**	f_3	5.64E+04	**2.58E+03**	f_4	5.24E+02	**4.85E+02**
Std.		3.34E+03	**2.03E+03**		1.14E+04	**7.04E+02**		4.38E+01	**5.03E+01**
Rank		2	1		2	1		2	1
Mean	f_5	6.86E+02	**5.97E+02**	f_6	6.13E+02	**6.00E+02**	f_7	1.04E+03	1.08E+03
Std.		2.69E+01	**6.34E+01**		3.82E+00	**4.16E−02**		4.40E+01	1.22E+01
Rank		2	1		2	1		1	2
Mean	f_8	9.89E+02	**8.97E+02**	f_9	2.79E+03	**9.00E+02**	f_{10}	6.77E+03	**4.82E+03**
Std.		2.52E+01	**6.32E+01**		6.30E+02	**5.95E−01**		6.11E+02	**1.05E+03**
Rank		2	1		2	1		2	1
Mean	f_{11}	1.32E+03	**1.21E+03**	f_{12}	1.51E+06	**2.62E+05**	f_{13}	5.66E+03	**2.70E+03**
Std.		5.30E+01	**2.20E+01**		9.71E+05	**9.62E+04**		2.41E+03	**1.97E+03**
Rank		2	1		2	1		2	1
Mean	f_{14}	4.53E+04	**1.58E+04**	f_{15}	**3.60E+03**	6.37E+03	f_{16}	2.96E+03	**2.38E+03**
Std		3.80E+04	**7.53E+03**		**2.50E+03**	3.46E+03		2.63E+02	**2.83E+02**
Rank		2	1		1	2		2	1
Mean	f_{17}	2.90E+03	**2.42E+03**	f_{18}	**3.84E+05**	4.82E+05	f_{19}	**3.81E+03**	1.64E+04
Std.		2.37E+02	**2.33E+02**		**2.18E+05**	1.86E+05		**2.99E+03**	4.30E+03
Rank		2	1		1	2		1	2
Mean	f_{20}	2.83E+03	**2.31E+03**	f_{21}	2.47E+03	**2.38E+03**	f_{22}	8.20E+03	**3.57E+03**
Std.		2.04E+02	**1.54E+02**		1.80E+01	**4.31E+01**		1.38E+03	**2.21E+03**
Rank		2	1		2	1		2	1
Mean	f_{23}	2.91E+03	**2.80E+03**	f_{24}	3.03E+03	**2.96E+03**	f_{25}	**3.03E+03**	3.05E+03
Std.		2.68E+01	**2.30E+01**		2.48E+01	**2.19E+01**		**1.99E+01**	2.96E+01
Rank		2	1		2	1		1	2
Mean	f_{26}	5.70E+03	**2.90E+03**	f_{27}	3.39E+03	**3.31E+03**	f_{28}	3.30E+03	**3.29E+03**
Std.		2.88E+02	**4.17E−13**		6.48E+01	**5.00E+01**		1.66E+01	**2.24E+01**
Rank		2	1		2	1		2	1
Mean	f_{29}	4.19E+03	**3.75E+03**	f_{30}	1.04E+06	**8.18E+05**			
Std.		2.42E+02	**1.97E+02**		2.23E+05	**4.86E+04**			
Rank		2	1		2	1			
A.R.	Overall	1.83	1.17						

the trap of local optima and converge to a better optimum with faster speed. During the late search stage, a smaller K improves particles' local search ability. Hence, particles can achieve higher accuracy. With a suitable combination of K and M, CfPSO outperforms SPSO by a good balance between the global and local search.

5   Conclusion

A novel PSO variant without communications called Communication-free PSO (CfPSO) is proposed in this work. It has two distinct features. First, CfPSO is suitable to search for a target/solution in environments without communications among particles. Second, CfPSO can be applied to high-dimensional

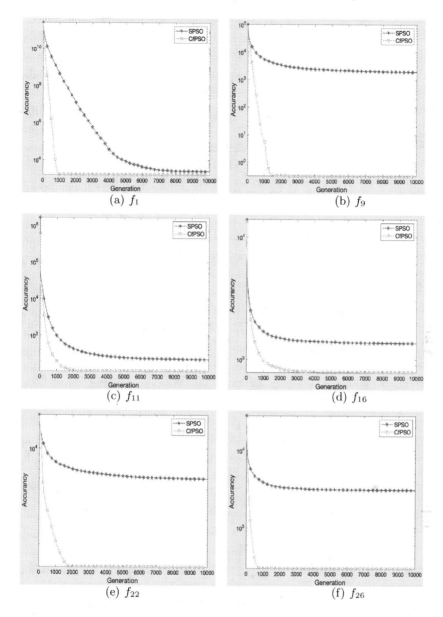

Fig. 2. Convergence curves of SPSO and CfPSO on CEC2017 benchmark functions.

and multimodal problems. Compared with the peer competitor of SPSO with communications, CfPSO shows its higher convergence accuracy and faster convergence speed on CEC2017 benchmark functions Our next work is try to apply the proposed ideas to other evolutionary algorithms, e.g., [18–34].

References

1. Al-Abri, S., Zhang, F.: A distributed active perception strategy for source seeking and level curve tracking. IEEE Trans. Autom. Control **67**(5), 2459–2465 (2022)
2. Awad, N., Ali, M., Liang, J., Qu, B., Suganthan, P.: Problem definitions and evaluation criteria for the CEC 2017 special session and competition on single objective real-parameter numerical optimization. Nanyang Technological University, Singapore and Computational Intelligence Laboratory, Zhengzhou University, Zhengzhou, China, Technical report 10 (2017)
3. Azuma, S.I., Sakar, M.S., Pappas, G.J.: Stochastic source seeking by mobile robots. IEEE Trans. Autom. Control **57**(9), 2308–2321 (2012)
4. Blackwell, T., Kennedy, J.: Impact of communication topology in particle swarm optimization. IEEE Trans. Evol. Comput. **23**(4), 689–702 (2019)
5. Bratton, D., Kennedy, J.: Defining a standard for particle swarm optimization. In: IEEE Swarm Intelligence Symposium, pp. 120–127 (2007)
6. Ceberio, J., Irurozki, E., Mendiburu, A., Lozano, J.A.: A review on estimation of distribution algorithms in permutation-based combinatorial optimization problems. Progr. Artif. Intell. **1**(1), 103–117 (2012)
7. Clerc, M., Kennedy, J.: The particle swarm-explosion, stability, and convergence in a multidimensional complex space. IEEE Trans. Evol. Comput. **6**(1), 58–73 (2002)
8. Colorni, A., Dorigo, M., Maniezzo, V., et al.: Distributed optimization by ant colonies. In: Proceedings of the First European Conference on Artificial Life, vol. 142, pp. 134–142 (1991)
9. Eberhart, R.C., Kennedy, J.: A new optimizer using particle swarm theory. In: Proceedings of the Sixth International Symposium on Micro Machine and Human Science, vol. 1, pp. 39–43 (1995)
10. Ho, S.Y., Lin, H.S., Liauh, W.H., Ho, S.J.: OPSO: orthogonal particle swarm optimization and its application to task assignment problems. IEEE Trans. Syst. Man Cybern. Part A Syst. Hum. **38**(2), 288–298 (2008)
11. Keles, C., Alagoz, B.B., Kaygusuz, A.: Multi-source energy mixing for renewable energy microgrids by particle swarm optimization. In: International Artificial Intelligence and Data Processing Symposium, pp. 1–5 (2017)
12. Kennedy, J., Eberhart, R., et al.: Particle swarm optimization. In: Proceedings of IEEE International Conference on Neural Networks, vol. 4, pp. 1942–1948 (1995)
13. Khalil, R.A., Saeed, N., Babar, M.I., Jan, T., Din, S.: Bayesian multidimensional scaling for location awareness in hybrid-internet of underwater things. IEEE/CAA J. Autom. Sinica **9**(3), 496–509 (2022)
14. Kinsey, J.C.R., Eustice, M., Whitcomb, L.: A survey of underwater vehicle navigation: Recent advances and new challenges. In: Proceedings of the IFAC Conference on Manoeuvering Control Marine Craft, vol. 88, pp. 1–12 (2006)
15. Robinson, J., Sinton, S., Rahmat-Samii, Y.: Particle swarm, genetic algorithm, and their hybrids: optimization of a profiled corrugated horn antenna. In: IEEE Antennas and Propagation Society International Symposium, vol. 1, pp. 314–317 (2002)
16. Shi, Y., Eberhart, R.: A modified particle swarm optimizer. In: Proceedings of IEEE World Congress on Computational Intelligence, pp. 69–73 (1998)
17. Wachowiak, M.P., Smolíková, R., Zheng, Y., Zurada, J.M., Elmaghraby, A.S.: An approach to multimodal biomedical image registration utilizing particle swarm optimization. IEEE Trans. Evol. Comput. **8**(3), 289–301 (2004)

18. Wang, J., Liu, C., Zhou, M.: Improved bacterial foraging algorithm for cell formation and product scheduling considering learning and forgetting factors in cellular manufacturing systems. IEEE Syst. J. **14**(2), 3047–3056 (2020)
19. Wang, X., Xing, K., Yan, C., Zhou, M.: A novel MOEA/D for multi-objective scheduling of flexible manufacturing systems. Complexity **2019** Article ID 5734149, 14 p (2019) https://doi.org/10.1155/2019/5734149
20. Wang, Y., Gao, S., Zhou, M., Yu, Y.: A multi-layered gravitational search algorithm for function optimization and real-world problems. IEEE/CAA J. Autom. Sin. **8**(1), 94–109 (2021)
21. Wang, Z., Gao, S., Zhou, M., Sato, S., Cheng, J., Wang, J.: Information-theory-based nondominated sorting ant colony optimization for multiobjective feature selection in classification, IEEE Trans. Cybern. (2022). https://doi.org/10.1109/TCYB.2022.3185554
22. Wu, Q., Zhou, M., Zhu, Q., Xia, Y., Wen, J.: MOELS: multiobjective evolutionary list scheduling for cloud workflows. IEEE Trans. Autom. Sci. Eng. **17**(1), 166–176 (2020)
23. Yu, Y., et al.: Scale-free network-based differential evolution to solve function optimization and parameter estimation of photovoltaic models. Swarm Evol. Comput. **74**, 101142 (2022). https://doi.org/10.1016/j.swevo.2022.101142
24. Zhang, J., Lu, Y., Che, L., Zhou, M.: Moving-distance-minimized PSO for mobile robot swarm, IEEE Trans. Cybern. **52**(9), 9871–9881 (2022)
25. Zhang, J., et al:. PSO-based sparse source location in large-scale environments with a uav swarm, IEEE Trans. Intell. Transp. Syst. **24**(5), 5249–5258 (2023)
26. Zhang, J., Zhu, X., Wang, Y., Zhou, M.: Dual-environmental particle swarm optimizer in noisy and noise-free environments. IEEE Trans. Cybern. **49**(6), 2011–2021 (2019)
27. Zhang, P., Zhou, M., Wang, X.: An intelligent optimization method for optimal virtual machine allocation in cloud data centers, IEEE Trans. Autom. Sci. Eng. **17**(4), 1725–1735 (2020)
28. Zhao, J., Liu, S.X., Zhou, M., Guo, X.W., Qi, L.: Modified cuckoo search algorithm to solve economic power dispatch optimization problems. IEEE/CAA J. Autom. Sin. **5**(4), 794–806 (2018)
29. Zhao, Z., Zhou M., Liu, S.: Iterated greedy algorithms for flow-shop scheduling problems: a tutorial. IEEE Trans. Autom. Sci. Eng. **19**(1) 251–261 (2022)
30. Zhou, J., Wu, Q., Zhou, M., Wen, J., Al-Turki, Y., Abusorrah, A.: LAGAM: a length-adaptive genetic algorithm with markov blanket for high-dimensional feature selection in classification, IEEE Trans. Cybern. (2022) https://doi.org/10.1109/TCYB.2022.3163577
31. Zhou, Y., Xu, W., Fu, Z.-H., Zhou, M.: Multi-neighborhood simulated annealing-based iterated local search for colored traveling salesman problems. IEEE Trans. on Intell. Transp. Syst. **23**(9), 16072–16082 (2022)
32. Zhu, H., Liu, G., Zhou, M., Xie, Y., Kang, Q.: Dandelion algorithm with probability-based mutation. IEEE Access **7**, 97974–97985 (2019)
33. Zhu, X., Li, J., Zhou, M.: Target coverage-oriented deployment of rechargeable directional sensor networks with a mobile charger. IEEE Internet Things J. **6**(3), 5196–5208 (2019)
34. Zuo, X., et al.: Optimizing hospital emergency department layout via multiobjective tabu search. IEEE Trans. Autom. Sci. Eng. **16**(3), 1137–1147 (2019)

An Atomic Retrospective Learning Bare Bone Particle Swarm Optimization

Guoyuan Zhou[1] , Jia Guo[1,2(✉)] , Ke Yan[3], Guoao Zhou[4], and Bowen Li[5]

[1] School of Information Engineering, Hubei University of Economics, Wuhan 430205, China
`zhouguoyuan@email.hbue.edu.cn, guojia@hbue.edu.cn`
[2] Hubei Internet Finance Information Engineering Technology Research Center, Wuhan 430205, China
[3] China Construction Third Engineering Bureau Installation Engineering Co., Ltd., Wuhan 430075, China
[4] School of Automation, Wuhan University of Technology, Wuhan 430070, China
[5] School of International Education, Wuhan University of Technology, Wuhan 430070, China

Abstract. In order to increase the diversity of bare-bone particle swarm optimization (BBPSO) population search range, enhance the ability to jump out of local optimum, we propose an atomic retrospective learning bare-bone particle swarm optimization (ARBBPSO) algorithm based on BBPSO. Different from the renewal strategy of BBPSO, inspired by electron motion around protons in ARBBPSO, we use a strategy of motion around nuclei to increase population diversity. At the same time, the retrospective learning strategy is used to allow the proton particles to have a chance to correct errors during the process of updating, thus allowing the population to have a chance to evade falling into a local optimum. To verify the performance of the proposed algorithm, 29 benchmark functions of CEC2017 are chosen to compare with four well-known BBPSO-based algorithms. The experimental results indicate that ARBBPSO is superior to several other algorithms for improving BBPSO from a comprehensive consideration.

Keywords: Retrospective learning · PSO · Evolutionary computation

1 Introduction

In recent years, the challenge of optimization problems has grown rapidly with the growing of artificial intelligence. More and more scholars are devoting themselves to the research of optimization algorithms and applying them to the fields of robot [1,2], energy [3] and medicine [4,5].

The particle swarm optimization [6](PSO) algorithm was proposed by Kennedy in 1995 inspired by the social behavior of birds for solving optimization problems. In PSO, the velocity and orbit of a particle is associated with both the global optimal position and the individual optimal position. However, the single evolutionary approach and simple formula of PSO lead to limited population diversity and difficulty in breaking through the local optimum.

In addressing the insufficiencies in PSO, many researchers proposed a wide range of improvements to be used in PSO. Beheshti [7] developed a novel x-shaped binary particle swarm optimization (XBPSO) for solving high-dimensional optimization problems. In XBPSO, the author proposed a new x-type transfer function for increasing the exploration and utilization of the binary search space. And the proposed x-shaped transfer function applies to all PSO algorithms to eliminate the disadvantage of existing transfer functions. Lu [8] employed delayed activation strategy, repulsive mechanism and multiple groups approach to improve PSO and hence prevent premature convergence. Zhang [9] proposed objective-constraint mutual-guided surrogate-based PSO. Zeng [10] et al. introduced dynamic-neighborhood strategy and switching strategy in traditional PSO.

The above mentioned PSO-based algorithms are commonly parameterized. Facing a new optimization problem, researchers need to adjust the parameters of PSO to guarantee the optimal solution of the problem. To address the issue, Kennedy [11] proposed bare-bone particle swarm optimization (BBPSO) in 2003. BBPSO eliminates the velocity concept from PSO and does not use inertial weight and acceleration coefficients. Relying on a random but guided search, PSO is vastly simplified by this methodology. To improve the efficiency of the BBPSO, Zhao [12] combines the characteristics of PSO and BBPSO to propose a new modified bare-bone particle swarm optimization. PSO suffers from the problems of premature convergence and loss of diversity. BBPSO is updated in a stochastic process. The choice of two update strategies is controlled by a stochastic parameter, which increases the search range and improves the possibility of falling into a local optimum. To enhance the ability of BBPSO to solve high-dimensional optimization problems and to jump out of local optima, Tian [13] proposed a kind of a bare-bone particle swarm optimization based on electronic transition (ETBBPSO). In ETBBPSO, the particles are separately partitioned into different orbits. The orbit merging operator is utilised to combine the orbits which have a low search ability to the orbits that have a high search ability in each iteration. To improve the capability of BBPSO, Guo [14] incorporates fission and fusion strategies into the BBPSO. The fission strategy is employed to enhance the searching range by distributing particles into different local groups. A fusion strategy is used to fuse several different local groups to a single group to speed up convergence.

However, these improved BBPSO algorithms suffer from slow global convergence and tend to miss population diversity. To address these shortcomings we propose an atomic retrospective learning bare bone particle swarm optimization (ARBBPSO) inspired by atomic motion around the nucleus and backtracking strategies. The rest of this paper is as follows: Sect. 2 gives a description of the proposed algorithm; Section 3 provides experiments to demonstrate the performance of the proposed algorithm; Section 4 gives the conclusion of the paper.

2   An Atomic Retrospective Learning Bare Bone Particle Swarm Optimization

In order to solve the shortage of BBPSO, improve the diversity of the search of the population, and avoid premature convergence into local optimum, we propose an atomic backtracking learning bare-bone particle swarm optimization (ARBBPSO) algorithm. In ARBBPSO we primarily employ an electron motion around the nucleus strategy and a retrospective learning strategy.

2.1   Motion Around the Nucleus Strategy

One reason why BBPSO losing population diversity is that the motion of each particle is only related to the optimal position of the period itself and an identical global optimal position. To reduce the impact of this drawback on the optimisation algorithm, inspired by the motion around the nucleus in the hydrogen atom. A negatively charged electron inside the atom is attracted to a positively charged proton and constantly works its way around the nucleus. The strategy of motion around the nucleus provides an additional way of updating the position of the particles, getting rid of the drawback of having only one updating strategy for the particles in PSO and BBPSO. This helps the particles to increase the diversity of the population search range to a certain extent.

Prior to any iteration, two particles from the population which have not been selected above are chosen at random and ranked based on their fitness. Particles with high fitness we call proton (PT) particles and those with relatively lower fitness we call electron (ET) particles. In ARBBPSO, the PT particles have a better fitness, therefore any ET particles move around their PT particles. In this way it is possible for the ET particles to search for a better adapted position. The equation for updating the position of an ET particle is displayed in the Eq. 1.

$$v = (PT^t(i) + ET^t(i))/2$$
$$\omega = |PT^t(i) - ET^t(i)| \tag{1}$$
$$ET^{t+1}(i) = Best(ET^t(i), GD(v, \omega))$$

where $PT^t(i)$ is the optimal position of $(i)th$ proton in the $t - th$ generation, $ET^t(i)$ denotes the optimal position of the $t - th$ iteration of the $(i)th$ electron. $GD()$ is a Gaussian distribution with v as mean and ω as standard division. $Best()$ is a function which selects the optimal of two inputs as an output.

2.2   Retrospective Learning Strategy

There is an old Chinese saying that there is no turning back from the opening bow. In BBPSO, the particles are renewed in the same irreversible way as this ancient saying. This leads to the fact that the particles have almost no chance to correct their mistakes after searching for the wrong position. To cope with the problem that particles in BBPSO converge too early and fall into local optimum, we propose a retrospective learning strategy for proton particles. In this retrospective strategy, when a proton particle makes an error, it returns to the previous step thus avoiding the error on top of the error.

At each iteration, each proton particle learns not only from the current global best position, but also from the global best position of the previous iteration, and the two learnings generate two candidate positions to choose the better one for updating. It provides the possibility to jump out of the local optimum when the population traps in the local optimum. The equation for candidate position generation is shown in Eq. 2.

$$\alpha = (PT^t(i) + Gbest(\eta))/2$$
$$\delta = |PT^t(i) - Gbest(\eta)| \tag{2}$$
$$PT^{t+1}_{candidate}(i, \eta) = GD(\alpha, \delta)$$

where $PT^t_{best}(i)$ is the position of $(i)th$ proton particle in $t-th$ generation,$Gbest(\eta)$ indicates the position of the global best particle, when $\eta = 1$ it is the global best position of the current generation, when $\eta = 2$ it indicates the global best position of particles of the previous generation. $PT^{t+1}_{candidate}$ is the nth candidate position of the $(i)th$ particle in the $t+1-th$ generation. $GD(\alpha, \delta)$ is a Gaussian distribution with α as mean and δ as standard division.

In ARBBPSO, PT particles learn the global best particle for searching and ET particles follow the proton particles for searching. ET particles also learn indirectly from the global best particles, and the multiple learning methods help ARBBPSO to further increase the diversity of population search.

2.3   Pseudo Code of ARBBPSO

To show the workflow of ARPBBPSO, the pseudo-code is in Algorithm 1.

3   Results

3.1   Experimental Methods

To demonstrate the improved performance of the proposed algorithms, four well-known BBPSO-based improvement algorithms PBBPSO [15], DLSBBPS [16], ETBBPSO [13], TBBPSO [17], are used to serve as the control group. To ensure uniformity of comparison, we employ the 29 benchmark functions in CEC2017 [18], as the test functions. All algorithms had a population size of 100, a dimension of 100 and a maximum number of iterations of 1.00E+4 a runtime of 37.

3.2   Experimental Results

To present the experimental results as optimally as possible for analysis, we employ AT to estimate the performance of each algorithm. the definition of AT is $|actualoptimum - theoreticaloptimum|$.

Algorithm 1 ARBBPSO

Require: Max iteration time, T
Require: Fitness function, f
Require: Particles position, $X = x_1, x_2, ..., x_n$
Require: Personal best position, $Pbest\_postion$
Require: Personal best value, $Pbest\_value$
Require: Global best position, $Gbest\_postion$
Require: Global best value, $Gbest\_value$
Require: Candidate position, $Candidate\_posi = c_1(1), c_1(2), c_2(1), c_2(2)...c_n(1), c_n(2)$
Require: Candidate value, $Candidate\_value$
  1: Randomly generate the initial position of X
  2: $t = 1$
  3: **while** $t < T$ **do**
  4:     **while** $x \neq \emptyset$ **do**
  5:         Select two particles,x_i and x_j from X
  6:         **if** $f(x_j) < f(x_i)$ **then**
  7:             Update x_i with Eq. (3)
  8:             Generate $c_j(w)$ by Eq. (4)
  9:             $x_j = best(c_j(1), c_j(2))$
 10:         **else**
 11:             Update x_j with Eq. (3)
 12:             Generate $c_i(w)$ by Eq. (4)
 13:             $x_i = best(c_i(1), c_i(2))$
 14:         **end if**
 15:         $t = t + 1$
 16:         Update $Gbest\_position, Gbest\_value, Pbest\_postion, Pbest\_value$
 17:     **end while**
 18: **end while**

To provide an intuitive representation of ARBBPSO's performance, the AT results of the five algorithms with the 29 benchmark functions of the CEC2017 are also shown in the Fig. 1, 2 and 3. The horizontal indicate the number of evolutions and the vertical denote the value of AT.

The specification of the results including the Mean and standard deviation (Std) are shown in Tables 1 and 2. In particular, an undifferentiated statistic is applied to all algorithms by the Friedman test. The rank of all algorithms in a certain benchmark function signifies the score obtained in a certain benchmark function. Finally, the rank scores are obtained by averaging the total rank scores of all algorithms. The algorithm ranks in a certain benchmark function are shown in the Table 1 and Table 2. The average ranks of all algorithms are on the bottom of Table 2.

After five algorithms are compared and measured, ARBBPSO ranks first in 17 of the 29 CEC2017 benchmarking functions, second in four, third in two, fourth in none and fifth in six. According to Friedman test statistics, the average rank of ARBBPSO is 2.1034 which is 26.509% better than the second ranked algorithm, ETBBPSO.

Table 1. Experimental Results, of ARBBPSO, DLSBBPSO, TBBPSO, BBPSO, ETBBPSO and PBBPSO for f_1–f_{16}.

Function Number	Data Type	ARBBPSO	DLSBBPSO	TBBPSO	ETBBPSO	PBBPSO
f_1	Mean	1.510E+04	1.914E+04	2.523E+04	2.370E+04	1.886E+04
	Std	2.570E+04	2.188E+04	2.296E+04	3.312E+04	2.165E+04
	Rank	1	3	5	4	2
f_2	Mean	6.339E+126	4.236E+118	8.169E+136	1.091E+121	5.857E+128
	Std	3.856E+127	2.567E+119	4.969E+137	6.579E+121	3.562E+129
	Rank	3	1	5	2	4
f_3	Mean	8.232E+05	3.883E+06	2.023E+06	4.098E+06	4.345E+06
	Std	4.023E+05	2.406E+06	8.864E+05	3.393E+06	3.079E+06
	Rank	1	3	2	4	5
f_4	Mean	1.719E+02	1.917E+02	1.764E+02	1.618E+02	1.630E+02
	Std	4.631E+01	5.500E+01	5.236E+01	5.148E+01	5.200E+01
	Rank	3	5	4	1	2
f_5	Mean	7.740E+02	8.250E+02	9.262E+02	9.228E+02	9.528E+02
	Std	1.459E+02	1.821E+02	1.596E+02	1.507E+02	1.412E+02
	Rank	1	2	4	3	5
f_6	Mean	2.862E+01	3.914E+01	3.568E+01	3.847E+01	3.800E+01
	Std	5.970E+00	9.147E+00	6.785E+00	8.184E+00	8.021E+00
	Rank	1	5	2	4	3
f_7	Mean	9.788E+02	8.491E+02	9.577E+02	8.849E+02	9.298E+02
	Std	1.353E+02	1.504E+02	1.659E+02	1.333E+02	1.518E+02
	Rank	5	1	4	2	3
f_8	Mean	8.022E+02	8.047E+02	9.257E+02	8.882E+02	9.043E+02
	Std	1.211E+02	1.610E+02	1.445E+02	1.562E+02	1.610E+02
	Rank	1	2	5	3	4
f_9	Mean	2.423E+04	2.741E+04	3.587E+04	3.121E+04	3.695E+04
	Std	8.315E+03	1.166E+04	8.435E+03	8.049E+03	1.515E+04
	Rank	1	2	4	3	5
f_{10}	Mean	2.669E+04	2.999E+04	2.541E+04	2.669E+04	3.142E+04
	Std	3.480E+03	5.856E+03	5.881E+03	7.885E+03	4.329E+03
	Rank	2	4	1	3	5
f_{11}	Mean	6.370E+02	6.658E+03	3.369E+03	7.839E+03	6.120E+03
	Std	2.388E+02	1.036E+04	2.988E+03	1.837E+04	6.103E+03
	Rank	1	4	2	5	3
f_{12}	Mean	2.045E+07	5.553E+07	4.725E+07	4.365E+07	4.645E+07
	Std	1.196E+07	2.932E+07	2.384E+07	2.169E+07	2.094E+07
	Rank	1	5	4	2	3
f_{13}	Mean	7.431E+03	1.165E+04	1.185E+04	1.304E+04	1.491E+04
	Std	8.781E+03	1.877E+04	1.426E+04	1.887E+04	2.138E+04
	Rank	1	2	3	4	5
f_{14}	Mean	4.262E+05	1.471E+06	1.103E+06	1.050E+06	1.246E+06
	Std	3.000E+05	1.172E+06	4.947E+05	6.805E+05	8.042E+05
	Rank	1	5	3	2	4
f_{15}	Mean	6.028E+03	1.180E+04	7.798E+03	1.393E+04	5.965E+03
	Std	8.467E+03	1.083E+04	1.075E+04	1.707E+04	8.146E+03
	Rank	2	4	3	5	1

Table 2. Experimental Results, of ARBBPSO, DLSBBPSO, TBBPSO, BBPSO, ETBBPSO and PBBPSO for f_{16}–f_{29}.

Function Number	Data Type	ARBBPSO	DLSBBPSO	TBBPSO	ETBBPSO	PBBPSO
16	Mean	5.190E+03	9.216E+03	5.901E+03	6.214E+03	9.146E+03
	Std	1.592E+03	2.680E+03	1.978E+03	2.520E+03	2.678E+03
	Rank	1	6	2	3	5
17	Mean	4.713E+03	5.788E+03	4.685E+03	4.911E+03	6.363E+03
	Std	8.945E+02	1.355E+03	1.003E+03	1.088E+03	1.323E+03
	Rank	2	5	1	3	6
18	Mean	2.188E+06	8.874E+06	7.049E+06	4.171E+06	6.655E+06
	Std	1.146E+06	6.355E+06	5.101E+06	2.234E+06	5.161E+06
	Rank	1	6	5	2	4
19	Mean	7.324E+03	7.230E+03	7.085E+03	9.547E+03	8.814E+03
	Std	8.259E+03	8.287E+03	9.939E+03	1.962E+04	1.028E+04
	Rank	3	2	1	5	4
20	Mean	3.337E+03	4.560E+03	3.720E+03	3.565E+03	4.878E+03
	Std	7.965E+02	1.502E+03	1.178E+03	1.243E+03	1.643E+03
	Rank	1	5	3	2	6
21	Mean	1.116E+03	1.026E+03	1.100E+03	1.106E+03	1.086E+03
	Std	1.817E+02	1.274E+02	1.494E+02	1.592E+02	1.576E+02
	Rank	6	1	3	5	2
22	Mean	2.583E+04	3.115E+04	2.758E+04	2.527E+04	3.280E+04
	Std	4.526E+03	4.865E+03	7.401E+03	8.602E+03	2.818E+03
	Rank	2	5	4	1	6
23	Mean	1.349E+03	1.250E+03	1.245E+03	1.280E+03	1.275E+03
	Std	1.265E+02	1.058E+02	9.325E+01	9.663E+01	1.043E+02
	Rank	6	2	1	4	3
24	Mean	2.008E+03	1.800E+03	1.876E+03	1.891E+03	1.858E+03
	Std	2.345E+02	1.769E+02	1.509E+02	1.376E+02	1.958E+02
	Rank	6	1	3	4	2
25	Mean	7.654E+02	7.619E+02	7.617E+02	7.514E+02	7.521E+02
	Std	6.750E+01	5.549E+01	4.056E+01	6.999E+01	6.757E+01
	Rank	6	5	4	2	3
26	Mean	1.477E+04	1.355E+04	1.388E+04	1.453E+04	1.456E+04
	Std	1.799E+03	1.755E+03	1.867E+03	1.661E+03	1.937E+03
	Rank	6	1	2	4	5
27	Mean	5.000E+02	5.000E+02	5.000E+02	5.000E+02	5.000E+02
	Std	5.245E−04	4.791E−04	4.914E−04	4.458E−04	2.992E−04
	Rank	1	5	2	4	6
28	Mean	5.000E+02	5.000E+02	5.000E+02	5.000E+02	5.000E+02
	Std	4.376E−04	3.304E−04	5.004E−04	4.254E−04	4.277E−04
	Rank	2	5	1	3	6
29	Mean	4.080E+03	4.083E+03	4.520E+03	4.325E+03	4.449E+03
	Std	7.656E+02	6.593E+02	9.597E+02	8.301E+02	9.325E+02
	Rank	1	2	6	3	5
Average Rank		2.1034	3.0690	3.2414	2.8621	3.7241

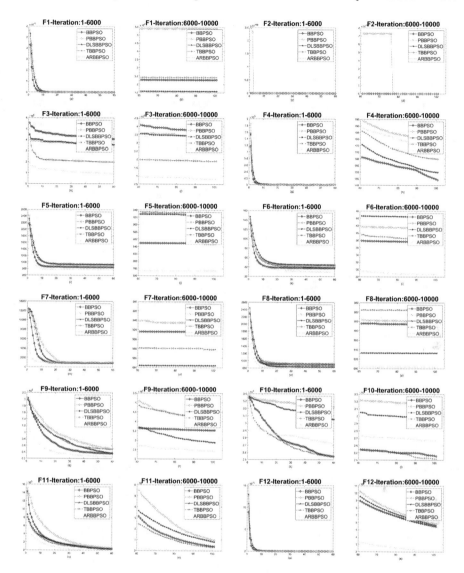

Fig. 1. Convergence diagram, f_1-f_{12}.

From the experimental results, ARBBPSO generally performs better than the control group on single-peaked functions (f_1–f_3), simple multi-peaked functions (f_4–f_{10}) and mixed functions (f_{11}–f_{20}), where the first ranking rates are 66.67%, 57.14% and 70.00%, respectively. However, the performance on the combined function problem (f_{21}–f_{29}) and especially f_{23}-f_{26} are inferior to the control group algorithms.

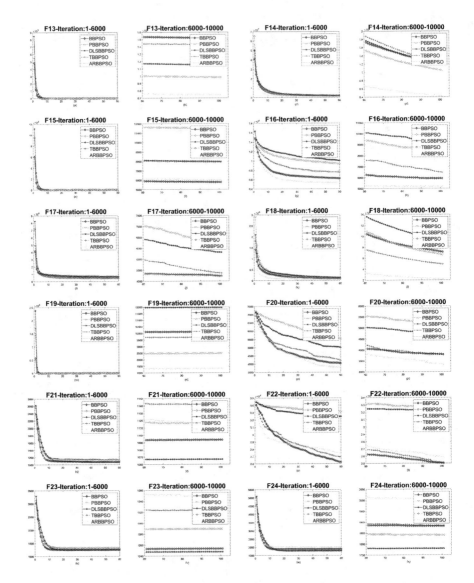

Fig. 2. Convergence diagram, $f_{13} - f_{24}$.

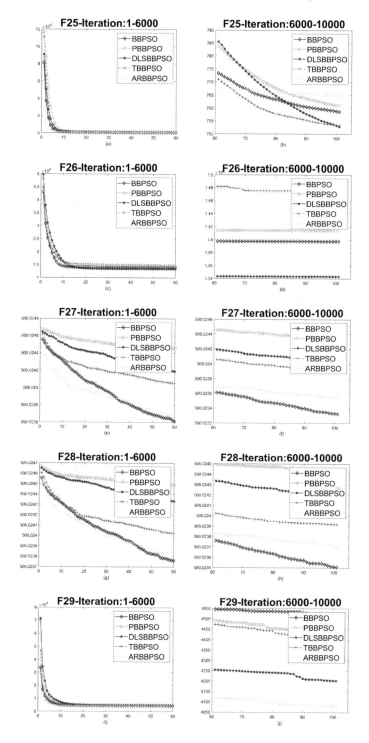

Fig. 3. Convergence diagram, $f_{25}-f_{29}$.

4   Conclusion

In this paper, we propose an atomic retrospective learning particle swarm optimization (ARBBPSO), which is proposed to address the optimization problem in a more efficient way. ARBBPSO is based on BBPSO, which succeeds the simplicity and parameter-free features of BBPSO. To improve the performance of BBPSO we propose two new strategies to improve BBPSO. Firstly, we present a strategy for the motion around the nucleus, which is inspired by the motion of negatively charged particles around positively charged protons in the nucleus of a hydrogen atom. Through this strategy particles can have more learning subjects, providing the possibility to solve the losing diversity in BBPSO. Secondly, we introduce a retrospective learning strategy, which enables the particles to have the ability of error correction in the process of searching for the optimum. By retrospective learning, particles have the opportunity to avoid falling into local optima in the early iterations.

To test the performance of ARBBPSO, the CEC2017 benchmark function is used in the experiments. And, five well-known BBPSO variant algorithms are chosen to be used as a control group. Finally, the experimental results and statistical analysis prove the performance of ARBBPSO in tackling the optimal problem.

Although ARBBPSO had the best overall position at CEC2017, there is still much to improve. For example, ARBBPSO performed worse on some Composition Functions at CEC2017. This is an important task for the future.

References

1. Nonoyama, K., Liu, Z., Fujiwara, T., Alam, M.M., Nishi, T.: Energy-efficient robot configuration and motion planning using genetic algorithm and particle swarm optimization. Energies **15**(6), 2074 (2022). https://doi.org/10.3390/en15062074
2. Rossides, G., Metcalfe, B., Hunter, A.: Particle swarm optimization-an adaptation for the control of robotic swarms. Robotics **10**(2), 58 (2021). https://doi.org/10.3390/robotics10020058
3. Gbadega, P.A., Sun, Y.: A hybrid constrained particle swarm optimization-model predictive control (CPSO-MPC) algorithm for storage energy management optimization problem in micro-grid. Energy Rep. **8**, 692–708 (2022). https://doi.org/10.1016/j.egyr.2022.10.035
4. Li, W., Li, B., Guo, H.L., Fang, Y.X., Qiao, F.J., Zhou, S.W.: The ECG signal classification based on ensemble learning of PSO-ELM algorithm. Neural Netw. World **30**(4), 265 (2020). https://doi.org/10.14311/NNW.2020.30.018
5. Beheshti, Z., Shamsuddin, S.M.H., Beheshti, E., Yuhaniz, S.S.: Enhancement of artificial neural network learning using centripetal accelerated particle swarm optimization for medical diseases diagnosis. Soft. Comput. **18**(11), 2253–2270 (2013). https://doi.org/10.1007/s00500-013-1198-0
6. Kumar, M., Raman, J., Priya, P.: Particle swarm optimization. Int. J. Mater. Forming Mach. Process. **2**(1), 54–87 (2015). https://doi.org/10.4018/ijmfmp.2015010104

7. Beheshti, Z.: A novel x-shaped binary particle swarm optimization. Soft. Comput. **25**(4), 3013–3042 (2020). https://doi.org/10.1007/s00500-020-05360-2

8. Lu, J., Zhang, J., Sheng, J.: Enhanced multi-swarm cooperative particle swarm optimizer. Swarm Evol. Comput. **69**(October 2021), 100989 (2022). https://doi.org/10.1016/j.swevo.2021.100989

9. Zhang, Y., Ji, X. F., Gao, X.Z., Gong, D.W., Sun, X.Y.: Objective-constraint mutual-guided surrogate-based particle swarm optimization for expensive constrained 'multimodal problems. IEEE Trans. Evol. Comput. (2022). https://doi.org/10.1109/TEVC.2022.3182810

10. Zeng, N., Wang, Z., Liu, W., Zhang, H., Hone, K., Liu, X.: A dynamic neighborhood-based switching particle swarm optimization algorithm. IEEE Trans. Cybern. **52**(9), 9290–9301 (2022). https://doi.org/10.1109/TCYB.2020.3029748

11. Kennedy, J.: Bare bones particle swarms. In: Proceedings of the 2003 IEEE Swarm Intelligence Symposium, SIS 2003 (Cat. No.03EX706), vol. 729, pp. 80–87 (2003). https://doi.org/10.1109/SIS.2003.1202251

12. Zhao, X., Liu, H., Liu, D., Ai, W., Zuo, X.: New modified bare-bones particle swarm optimization. In: 2016 IEEE Congress on Evolutionary Computation, CEC 2016, pp. 416–422 (2016). https://doi.org/10.1109/CEC.2016.7743824

13. Tian, H., Guo, J., Xiao, H., Yan, K., Sato, Y.: An electronic transition-based bare bones particle swarm optimization algorithm for high dimensional optimization problems. PLoS ONE **17**(7 July), 1–23 (2022). https://doi.org/10.1371/journal.pone.0271925

14. Guo, J., Sato, Y.: A fission-fusion hybrid bare bones particle swarm optimization algorithm for single-objective optimization problems. Appl. Intell. **49**(10), 3641–3651 (2019). https://doi.org/10.1007/s10489-019-01474-9

15. Guo, J., Sato, Y.: A pair-wise bare bones particle swarm optimization algorithm for nonlinear functions. Int. J. Netw. Distrib. Comput. **5**(3), 143–151 (2017). https://doi.org/10.2991/ijndc.2017.5.3.3

16. Guo, J., Sato, Y.: A bare bones particle swarm optimization algorithm with dynamic local search. In: Tan, Y., Takagi, H., Shi, Y. (eds.) ICSI 2017. LNCS, vol. 10385, pp. 158–165. Springer, Cham (2017). https://doi.org/10.1007/978-3-319-61824-1_17

17. Guo, J., Shi, B., Yan, K., Di, Y., Tang, J., Xiao, H., Sato, Y.: A twinning bare bones particle swarm optimization algorithm. PLoS ONE **17**(5 May), 1–30 (2022). https://doi.org/10.1371/journal.pone.0267197

18. Awad, N.H., Ali, M.Z., Liang, J., Qu, B.Y., Suganthan, P.N.: Problem definitions and evaluation criteria for the CEC 2017 special session and competition on real-parameter optimization. Nanyang Technology University, Singapore, Technical report, pp. 1–34 (2016)

A PSO-Based Product Design Tolerance Optimization Method Considering Product Robustness

Shuai Li[✉], Ruizhao Zheng, Yang Yang, Chunlin He, and Yong Zhang

School of Information and Control Engineering, China University of Mining and Technology,
Xuzhou 221116, China
lishuaixxls@163.com

Abstract. The tolerance design of product affects the cost of the product, as well as its reliability. To address the problem that traditional tolerance design models only focus on product value factors but neglect the product robustness, first this paper introduces a product robustness evaluation index into the tolerance optimization allocation. Taking the product robustness and product quality loss cost as the objectives, and letting the processing capability as constraint, a new multi-objective optimization model of tolerance allocation is established, and a solution algorithm based on particle swarm algorithm is given. The applicability and effectiveness of the model are verified by using the DC-DC circuit system of a television mainboard as an example.

Keywords: Tolerance optimization · particle swarm optimization · cascading failures · product robustness

1 Introduction

Reserving certain component tolerances in product design can ensure product performance while reducing production costs. Tolerance allocation refers to the allocation of certain tolerances to each component in production manufacturing based on the function and assembly requirements of the parts, to ensure that the parts can meet the corresponding design requirements after assembly.

Traditional tolerance optimization design usually takes the product quality loss and the manufacturing cost as objective functions [1], and establishes an optimization model. Furthermore, Muthu et al. [2] proposed a tolerance optimization model based on the Taguchi quality loss function and processing cost function, and used particle swarm optimization algorithm to solve the model. Xiao et al. [3] established a multi-objective tolerance design model considering product quality based on the Taguchi quality view and Pareto optimal solution set. Kuang et al. [4] proposed a mixed optimization allocation model to solve the size and position of tolerance. Jawahar et al. [5] established a multi-objective model for the tolerance allocation of pin-hole interchangeable assembly with total assembly cost and minimum clearance variation, and obtained its optimal solution

© The Author(s), under exclusive license to Springer Nature Switzerland AG 2023
Y. Tan et al. (Eds.): ICSI 2023, LNCS 13968, pp. 180–191, 2023.
https://doi.org/10.1007/978-3-031-36622-2_15

set by using iterative search algorithm. Liu et al. [6] established a multi-objective tolerance optimization model with cost and tolerance sensitivity as objective functions, and
used a small habitat particle swarm algorithm to obtain a uniformly distributed Pareto
front. Guo et al. [7] introduced the product performance deviation to constrain the tolerance based on the traditional economic processing cost constraint, and used genetic
algorithm to solve the model with minimizing manufacturing cost and quality loss cost
as the optimization objective. Chen et al. [8] proposed a remanufacturing tolerance
optimization design method considering process condition constraints, and established
a remanufacturing tolerance design multi-objective optimization model with manufacturing cost, quality loss and process capability as optimization objectives. Xing et al.
[9] proposed a multi-objective matching optimization model for the cost and quality co-
optimization of remanufactured products, and used a co-evolutionary algorithm to obtain
the final matching optimization solution. Hassani et al. [10] proposed a reliability-based
tolerance design method for addressing the degradation effects during the service life of
mechanical components. The method reformulates the tolerance allocation problem as
a bi-objective optimization problem with probability constraints. Non-dominated optimal tolerances and solutions are obtained using NSGA-II and entropy-based TOPSIS
methods, respectively. Ma et al. [11] proposed a total cost function model based on
a two-stage Bayesian sampling method, where the total cost consists of tolerance cost,
scrap cost, and quality loss. The constructed total cost model is optimized using a genetic
algorithm.

However, the above tolerance optimization models only focus on product value factors but ignore the issue of product robustness. During the development or production
process of a product, failures of components often occur due to various factors, and these
failures can spread internally within the product due to the interrelatedness of components [12]. Allocating reasonable tolerances to key components can effectively reduce
the impact of failure change propagation [13]. In order to remedy these defects, this
paper investigates a new multi-objective optimization model for tolerance assignment,
taking the product robustness and the product quality loss cost as two objectives, and a
solution algorithm based on particle swarm algorithm is given.

2 Product Design Optimization Model Considering Tolerance

2.1 Objective 1: Product Robustness Indicators

Currently, researchers usually employ system design, parameter design, and tolerance
design to enhance the robustness of products and reduce their sensitivity to noise factors in terms of quality and performance. This paper aims to investigate the network
robustness of products and analyze how to allocate the design tolerances of product
components in a reasonable manner, so that the product can maintain its normal structure and function when facing design changes. This section adopts complex network
technology to represent the correlation and characteristics among product parts and constructs a network model for complex products. A product network mainly comprises
many parts and various types of correlation. The parts are mapped to network nodes to
form a node-set $VP = (vp_1, vp_2, \ldots, vp_n)$. The correlation among the parts is mapped

to network edges to form an edge set $EP = \{ep_{ij} | i, j = 1, 2, \ldots, n, i \neq j\}$. The correlation strength between nodes forms a weight set $WP = \{wp_{ij} | i, j = 1, 2, \ldots, n, i \neq j\}$. Therefore, a product part network can be represented as $GP = (VP, EP, WP)$.

Existing research mainly evaluates the robustness of products from the perspective of the number of failed nodes after change propagation and their impact on the network structure. This definition only considers whether the nodes fail or not from a macroscopic perspective, without considering the changes in the node states from a microscopic perspective. In view of this, this section proposes a new evaluation index for the robustness of product networks, which considers both the impact of changes on the entire product network (structural robustness) and the degree of impact on certain component nodes (performance robustness). When a design change occurs, some nodes in the product part network may fail and become unable to work. The ratio of failed nodes can be used to indicate the product's structural robustness; the smaller the ratio, the better the structural robustness of the product. The definition of product structural robustness is the ability of the product to maintain its normal structure and function. The product's structural robustness(ϕ) is defined as:

$$\phi = \frac{1}{n} \sum_{i=1}^{n} s(i) \tag{1}$$

where n represents the number of part nodes and $s(i)$ represents the ratio of failed nodes after the end of the cascading failure propagation caused by the initial change node i.

The above definition can evaluate the structural robustness of the product network from a macro perspective. However, according to the risk load distribution strategy, when a node changes and fails, its risk load will be distributed to neighboring nodes. If the neighboring nodes can absorb the additional load the failed node brings, they will not fail. Usually, each neighboring node has its own maximum capacity. When the load borne by a node approaches its capacity limit, the performance of the part node will be affected, and the node becomes very fragile, which will significantly reduce the robustness of the entire network. Therefore, the average load change of nodes can be used to represent the performance robustness of the product:

$$\gamma = \frac{1}{n} \sum_{i=1}^{n} LV_i \tag{2}$$

where n represents the number of part nodes, and LV_i represents the average load change of neighboring nodes after the end of the cascading failure propagation caused by the initial change node i.

The above ϕ value reflects the average impact range of the change on the entire product network, while γ characterizing the average impact degree of the change on all neighboring nodes. Based on this, the comprehensive robustness of the product network can be defined as:

$$R = \sqrt{\phi(1 - \gamma)} \tag{3}$$

2.2 Objective 2: Loss Cost of Product Quality

According to Taguchi's quality theory, when there is a dimensional deviation in the machining of parts, these deviations will eventually affect the product quality [14]. Assuming that the product quality characteristic is y, and the designed target value is m, the quality loss function $L(y)$ can be represented as:

$$L(y) = K(y - m)^2 \tag{4}$$

where y is the product quality characteristic, m is the target value, and K is the quality loss coefficient that is independent of y. During manufacturing, $(y - m)$ presents the design tolerance. In tolerance design, the tolerance is generally a two-way symmetric distribution, that is, $y - m = T/2$, then the function relationship between quality loss cost and tolerance can be expressed as:

$$L(T_i) = k\frac{T_i^2}{4} \tag{5}$$

where $k = 4A/T^2$ and A represents the loss caused by a non-conforming product.

Simultaneously considering n affected parts, the total quality loss cost of product part tolerances is given by:

$$C(T) = \sum_{i=1}^{n} L(T_i) \tag{6}$$

2.3 Constraints

When optimizing the allocation of tolerances for product parts, some constraints must be taken into consideration, as the tolerances must be within a reasonable range. If the constraints are too strict, the optimized tolerances may not be practical, while overly loose constraints may result in excessive precision during product manufacturing and increased costs. The tolerance range is limited to:

$$T_{imin} \leq T_i \leq T_{imax} \tag{7}$$

where T_{imin} is the minimum tolerance of the i-th part, and T_{imax} is the maximum tolerance of the i-th part.

3 Model Solving Based on PSO

The multi-objective optimization model for design tolerance allocation established in this paper is a typical constrained multi-objective optimization problem. When solving such problems, traditional optimization algorithms convert them into single-objective problems first, and then use intelligent optimization algorithms to obtain an optimized solution to the problem. In this paper, an improved multi-objective particle swarm optimization algorithm is used to solve the above model and obtain multiple Pareto optimal solutions for decision-makers to choose from.

Particle Swarm Optimization (PSO) is a random search algorithm that simulates the behavior of organisms in nature and is an important branch of intelligent optimization algorithms. In the PSO algorithm, each particle represents a potential solution to the problem, and a group of particles work together to find the global optimal solution to the problem through continuous cooperation. Due to its simplicity, ease of implementation, and the fact that it does not require the use of feature information of the problem, PSO has received attention from many scholars, and new achievements have been continuously made in improving and analyzing its performance, and it has been widely used in multiple fields [15–18].

3.1 Encoding

The real number coding method is applied to the tolerance assignment optimization problem of parts. For the tolerance assignment problem containing n tolerance parameters, each particle corresponds to an n-dimensional position vector, where the value of each element represents the tolerance value of each part, that is:

$$x_i = \{x_{i1}, x_{i2}, \cdots, x_{in}\} \rightarrow T = \{t_1, t_2, \cdots, t_n\} \tag{8}$$

where, $x_i = \{x_{i1}, x_{i2}, \cdots, x_{in}\}$ represents the position vector of the i-th particle, that is, the solution vector of tolerance assignment; x_{ij} represents the tolerance parameter value of part node j in particle i.

3.2 Adaptive Adjustment of Inertia Weight and Learning Factor

This paper adopts the method proposed in [19] to set the inertia weight:

$$\omega(t) = \omega_{max} - \frac{\omega_{max} - \omega_{min}}{iterNum}t \tag{9}$$

where, $\omega(t)$ is the value of particle i in the t-th iteration; t and $iterNum$ are the preset maximum number of iterations and the current number of iterations, respectively; ω_{max} and ω_{min} are the maximum and minimum inertia weights respectively. In general, their values are 0.9 and 0.4 respectively.

The values of learning factors c_1 and c_2 are adjusted linearly. Reference [20] analyzed the different influences of the cognitive and social part on the algorithm, and proposed a learning factor that changes asynchronously and linearly with time:

$$c_1 = c_{1s} + \frac{c_{1e} - c_{1s}}{iterNum}t \tag{10}$$

$$c_2 = c_{2s} + \frac{c_{2e} - c_{2s}}{iterNum}t \tag{11}$$

where, c_{1s} and c_{1e} are the initial and termination learning factor of c_1, respectively. In general, their values are 2.5 and 0.5, respectively. c_{2s} and c_{2e} are the initial and termination learning factor of c_2, respectively. In general, their values are 0.5 and 2.5, respectively.

3.3 Algorithm Step

The steps for solving the tolerance assignment of product parts based on a multi-objective particle swarm optimization algorithm are as follows:

Step 1: Population initialization. Initializing the position of particles within the specified range randomly, and setting the individual optimal position of each particle. Formulas (3) and (6) were used to calculate the fitness value of each particle and initialize the external archive set A.

Step 2: Select the globally optimal position for each particle from the calculation.

Step 3: Updating the velocity and position of the particles, and calculating the fitness value of the particles by the method in [20].

Step 4: Updating individual optimal position, according to Pareto dominance relation. If the updated particle dominates the current individual optimal position, the new particle is taken as its individual optimal position; if they do not dominate each other, one of them is randomly selected as the individual optimal position.

Step 5: Updating the archive set A by the method proposed in [21].

Step 6: Determining whether the maximum number of iterations has been reached. If so, outputting the optimal solution set in A. Otherwise, go to Step 3.

4 Case Study

4.1 Model Introduction

The motherboard is one of the main components of the TV, integrating the central processing unit (CPU), memory, graphics processing unit (GPU), storage, various chips, capacitors, resistors and inductor circuits, etc. If these components are damaged, it will cause the entire motherboard to not work properly. Electronic components generally have a certain tolerance margin to ensure that the motherboard is able to maintain normal operation in the event of interference. Therefore, in order to improve the motherboard's anti-interference ability and reduce the user's loss, it is necessary to make a reasonable selection of the tolerance margin of the electronic components on the motherboard to improve the robustness of the product and ensure the quality loss within a certain range.

To verify the effectiveness of the proposed method, we take a certain DC-DC circuit on the motherboard of a model TV as an example and construct its multi-objective model for part tolerance allocation. The DC-DC circuit consists of 14 components, as shown in Fig. 1, and the network topology between its components is shown in Fig. 2. The relationship between the tolerance value and the scrap cost of each component counted by the enterprise is shown in Table 1.

According to the method in Sect. 2, the tolerance optimization multi-objective model for this problem is as follows:

$$minR(T) = \sqrt{\phi(T)(1 - \gamma(T))} \tag{12}$$

$$minC(T) = \sum_{i=1}^{10} L(T_i) \tag{13}$$

where, the decision variables are $T = \{T_1, T_2, \cdots, T_{10}\}$ and the following constraints are satisfied:

$0.10 \leq T_1 \leq 0.22$, $0.20 \leq T_2 \leq 0.42$, $0.05 \leq T_3 \leq 0.13$, $0.10 \leq T_4 \leq 0.22$, $0.20 \leq T_5 \leq 0.43$, $0.20 \leq T_6 \leq 0.41$, $0.10 \leq T_7 \leq 0.21$, $0.20 \leq T_8 \leq 0.40$, $0.05 \leq T_9 \leq 0.12$, $0.20 \leq T_{10} \leq 0.43$.

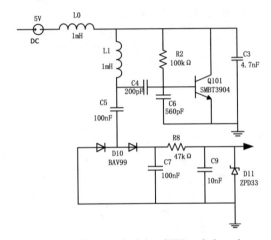

Fig. 1. Circuit principle of TV main board

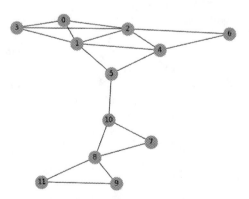

Fig. 2. Component network model

Table 1. Component design tolerances and end-of-life costs

Component name	Nominal value	Deviation	End-of-life costs A_i	Component name	Nominal value	Deviation	End-of-life costs A_i
Inductor L0	1 mH	±10%	1.25	Capacitance C5	100 nF	±20%	0.22
Inductor L1	1 mH	±20%	0.86	Capacitance C6	560 pF	±10%	0.23
Resistance R2	100 kΩ	±5%	0.16	Capacitance C7	100 nF	±20%	0.22
Capacitance C3	4.7 nF	±10%	0.20	Resistance R8	47 kΩ	±5%	0.14
Capacitance C4	200 pF	±20%	0.22	Resistance C9	10 nF	±20%	0.16

4.2 PSO Parameter Analysis

To verify the effectiveness of the improved particle swarm optimization algorithm (IPSO) used in this paper, it was compared with the standard particle swarm optimization algorithm (SPSO), with the following parameter settings for SPSO: inertia weight ω=0.6, acceleration coefficients c_1=c_2=2.5, and other parameters consistent with the algorithm used in this paper, where the population size is N = 100 and the number of iterations is *iterNum* =100. The Pareto fronts obtained by the two algorithms are shown in Fig. 3.

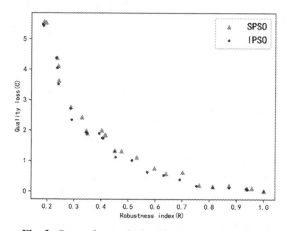

Fig. 3. Pareto fronts obtained by the two algorithms

To further demonstrate the superiority of the proposed algorithm, the hypervolume (HV) measure [22] is used to evaluate the algorithm's performance. Compared with other measures in multi-objective optimization, the calculation of this measure does not require knowing the true optimal solution set of the problem in advance, and can simultaneously evaluate the distribution and convergence of the Pareto optimal solution set. The larger the HV value, the better the distribution and convergence of the obtained solution set. Considering the Pareto optimal solution set obtained by a certain algorithm, the HV measure value is the volume of the objective space region enclosed by this solution set and the reference point. Typically, the worst results for each objective function in the Pareto optimal solution set obtained by the algorithm are used as reference points. The two algorithms were independently run 20 times, and the average value of the HV measure for each algorithm was calculated and is shown in Table 2.

Table 2. Average HV values obtained by the two algorithms

Two algorithms	Average HV value
Proposed method (IPSO)	1.825
Standard particle swarm optimization (SPSO)	1.619

As shown in Table 2, the average HV value of the proposed method is greater than that of the standard particle swarm optimization algorithm. This indicates that under the same number of iterations and population size, the proposed method obtains a Pareto front with better distribution and convergence.

4.3   Result Analysis

Using the algorithm proposed in this paper to solve the above model, Fig. 4 shows a set of Pareto fronts obtained, and Table 3 lists the information of some optimal solutions in the Pareto front. It can be seen that the algorithm in this paper obtains multiple complementary Pareto optimal solutions that satisfy the constraints. Taking one of the representative solutions as an example, when the system robustness indicator is 0.288 and the total mass loss cost is 2.717 yuan, the tolerances of component are $T_1=0.197$, $T_2=0.287$, $T_3=0.126$, $T_4=0.213$, $T_5=0.216$, $T_6=0.245$, $T_7=0.131$, $T_8=0.324$, $T_9=0.109$, $T_{10}=0.315$, which fully meet the requirements.

Furthermore, by observing the obtained Pareto front, the following regularities can be found between robustness and quality loss cost: when the robustness index is less than 0.4, the improvement of the robustness index has a significant impact on the quality loss cost, and the quality loss cost will sharply increase; when the robustness index is in the range of 0.4 to 0.6, the robustness index and the quality loss cost have a roughly linear relationship; when the robustness index is greater than 0.6, the change in the quality loss cost is small, which indicates that the improvement of machining accuracy has no significant effect on the improvement of product quality at this point, but it will cause a sharp decline in the robustness of the product. Decision makers or users can choose a suitable tolerance allocation scheme based on their own needs. If they value the robustness of the product more, they can choose the optimal configuration scheme from the left half of the Pareto front; if they value the quality loss cost more, they can choose the optimal configuration scheme from the right half of the Pareto front.

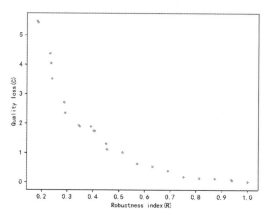

Fig. 4. Pareto optimal front obtained by the proposed algorithm

Table 3. Partial Pareto solutions obtained by the algorithm

T_1	T_2	T_3	T_4	T_5	T_6	T_7	T_8	T_9	T_{10}	R	C
0.189	0.243	0.096	0.218	0.263	0.348	0.207	0.318	0.114	0.233	0.392	1.899
0.178	0.367	0.125	0.207	0.339	0.361	0.179	0.298	0.070	0.315	0.233	4.375
0.168	0.408	0.074	0.194	0.280	0.323	0.185	0.386	0.103	0.344	0.185	5.435
0.178	0.343	0.062	0.215	0.284	0.246	0.199	0.299	0.074	0.384	0.236	4.045
0.213	0.254	0.095	0.183	0.234	0.240	0.168	0.315	0.091	0.229	0.403	1.750
0.103	0.206	0.064	0.210	0.218	0.286	0.200	0.289	0.053	0.254	0.514	1.020
0.212	0.419	0.123	0.209	0.309	0.330	0.191	0.344	0.115	0.324	0.182	5.474
0.131	0.236	0.069	0.267	0.229	0.239	0.210	0.300	0.080	0.219	0.450	1.309
0.131	0.219	0.065	0.181	0.232	0.235	0.121	0.208	0.068	0.238	0.753	0.184
0.110	0.210	0.052	0.174	0.210	0.200	0.118	0.207	0.053	0.020	0.873	0.127
0.100	0.226	0.050	0.173	0.200	0.200	0.138	0.200	0.050	0.200	0.937	0.095
0.127	0.200	0.050	0.111	0.200	0.200	0.123	0.228	0.050	0.205	0.940	0.064
0.186	0.273	0.114	0.163	0.215	0.228	0.170	0.291	0.111	0.219	0.407	1.740
0.207	0.250	0.118	0.187	0.215	0.243	0.148	0.288	0.100	0.216	0.345	1.930
0.197	**0.287**	**0.126**	**0.213**	**0.216**	**0.245**	**0.131**	**0.325**	**0.109**	**0.315**	**0.288**	**2.717**
0.109	0.218	0.079	0.219	0.215	0.213	0.117	0.210	0.053	0.210	0.691	0.394
0.213	0.350	0.125	0.213	0.219	0.284	0.144	0.300	0.118	0.315	0.241	3.518
0.105	0.200	0.086	0.195	0.205	0.207	0.114	0.216	0.082	0.210	0.631	0.535
0.134	0.214	0.052	0.213	0.211	0.218	0.181	0.293	0.054	0.220	0.454	1.122

(continued)

Table 3. (*continued*)

T_1	T_2	T_3	T_4	T_5	T_6	T_7	T_8	T_9	T_{10}	R	C
0.202	0.248	0.117	0.203	0.214	0.234	0.140	0.289	0.085	0.218	0.348	1.901
0.106	0.216	0.094	0.201	0.206	0.208	0.121	0.217	0.089	0.211	0.571	0.637
0.010	2.081	0.050	0.176	0.200	0.200	0.107	0.207	0.056	0.210	0.813	0.145
0.100	0.200	0.050	0.100	0.200	0.200	0.100	0.200	0.050	0.210	1.000	0.016
0.213	0.287	0.117	0.213	0.214	0.219	0.115	0.297	0.112	0.321	0.291	2.352

5   Conclusions

This paper proposed a novel multi-objective tolerance allocation optimization model, which considers both product robustness and quality loss cost, and designed a multi-objective particle swarm optimization algorithm to solve it. Taking the design tolerance optimization problem of the DC-DC circuit on the motherboard of a certain model of TV as an example, the experimental results show that: (1) the introduced automatic adaptive parameter adjustment strategy improves the ability of the particle swarm optimization algorithm to handle the design tolerance optimization problem; (2) the established multi-objective tolerance optimization model considering product robustness can provide decision-makers with multiple complementary and constraint-satisfying optimal solutions. Decision-makers or users can choose a suitable tolerance configuration scheme based on their own preferences. This paper also provides a new idea and direction for tolerance optimization.

References

1. Wu, Z.T.: Computer Aided Tolerance Optimization Design. Zhejiang University Press (1999)
2. Muthu, P., Dhanalakshmi, V., Sankaranarayanasamy, K.: Optimal tolerance design of assembly for minimum quality loss and manufacturing cost using metaheuristic algorithms. Int. J. Adv. Manuf. Technol. **44**(11–12), 1154–1164 (2009)
3. Xiao, R.B., Zou, H.F., Tao, Z.W.: Multi-objective model of tolerance design and its solution with particle swarm optimization algorithm. Comput. Integr. Manuf. Syst. **07**, 976–980+989 (2006)
4. Kuang, B., Huang, M.F., Zhong, Y.R.: Optimal tolerance allocation for composite dimensional and geometric tolerances. Comput. Integr. Manuf. Syst. **118**(02), 398–402 (2008)
5. Jawahar, N., Sivasankaran, R., Ramesh, M.: Optimal Pareto front for manufacturing tolerance allocation model. Proc. Inst. Mech. Eng. Part B: J. Eng. Manuf. **231**(7), 1190–1203 (2017)
6. Liu, H.B., Liu, J.H., He, Y.X., Guo, C.Y., Jiang, K.: Optimal tolerance allocation for composite dimensional and geometric tolerances. Comput. Integr. Manuf. Syst. **21**(03), 585–592 (2015)
7. Guo, Y.F., Luo, Y.W., Zhao, Y.M., He, H.X., Tan, H.J.: Constraint tolerance optimization design based on performance deviation. China Mech. Eng. **28**(17), 2069–2074 (2017)
8. Chen, Y.X., Jiang, Z.G., Zhu, S.: Optimization method for remanufacturing tolerance of electromechanical products with process condition constraint. Computer Integrated Manufacturing Systems (2023). http://kns.cnki.net/kcms/detail/11.5946.tp.20210918.1146.010.html

9. Xing, S.X., Jiang, Z.G., Zhu, S., Zhang, H.: Research on optimization method for parts selection of remanufactured products under dimensional accuracy constraints. J. Mech. Eng. **58**(19), 221–228 (2022)
10. Hassani, H., Khodaygan, S.: Reliability-based optimal tolerance design of mechanical systems including epistemic uncertainty. Int. J. Mech. Mater. Des. 1–18 (2022)
11. Ma, Y., Wang, J., Tu, Y.: Concurrent optimization of parameter and tolerance design based on the two-stage Bayesian sampling method. Q. Technol. Quant. Manag. 1–23 (2023)
12. Tang, D.B., Xu, R.H., Tang, J.C.: Analysis of engineering change impacts based on design structure matrix. J. Mechan. Eng. **46**(01), 154–161 (2010)
13. Liu, J.H., Hou, Y.Z.: MBD Model parametric method based on hybrid-attributed adjacency graph. J. Comput.-Aided Des. Comput. Graph. **30**(07), 1329–1334 (2018)
14. Jiang, Y., Jiang, Z.G., Zhang, H., Cheng, H.: Optimization study of remanufacturing reconditioning scheme for used parts based on failure features. Mach. Tool Hydraulics **44**(21), 168–172 (2016)
15. Xiao, T.Y., Zhang, Z.H.: Particle swarm optimization algorithm for solving large-scale function optimization. Comput. Eng. Des. **42**(06), 1614–1622 (2021)
16. Alsaidy, S.A., Abbood, A.D., Sahib, M.A.: Heuristic initialization of PSO task scheduling algorithm in cloud computing. J. King Saud Univ.-Comput. Inf. Sci. **34**(6), 2370–2382 (2022)
17. Xue, Y., Xue, B., Zhang, M.: Self-adaptive particle swarm optimization for large-scale feature selection in classification. ACM Trans. Knowl. Discov. Data **13**(5), 1–27 (2019)
18. Song, X.F., Zhang, Y., Gong, D.W., et al.: A fast hybrid feature selection based on correlation-guided clustering and particle swarm optimization for high-dimensional data. IEEE Trans. Cybern. **52**(9), 9573–9586 (2021)
19. Hu, Y., Zhang, Y., Gong, D.W.: Multiobjective particle swarm optimization for feature selection with fuzzy cost. IEEE Trans. Cybern. **51**(2), 874–888 (2021)
20. Ratnaweera, A., Halgamuge, S.K., Watson, H.C.: Self-organizing hierarchical particle swarm optimizer with time-varying acceleration coefficients. IEEE Trans. Evol. Comput. **8**(3), 240–255 (2004)
21. Zhang, Y., Yuan, L.J., Zhang, Q., Sun, X.Y.: Multi-objective optimization of building energy performance using a particle swarm optimizer with less control parameters. J. Build. Eng. **32**, 101505 (2020)
22. Zitzler, E., Thiele, L.: Multiobjective evolutionary algorithms: a comparative case study and the strength Pareto approach. IEEE Trans. Evol. Comput. **3**(4), 257–271 (1999)

Genetic Algorithms

Competitive Learning and Dynamic Genetic Algorithms for Robust Layout Designs Under Uncertainties

Srisatja Vitayasak and Pupong Pongcharoen[✉]

Centre of Operations Research and Industrial Applications (CORIA), Department of Industrial Engineering, Faculty of Engineering, Naresuan University, Phitsanulok 65000, Thailand
{srisatjav,pupongp}@nu.ac.th

Abstract. Machine layout design is a crucial part of company's operations and commonly aimed at maximising the effectiveness of resource utilisation within the manufacturing process to meet the uncertain needs of customers. Previous studies on Genetic Algorithm have rarely applied both adaptive and learning mechanisms simultaneously. This paper presents the development of a new competitive learning Genetic Algorithm (CLGA) and dynamic Genetic Algorithm (DGA) for layout design under machine availability and demand uncertainty scenario. The internal logistics of raw materials within the shop floor (minimum flow distance) was considered. The computational programme was carried out using eight benchmarking datasets. The analysis on the computational results indicated that the proposed methods statistically outperformed the conventional Genetic Algorithm in solution quality and convergent speed.

Keywords: Adaptive parameter · Machine learning · Facility layout · Dynamic demand · Unavailable machine

1 Introduction

Computational Intelligences (e.g. nature-inspired search algorithms or metaheuristics) have been extensively applied to solve optimisation problems. Within an algorithm, a balance between intensification and diversification strategies always plays a crucial role in the algorithm performance. The balance can be achieved from an optimal parameter setting, an adjustment of mechanism within algorithm, or a combination of algorithm and other approaches [1]. The values of the parameters considerably influence whether the algorithm will find local optimums or optimal solution and whether it will efficiently find such a solution [2], whilst the universal optimal parameter setting for all problem domains does not exist [3].

The setting of algorithm's parameters can be either statically or dynamically carried out during a computational run. Static parameter setting usually adopts its values in ad hoc fashion, best guess method or from previous research that may be often conducted in other problem domains. Although those settings are time-saving approaches, they do not guarantee the best performance of the searching process. An alternative approach may be trial and error, in which the experience of researchers is needed. Otherwise, the parameter tuning process could be very tedious and time consuming [4]. The adaptation of non-tuned parameter setting for particular problem usually results in local optimum due to a premature convergence [5]. However, the parameter setting may be inefficient during the computational run because of unbalance between exploitation and exploration. This suggests that the search process may not be able to escape from the local optimum [1]. Adjustable or adaptive parameter setting during the computational run or dynamic parameter setting (or parameter control) can overcome this shortcoming [2].

Facility layout problem (FLP) as one of quadratic assignment problems [6, 7] and non-deterministic polynomial complete problem [8, 9] refer to the location of facilities, machines or blocks (see Fig. 1) on the production area [10]. Layout can be configured as single or multiple rows, loop, or cluster. An effective facilities' arrangement leads to the minimum flow of handling materials, which can account for the total operating costs. The reduction of handling flow sequentially leads to quicker transfer times; higher productivity rate; lower unit cost of product; and better competitiveness. Many optimisation algorithms have been successfully applied to solve the FLPs. After performing a systematic literature survey on databases including the ISI Web of Science and Scopus during 2007 to 2022, a total number of FLP related research works were 487 papers. Genetic Algorithm (GA) has been extensively applied to solve the problems mostly using fixed values of GA parameters [11]. Recently, research has been devoted to improve GA effectiveness by integrating GA with another algorithm [12–14] to accelerate the search to find the global optimum [15].

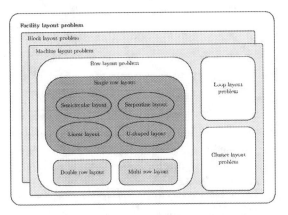

Fig. 1. Context of layout problem [16]

Uncertainty issues faced by a manufacturing company usually have an influence from both external and internal environment [17]. External variations (e.g., customer demand) cause changes in production quantity, directly affecting production mix and the intensity of material flow between machines [18]. Based on the literature review, there were only 22 of 487 papers (about 4.5%) applying GA to solve FLP with a consideration on demand uncertainty. Internal variations (e.g., unavailable machines [19]) can disrupt the production flow and sequentially cause uneven workload, slow manufacturing flow, poor productivity rate and higher unit cost of product. When a machine is under maintenance, a manufacturing system with alternative machine flexibility routing can maintain production performance. Nevertheless, the alternative routes may generate longer flow distance.

The objective of the present work was aimed at developing dynamic and competitive learning Genetic Algorithm (GA) to solve machine layout problem based on robust design approaches with subjected to demand uncertainty and machine availability. The present paper is organised as follows. Section 2 deals with the literature reviews on dynamic Genetic Algorithm parameters and layout design with uncertain conditions followed by the proposed methods for solving the problem in Sect. 3. Section 4 shows the results obtained from the proposed methods. Lastly, the discussion and conclusions of the work are presented in Sect. 5.

2   Comprehensive Literature Reviews

The related literature reviews of parameters in stochastic search algorithms, facility layout design under dynamic demand, and machine layout design in uncertain production are presented in the following subsections.

2.1   Parameters in Stochastic Search Algorithms

Parameter setting can be classified into two major forms: parameter tuning and control [2]. Parameter tuning ensures that good values are identified before the computational run and fixed throughout the run. Tuning methods can be classified as sampling methods (e.g. Latin-square, Taguchi orthogonal arrays), model-based methods (e.g. logistic regression, design of experiment), screening methods (e.g. multiple comparison procedure), and meta-evolutionary algorithms. These tuning methods differ in balancing between exploitation and exploration that affects the number of experiments. The drawbacks of parameter tuning has been mentioned that its computational time and resources could be highly expensive especially when considering high numbers of parameters and its values and adopting full factorial design; parameters may not be independent; and the selected parameter values need not be optimal [2]. Although the parameter setting is optimised, the results can prematurely get trapped because of changes in the effectiveness of a parameter setting during iterative search [1]. This problem can be solved using parameter control [20].

Parameter control means that the parameter can be adjusted throughout the run so called dynamic parameter setting. Parameter values can be modified without taking into account the random or deterministic update [1]. This does not involve the feedback from the search process for reviewing the balance between intensification and diversification. Alternatively, evolving parameter value during the search can be called self-adaptive parameter setting. The parameters are encoded into the representation and are subject to changes according to the solutions found [1]. Although this approach results in an increased search space, a more computational time is required [21].

Adaptive parameter setting is another strategy for determining the direction or magnitude of parameters using the memory of search. It does not alter the amount of search and makes it more effective via adjusting strategy [22]. Concept of adjustable parameter setting has received considerable interest with the intention to improve the algorithm's performance, so that the solution quality is better [23].

2.2   Facility Layout Design Under Dynamic Demand

A robust layout design considers a wide variety of demand scenarios [8] and attempts to minimise material flow costs over planning horizon, by which the layout may not be optimal under any specific time horizon. The layout can be periodically adapted to optimise total cost including material flow and rearrangement costs. If the rearrangement (shifting) costs are fairly low, the layout may be redesigned to suit the changed material flow [24]. Frequently, machine relocation is expensive and unfeasible with a short notice, so the machines are not relocated.

Based on the reviewed FLP-related papers as mentioned in Sect. 1, 96 papers have applied GA to solve FLD problem. There are 22 of 96 papers (22.3%) that proposed the GA-based layout design under fluctuated demand scenario. Only 2 of 96 papers have examined with dynamic parameter. Datta et al. [25] have introduced the self-adaptive mutation probability in single-row layout problem. Peng et al. [26] have developed adaptive mutation and crossover based on the fitness value of the chromosome and applied to solve the equal-size layout design problem under varied demand level. However, there has been no report on applying GA with competitive learning crossover mechanism and adjustable parameter for optimising layout design using robust design approaches by simultaneously considering dynamic demand and machine availability.

2.3   Facility Layout Design in Uncertain Conditions

Product demand is varied in the form of deterministic or probabilistic profiles. The demand can be presented by the flow matrices among facilities [27, 28]. Deterministic demand can be pre-defined in advance via forecasting and/or pre-order systems. Probabilistic (stochastic) demand can be determined using empirical data [29], or probability distributions [30], such as Uniform, Normal, or Exponential distribution. The demand profiles can also be generated using different probabilities [31].

Machine unavailability e.g. breakdown is a stochastic event, which can interrupt the operations leading to a halt and delay of production. The schedule may be revised to re-design the remaining tasks concerning the machine utilisation. A number of machine breakdown are commonly signified by the Poisson distribution whilst machine failures can be generated randomly [32]. The lifetime is usually modelled with the Weibull distribution [33]. However, mean failure time is usually represented by Normal or Exponential distributions and Uniform distribution for broken-down machine [34].

To mitigate problems due to unavailable machine, production should be flexible, e.g., alternative machines or routings. Alternative machines should be able to perform the same operation as the unavailable machines, whereas alternative routings provide the same operation sequence. Material flow time and distances can be changed because of alterations of the route [35]. Unavailable machines have not been considered in the layout design problem.

3 Proposed Methods for Dynamic Layout Design

In this work, Dynamic Genetic Algorithm (DGA) and Competitive Learning Genetic Algorithm (CLGA) were developed and employed to solve dynamic layout design problems subjected to demand fluctuation and machine availability using TCL/TK programming language [36]. The assumptions in this work were made as follows: i) material flow is functioned on pick-up and drop-off (P/D) stations; ii) the flow distance was calculated from machine-to-machine; iii) machine were placed in row-by-row; iv) a set of alternate machines was implemented in case of machine unavailability; v) the automated guided vehicles were travelled on the guided lines; vi) the passageway between machines was comparable; and vii) the capacity of machine was not considered. Descriptions of the proposed methods are provided in the following subsections.

3.1 Dynamic GA to Design Machine Layout

The probabilities of crossover and mutation parameters of DGA were dynamic using the self-adjustment. The robust layout design using the DGA can be divided into thirteen following steps:

- i) encoding - each gene represents machine to be placed;
- ii) recorded data - numbers of products (N) and machines (M); machine width (MW) and length (ML); machine sequences (MS); and alternative routes for each machine;
- iii) itemise parameters - probabilities of crossover (P_c) and mutation (P_m); population size (Pop); number of generations (Gen); floor width (F_W) and length (F_L); passageway gap (G) between machines; time-periods (P); and probability of machine unavailability (%UM) for repairing or maintenance;
- iv) forecasting demand levels on any period (D_{gk}) of all products;
- v) aimlessly create a list of unavailable machines based on the %UM;
- vi) randomly initialise chromosomes based on the Pop pre-defined in step iii. Each solution is composed of a sequence of machines considered for layout design, for example in case of ten machines, a generated solution can be 1–8–5–4–3–9–6–10–2–7;
- vii) redirect the material flow to alternative machines in case machines are unavailable (e.g. maintenance). For example, the normal processing route for product A is M2-M5-M8 but M8 was planned for maintenance in the third time-period. If machine 1 (M1) is defined as alternative machine due to the unavailability of machine 8, the redirecting manufacturing route for the third time-period will be 2–5–1;
- viii) perform evolutional operations to produce offspring according to P_c and P_m;
- ix) arrange machines based on multiple rows arrangement as shown in Fig. 2. The placement algorithm locates rectangular machines into a restricted manufacturing area with a gap between adjacent machines. The red dash line in Fig. 2 presents the route of vehicle's movement between rows. The pre-define vehicle can flow right or left of the row, followed by down or up to the target row;

x) calculate total material flow distance using (1) as the fitness function.

$$Total\ material\ flow\ distance\ =\ \sum_{i=1}^{M}\sum_{j=1}^{M}\sum_{g=1}^{N}\sum_{k=1}^{P}d_{ijk}f_{ijgk}D_{gk} \qquad (1)$$

where i and j indicate the indexes (i and j = 1, 2, 3, ..., M) (i ≠ j); M denotes machine number; g is the index of any product (g = 1, 2, 3, ..., N); N denotes the number of product types; k is the index of any time-period (k = 1, 2, 3, ..., P); P denotes the number of time periods; d_{ijk} is the material handling distance from machine i^{th} to j^{th} in period k^{th}; f_{ijgk} denotes the amount of material flow of product g^{th} from machine i^{th} to j^{th} in period k^{th}; D_{gk} refers to the customer demand of product g^{th} in time-period ^{th}k.

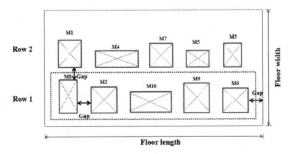

Fig. 2. Example of multi-row configuration

xi) select the best chromosome based on the elitist selection mechanism;
xii) compare the best solution to the best-so-far solution and adjust P_c and P_m based on the predetermined constraint. If the fitness value of the best solution in the current generation is worse than the fitness value of the best-so-far solution, the dynamic rate for adjusting P_c and P_m were formulated as in (2) and (3), respectively. The P_c and P_m for the next generation are adjusted by (4) and (5);

$$Dynamic\ rate\ for\ P_c = random\ number\ *\ the\ current\ P_c \qquad (2)$$

$$Dynamic\ rate\ for\ P_m = random\ number\ *\ the\ current\ P_m \qquad (3)$$

$$P_c = P_c + dynamic\ rate\ for\ P_c \qquad (4)$$

$$P_m = P_m + dynamic\ rate\ for\ P_m \qquad (5)$$

xiii) randomly choose chromosomes; and terminate the GA process if stopping criteria is satisfied. Once it is completed, then reporting the best-so-far solution.

3.2 Competitive Learning Genetic Algorithm for Dynamic Machine Layout Design

In this work, a competitive learning mechanism was embedded within the GA crossover operations with an aim to enhance the evolutionary search direction. Unlike the crossover mechanism in classical GA, the selection of parent chromosomes was based on the best solutions. The number of the new offspring depend on the probability of crossover. The competitive learning mechanism consists of four steps: i) The best-so-far chromosome ($CHRO_{best}$) was recognised; ii) The new chromosome ($CHRO_{new,a}$) was produced based on the $CHRO_{best}$ using the following equation:

$$CHRO_{new,a} = CHRO_{old,a} + \theta_a(CHRO_{best} - SLF * CHRO_{old,a})$$

$CHRO_{new}$ was a new chromosome. $CHRO_{old}$ was a current chromosome. a was a chromosome index ($a = 1, 2, 3, ..., cross$). θ_a was a scaling value ranged between [0,1] at a^{th} chromosome. SLF was a self-learning factor. The value of SLF can be either 1 or 2. The current chromosome ($CHRO_{old}$) was transformed on its own and by the best chromosome. Modifying the offspring according to the $CHRO_{best}$ is demonstrated by the examples in Fig. 3 and Fig. 4. If the SLF was equal to 2, meaning that two swap operations were required, the $SLF*CHRO_{old}$ was 1–4–5–8–3–6–9–10–2–7 (Fig. 3). The learning process between the $CHRO_{best}$ and the $CHRO_{old}$ was expressed in Fig. 4.

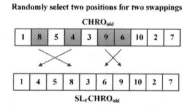

Fig. 3. Swap operation in case of $SLF = 2$

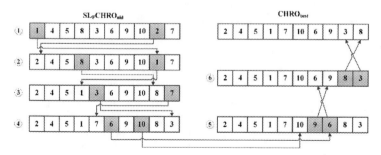

Fig. 4. Modifying the chromosome according to the best chromosome

The swapping procedure was repeated until the $SL_F * CHRO_{old}$ will be similar to $CHRO_{best}$. In this case, six swapping operations were (1,2) (8,1) (3,7) (6,10) (9,6) and (3,8). The number of swapping operations was calculated by multiplying with a scaling value (θ_a). For example, if θ_a was 0.78, the number of swapping operations was 6*0.78 = 4.68 (then rounded up to 5). The $CHRO_{old}$ was then performed five swaps: (1,2) (8,1), (3,7), (6,10), and (9,6). The $CHRO_{new}$ is 2–4–5–1–7–10–6–9–8–3.

iii) The $CHRO_{new,a}$ competes with $CHRO_{old,a}$ by calculating the material flow distance of $CHRO_{new,a}$; and iv) The $CHRO_{new,a}$ was accepted if it was better than $CHRO_{old,a}$.

4   Design and Analysis of Experiments

The numerical experiments carried out on eight datasets were different in numbers of unequal-size machines and product types [37], in which the number of products were ranged between 5–25 to be manufactured on 10–50 machines. For example, the biggest dataset considered in this work was based on 50 products to be sequentially manufactured on 25 machines (50N25M). The initial P_c and P_m for DGA were set at 0.5 and 0.1. The probability of crossover has been recommended to be 0.9 whilst 0.5 for mutation based on population size of 50 and 50 generations. The two-point centre crossover (2PCX) and two operation random swap (2ORS) were adopted as suggested by [38].

Demand profiles were obtained from either empirical data or probability distributions (exponential, normal, or uniform). Unavailable machines were randomly chosen and the percentage of unavailable machines in each period was considered at 10%. The number of time periods were set at ten. The alternative machines were used during the periods when changing the processing route. Thirty random seeds were used to perform the experiment based on the full factorial design. All computational runs were based on a computer with CPU Intel Core i7 3.4 GHz with 16 GB of DDR3 RAM.

The average and standard deviation (SD) of the material flow distance (MFD) (unit: metre), executing time (unit: second) are shown in Table 1. The average MFD obtained from GA was higher than those obtained from DGA in all datasets. The self-adjustment probabilities of crossover and mutation operations in DGA were useful to increase the diversity of chromosomes throughout search process as well as to escape from the local optimum. The average MFD was minimised by the CLGA in most datasets except the 15N30M dataset. Modifying the crossover operation by adopting the proposed competitive learning mechanism can improve the efficiency of the evolutionary search mechanism. The mean differences of MFD between GA and DGA, GA and CLGA, and DGA and CLGA were analysed by using Student's t-test. The results shown in Table 1 suggested that two approaches of GA parameter settings (between GA and DGA) were significantly differed for 15N30M and 20N40M datasets. There were statistically insignificant in the differences of the mean MFD between CLGA and DGA for 15N30M and 40N40M datasets. However, the CLGA took more execution time than the DGA and the conventional GA. This resulted from the number of iterations and selection process of new chromosomes. In 25N50M case, the time for CLGA in robust layout was about 31% and 61% longer than the time of DGA and GA, respectively. The computational time for DGA was longer than the GA in some datasets because of dynamic parameter setting being related to the quality of the current best and the best-so-far (BSF) solutions.

The convergences of the CLGA, DGA and GA in 20N40M case were proposed by plotting the average BSF solutions for each generation reported from 30 repetitive computational runs (Fig. 5). The convergence rate of DGA during 50 generations was faster than the classical GA. By using Student's t-test, the average BSF solutions obtained from GA and DGA in each generation were statistically significant after the seventeenth generation. During the run, the GA parameters' setting was inefficient in that the GA were trapped in the local optimum solution. The exploitative and explorative features are unbalanced. Adjusting the probabilities of mutation and crossover with a comparison of the best solution of the current generation to the BSF solution leads to more efficiency of searching process. Increment of crossover and mutation probabilities can lead to a more dissimilarity in the solutions, so the standard deviation of solutions is increased. Because of this procedure, the GA can avoid premature convergence and can escape from local optimum. However, the dynamic probabilities may deteriorate capabilities of searching and learning which is resulted from an amount of search is increased with the adjusted probabilities. Solution quality development of DGA was slower than CLGA. The convergent lines of both methods are shown in Fig. 5. In case of CLGA, the solution was better than the DGA and the conventional GA from the beginning as well as the convergence was continuously conducted. These issues resulted from the self-learning operation of the current chromosome and the searching mechanism within the best chromosome for discovering the outclass search space.

The values of standard deviation (SD) for the BSF solutions (CLGA, DGA and GA) in each generation were plotted in bar chart (Fig. 5). The SD values of the CLGA and DGA were more than those of the GA especially in the last twenty-five generations. The BSF solutions of CLGA and DGA were more diverse than those of GA and as such providing more opportunity to get better solutions. A diversity of solutions affected the standard deviation. However, high diversity might lead to the loss of computational time and did not guarantee the quality of solution. The SD values in Table 1 were calculated from the BSF solutions with 30 replications (30 seed numbers). The BSF solutions' SD of DGA was lower than that of GA in some datasets but the best solution can be found in the solutions having less deviation. The SD of the CLGA was higher than that of DGA and GA in most datasets and the solution quality was also better.

Table 1. Experimental results

Dataset	Value	MFD based on			P-value	
		GA	DGA	CLGA	GA-CLGA	DGA-CLGA
5N10M	Mean	590,753.3	587,237.0	**555,559.0**	0.000	0.000
	SD	16,296.5	16,892.5	12,923.8		
	Time	120.9	109.6	128.0		
10N20M	Mean	3,502,114.8	3,473,331.9	**3,243,174.2**	0.000	0.000
	SD	65,178.1	46,221.5	105,961.5		
	Time	424.8	398.3	648.2		
20N20M	Mean	9,975,829.5	9,933,818.9	**9,429,173.9**	0.000	0.000
	SD	150,438.9	154,803.6	290,973.0		
	Time	660.0	842.6	1,187.2		
40N20M	Mean	20,055,976.8	19,976,663.6	**17,814,658.9**	0.000	0.000
	SD	256,549.6	179,713.0	491,625.3		
	Time	1,564.4	1,592.6	2,216.2		
15N30M	Mean	8,112,260.9	**8,024,490.8**	8,077,962.8	0.459	0.237
	SD	160,677.7	149,243.5	193,970.8		
	Time	663.8	777.9	1,139.8		
20N40M	Mean	16,963181.9	16,558,455.4	**16,137,934.0**	0.000	0.000
	SD	450,898.8	398,220.0	377,207.6		
	Time	1,169.8	1,114.0	1,684.6		
40N40M	Mean	30,014,644.7	29,883,421.8	**29,550,914.4**	0.017	0.064
	SD	668,480.4	581,028.5	768,120.3		
	Time	2,032.8	2,427.2	2,965.6		
25N50M	Mean	27,092,641.4	26,810,841.4	**26,497,369.1**	0.000	0.041
	SD	568,595.4	602,286.1	558,800.7		
	Time	1,287.2	1,588.8	2,077.0		

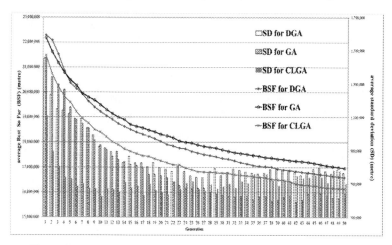

Fig. 5. Convergences on the proposed methods for the case of 20N40M

5   Conclusions

In this present paper, the development of the dynamic Genetic Algorithm (DGA) and competitive learning Genetic Algorithm (CLGA) for designing robust unequal-sized machine layouts with dynamic demand and machine availability was described. The proposed CLGA and DGA with self-adjustment produced better solutions than the conventional GA in all datasets Adjusting GA mechanism with parameter setting approach and the competitive learning in crossover operation resulted in higher performance algorithm. Further studies may focus on the trade-off cost associated with material flow and machine re-location for periodic layout redesigning; and a combination of new parameter setting and learning approaches.

Acknowledgements. This work was a part of research projects under grant numbers R2565C006 and R2566B066. The work was also co-funded by the National Research Council of Thailand and Naresuan University under grant number N42A650329.

References

1. Talbi, E.G.: METAHEURISTICS from Design to Implementation. Wiley, Hoboken (2009)
2. Eiben, A.E., Michalewicz, Z., Schoenauer, M., Smith, J.E.: Parameter control in evolutionary algorithms. In: Lobo, F.G., Lima, C.F., Michalewicz, Z. (eds.) Parameter Setting in Evolutionary Algorithms. Studies in Computational Intelligence, vol. 54, pp. 19–46. Springer, Berlin (2007). https://doi.org/10.1007/978-3-540-69432-8_2
3. Chew, E.P., Ong, C.J., Lim, K.H.: Variable period adaptive genetic algorithm. Comput. Ind. Eng. **42**, 353–360 (2002)
4. Michalewicz, Z., Fogel, D.V.: How to Solve It: Modern Heuristics. Springer, Cham (2010)
5. Bingul, Z.: Adaptive genetic algorithms applied to dynamic multiobjective problems. Appl. Soft Comput. **7**, 791–799 (2007)

6. Matousek, R., Dobrovsky, L., Kudela, J.: How to start a heuristic? Utilizing lower bounds for solving the quadratic assignment problem. Int. J. Ind. Eng. Comput. **13**, 151–164 (2022)
7. Hameed, A.S., Aboobaider, B.M., Mutar, M.L., Choon, N.H.: A new hybrid approach based on discrete differential evolution algorithm to enhancement solutions of quadratic assignment problem. Int. J. Ind. Eng. Comput. **11**, 51–72 (2020)
8. Lashgari, M., Kia, R., Jolai, F.: Robust optimisation to design a dynamic cellular manufacturing system integrating group layout and workers' assignment. Eur. J. Ind. Eng. **15**, 319–351 (2021)
9. Sooncharoen, S., Vitayasak, S., Pongcharoen, P., Hicks, C.: Development of a modified biogeography-based optimisation tool for solving the unequal-sized machine and multi-row configuration facility layout design problem. ScienceAsia **48**, 12–20 (2022)
10. Nagarajan, L., Mahalingam, S.K., Gurusamy, S., Dharmaraj, V.K.: Solution for bi-objective single row facility layout problem using artificial bee colony algorithm. Eur. J. Ind. Eng. **12**, 252–275 (2018)
11. Dapa, K., Loreungthup, P., Vitayasak, S., Pongcharoen, P.: Bat algorithm, genetic algorithm and shuffled frog leaping algorithm for designing machine layout. In: Ramanna, S., Lingras, P., Sombattheera, C., Krishna, A. (eds.) MIWAI 2013. LNCS (LNAI), vol. 8271, pp. 59–68. Springer, Heidelberg (2013). https://doi.org/10.1007/978-3-642-44949-9_6
12. Hosseini, S.S., Azimi, P., Sharifi, M., Zandieh, M.: A new soft computing algorithm based on cloud theory for dynamic facility layout problem. RAIRO Oper. Res. **55**, S2433–S2453 (2021)
13. Guo, W., Jiang, P., Yang, M.: Unequal area facility layout problem-solving: a real case study on an air-conditioner production shop floor. Int. J. Prod. Res. **61**, 1479–1496 (2023)
14. Pourvaziri, H., Salimpour, S., Akhavan Niaki, S.T., Azab, A.: Robust facility layout design for flexible manufacturing: a doe-based heuristic. Int. J. Prod. Res. **60**, 5633–5654 (2022)
15. El-Mihoub, T.A., Hopgood, A.A., Nolle, L., Battersby, A.: Hybrid genetic algorithms: a review. Eng. Lett. **13**, 124–137 (2006)
16. Keller, B., Buscher, U.: Single row layout models. Eur. J. Oper. Res. **245**, 629–644 (2015)
17. Wahab, M.I.M., Stoyan, S.J.: A dynamic approach to measure machine and routing flexibilities of manufacturing systems. Int. J. Prod. Econ. **113**, 895–913 (2008)
18. La Scalia, G., Micale, R., Enea, M.: Facility layout problem: bibliometric and benchmarking analysis. Int. J. Ind. Eng. Comput. **10**, 453–472 (2019)
19. Deep, K.: Machine cell formation for dynamic part population considering part operation trade-off and worker assignment using simulated annealing-based genetic algorithm. Eur. J. Ind. Eng. **14**, 189–216 (2020)
20. Alam, M.S., Islam, M.M., Yao, X., Murase, K.: Diversity guided evolutionary programming: a novel approach for continuous optimization. Appl. Soft Comput. **12**, 1693–1707 (2012)
21. Eiben, Á.E., Hinterding, R., Michalewicz, Z.: Parameter control in evolutionary algorithms. IEEE Trans. Evol. Comput. **3**, 124–141 (1999)
22. Aleti, A., Grunske, L.: Test data generation with a Kalman filter-based adaptive genetic algorithm. J. Syst. Softw. **103**, 343–352 (2015)
23. Senthil Babu, S., Vinayagam, B.K.: Surface roughness prediction model using adaptive particle swarm optimization (APSO) algorithm. J. Intell. Fuzzy Syst. **28**, 345–360 (2015)
24. Balakrishnan, J.D., Cheng, C.H., Conway, D.G., Lau, C.M.: A hybrid genetic algorithm for the dynamic plant layout problem. Int. J. Prod. Econ. **86**, 107–120 (2003)
25. Datta, D., Amaral, A.R.S., Figueira, J.R.: Single row facility layout problem using a permutation-based genetic algorithm. Eur. J. Oper. Res. **213**, 388–394 (2011)
26. Peng, Y.F., Zeng, T., Fan, L.Z., Han, Y.J., Xia, B.X.: An improved genetic algorithm based robust approach for stochastic dynamic facility layout problem. Discrete Dyn. Nat. Soc. **2018**, 1–8 (2018)

27. Gong, J., Zhang, Z., Liu, J., Guan, C., Liu, S.: Hybrid algorithm of harmony search for dynamic parallel row ordering problem. J. Manuf. Syst. **58**, 159–175 (2021)
28. Zouein, P.P., Kattan, S.: An improved construction approach using ant colony optimization for solving the dynamic facility layout problem. J. Oper. Res. Soc. **73**, 1517–1531 (2022)
29. Khajemahalle, L., Emami, S., Keshteli, R.N.: A hybrid nested partitions and simulated annealing algorithm for dynamic facility layout problem: a robust optimization approach. Infor **59**, 74–101 (2021)
30. Salimpour, S., Pourvaziri, H., Azab, A.: Semi-robust layout design for cellular manufacturing in a dynamic environment. Comp. Oper. Res. **133** (2021)
31. Yang, T., Brett, A.P.: Flexible machine layout design for dynamic and uncertain production environments. Eur. J. Oper. Res. **108**, 49–64 (1998)
32. Siddique, P.J., Luong, H.T., Shafiq, M.: An optimal joint maintenance and spare parts inventory model. Int. J. Ind. Syst. Eng. **29**, 177–192 (2018)
33. Yeh, R.H., Kao, K.-C., Chang, W.L.: Preventive-maintenance policy for leased products under various maintenance costs. Expert Syst. Appl. **38**, 3558–3562 (2011)
34. Lu, Z.Q., Cui, W.W., Han, X.L.: Integrated production and preventive maintenance scheduling for a single machine with failure uncertainty. Comput. Ind. Eng. **80**, 236–244 (2015)
35. Parika, W., Seesuaysom, W., Vitayasak, S., Pongcharoen, P.: Bat algorithm for designing cell formation with a consideration of routing flexibility. In 2013 IEEE International Conference on Industrial Engineering and Engineering Management, IEEM 2013, pp. 1353–1357 (2014)
36. Thioulouse, J., Dray, S.: Interactive multivariate data analysis in R with the ade4 and ade4TkGUI packages. J. Stat. Softw. **22**, 1–14 (2007)
37. Vitayasak, S., Pongcharoen, P., Hicks, C.: Robust machine layout design under dynamic environment: Dynamic customer demand and machine maintenance. Expert Syst. Appl. X **3**, 100015 (2019)
38. Vitayasak, S., Pongcharoen, P.: Interaction of crossover and mutation operations for designing non-rotatable machine layout. In: Operations Research Network Conference, Thailand (2011)

A Bilevel Genetic Algorithm for Global Optimization Problems

Ziru Lun, Zhanyu Ye, and Qunfeng Liu[✉]

Dongguan University of Technology, Dongguan 532000, China
liuqf@dgut.edu.cn

Abstract. Genetic algorithm is an important intelligent optimization algorithm that operates on specific population by simulating the natural evolution process and using artificial evolution to continuously optimize the population so as to search for the optimal solution. At present, there are a large number of methods focus on improving genetic algorithms, but the current stage of genetic algorithm tends to have the problems of falling into local optimal premature and slow convergence. In this paper, we try to design a bilevel evolutionary particle swarm optimization algorithm based on the idea of genetic algorithm within the framework but without increasing the complexity, using a data-driven idea, and verify it by the genetic algorithm in the commercial software MATLAB. Numerical experiments show that the data-driven a bilevel genetic algorithm-based algorithm significantly improves the algorithm performance.

Keywords: intelligent optimization algorithm · genetic algorithm · bilevel evolution.s

1 Introduction

Genetic algorithm is one of the most popular swarm-based optimization algorithms for global optimization problem as follows [1].

$$\min_{x} f(x), \ x \in \Omega \subseteq R^{n}, \tag{1}$$

where $f(x)$ is the objective function, n-dimensional vector x is the decision variable, and Ω is the feasible region. However, comparing with other swarm-based optimization algorithms, such as particle swarm optimization and differential evolution, genetic algorithm's numerical performance is often not satisfactory [8].

Recently, a recursive deep swarm search (RDSS) framework is proposed to improve swarm-based optimization algorithms' performance, and it has been applied successfully on the particle swarm optimization [9] and partition-based global optimization problems [6], especially the DIRECT algorithm [3–5, 7]. The RDSS framework adopts the swarm-based algorithm itself to explore in one level (e.g., the original search space) and exploit in another level (e.g., a promising subregion of the search space). Adopting the above two-level search approach in the second level, then a recursive deep swarm search is obtained naturally. The most important feature of the RDSS framework is the global search (exploration) and the local search (exploitation) are implemented effectively by the same swarm-based optimization algorithm. In other words, no need to employ another local optimization algorithm.

Encouraged by the success of the RDSS framework, in this paper, we apply the RDSS to genetic algorithm and hope to improve the numerical performance significantly. A bilevel genetic algorithm is proposed, where genetic algorithm searches in the original search space for several iterations and then go deep into a promising subregion defined by the elite individuals. Repeat such process until the stopping condition satisfied. Numerical results show that the bilevel genetic algorithm improve significantly the original genetic algorithm's performance.

The rest of this paper is organized as follows. A brief review of genetic algorithm is presented in Sect. 2, and then the bilevel genetic algorithm is proposed detailly in Sect. 3. Numerical experiments are reported in Sect. 4, and the conclusion is summarized in the final section.

2 Brief Review of Genetic Algorithm

In this section, we introduce the traditional genetic algorithm, then analyzes the characteristics of the traditional genetic algorithm, and finally describes the basic process of the genetic algorithm.

2.1 The Idea of Genetic Algorithm

The genetic algorithm is a typical intelligent optimization algorithm proposed by Professor Holland, which integrates the law of "survival of the fittest" in nature, and is often used to solve optimization problems [1]. It has powerful search capabilities and occupy a high position among many intelligent algorithms.

Genetic algorithm simulate genetic inheritance and species evolution in nature, each individual represents an efficient solution, feasible solutions are limited to a specific search area, populations represent sets of feasible solutions, and feasible solutions to problems are represented by genetic coding. The algorithm initializes a population, encodes each individual in the population, calculates the adaptation value of each individual, and evolves the population through heredity, mutation and selection, and finally converges the population to the chromosomes that are most adapted to the environment [2].

2.2 Basic Process of Genetic Algorithm

The genetic algorithm is an adaptive global search algorithm, which can continuously perform iterative search according to the fitness function, and continuously optimize the solution space of the problem to be solved during the search process until the stopping condition position is reached[]. The flowchart of the traditional genetic algorithm is shown in Fig. 1:

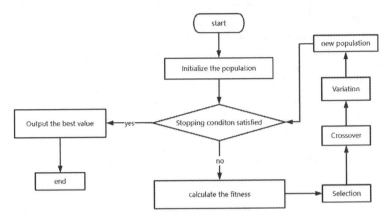

Fig. 1. The flow of genetic algorithm

2.3 Features of Genetic Algorithm

Genetic algorithm is a kind of global probability search optimal solution algorithm, with the genetic calculation is continuously optimized and improved, genetic algorithm is widely used in all walks of life, such as image processing optimization, network optimization, engineering management process optimization system dynamics optimization, etc.When solving practical problems, genetic algorithm present the following characteristics:

Parallelism. The genetic algorithm can use multiple computers to search the population online during the search, so that the search efficiency can be improved when the population is large. And the genetic algorithm not only searches for single-target problems, but can search for multi-target problems, and can solve complex problems with multiple parameters, multiple variables, multiple targets, and multiple regions. [11].

Robustness. The genetic algorithm is based on the solution set as the initial population, and the individuals who are far from the optimal solution of the objective function are quickly found through genetic operations, and the adaptive function is used as the basis for searching in the search process, without derivation, the discrete solution can be processed, and there is no limit to solving the problem, which can be a discrete function or a continuous function; It can be explicit or recessive, can be differentiable or non-differentiable. Genetic algorithm can mix with other algorithms [12].

Adaptability. The genetic algorithm adopts adaptive search technology to improve the probability of the next generation of highly adapted individuals in the evolutionary process, and can retain better individuals, so that the screened individuals can better adapt to the living environment [12].

It is easy to fall into local optimum. Due to the limited size of the initial population, a large number of highly adapted individuals survive, the chromosomal coincidence rate in the population is high, and the population lacks diversity, resulting in the selection, crossover, and mutation of genetic algorithm, the population no longer changes, resulting in the inability to find the global optimal solution, but falling into the local optimal situation [11].

3 Bilevel Evolution of Genetic Algorithm

As reviewed in Sect. 2, genetic algorithm has good optimization effect, but the convergence speed is slow, and it is easy to fall into the local optimal solution. At present, most of the genetic algorithm is mathematical optimization for the algorithm process, and the innovation idea is limited, it is difficult to jump out of the parameter optimization of the algorithm in the province, and there is no macro level evolution involved. Aiming at the defects of the existing optimization strategies, this paper proposes a new optimization idea, which mainly optimizes the algorithm at the macro level, so that the accuracy of the algorithm is higher.

The algorithm is mainly divided into three parts. In the first part, the elite strategy is used to find the elite subgroup to iterate separately. In the second part, the elite retention strategy was used to select the elite individuals. In the third part, evolution is carried out with the Bilevel framework [8].

3.1 Bilevel Evolution Framework

Single-level evolution always uses all individuals in the population to evolve, which is easy to fall into local optimum due to the fixity of the solution space. Bilevel evolution means that the population is divided into multiple level Spaces based on the initial population, and each level adopts different selection strategy or evolution strategy. The Bilevel evolution strategy adopted in this paper constructs two levels of biological evolution and cultural evolution to perform genetic operations on the original population [9].

Biological evolution refers to the mimicry or imitation of genetic selection and biological behavior of organisms in nature, with a lower probability of genetic mutation and a slower rate of population evolution. In contrast, cultural evolution refers to the evolution in a new solution space based on the experience and laws obtained from biological evolution, which has a faster update iteration and shorter period compared to biological evolution. By applying cultural evolution to genetic algorithm, better solutions can be found more quickly [6].

3.2 Elite Selection Strategy

The selection strategy of the traditional genetic algorithm, such as roulette wheel and bidding competition, is randomly selected, which has slow convergence speed, the cost of each evolution is the whole population, and the calculation cost is also high. Due to the single fitness function and the single system of evaluating individuals, it is easy to cause a single chromosome and fall into the local trap. In order to save the computational cost and increase the diversity of the population, an elite selection strategy can be used to select the offspring [5].

Elite selection strategy is to select several elite individuals with better evaluation from the entire population to form the initial elite population, and to exclusively implement genetic operations on the elite population to cultivate elite offspring that are closer to the optimal value of the target function.

3.3 Elite Retention Strategy

After obtaining the offspring of the elite population, the elite individuals of the population should be put back into the initial population according to the elite retention strategy, and the elite group and other individuals should be cultivated together for several generations, so that the information can be exchanged between biological evolution and cultural evolution, and the excellent individuals in the parent generation can be retained in the population to a greater extent. And promote the convergence of the population to the direction of the optimal solution of the objective function.

After executing the elite selection strategy in the cultural evolution layer, the population individuals can be divided into two types, elite individuals and ordinary individuals. Elite retention strategy is a strategy of gene exchange between biological evolution layer and cultural evolution layer, which can help the genes of elite offspring play a greater role in the population. Elite retention strategy refers to selecting a certain proportion of individuals in the elite population to replace their parents and put them into the population. At this time, two situations will occur: In the first case, the elite offspring is closer to the optimal value of the objective function than the parent. In this case, the elite offspring replaces the parent to evolve, which can retain excellent genes, reduce the calculation time of the intelligent optimization algorithm, reduce the calculation cost, and increase the convergence speed. In the second case, the parent generation is closer to the optimal value of the objective function than the elite offspring, which can increase the diversity of species and avoid falling into local optimum. In either case, the genetic algorithm will be improved to a certain extent and the effect of the algorithm can be optimized.

3.4 A Bilevel Genetic Algorithm

Firstly, introduce elite culture based on GA in MATLAB. Then, use elite selection strategy to select the best 30% of individuals in the population to build an elite subpopulation for cultural evolution. Next, use elite retention strategy to put the elite individuals back into the original population after evolution [14]. Last, repeat the above process until the computational cost is used up. The pseudo-code of the algorithm is expressed as follows.

Algorithm (Genetic algorithm based on two-level evolution, GA-2Level)

Input: Maximum calculated cost MaxNF, population size N
Output: the best function value.
Initialize the population P;
While the function value is evaluated <MaxNF , do the following:
1. Biological evolution: execute the GA and iterate 30 times.
2. Cultural evolution: Select 30 best individuals from population P, build subgroup, and iterate GA code 1 times based on the subgroup, spending 30 times of calculation cost;
3. Two-level communication: Put 30 individuals back into the initial population and continue the iteration.
End.

---

4 Numerical Experiments

In this section, we will compare the following two algorithms: GA: (1) The genetic algorithm built into MATLAB; (2) GA-2level: Genetic algorithm based on a two-level framework.

4.1 Test Data Collection and Processing

In this section, we adopt the Hedar dataset and CEC2020 dataset to test the performance of the algorithm [10]. Hedar dataset has 27 basic functions. If the original search space belongs to different dimensions or the variables are different, it is regarded as two functions, so there are a total of 68 functions to test. The CEC2020 dataset contains 10 unconstrained test functions, which are divided into multimodal functions and combinatorial functions. With a large number of functions and a high number of dimensions, the algorithm performance can be well measured.

In the experiment, GA-2level and GA tested 68 problems in hedar dataset and 10 problems in CEC2020 dataset respectively until the consumption of 20,000 function calculations stopped [13]. Each test function in the Hedar dataset and the CEC2020 dataset is independently tested 30 times. After 30 independent tests, the change in population after each iteration was recorded. Take the average of 30 calculations as the calculation result of the algorithm under this function. By recording the history of function value changes during evolution, a high-dimensional matrix can be obtained as follows.

$$H(MaxNF, 30, M , 2)$$

Matrix H indicates that two algorithms (MATLAB-GA, GA-2Level) were tested, each of which tested M functions, each solved 30 times independently to eliminate the noise of randomness as much as possible. The calculation cost of each test is the number of MaxNF function value calculations. With the help of the H matrix, the history of the decline of the function values in each test can be described. Due to the large number of test functions, in order to facilitate the presentation of numerical performance comparison results, we adopt the data profile technology to analyze process data [10], which has been proved to be a paradox-free data analysis method recently [13], proposed by John Moré and Stefan Wild in 2009, employs the following inequality as a convergence condition for the algorithm to solve the problem.

$$f(x) \leq f_L + \tau(f(x_0) - f_L). \tag{2}$$

where x_0 is the initial iteration point, and τ is the accuracy parameter ($\tau = 10\text{--}7$ in this paper). f_L is the minimum function value obtained when solving the test function for all optimization algorithms tested within a given computational cost, each corresponding to a f_L. $t_{p,s}$ is the lowest computational cost for the algorithm $s \in S$ to solve the function $p \in P$ such that condition (2) is satisfied, and if condition (2) is never met, then $t_{p,s}$ is infinity. Among them, the sets S and P are the set of algorithms and test functions to be compared, respectively. Then the data profile curve of the algorithm $s \in S$ is defined as:

$$d_s(\alpha) = \frac{1}{|P|} \left| \left\{ p \in P : \frac{t_{p,s}}{n_p + 1} \leq \alpha \right\} \right|. \tag{3}$$

where n_p is the number of dimensions of the test function p, $\frac{t_{p,s}}{n_p+1}$ roughly represents the relative computational cost of averaging each dimension. Thus, the data profile curve describes the proportion of problems that the algorithm s can solve within α relative computational costs.

4.2 Algorithm Parameter Design

In the test of GA-2level, the parameter design suggestions are as follows: the initial population size is set to 100, the calculation cost is set to 20000, other parameters of GA are set to default values, and the termination condition is set to all use up the computational cost.

4.3 Sensitivity Analysis of Parameters

Firstly, we test the sensitivity analysis of GA-2level to the number of iterations of biological evolution, and the number of iterations of biological evolution was set to 10, 20, 30, and 40 generations in turn. Figure 3 shows the comparison of GA and GA-2level on the Hedar test dataset, both algorithms terminated after 20,000 function computations were exhausted. The horizontal axis is the relative calculation cost calculated by adding one of the number of calculations and the dimension of the problem, and the vertical axis is the proportion of the problem solved. Given the same computational cost, the higher the data profile curve, the better the performance of the algorithm. The iteration times corresponding to biological evolution of the four subgraphs are set to 10, 20, 30, and 40 generations. The blue line is the curve corresponding to the original genetic algorithm solution problem, and the red line is the curve obtained by two levels of genetic algorithm [8].

As can be seen from Fig. 2, all two levels of genetic algorithm perform better on the Headar dataset than the genetic algorithm built into MATLAB. At the same computational cost, four GA-2level can solve 95% of the test functions, but GA can only

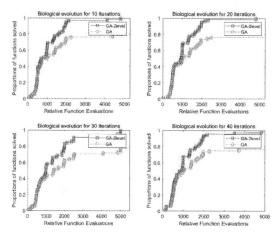

Fig. 2. Sensitivity analysis of number of biological evolutions.

solve 70% of the test functions [10]. Therefore, the performance gap between the four GA-2level and GA is about 25%, which is significant. That is, regardless of whether the number of biological iterations is set to 10, 20, 30, or 40, both GA-2level perform better on the Hedar test set than GA. Therefore, we can measure that the GA-2level are insensitive to number of biological evolutions.

Next, we test the sensitivity analysis of GA-2Level to the number of iterations of cultural evolution, and the number of iterations of cultural evolution was set to 1 generation, 2 generations, 3 generations, and 4 generations, and the experimental results were shown in Fig. 4.

As can be seen from Fig. 3, all GA-2level perform better on the Hedar dataset than the GA. However, when the number of biological evolutions reaches 4 generations, the performance of GA-2level decreases, that is, when the number of the biological evolutions is 3 generations, two-level genetic algorithm has the best performance, so the number of cultural evolution is usually set to 30.

4.4 Comparison of GA-2level and GA

To determine the performance of GA-2level, this section compares the performance of two algorithms, GA and GA-2level, on the Hedar test set and on the CEC2020 test set when the number of the biological evolution is 10 and the number of cultural evolutions is 3. In order to ensure that the number of calculations for a given function can be exhausted, only the Generations and StallGenLimi parameters are modified and the other parameters are left at default values.

Figure 4 shows the comparison results of GA and GA-2level on the Hedar test dataset, which is the data profile curve of GA and GA-2level. A significant law can be seen from the figure, and the performance of the improved algorithm GA-2Level proposed by us is superior. At the cost of 20,000 function computations, GA solved 45% of the functions and the GA-2level solved 98% of the functions, with a performance difference of more than 53%, and the difference was significant.

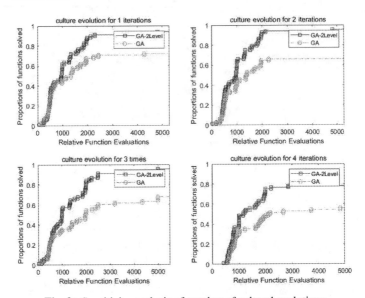

Fig. 3. Sensitivity analysis of number of cultural evolutions

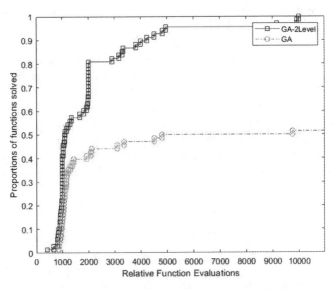

Fig. 4. Comparison of GA-2level and GA on Hedar

Figure 5 shows the comparison results of GA and GA-2level on the CEC2020 test dataset, which is the data profile curve of GA and GA-2level. A significant law can be seen from the figure, and the performance of the improved algorithm GA-2Level proposed by us is superior. At the cost of 20,000 function computations, GA solved 75%

of the functions and the GA-2level solved 98% of the functions, with a performance difference of more than 23%, and the difference was significant.

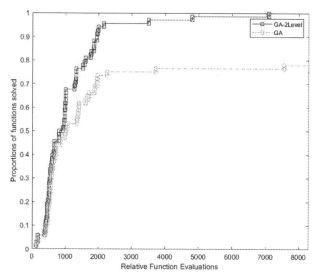

Fig. 5. Comparison of GA-2level and GA on CEC2020

5   Conclusion and Future Work

Encouraged by the successful applications of the recursive deep swarm search (RDSS) framework on some popular swarm-based optimization algorithms, in this paper, the RDSS framework is applied on GA, and a bilevel GA algorithm is designed. Through repeating the GA's biological evolutions in the original search space and a kind of cultural evolutions in the brief space inspired by the elites, the proposed bilevel GA performs much better than GA in MATLAB. Specifically, our numerical results show that more than 20% performance difference is improved by the bilevel GA.

Futural work includes more experimental evidences of the bilevel GA, designing of the multilevel GA, and more applications of the RDSS framework.

References

1. Holland John, H.: Adaptation in natural and artificial systems. University of Michigan Press, Ann Arbor (1975)
2. Li, L., Saldivar, A.A.F., Bai, Y., Chen, Y., Liu, Q., Li, Y.: Benchmarks for evaluating optimization algorithms and benchmarking MATLAB derivative-free optimizers for practitioners' rapid access. IEEE Access **7**, 79657–79670 (2019)
3. Liu, Q., Zeng, J., Yang, G.: MrDIRECT: a multilevel robust DIRECT algorithm for global optimization problems. J. Global Optim. **62**(2), 205–227 (2015)

4. Liu, Q., Yang, G., Zhang, Z., Zeng, J.: Improving the convergence rate of the DIRECT global optimization algorithm. J. Global Optim. **67**(4), 851–872 (2017)
5. Liu, Q., Zeng, J.: Global optimization by multilevel partition. J. Global Optim. **61**(1), 47–69 (2015)
6. Liu, Q., Cheng, W.: A modied DIRECT algorithm with bilevel partition. J. Global Optim. **60**(3), 483–499 (2014)
7. Liu, Q., et al.: Benchmarking stochastic algorithms for global optimization problems by visualizing confidence intervals. IEEE Trans. Cybern. **47**(9), 2924–2937 (2017)
8. Yan, Y., Zhou, Q., Cheng, S., Liu, Q., Li, Y.: Bilevel-search particle swarm optimization for computationally expensive optimization problems. Soft Comput. **25**, 14357–14374 (2021)
9. Moré, J.J., Wild, S.M.: benchmarking derivative-free optimization algorithms. SIAM J. Optim. **20**(1), 172–191 (2009)
10. Sun, J., Li, J., Wang, D., et al.: Thinned array optimization based on genetic model improved artificial bee colony algorithm. High Power Laser Part. Beams **33**(12), 43–50 (2021)
11. Euziere, J., Guinvarc'h, R., Uguen, B., Gillard, R.: Optimization of sparse time-modulated array by genetic algorithm for radar applications. IEEE Antennas Wirel. Propag. Lett. **13**, 161–164 (2014)
12. Yan, Y., Liu, Q., Li, Y.: Paradox-free analysis for comparing the performance of optimization algorithms. IEEE Trans. Evol. Comput. (2022)
13. Jing, L., Cai, T.: Problem Definitions and Evaluation Criteria for the CEC Special Session on Evolutionary Algorithms for Sparse Optimization Technical Report. Nanyang Technological University, Singapore (2020)
14. Herrera, F., Herrera-Viedma, E., Lozano, M.: Fuzzy tools to improve genetic algorithms. Dept. Comput. Sci. Artif. Intell. (1994)

Genetic Algorithm for Product Design Optimization: An Industrial Case Study of Halo Setting for Jewelry Design

Somlak Wannarumon Kielarova[1,2,3]([⊠]) [iD] and Prapasson Pradujphongphet[2,3]

[1] Centre of Excellence on Energy Technology and Environment, Naresuan University, Phitsanulok 65000, Thailand
somlakw@nu.ac.th
[2] iD3-Industrial Design, Decision and Development Research Unit, Faculty of Engineering, Naresuan University, Phitsanulok 65000, Thailand
[3] Department of Industrial Engineering, Faculty of Engineering, Naresuan University, Phitsanulok 65000, Thailand

Abstract. This paper proposes a new approach for the optimization of design parameters and the generation of design alternatives in product design process by using the generative design method (GDM). The proposed GDM is comprised of two stages: design optimization and design variation. This work focuses on developing a prototype system for jewelry design parameter optimization and shape variation. Using Genetic Algorithm (GA), the system will assist designers in creating halo ring designs with optimization of the number of halo gemstones and the gap size between them in the halo setting. Therefore, this system helps designers to avoid undesirable design iterations that cause waste of a great amount of designing time and effort. In GA the properties of the objects are described using chromosomes, where each chromosome is encoded with a set of jewelry design parameters. Uniform crossover and point mutation are utilized as genetic operation. Fitness function is derived from the mathematical relationships between the design parameters to calculate the gap size between halo stones. The empirical results indicate that the proposed approach is applicable for product design optimization. The proposed design system can reduce time for setting halo gemstones surrounding a center gemstone in comparison to manual setting using the trial-and-error method. Using this generative design technique, the proposed design system can automatically generate various design alternatives of halo rings with the set of optimized design parameters, such as center gemstone size, halo gemstone size, number of halo gemstones, and gap size. The experimental results of halo ring setting are also provided in this paper.

Keywords: Jewelry Design · Genetic Algorithm · Evolutionary Design · Design Optimization · Generative Design Method

1 Introduction

Nowadays, a competitive business environment forces enterprises to speed up their abilities in creating new products. Quick response to customer needs and high flexibility in product design are needed. The proposed generative design methodology helps to avoid redesign iterations, which can shorten the design cycle time to the market.

This paper proposes a new GDM for optimizing design parameters in product design process by using GA and generation of various design alternatives. The paper presents the proposed method through the development of a prototype system for jewelry design parameter optimization and design variation. Halo gemstone rings are one of the most popular types of jewelry rings that enjoy a high demand in the jewelry market. Therefore, it was chosen to be an industrial showcase for this work.

At present, various CAD packages are available in the market and play important roles in the jewelry sector. Even though the function of automatic arrangement of objects along curves, which could be applied to arrange the halo gemstones, is provided in the CAD packages, the users still need to experiment with parameters such as the number of gemstones or the equal gap size between them.

In our previous research, the mathematic models of relationship between parameters in halo gemstone arrangement were derived to be used in the computer-aided design module for automatic gemstone setting on a halo ring [1]. The design module has the non-trivial task of deriving the mathematic models, which are varied by the shape of the center gemstone and can provide only a single solution to the problem.

Given these limitations, this paper proposes a generative design system that uses Genetic Algorithm (GA) for optimizing the design parameters used in the generation of halo ring designs. With this GA, users can obtain a set of solutions generating a variety of halo rings according to the size and cut of the center gemstone and halo gemstones in hand.

This paper is organized into five sections. The next section provides the related theories and related published reports, which applied GA to the optimization of products. Section 3 describes the framework of the proposed generative design system based on GA. An industrial case study, as well as the experimental results are described and discussed in Sect. 4. Finally, this research work and its future directions are summarized in Sect. 5.

2 Literature Review

Evolutionary art [2] and design systems [3] offer an effective approach for the creation of pleasing art forms. The evolutionary process in the evolutionary art and design systems works as a design parameter optimizer and form generator. Using the advantages of such systems, designers or users can explore more solutions or design alternatives. Evolutionary design systems typically generate new forms based on random initial populations. Each individual item of a population is evaluated for its fitness by a user or the computer-based design system. The evolutionary process then generates new design solutions based on the highest fitness ranking. This process employs the advantage of evolution for the improvement of shape generation.

In GA a population of individuals or phenotypes is evolved toward better solutions as part of the optimization problem. Each individual item has a set of properties called chromosomes or genotypes, which can be mutated and crossed over. The solutions can be encoded in binary code or real numbers [4]. The evolutionary process starts from a population of randomly generated individuals. It is an iterative process with the population of each iteration being called a generation. In each generation, the fitness of every individual in the population is evaluated. The fitness is usually the value of the objective function in the optimization problem being solved. The more fit individuals are stochastically selected from the current population, and everyone's genome is modified by recombination and mutation, to form a new generation. The new generation of candidate solutions is then used in the next iteration of the algorithm. GA typically terminates when either a maximum number of generations has been reached, or a satisfactory fitness level has been achieved for the population.

A typical GA requires a genetic representation of the solution domain and a fitness function to evaluate the solution domain. The major elements of a GA [3] are described as follows. Genotype is a genetic representation that is mostly encoded in the string of chromosomes considered as the basic units of evolution. More suitable chromosome structures and more suitable genotype representations are chosen, meaning easier genetic operators can be applied. Genotype can be encoded either in binary or real numbers. In evolutionary design process, before the quality of an individual will be evaluated, genotypes are mapped onto phenotypes. In general, phenotypes consist of sets of parameters that represent the shape or form of the studied product. Phenotypes have been represented by various techniques depending on the objectives of the system. The new offspring are typically reproduced by genetic operators: crossover and mutation. Crossover is an event where parts of the chromosomes of the two selected parents are randomly recombined to create a new set of offspring. These offspring then inherit the characteristics of each parent. Mutation changes an arbitrary part in a genetic sequence from its original state. Mutation is used to maintain population diversity during evolution.

Fitness function represents a heuristic estimation of the solution quality. It is derived from the objective functions, to measure the phenotypes' abilities or properties. It is a key method for leading the individuals' evolution. For every phenotype, its fitness, or a level of goodness for each solution must be evaluated. In an evolutionary design, almost all computational time is spent in the evaluation process [3], which can take from few minutes up to several hours to evaluate a single solution. The process can be improved by reducing the number of evaluations during the evolutionary process. Population size is often less than ten individuals [3], which are then judged rapidly in each generation. Selection is a process of choosing suitable phenotypes according to their fitness. The selection scheme can determine the generating direction of the process.

Recently, there have been several reports that applied GA in design applications and creation of artworks e.g., [5–12]. Some of these works integrated GA with other techniques such as artificial neural network, fuzzy set, etc., depending on the problem domains.

3   Genetic Algorithm for Product Design Optimization: An Industrial Case Study of Halo Setting for Jewelry Design

This paper proposes a framework of a generative product design system, which consists of two stages. The first stage is the optimization of the design parameters. The second stage is the generation of the 3D models of design solutions generated according to the set of design parameters optimized in the previous stage. The framework of the proposed generative design of the industrial case study of jewelry design system is illustrated in Fig. 1.

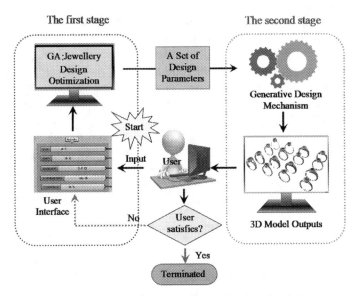

Fig. 1. The framework of the proposed generative jewelry design system.

During this research project, we collaborated with a jewelry entrepreneur in Thailand. An industrial case study of halo gemstone ring design was chosen as the showcase problem. We developed a new approach for the optimization of gem arrangement on halo rings by using GA. This work was undertaken to support CAD designers to automatically generate halo ring designs with optimized number of halo gemstones used and the gap size between them using input parameters: size of the center gemstone and size of the halo gemstones.

In this work, direct value encoding is used for encoding chromosomes to represent the head of the halo ring. The chromosome is encoded as a set of integer and real numbers with the same length of the number of design parameters as shown in Fig. 2. An example of chromosome encoding is shown in Fig. 3.

In the GA process, uniform crossover and point mutation are used to create new various solutions to maintain diversity in a population. Isotropic selection [13] is used to protect a premature colonization of a local optimum.

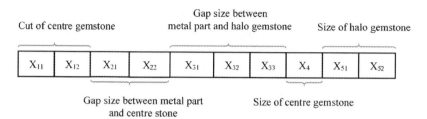

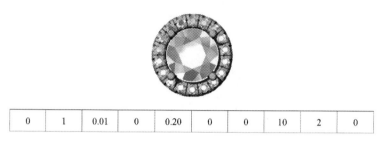

Fig. 2. Example of chromosome design used in the evolutionary design process.

0	1	0.01	0	0.20	0	0	10	2	0

Fig. 3. An example of chromosome encoding.

4   An Industrial Case Study of Halo Gemstone Setting: Experimental Results and Discussions

4.1 Problem Definition

Halo ring is a ring in which a center gemstone is surrounded with a set of halo stones (a set of round pave diamonds or faceted color gemstones). The function of the halo stones is to bring attention to the center stone.

We have identified a bottleneck in the design process of halo rings, which is the setting of halo gemstones surrounding the center stone with suitable equal gaps. Changing the gemstone size and/or changing the ring size requires rearrangement or re-setting of the halo gemstones on the head of the ring to obtain the proper gaps between the halo gemstones to achieve the desired aesthetics. This necessitates further metalsmith's work. The rearrangement of halo gemstones to optimize the gaps between them by manual approach with the trial-and-error method is therefore relatively time-consuming.

The industrial case study undertaken during this project has aimed to automate the setting of halo gemstones on the head of a ring using two input parameters: size of the center gemstone and size of the halo gemstones. The anatomy of a halo ring is illustrated in Fig. 4.

The work procedure and the conditions of halo setting in our study are summarized as follows:

In this study, the center gemstone cuts are limited to round cut and cushion cut, while only round cut is used for the halo gemstones. The center gemstone cut dominates the setting pattern of the halo setting defining the number of halo gems and gap size between the halo gemstones.

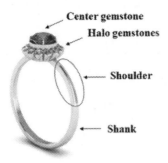

Fig. 4. Anatomy of a halo ring.

Examples of acceptable halo setting patterns are shown in Fig. 5. The concept of arrangement of gemstones on the head of a halo ring is based on trying to set the gemstones in the positions at quarters and angles as illustrated in Fig. 5. This arrangement will result in the halo appearing to look round. Figure 6 compares examples of acceptable and unacceptable halo gemstone setting.

Fig. 5. Illustration of acceptable patterns of halo gem setting.

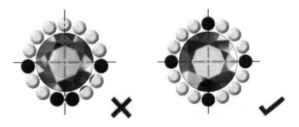

Fig. 6. Comparison of acceptable and unacceptable halo gemstone settings.

The popular sizes of the center gemstones range from 0.15 carat to 4.0 carat, while the popular sizes of the halo gems range from 0.002 carat to 0.035 carat. The gap between halo gems is typically between 0.20 and 0.35 mm. Figure 7 illustrates the gap size between halo gems.

Kielarova et al. [1] studied the relationships between the design parameters and derived a mathematical model for halo gemstone setting for round-cut center gemstone

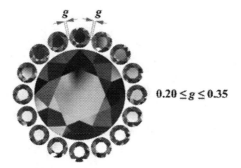

Fig. 7. Illustration of the gap size between halo gemstones.

by surrounding it with a single layer. The mathematical relationships are shown in Eq. (1)–Eq. (3) and the parameters used are illustrated in Fig. 8.

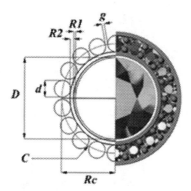

Fig. 8. Illustration of the design parameters used in halo gemstone setting.

The number of halo gems (*Num*) and gap size between the halo gemstones (*gap*) are calculated from the radius of the center gemstone (R_{centre}). Additional parameters used are the diameter of the halo gem (d_{halo}), the length of the curve for setting halo gems (*LC*), the radius of curve (R_{curve}), the gap size between the center gemstone and the metal part (gap_{cm}), and the gap size between the metal part and the halo gemstone (gap_{mh}). The number of halo gems is calculated by dividing the length of the curve for setting halo gems by the summation of the radius of the halo gem and the gap size between the halo gemstones as shown in Eq. (1). The length of the curve for setting halo gems (*LC*) is calculated by Eq. (2). Finally, the radius of curve (R_{curve}) is calculated by the summation of the radius of the center gemstone (R_{centre}), the radius of the halo gem (r_{halo}), the metal part (gap_{cm}), and the gap size between the metal part and the halo gemstone (gap_{mh}) as shown in Eq. (3).

$$Num = LC/(d_{halo} + gap) \tag{1}$$

$$LC = 2\pi R_{curve} \tag{2}$$

$$R_{curve} = R_{centre} + r_{halo} + gap_{cm} + gap_{mh} \qquad (3)$$

In this research, GA is used to optimize the number of halo gems (*Num*) and the gap size between them (*gap*). Minimum number of halo gems is required, while the gap size between them is minimized within 0.20 and 0.35 mm.

4.2   Experimental Results and Discussions

The prototype system was developed using Grasshopper in Rhinoceros. We employed Galapagos [14], a plug-in in Grasshopper [15], to build the proposed GDM. In the first stage, the generative design system starts when the user provides inputs of the size of the center gemstone and the size of the halo gemstones to the system via the user interface.

An illustrative test of designing a halo ring was performed with the following details:

- Ring size: US standard size 8.5 with the diameter of 18.57 mm,
- A center gemstone with the size 3.87 carat or 10.00 mm. in round cut or brilliant,
- Halo gemstone: size 0.015 carat (1.50 mm.) in round cut or brilliant; number of halo gemstones is even, and the target gap size is 0.20 mm.

In this example, the number of halo gemstones is unknown, but it is requested to be even. The results of outputs from the generative jewelry design system are shown in Fig. 9, which provides the solution as follows:

- The number of halo gemstones used for setting around the center gemstone is 24.
- The gap size between the halo gems is 0.202 mm.

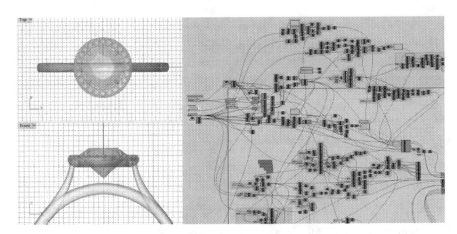

Fig. 9. Sample experimental results of the proposed generative jewelry design system.

After obtaining the results from the first stage, the system uses these results of design parameters to generate various 3D models of the resulting halo rings as shown in Fig. 10.

Comparison of the proposed GA used for optimizing the halo setting examples and automatically setting gemstones on the halo ring with the manual approach using a

Fig. 10. Selected 3D models of halo rings generated in the second stage.

CAD package was performed by assigning the design of the halo rings as explained in the previous section. The process began with the participants, who are CAD designers, drawing halo ring models. Subsequently, the participants performed setting of the center stone and setting of the halo gems onto the models. They usually set the halo gems by manual trial-and-error method with different number of halo gems until obtaining the gap size between the halo gems within the preferred range from 0.20 to 0.35 mm. Moreover, they needed to prepare the cutters to cut the channels for the settings and model the prongs. These tasks are very time consuming. For the example, in the case of setting a round cut center stone with halo gems in one layer, the CAD designers spent on average 55 min to complete the task. In comparison the proposed GA spent only approximately 15 min on the task, which reduced production time by about 72.7%. Comparing the proposed GA to the mathematic halo setting module, the proposed GA can generate more varied solutions and can be easily extended to other gemstone cuts.

5   Conclusions and Future Direction

A generative design system based on GA was developed for jewelry design applications. An industrial case study focused on setting halo gemstones surrounding a center gemstone to achieve an aesthetic appearance through the suitable gap size between the halo gems was performed in collaboration with a jewelry entrepreneur in Thailand. The paper proposes a new generative design method, which is comprised of two stages: design optimization and design variation. A generative design system based on GA for jewelry design applications was developed to present the methodology of the proposed GDM.

The generative design system for automatic setting and arranging of halo gemstones on the head of halo rings is proposed. Based on the experimental results, the proposed system is capable of automatically setting the center stone and halo gemstones by using a set of inputs from the user. The inputs that the system requires are the size and cut of the center gemstone and the size of the halo gemstone. Those parameters are considered as components of the fitness function in the GA-based generative design system. The proposed system was developed for supporting jewelry CAD designers in optimizing the number of halo gemstones and the gap sizes between them and to be used for automatically generating halo rings. It was developed by using data and information about jewelry ring design from jewelry designers and a jewelry manufacturer. Testing of the system by the manufacturer showed that the proposed system can aid the CAD designers and reduce halo setting and arrangement time by about 72% in comparison with the standard manual method. In this case study, round cut and cushion cut of the center gemstone were studied. In future, we plan to include other gem cuts in our system. Moreover, accent gemstones on the shoulder of the ring will be included into the halo ring model.

Acknowledgements. This work was supported by Naresuan University (NU) and Faculty of Engineering, and National Science, Research and Innovation Fund (NSRF) Grant No. R2565B041. The authors would like to gratefully thank all participants for their collaborations in this research.

References

1. Kielarova, S.W., Pradujphongphet, P., Nakmethee, C.: Development of computer-aided design module for automatic gemstone setting on halo ring. KKU Eng. J. **43**(S2), 239–243 (2016)
2. Sims, K.: Artificial evolution for computer graphics. Comput. Graph. **25**(4), 319–328 (1991)
3. Bentley, P.: Evolutionary Design by Computers. Morgan Kaufmann, San Francisco (1999)
4. Whitley, D.: A genetic algorithm tutorial. Stat. Comput. **4**(2), 65–85 (1994)
5. Lourenço, N., Assunção, F., Maçãs, C., Machado, P.: EvoFashion: customising fashion through evolution. In: Correia, J., Ciesielski, V., Liapis, A. (eds.) EvoMUSART 2017. LNCS, vol. 10198, pp. 176–189. Springer, Cham (2017). https://doi.org/10.1007/978-3-319-55750-2_12
6. Tabatabaei Anaraki, N.A.: fashion design aid system with application of interactive genetic algorithms. In: Correia, J., Ciesielski, V., Liapis, A. (eds.) EvoMUSART 2017. LNCS, vol. 10198, pp. 289–303. Springer, Cham (2017). https://doi.org/10.1007/978-3-319-55750-2_20
7. Cohen, M.W., Cherchiglia, L., Costa, R.: Evolving mondrian-style artworks. In: Correia, J., Ciesielski, V., Liapis, A. (eds.) EvoMUSART 2017. LNCS, vol. 10198, pp. 338–353. Springer, Cham (2017). https://doi.org/10.1007/978-3-319-55750-2_23
8. Byrne, J., Cardiff, P., Brabazon, A., O'Neill, M.: Evolving parametric aircraft models for design exploration and optimisation. Neurocomputing **142**(Supplement C), 39–47 (2014)
9. Martín, A., Hernández, A., Alazab, M., Jung, J., Camacho, D.: Evolving generative adversarial networks to improve image steganography. Expert Syst. Appl. **222**, 119841 (2023)
10. Mok, P.Y., Xu, J., Wang, X.X., Fan, J.T., Kwok, Y.L., Xin, J.H.: An IGA-based design support system for realistic and practical fashion designs. Comput. Aided Des. **45**(11), 1442–1458 (2013)
11. Tang, C.Y., Fung, K.Y., Lee, E.W.M., Ho, G.T.S., Siu, K.W.M., Mou, W.L.: Product form design using customer perception evaluation by a combined superellipse fitting and ANN approach. Adv. Eng. Inform. **27**(3), 386–394 (2013)
12. Starodubcev, N.O., Nikitin, N.O., Andronova, E.A., Gavaza, K.G., Sidorenko, D.O., Kalyuzhnaya, A.V.: Generative design of physical objects using modular framework. Eng. Appl. Artif. Intell. **119**, 105715 (2023)
13. Evolutionary Principles applied to Problem Solving using Galapagos. https://ieatbugsforbreakfast.wordpress.com/2011/03/04/epatps01/. Accessed 09 Jan 2023
14. Galapagos Evolutionary Solver. http://www.grasshopper3d.com/group/galapagos. Accessed 09 Jan 2023
15. Grasshopper-Algorithmic Modelling for Rhino. http://www.grasshopper3d.com/. Accessed 20 Jan 2023

Quad-Criteria Task Scheduling in Multicores Based on NSGAIII

Jiaxuan Wang[1,2], Zheng Tan[1,3], and Bo Yuan[1(✉)]

[1] Department of Computer Science and Engineering, Southern University of Science and Technology, Shenzhen, China
yuanb@sustech.edu.cn
[2] Academy for Advanced Interdisciplinary Studies, Southern University of Science and Technology, Shenzhen, China
[3] School of Computer Science, University of Birmingham, Birmingham, UK

Abstract. It is very challenging to schedule multiple tasks in multicore systems, as there are multiple antagonistic optimization objectives, such as makespan, energy consumption, reliability, and peak temperature. To produce Pareto front in this four-dimensional space, the previous $\epsilon - constraint$ method transforms the original multi-objective optimization problem into multiple constrained single objective optimization problems, which relies heavily on the thresholds chosen for the constraints. As population-based global optimization methods, Multi-Objective Evolutionary Algorithms (MOEA) have been successfully applied in solving real-world multi-objective scheduling problems. In this work, we propose a problem-specific encoding strategy for the quad-criteria task scheduling problem and use the Non-dominated Sorting Genetic Algorithm-III (NSGAIII) as the search engine. Compared with the $\epsilon - constraint$ method, experimental results over a large set of benchmarks demonstrate that the proposed NSGAIII based method can provide a better solution set within approximately the same running time, especially when the number of cores is large.

Keywords: Task scheduling · Multicore systems · Multi-objective optimization · Evolutionary algorithm

1 Introduction

The multicores are widely used in embedded systems due to their ability to contain multiple independent processors [1,2]. The design paradigm of multicore embedded systems can be roughly divided into two different parts [3]. One is the basic hardware facilities of the systems, and the other is the application scheduling on the multicore structures [4]. The scheduling strategy affects how the application runs, so it is crucial for the system performance. The application

This work was supported by the National Natural Science Foundation of China under Grant 61976111 and Grant 62250710682.

scheduling problem is one of the most urgent problems to be solved for implementing embedded systems [5]. Our work mainly focuses on the task scheduling problem, while optimizing multiple objectives.

Multicore embedded systems have to balance many performance objectives in their design phase, such as the total execution time (makespan, which affects the response time to inputs), reliability, energy consumption (important for battery powered systems), and peak temperature (which can affect processing speed, reliability, and power consumption). However, these objectives are often antagonistic. Some antagonistic relationships are obvious, for example, between the total execution time and reliability. Improving reliability requires some redundancy (repeating tasks in space or time), which increases the total execution time. Some antagonistic relationships are less obvious, for example, between energy consumption and reliability. Lowering the core voltage reduces energy consumption by lowering power consumption, but also reduces reliability by increasing the failure rate of the core (lower voltage makes the circuit more vulnerable to particle interference [6]), so energy consumption and reliability are also antagonistic.

Several related works optimize the makespan, reliability, power consumption, peak temperature, etc., for applications on multicores, but they usually focus on only two or three objectives. ESRG [7] is a bi-objective list scheduling algorithm which minimizes energy consumption while satisfying the reliability constraint. The key idea of ESRG is to assign a reliability constraint to each task based on the application's reliability constraint. ESRG schedules tasks in priority order and chooses the processor and frequency combination that consumes the least energy while satisfying the task's reliability constraint. UMOTS [8] uses MOEA to find the Pareto front in the objective space of makespan, energy consumption, and reliability. It also uses a Monte Carlo (MC) based approach to deal with the uncertainties in both hardware and software components. HiMap [9] is a list scheduling method to maximize the lifetime reliability while satisfying makespan, power, and temperature constraints. HiMap uses a two-stage mapping approach. The first stage is a Simulated Annealing (SA) based mapping algorithm that assigns each task to a core set. The second stage is a greedy selection algorithm that chooses the best frequency for each core to meet the performance requirement.

All of these four objectives (makespan, energy, temperature, and reliability) are firstly studied in ERPOT [10] via the $\epsilon - constraint$ method, which transforms the original multi-objective optimization problem into multiple constrained single-objective optimization problems. Specifically, the makespan is minimized by a list scheduling method under preset constraints of the other three objectives.

However, the $\epsilon - constraint$ method has some limitations. First, it is hard to use energy or reliability as constraints directly because of the "funnel effect" [11] which means that the reliability of the application decreases with the number of tasks to be scheduled. For example, as shown in Fig. 1, if tasks 1–4 are executed only once, the whole reliability meets the constraint R_{obj}. But if more tasks are added, they need to be executed multiple times to ensure the whole reliability does not violate R_{obj} (the number of replicas of task 5, 6, and etc. increases to 4). Therefore, instead of reliability and energy, the global system failure rate (GSFR) and power consumption are considered as constraints in EROPT. The GSFR of a partial schedule is the failure rate of this schedule as if it was a single task mapped on a single core [12]. Using GSFR as a constraint for each task is a conservative estimate of the failure rate of the entire application, and so is the power consumption. Second, the performance of the $\epsilon - constraint$ method depends heavily on the preset values of the constraints.

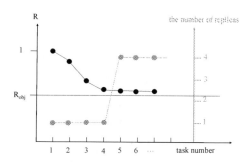

Fig. 1. Funnel effect [10]

In this paper, we propose to use MOEA as the search framework for this quad-criteria scheduling problem. Unlike ERPOT, MOEA can optimize all four objectives (makespan, reliability, energy consumption and peak temperature) directly without converting the multi-objective problem into single-objective problems with constraints or using GSFR and power consumption as proxies for reliability and energy consumption.

The main contribution of this work is that a problem-specific encoding strategy is designed for this quad-criteria task scheduling problem (makespan, reliability, energy consumption and peak temperature). Particularly, each chromosome consists of three parts which indicate the assignments of the cores and frequencies, the number of replicas, and the reference temperature, respectively. Since Non-dominated Sorting Genetic Algorithm-III (NSGAIII) [13] is one of the most widely used MOEAs, we present our quad-criteria task scheduling method based on NSGAIII. Experiments show that our method outperforms ERPOT [10] in two performance indicators, especially when the number of cores is large.

The rest of the paper is structured as follows. In Sect. 2, the background knowledge of quad-criteria scheduling is introduced. The details of the proposed encoding strategy and the NSGAIII framework are presented in Sect. 3. Section 4 provides the experimental results on benchmarks and case study. Finally, Sect. 5 concludes this work.

2   Problem Description

The multicore system can be modeled as multiple heterogeneous cores sharing multiple data paths. It can be described as $Arc = (C, B, L)$, where C is the set of heterogeneous cores, B is the set of communication buses, and L is the set of connections. A connection $l = (c, b), l \in L$, indicates that the core c is connected to the bus b. An application can be modeled as a directed acyclic graph (DAG), $DAG = (V, E)$, where V is the set of nodes and E is the set of edges. A computational task is represented by each node and the data dependency between two tasks is represented by each edge.

The task scheduling problem is to map the tasks to the multicores and determine their execution orders on each core. This means assigning a set of cores and corresponding running frequencies for each task of the application. The mapping of task t can be expressed as $M_t = \{(c_1, F_1), ..., (c_j, f_j), ...\}^{NR_t}$, where NR_t is the number of replicas for task t, (c_j, f_j) is a pair of core and frequency assigned for j_{th} replica of task t. Then determine the execution order and start time of each task t according to the endtime and cooling time of the previous tasks of t.

The scheduling result can be evaluated by four performance metrics: makespan ($makespan$), energy consumption (E), reliability (R), and peak temperature (T_{max}). The optimization objectives are defined as follows:

Makespan: The makespan of an application is the maximum endtime of all tasks. The execution time of a task executing on a core is affected by two factors: 1) the nominal (corresponding to the highest frequency) worst case execution time (WCET) of the task on the core; and 2) the selected frequency.

It is generally believed that the execution time of a task on a core is inversely proportional to the frequency selected by the core [10, 11], that is, the higher the frequency, the shorter the execution time. The execution time $exe(t, c, f)$ of task t executing on core c with frequency f is calculated as follows. To ease the computations, we transform the frequencies into normalized factors $f_{norm} \in \{f_{max} = f_3 = 1, f_2 = 2/3, f_{min} = F_1 = 1/3\}$.

$$exe(t, c, f) = WCET(t, c)/f_{norm} \tag{1}$$

Reliability: The reliability is the probability that the system will operate without faults after a specified period of time. Generally, the fault rate is used to model the reliability. Faults related to integrated circuits can be divided into two types. One is transient failure, which is a single-event disturbance and occurs in an unpredictable manner [14]. The other one is permanent failure, which is an irrecoverable hardware failure seen as a lifetime reliability threat [12].

Because the frequency of transient faults is much higher than that of permanent faults [15], this paper only considers transient faults. Transient faults are random errors modeled by a Poisson distribution with a failure rate denoted by λ. The reliability $R(t, c, f)$ of one task t executing on core c at frequency f is calculated as follows.

$$R(t, c, f) = e^{-\lambda \cdot exe(t,c,f)} \tag{2}$$

$$\lambda = \lambda_0 \cdot \lambda(f) \cdot \lambda(T) \tag{3}$$

$$\lambda(f) = 10^{\frac{b(1-f_{norm})}{1-f_{min}}} \tag{4}$$

$$\lambda(T) = e^{\frac{-Ea}{K} \cdot (\frac{1}{T_{end}} - \frac{1}{T_{start}})} \tag{5}$$

where $b \geq 0$ is a constant [11]. In fact, the lower the voltage, the smaller the critical energy, and thus the more sensitive the system is to low-energy particles. Ea is the activation energy, K is the Boltzmann's constant, T_{start} is the core temperature at the beginning of task execution, and T_{end} is the core temperature at the end of task execution. The reliability of a task t is calculated as follows, where $(c_j, f_j) \in M_t$.

$$R(t, M_t) = 1 - (\prod_{j=1}^{NR_t} (1 - R(t, c_j, f_j))) \tag{6}$$

Energy Consumption: Energy consumption is the integral of power consumption P over time. Power consumption mainly comes from two sources [16]: 1) Static power is due to the leakage current, and is proportional to the instantaneous temperature; 2) Dynamic power comes from switching activity, such as transistors, wires and other small capacitors that are filled and emptied when the corresponding value switches from 0 to 1 or 1 to 0. Therefore, the consumption depends on the capacitance of the physical components and the frequency of the change [16]. The energy consumption $E(t, c, f)$ of a task t is calculated as follows.

$$P(t, c, f) = \alpha * T(t, c) + \beta + \gamma \cdot C_{ef} \cdot V^2 \cdot f \tag{7}$$

$$E(t, c, f) = P(t, c, f) * exe(t, c, f) \tag{8}$$

where $T(t, c)$ is the temperature during execution, α and β are the coefficients of linear relationship between temperature and static power consumption, which are related to hardware. C_{ef} is the switching capacitance. V stands for voltage, and f stands for frequency. γ is the coefficient of dynamic power and static power. For simplicity, the energy consumed during task execution $E(t, c, f)$ is roughly the product of power consumption and execution time.

Temperature: Many models have studied the temperature of a core during the task execution [10,17,18]. The core temperature depends on several factors, such as power consumption, initial core temperatures, ambient temperatures and temperatures of other cores.

$$C \cdot \left(\frac{dT_c(\tau)}{d\tau}\right) + G(T_c(\tau) - T_{amb}) = P \tag{9}$$

$$T_{steady} = \frac{G \cdot T_{amb} + \sum_{c' \in nbr(c)} \kappa\left(c, c'\right) \cdot T_{c'}(\tau) + P}{G - \alpha + \sum_{c' \in nbr(c)} \kappa\left(c, c'\right)} \tag{10}$$

$$T(t, c) = T_{steady} + (T_{start} - T_{steady}) * e^{-A \cdot exe(t,c,f)} \tag{11}$$

Equation (9) shows how the core temperature changes over time [10], where T_{amb} is the ambient temperature, P is the power, $T_c(\tau)$ is the temperature of core c at time τ, and C and G are architecture-based constants for heat conductivity. By solving this equation and taking into account the effect of the neighboring cores $nbr(c)$ on its temperature, we can obtain the steady-state temperature of the core T_{steady} as Eq. (10), where $\kappa\left(c, c'\right)$ is the thermal conductivity between c and c', and α is the same as in Eq. (6). Then we can calculate the temperature $T(t, c)$ of the core after executing task t on core c using Eq. (11), where A is related to heat conduction.

When a core is idle (cooling time), its temperature still follows the same formulas. The only difference is that the power P is set to 0, and the $exe(t, c, f)$ is set to cooling time CT_t.

$$T_{steady} = \frac{G \cdot T_{amb} + \sum_{c' \in nbr(c)} \kappa\left(c, c'\right) \cdot T_{c'}(\tau) + 0}{G - \alpha + \sum_{c' \in nbr(c)} \kappa\left(c, c'\right)} \tag{12}$$

$$T(t, c) = T_{steady} + (T_{start} - T_{steady}) * e^{-A \cdot CT_t} \tag{13}$$

3   The Proposed Method

3.1   Encoding Strategy

MOEA is a type of optimization algorithm that uses individuals (chromosomes) to represent possible solutions to a problem and uses a population to represent the set of explored solutions. The optimization process is achieved by iteratively updating the population. Therefore, the performance of the MOEA depends largely on how the solutions are represented, that is, the encoding strategy of the chromosomes.

We can express the population as $P = \{x_1, x_2, \ldots, x_i, \ldots\}$. Let NT be the number of tasks and NC be the number of cores. As shown in Fig. 2, we divide the chromosome into three parts: $x_i = \{TM_i, NR_i, RT_i\}$.

The first part specifies the cores and the frequencies assigned to each task, expressed as $TM_i = \{(C_{i,1}, F_{i,1}), (C_{i,2}, F_{i,2}), \ldots, (C_{i,t}, F_{i,t}), \ldots\}^{|NT|}$. For each task, we use NC binary codes $C_{i,t} = \{SelectTag\}^{|NC|}$ to indicate which cores are selected, where $SelectTag \in \{0, 1\}$ and $SelectTag = 1$ means that the core at that position is chosen to execute task t. We also use NC integer codes

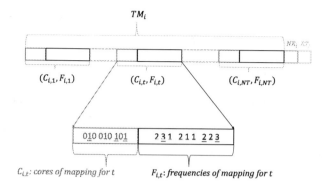

Fig. 2. Encoding example

$F_{i,t} = \{f^1, f^2, \ldots\}^{|NC|}$ to represent the frequencies of each core. The length of the first part is $2 \cdot NC \cdot NT$.

The second part determines the number of replicas for all tasks in this chromosome, that is, $NR_t = NR_i$ for every task. The overall error rate of a scheduling solution is approximately equal to the weighted average of the error rates of all tasks [12]. However, this may be unfair to tasks with lower error rates, since they are affected by the task with the highest error rate. Therefore, it is better to make the error rate of each task as similar as possible. One way to do this is to make the number of replicas for all tasks in a scheduling policy equal. If the number of replicas does not match the number of selected cores during decoding, we randomly select or discard cores as needed.

The third part sets the reference temperature (RT) of the tasks, where $RT \in [T_{lo}, T_{up}]$. We use RT to calculate the cooling time CT_t before executing task t according to the JUST strategy [19], because we want to optimize the peak temperature. Instead of encoding the cooling time for each task, we only encode the reference temperature in the chromosome.

3.2   Algorithm Structure

A multi-objective optimization problem can be defined as follows. x is the chromosome (individual or solution) in the decision space, which is an operational representation of the solution, such as using the encoding strategy above to represent a scheduling solution. $F(x)$ is the mapping of x in the decision space measuring the performance of x on m optimization objectives, which is called the fitness of solution x.

$$minimize \quad F(x) = (f_1(x), f_2(x), \ldots, f_m(x)) \tag{14}$$

As an effective method to deal with Multi-objective Optimization Problems (MOP), MOEA have been widely recognized. **NSGAIII** [13] is one of the most widely used MOEAs, especially for MOP with more than 3 objects.

The structure of NSGAIII is shown on the left side of Fig. 3, assuming that the size of the population is N. The first step of the algorithm is initialization, where the initial population P_0 is randomly generated according to the proposed encoding strategy in Sect. 3.1. Then in each generation, three operations are included: parent selection, reproduction, and survival selection. Parent selection is to select parents from P_i based on their fitness. Reproduction is to perform crossover and mutation operations on the selected parents to generate offspring Q_i. Survival selection is to select individuals from the set of P_i and Q_i as the population for the next iteration P_{i+1} based on their fitness. These three operations iterate until the termination condition is reached, and the population at this time is the final result of the algorithm.

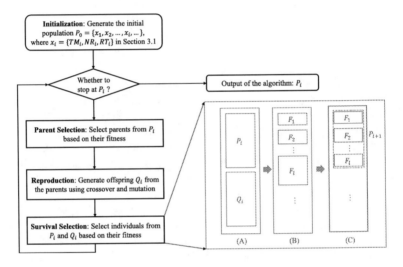

Fig. 3. Framework of NSGAIII [13]

Specially, the survival selection of NSGAIII mainly includes the following two steps, showed in the right side of Fig. 3:

The first step is to divide the population set consisting of P_i and Q_i into different nondominated levels based on the non-dominated sorting, as shown by the transformation from (A) to (B) in Fig. 3. The process is to sort all solutions according to the dominance relationship, and obtain the nondominated solution as the first layer of nondomination called F_1. Then all solutions in F_1 are deleted from the solution set, and the process above is repeated to find F_2 and so on. The iterations stop when all solutions are assigned.

The second step is to select each nondominated level to construct a new population set P_{i+1}, as shown by the transformation from (B) to (C) in Fig. 3. The selection starts from F_1, and ends when the size of P_{i+1} is equal to N or exceeds N for the first time, denoting this layer as the last accepted layer F_l. If it is equal to N, the survival selection is completed. If it exceeds N for the first time, only part of layer F_l is selected, and the selection includes five steps:

- Step 1: Choose several reference points to ensure diversity in selected solutions.
- Step 2: Normalize each objective based on the extent of members of F_l in the objective space.
- Step 3: Define a reference line corresponding to each reference point by joining the reference point with the origin.
- Step 4: Associate each solution in F_l with a reference point, according to the distance between the solution and the reference line corresponding to each reference point.
- Step 5: Iteratively perform the following two operations to select solutions from F_l to P_{i+1} until the size of P_{i+1} is equal to N: 1) select the reference point with the least number of corresponding individuals, and 2) randomly select one individual from the corresponding individuals of this reference point to join in P_{i+1}.

4 Experimental Results

4.1 Experimental Configuration

The task graphs of the applications used in all experiments are generated by TGFF [20]. The number of cores NC is set to 4 or 9, and the number of tasks in the task graph NT is 10, 20, 30, 40, or 50 for $NC = 4$; and 20, 40, 60, 80, or 100 for $NC = 9$. The nominal WCET of each task is randomly chosen from the interval $[5, 15]$. The parameter values are provided in Table 1 according to [10].

Table 1. Parameter Values

$f \in \{1.2\,V, 1.1\,V, 1.06\,V\}$
$f_{norm} \in \{f_{max} = f_3 = 1, f_2 = 2/3, f_{min} = F_1 = 1/3\}$
$V \in \{900\,MHz, 600\,MHz, 300\,MHz\}$
$\lambda_0 = 10^{-5}, b = 0.5$
$\alpha = 0.1\,WK^{-1}, \beta = -11\,W, \gamma = 0.2, C_{ef} = 10^{-8}\,JV^{-2}$
$C = 0.03\,JK^{-1}, G = 0.3\,WK^{-1}, \kappa(c, c') = 0.1\,WK^{-1}$

The following are the preset constraint values for ERPOT:

- $GSFR_{obj} \in \{10^{-9}, 10^{-8}, 10^{-7}, 10^{-6}, 10^{-5}, 10^{-4}\}$
- $P_{obj} \in \{30, 39, 48, 57, 66, 75\}$
- $T_{obj} \in \{340, 349, 358, 367, 376, 385\}$

A population size of 100 was used in our method. The maximum number of iterations for NSGAIII was set to 1500 when $NC = 9$ and to 500 when $NC = 4$. The experiment was repeated 30 times.

Two metrics were adopted to compare different algorithms. The first one is the *AveMakespanImpr* [10] which indicates how much our method improved compared to ERPOT in terms of *makespan*. The objective space is evenly divided into multiple cubes according to three constraints (objectives) other than *makespan*. We calculated the *makespan* improvement rate of NSGAIII compared to ERPOT in the cubes where both algorithms found solutions as follows. Then we averaged these values over all cubes to obtain the *AveMakespanImpr*.

$$\frac{makespan(ERPOT) - makespan(NSGAIII)}{makespan(ERPOT)} \tag{15}$$

The second one is the hypervolume (HV) [21]. HV of solution set A is calculated as follow, where r is the reference point, $\mathcal{L}$ denotes the Lebesgue measure of a set, and $\prec$ means dominating. HV measures the convergence and diversity of the solution set simultaneously and is widely used in the MOEA field.

$$HV(A, r) = \mathcal{L} \left(\bigcup_{a \in A} \{ \mathbf{b} \mid \mathbf{a} \prec \mathbf{b} \prec \mathbf{r} \} \right) \tag{16}$$

4.2   Experimental Results and Analysis

The experimental results is shown in Table 2. The first column is the description of the test cases, the second column is the average *AveMakespanImpr* value of 30 repeated experiments of NSGAIII compared to ERPOT on each testcase, the third column is the average HV value of 30 repeated experiments of the proposed method based on NSGAIII, the fourth column is the HV value of ERPOT, and the fifth column is the improvement ratio of NSGAIII compared to ERPOT in terms of HV.

It can be seen from Table 2 that there is no significant difference of the two methods on small scale problems, such as $NC = 4$, but when $NC = 9$, it is obvious that the proposed NSGAIII-based method is superior to EROPT in terms of both *AveMakespanImpr* and HV. Therefore, the proposed method is expected to be more promising for large scale problems. It should be noted that the average *AveMakespanImpr* is only 2.3379% (when $NC = 9$) which is not significant. This is because that ERPOT transforms the four-objective problem into multiple single optimization problems for the *makespan*, meaning EPORT prefers to minimize *makespan*. However, there is no bias for any of the four objectives in our NSGAIII based method. And HV value of our NSGAIII based method is 17.801% (when $NC = 9$) higher than that of ERPOT.

Table 2. Experiential results on synthetic benchmarks of ERPOT and our method

TestCase	$AveMakespanImpr$	$HV_{NSGAIII}$	HV_{ERPOT}	HV_{Impr}
$NC = 4, NT = 10$	0.26408%	1.4059	1.1802	19.129%
$NC = 4, NT = 20$	−0.023272%	1.3752	1.1348	21.189%
$NC = 4, NT = 30$	0.87815%	1.3125	1.3055	0.53897%
$NC = 4, NT = 40$	−0.26311%	1.3722	1.3931	−1.4984%
$NC = 4, NT = 50$	0.37175%	1.3656	1.4031	−2.6742%
Average on $NC = 4$	0.24552%	1.3663	1.2833	7.3369%
$NC = 9, NT = 20$	2.1395%	1.3144	1.1004	19.449%
$NC = 9, NT = 40$	3.6922%	1.2949	1.0291	25.836%
$NC = 9, NT = 60$	2.7575%	1.3013	1.0485	24.109%
$NC = 9, NT = 80$	1.9870%	1.2429	1.0988	13.119%
$NC = 9, NT = 100$	1.1132%	1.2662	1.1890	6.4910%
Average on $NC = 9$	2.3379%	1.2839	1.0931	17.801%

4.3 Case Study

We further verified the performance of our method on five real-world benchmarks from the E3S Suite [22]. Table 3 describes the benchmarks and presents the results of different methods. It can be seen that the average $AveMakespanImpr$ of our method compared to EPORT is about 0.481% and the HV is improved by about 12.113%.

Table 3. Experiential results on real-world benchmarks of ERPOT and our method

TestCase	NT	$AveMakespanImpr$	$HV_{NSGAIII}$	HV_{ERPOT}	HV_{Impr}
auto-indust	24	0.47243%	1.3712	1.2900	6.3588%
consumer	12	0.97572%	1.3956	1.2039	15.928%
networking	13	0.03752%	1.4104	1.1510	22.533%
office-automation	5	0.96211%	1.4360	1.2891	11.405%
telecom	30	−0.043016%	1.3943	1.3354	4.4091%
Average		0.48095%	1.4015	1.2539	12.113%

5   Conclusion

As far as we know, for the first time, we solved the task scheduling problem with four objectives (makespan, reliability, energy consumption and peak temperature) in multicores using multi-objective evolutionary algorithm from the perspective of global optimization, which can effectively avoid the indirect optimization and the requirement of preset constraints existing in the previous

$\epsilon - constraint$ method. The experiments show that our method can indeed find a better distributed solution set, especially when the number of cores is large. In particular, without prioritizing *makespan* optimization, the *AveMakespanImpr* of our method compared to ERPOT is greater than 0 indicating that our method slightly outperforms ERPOT in terms of makespan. Moreover, our method increases the *HV* by about 17%.

References

1. Jia, T., Mantovani, P., Dos Santos, M.C., et al.: A 12nm agile-designed SoC for swarm-based perception with heterogeneous IP blocks, a reconfigurable memory hierarchy, and an 800MHz multi-plane NoC. In: ESSCIRC 2022-IEEE 48th European Solid State Circuits Conference, pp. 269–272. IEEE (2022)
2. Saddik, A., Latif, R., El Ouardi, A., Elhoseny, M., Khelifi, A.: Computer development based embedded systems in precision agriculture: tools and application. Acta Agriculturae Scandinavica **72**(1), 589–611 (2022)
3. Sahu, P.K., Chattopadhyay, S.: A survey on application mapping strategies for network-on-chip design. J. Syst. Architect. **59**(1), 60–76 (2013)
4. Chakravarthi, V.S., Koteshwar, S.R.: Introduction to design of system on chips and future trends in VLSI. In: Chakravarthi, V.S., Koteshwar, S.R. (eds.) SoC Physical Design: A Comprehensive Guide, pp. 1–20. Springer, Cham (2022). https://doi.org/10.1007/978-3-030-98112-9_1
5. Singh, A.K., Shafique, M., Kumar, A., Henkel, J.: Mapping on multi/many-core systems: survey of current and emerging trends. In: 2013 50th ACM/EDAC/IEEE Design Automation Conference, pp. 1–10. IEEE (2013)
6. Andjelkovic, M., Krstic, M., Kraemer, R., Veeravalli, V.S., Steininger, A.: A critical charge model for estimating the SET and SEU sensitivity: a muller c-element case study. In: 2017 IEEE 26th Asian Test Symposium, pp. 82–87. IEEE (2017)
7. Xie, G., Chen, Y., Xiao, X., Xu, C., Li, R., Li, K.: Energy-efficient fault-tolerant scheduling of reliable parallel applications on heterogeneous distributed embedded systems. IEEE Trans. Sustain. Comput. **3**(3), 167–181 (2018)
8. Raji, M., Nikseresht, M.: UMOTS: an uncertainty-aware multi-objective genetic algorithm-based static task scheduling for heterogeneous embedded systems. J. Supercomput. **78**(1), 279–314 (2022)
9. Rathore, V., Chaturvedi, V., Singh, A.K., Srikanthan, T., Shafique, M.: Longevity framework: leveraging online integrated aging-aware hierarchical mapping and VF-selection for lifetime reliability optimization in manycore processors. IEEE Trans. Comput. **70**(7), 1106–1119 (2020)
10. Girault, A., Zarandi, H.R., et al.: EPORT: a quad-criteria scheduling heuristic to optimize execution time, reliability, power consumption and temperature in multicores. IEEE Trans. Parallel Distrib. Syst. **30**(10), 2193–2210 (2019)
11. Assayad, I., Girault, A., Kalla, H.: Tradeoff exploration between reliability, power consumption, and execution time for embedded systems. Int. J. Softw. Tools Technol. Transfer **15**(3), 229–245 (2013)
12. Girault, A., Kalla, H.: A novel bicriteria scheduling heuristics providing a guaranteed global system failure rate. IEEE Trans. Dependable Secure Comput. **6**(4), 241–254 (2009)
13. Deb, K., Jain, H.: An evolutionary many-objective optimization algorithm using reference-point-based nondominated sorting approach, part I: solving problems with box constraints. IEEE Trans. Evol. Comput. **18**(4), 577–601 (2014)

14. Dilek, S., Smri, R., Tosun, S., Dal, D.: A high-level synthesis methodology for energy and reliability-oriented designs. IEEE Trans. Comput. **71**(1), 161–174 (2022)
15. Zhu, D., Melhem, R., Moss'e, D.: The effects of energy management on reliability in real-time embedded systems. In: IEEE/ACM International Conference on Computer Aided Design, pp. 35–40. IEEE (2004)
16. Moy, M., Helmstetter, C., Bouhadiba, T., Maraninchi, F.: Modeling power consumption and temperature in TLM models. Leibniz Trans. Embed. Syst. **3**(1), 03 (2016)
17. Skadron, K., Stan, M.R., Huang, W., Velusamy, S., Sankaranarayanan, K., Tarjan, D.: Temperature-aware microarchitecture. ACM SIGARCH Comput. Archit. News **31**(2), 2–13 (2003)
18. Huang, W., Stan, M.R., Skadron, K., Sankaranarayanan, K., Ghosh, S., Velusam, S.: Compact thermal modeling for temperature-aware design. In: Proceedings of the 41st annual Design Automation Conference, pp. 878–883 (2004)
19. Kumar, P., Thiele, L.: Thermally optimal stop-go scheduling of task graphs with real-time constraints. In: 16th Asia and South Pacific Design Automation Conference, pp. 123–128. IEEE (2011)
20. Dick, R.: Task graphs for free. https://robertdick.org/projects/tgff/index.html. Accessed 4 Feb 2023
21. Tian, Y., Cheng, R., Zhang, X., Li, M., Jin, Y.: Diversity assessment of multi-objective evolutionary algorithms: performance metric and benchmark problems. IEEE Comput. Intell. Mag. **14**(3), 61–74 (2019)
22. Dick, R.: E3S. http://ziyang.eecs.umich.edu/dickrp/e3s/. Accessed 4 Feb 2023

An Improved NSGA-II for Solving Reentrant Flexible Assembly Job Shop Scheduling Problem

Xiuli Wu$^{(\boxtimes)}$, Yaqi Zhang, and Kunhai Zhao

College of Mechanical Engineering, University of Science and Technology Beijing, 30 Xueyuan
Road, Haidian District, Beijing, China
wuxiuli@ustb.edu.cn

Abstract. In the wafer manufacturing process of micro electro mechanical systems (MEMS), there are reentrant flow, parallel machines, and assembly operation. Therefore, this study models its scheduling problem as a reentrant flexible assembly job shop scheduling problem. First, a mathematical model is formulated to minimize the total tardiness and the total energy consumption. Second, an improved non-dominated sorting genetic algorithm II (INSGA-II) is proposed to solve this NP-hard problem. An encoding and decoding method are designed according to the problem characteristics. A rule-based initialization strategy is developed to improve the quality of the initialized population. Specific crossover, mutation and selection operators are designed. Finally, numerical experiments are carried out, and the result shows that the proposed algorithm can effectively solve the problem.

Keywords: MEMS Wafer Manufacturing · Reentrant Flexible Assembly Job Shop Scheduling · Non-dominated Sorting Genetic Algorithm II

1 Introduction

MEMS are devices have the ability to sense, control, and drive on a micro scale and have an impact on a macro scale [1], and are one of the components of the semiconductor industry. MEMS has the advantages of small size and light weight. Therefore, in recent years, MEMS has become the key component of many high-tech products. For example, accelerometers and gyroscopes are used in smart phones; blood glucose and blood pressure sensors are used in wearable devices [2]. The development of these emerging industries has driven the development of MEMS, but there are few studies on the scheduling problem of its manufacturing system. The manufacturing process of MEMS includes four steps: wafer manufacturing, testing, assembly, and final test, of which wafer manufacturing is the most expensive and complex part [3]. The level of management directly affects the production efficiency of the production system and poses great challenge to managers. Therefore, scheduling the MEMS wafer manufacturing effectively is an urgent problem for managers to face.

Supported by the National Natural Science Foundation of China (No. 52175449).

MEMS wafer manufacturing process has reentrant flow, parallel machines and assembly operation. Thus, it can be modeled as a reentrant flexible assembly job-shop scheduling problem (RFAJSP). Wu et al. [4, 5] have studied reentrant scheduling problem with batch processing machines and distributed flow-shop. But the study on RFAJSP is rare. To our best knowledge, there are two studies focused on RFAJSP. Francis et al. [1] studied the scheduling problem in the MEMS wafers manufacturing process with bond process, proposed scheduling rules, and discussed the performance of the rules aiming at minimizing cycle time and total work-in-process. Mittler [6] studied a problem in semiconductor manufacturing, in which prefabricated parts need to be assembled into modules synchronically, and proposed two rules to reduce the synchronization delay of prefabricated parts. To sum up, heuristic rules are mainly used to solve the problem, and few scholars develop meta-heuristic algorithm which have better optimization effect to optimize the problem.

Moreover, in recent years, with the increasingly serious problem of global warming, more and more studies have begun to pay attention to the green indicators and used multi-objective optimization algorithm to optimize the economic indicators and green indicators of scheduling problem jointly. Dong et al. [7] proposed an improved multi-verse optimizer algorithm to minimize the makespan, the total carbon emissions and the total tardiness of a reentrant hybrid flow shop scheduling problem with learning effect. Geng et al. [8] developed an improved multi-verse optimization algorithm to minimize the makespan, the maximum delay and the idle energy consumption of a reentrant hybrid flow-shop scheduling problem.

From the aforementioned literatures, we can see that: (1) It is necessary to develop a scheduling method for wafer manufacturing of MEMS. (2) Few scholars adopted meta-heuristic algorithms to solve the RFAJSP. (3) Scholars have begun to pay attention to optimize the economic and green indicators of scheduling problem jointly.

Therefore, this study focuses on the wafer manufacturing of MEMS, model its scheduling problem as a multi-objective RFAJSP (MO-RFAJSP) and proposes an INSGA-II to optimize the economic and green indicators jointly. The main contributions of this study are as follows: (1) A mathematical model of MO-RFAJSP with the goal of minimizing total delay and total energy consumption is formulated. (2) An INSGA-II algorithm is proposed to optimize MO-RFAJSP. (3) Special encoding and decoding method, crossover and mutation operators are designed. (4) A rule-based initialization strategy is proposed to improve the quality of the initialized population.

The rest of this paper is organized as follows: Sect. 2 describes and formulates model of MO-RFAJSP. Section 3 proposes the INSGA-II. In Sect. 4, numerical experiments are carried out. Section 5 summarizes this study and proposes future research directions.

2 Description and Formulation of MO-RFAJSP

2.1 Description of MO-RFAJSP

In MO-RFAJSP, the processing operation tree is used to describe the processing sequence of products, and the pruning search method [9] is used to identify virtual components (VCs) on the processing operation tree to simplify the difficulty of the problem.

VCs refers to the maximum continuous operations of the workpiece. An example of a processing operation tree of product A is shown in Fig. 1.

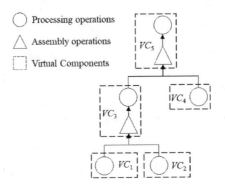

Fig. 1. An example of the processing operation tree of product A.

MO-RFAJSP can be described as follows. There are several workstations in the workshop, and a workstation contains multiple machines. The processing process of each product is composed of multiple VCs, and there is an assembly relationship between VCs. Each VC needs to go through L_h processes, and the processing process of the same VC need enter the same workstation many times. The task of the problem is to determine the processing sequence and the processing machine of each operation.

2.2 Formulation of MO-RFAJSP

The notations are shown in Table 1.

The following assumptions are made:

- All jobs and machines are available at time zero.
- The processes cannot be interrupted, once the process starts.
- Each operation can be processed on any parallel machine in the workstation.
- The jobs should be processed according to the processing process.

The mathematical model is formulated as follows:

$$f = \min(TT, TE) \tag{1}$$

$$K_j \geqslant 1 \, \forall j \tag{2}$$

$$S_{hl} \geqslant 0 \, \forall h \, \forall l \tag{3}$$

$$\sum_{k=1}^{K_{Ohl}} X_{hlk} = 1 \forall h \, \forall l \tag{4}$$

$$C_{hl} = S_{hl} + \sum_{k=1}^{K_{Ohl}} (X_{hlk} * P_{hlk}) \, \forall h \, \forall l \tag{5}$$

Table 1. Notations.

Type of Notation	Notations	Definition
Indices	i	The index of jobs, $i = 1, 2, 3, ..., I$
	j	The index of workstations, $j = 1, 2, 3, ..., J$
	h, b	The index of VCs, $h, b = 1, 2, 3, ..., H$
	l, v	The index of operations of h-th VC., $l, v = 1, 2, 3, ..., I_h$
	k	The index of machine of j-th workstation, $k = 1, 2, 3, ..., K_j$
	o	The index of operation performed on k-th machine of j-th workstation, $o = 1, 2, 3, ..., N_{jk}$
Parameters	F_{hlv}	If the l-th operation is the successor of the v-th operation of VC_h, $F_{hlv} = 1$, otherwise $F_{hlv} = 0$
	G_{hb}	If VC_h is the successor of the VC_b, $G_{hb} = 1$, otherwise 0
	O_{hl}	The l-th operation of the h-th VC
	D_i	The deadline of the i-th job
	V_{jk}	The idle energy consumption per unit of k-th machine of j-th workstation
	W_{lk}	The processing energy consumption per unit of k-th machine of j-th workstation
	P_{hlk}	The processing time of O_{hl} on k-th machine
Variables	S_{hl}	The start time of O_{hl}
	C_{hl}	The completion time of O_{hl}
	A_{jko}	The index of VC of the o-th operation processed on k-th machine of j-th workstation
	B_{jko}	The index of operation of the o-th operation on k-th machine of j-th workstation
	R_{jk}	The total idle energy consumption of k-th machine of j-th workstation
	T_h	The total processing energy consumption of h-th VC
	TT	Total tardiness of the solution
	TE	Total energy consumption of the solution
Decision variables	X_{hlk}	If the O_{hl} is processed on machine k, $X_{hlk} = 1$, otherwise 0
	Y_{hlbv}	If the O_{hl} is processed behind O_{bv}, $Y_{hlbv} = 1$, otherwise 0

$$Y_{hlbv} * S_{hl} \geqslant Y_{hlbv} * C_{bv} \ \forall h \ \forall l \ \forall b \ \forall v h \neq b l \neq v \qquad (6)$$

$$G_{hb} * S_{h1} \geqslant G_{hb} * C_{bL_b} \ \forall h \ \forall b h \neq b \qquad (7)$$

$$F_{hlv} * S_{hl} \geqslant F_{hlv} * C_{hv} \ \forall h \ \forall l \ \forall v l \neq v \qquad (8)$$

$$R_{jk} = \sum_{o=2}^{N_{jk}} ((C_{A_{jko}B_{jko}} - S_{A_{jk(o-1)}B_{jk(o-1)}}) * V_{jk}) \; \forall j \; \forall k \tag{9}$$

$$T_h = \sum_{l=1}^{L_h} \sum_{k=1}^{K_{O_{hl}}} (X_{hlk} * P_{hlk} * W_{lk}) \; \forall h \tag{10}$$

$$TT = \sum_{h=1}^{H} (max(max_{l \in L_h}(C_{hl} - D_i), 0)) \tag{11}$$

$$TE = \sum_{j=1}^{J} \sum_{k=1}^{K_j} R_{jk} + \sum_{h=1}^{H} T_h \tag{12}$$

$$X_{hlk} \in \{0, 1\} \; \forall h \; \forall l \; \forall k \tag{13}$$

$$Y_{hlbv} \in \{0, 1\} \; \forall h \; \forall l \; \forall b \; \forall vh \neq b \; l \neq v \tag{14}$$

Equation (1) indicates the objectives, i.e. minimizing the total tardiness and the total energy consumption. Equation (2) indicates there should be at least one machine in the workstation. Equation (3) represents the start time of each operation should be greater than 0. Equation (4) ensures each operation can only be processed by one machine. Equation (5) indicates the completion time of each operation is equal to the start time plus the processing time of the operation. Equation (6) ensures the start time of each operation should not be earlier than the completion time of its previous operation processed on the same machine. Equation (7) represents the completion time of each VC should not be later than the start time of its successor VC. Equation (8) ensures the start time of each operation of each VC should not be earlier than the completion time of its predecessor operations. Equation (9) indicates the calculation formula of idle energy consumption of each machine. Equation (10) indicates the calculation formula of idle and processing energy consumption of each VC. Equation (11)–(12) calculate the total tardiness and total energy consumption of the solution, respectively. Equation (13)–(14) define the decision variables.

3   INSGA-II for MO-RFAJSP

3.1   Framework of the Algorithm

Job-shop scheduling has been proved to be NP-hard problem by [10]. The MO-RFAJSP is a job-shop scheduling problem with many constraints, which increases the solving complexity of the scheduling problem and makes it strongly NP-hard. It is hard to use the exact solution method to get an optimal solution for the problem in limited time. Hence, a meta-heuristic algorithm is developed to optimize the problem.

NSGA-II [11] is a classical multi-objective optimization algorithm. However, the algorithm cannot use for this problem directly and exists some shortcomings. So, in order to effectively solve the MO-RFAJSP, an INSGA-II is proposed. (1) According to the characteristics of the problem, the coding, decoding, crossover and mutation methods of the algorithm are developed. (2) The randomly population initialization method of NSGA-II will lead to low quality of the initial population, resulting in premature convergence. So, a population initialization strategy is designed. The framework of the INSGA-II is shown in Algorithm 1.

Algorithm 1: The framework of the INSGA-II.

Input: stop condition, size of population Np, crossover probability CR, mutant probability MR

Output: E

1:   Generate the initial population **POP** by the population initialization strategy.
2:   Set the external population E to record the current non-dominated solutions.
3:   **While** the stopping criterion is not satisfied **do**
4:   **For** $i = \{1, ..., Np\}$ **do**
5:   Randomly select individuals **P1** and **P2** from **POP** without repetition.
6:   Execute the crossover operator on **P1** and **P2** to generate a new individual **C** with CR.
7:   Execute the mutation operator on **C** to generate a new individual **C'** with probability MR.
8:   **End for**
9:   Select the individuals to enter the next generation and set these individuals as population **POP**.
10: Update E.
11: **End while**

3.2 Encoding and Decoding

A dual chromosome encoding method is designed. The first chromosome represents the processing sequence of the operation, each gene represents the index of the VC, and the number of times the index of the VC appears indicates the number of operations in the VC. The second chromosome represents the index of the processing machine. The example of the product A in Fig. 1 is encoded as [[1, 1, 2, 4, 1, 3, 3, 5], [1, 2, 1, 1, 2, 1, 1]]. According to the structure of product A, VC_1 and VC_2 are assembled to obtain VC_3, VC_3 and VC_4 are assembled to obtain VC_5, so in the first chromosome, 3 appears after 1 and 2, and 5 appears after 3 and 4.

The decoding process is as follows. According to the operation processing sequence in the first chromosome, the process is successively assigned to the machine selected in the second chromosome. For the aforementioned example, the operation sequence of solution can be obtained: $O_{11}, O_{12}, O_{21}, O_{41}, O_{13}, O_{31}, O_{32}, O_{51}$, the index of the processing machines used for these operations are: 1, 2, 1, 1, 1, 2, 1, 1, 1.

3.3 Population Initialization Strategy

Three rules are adopted to generate the first chromosome. Two rules are adopted to generate the second chromosome. The rules are shown in Table 2 and Table 3. The associated parts of VC_h refer to the VCs which belonging to the same predecessor VC of VC_h.

Table 2. Rules for generating operation sequence.

Rules	Description
FCFS	Select the VC with the earliest arrival time in the queue
IR	Select the VC with the maximum important rate in the queue. The maximum important rate of a VC refers to the ratio of remaining operations of the VC to the maximum of remaining operations of its associated parts
LRPT	Select the VC with the longest remaining processing time in the queue

Table 3. Rules for generating machine selection.

Rules	Description
R1	Choose the machine with the earliest completion time. Then, choose the machine with the smallest energy consumption increment among the machines with the same completion time
R2	Choose the machine with the smallest energy consumption. Then, choose the machine with the earliest completion time among the machines with the same energy consumption

The combination of the rules is used to generate initial population, which is shown in Table 4, and the procedure of generating initial population is shown in Algorithm 2. All the initial solutions generated by this strategy are feasible solutions.

Table 4. Rules for initialization.

Index	Rule for generating operation sequence R_O	Rule for generating machine selection R_M
1	FCFS	R1
2	IR	R1
3	LRPT	R1
4	FCFS	R2
5	IR	R2
6	LRPT	R2

Algorithm 2: The procedure of generating initial population.

Input: total number of operations Lp, size of population Np

Output: POP

1:    Set initial population $POP=\varnothing$.

2:    **For** $i = 1$ to 6 **do**

3:        **For** $j = 1$ to floor($Np/6$) **do**

4:            Set $pop\_1=\varnothing$, $pop\_2=\varnothing$, and place VCs which do not has predecessor VC into L.

5:            **For** $l = 1$ to Lp **do**

6:                Select VC_h from L with i-th R_O, and set $pop\_1= pop\_1 \cup \{h\}$.

7:                Select a machine k to process the operation with i-th R_M, set $pop\_2= pop\_2 \cup \{k\}$.

8:                **If** VC_h completes all operations **do**

9:                    Delete h from L.

10:                  Add all the index of the successor VC of VC_h to L.

11:                **End if**

12:            **End for**

13:            Set $POP = POP \cup \{pop\_1+pop\_2\}$.

14:        **End for**

15:    **End for**

3.4 Crossover Operator and Mutant Operator

Crossover operator adopted in this study refers to the precedence-preserving order-based crossover operator [12]. First, select a VC at random. Then, set the VC and its predecessor VC as *V1*, and set the remaining VC as the *V2*. Next, keep the position of the gene in the chromosome unchanged and copy the gene belong to *V2* in *P1* to *C*. After that, keep the sequence of genes in the chromosome unchanged and copy the genes belonging to *V1* in *P2* to the empty position in *C*.

Mutation operator adopts a gene insertion mutation operator [13]. First, a number rand is generated between 1 and Lp randomly. Next, the *rand*-th gene h is deleted from the first chromosome, and the *rand*-th gene g is deleted from the second chromosome. After that, determine the leftmost position l in first chromosome of all successor VCs of VC_h, and the rightmost position r in first chromosome of all predecessor VCs of VC_h. Then, generate a number *rand'* between l and r randomly. Insert the deleted gene h and g into *rand'*-th position of first and second chromosome respectively, and record the chromosome as C'. After that, the machine selection of VC_h needs to be reset. Replace the genes of machine selection of VC_h in the second chromosome of C' with the genes of machine selection of VC_h of C.

The chromosome C' after using crossover and mutant operator is still a feasible solution, which ensures the chromosome in the algorithm is always feasible solution.

3.5 Select Operator

Deb et al. [11] adopted non-dominant levels and crowding distances to select individuals entering the next generation. The non-dominant level represents the superiority among individuals in the population. The crowding distance represents the relative distance between two individuals at the same level. The steps of selection are as follows: First, merge the child and the parent population. Calculate the non-dominant level and the crowding distance of each individual in the merged population. Then, according to the non-dominant level from high to low, the individuals are selected to enter the population of the next generation until the population exceeds the population size, at which time the non-dominant level is i. Finally, the individuals with larger crowding distance are preferentially selected from the individuals with non-dominant level i to join the next generation of population until reaching the population size.

4 Numerical Experiment

4.1 The Experiments Design

The algorithms are coded in Python 3.9, and run on a computer with 64-bit Windows 10, Intel (R) Core (TM) i7-8550uCPU@1.80 GHZ.

The parameters of the algorithm are set as follows: maximum iterations I: 100, population size Np: 100, crossover probability CR: 0.9, mutation probability MR: 0.1.

Experiments are carried out on randomly generated instances. The generated instance contains two identical jobs. The processing operation tree of the jobs is shown in Fig. 1. Table 5 shows the operations of each VC, and processing time and energy consumption per unit of each operation. Table 6 shows the idle energy consumption per unit of each machine. Table 7 shows the deadline. In the table, O_{hl} represents the l-th operation of h-th VC, and M_{jk} represents the k-th machine of j-th workstation.

Two experiments were designed as follows:

- A case study was carried out to show how the scheduling algorithm works.
- Comparison experiments were carried out to prove the excellent performance of proposed algorithm.

Table 5. Data for a MO-RFAJSP instance.

Index of VC	Operations	Processing time, processing energy consumption per unit						
		M_{11}	M_{12}	M_{21}	M_{22}	M_{31}	M_{32}	M_{41}
1	O_{11}	–	–	–	–	10, 43	7, 57	–
	O_{12}	9, 48	8, 58	–	–	–	–	–
	O_{13}	–	–	–	–	7, 53	7, 57	–
	O_{14}	–	–	8, 56	5, 56	–	–	–
	O_{15}	–	–	–	–	8, 56	6, 60	–
2	O_{21}	–	–	6, 53	10, 42	–	–	–
	O_{22}	–	–	7, 42	7, 52	–	-	–
	O_{23}	–	–	–	–	8, 57	10, 51	–
	O_{24}	5, 53	9, 43	–	–	–	–	–
3	O_{31}	–	–	–	–	–	–	8, 52
	O_{32}	–	–	–	–	7, 54	7, 60	–
	O_{33}	6, 57	10, 53	–	–	–	–	–
	O_{34}	–	–	10, 46	9, 55	–	–	–
	O_{35}	–	–	8, 49	6, 49	–	–	–
	O_{36}	5, 52	7, 47	–	–	–	–	–
4	O_{41}	–	–	6, 60	10, 52	–	–	–
	O_{42}	–	–	7, 58	7, 50	–	–	–
	O_{43}	–	–	7, 49	5, 60	–	–	–
	O_{44}	–	–	–	–	5, 59	9, 47	–
	O_{45}	–	–	6, 42	6, 57	–	–	–
	O_{46}	6, 53	8, 57	–	–	–	-	–
5	O_{51}	–	-	–	–	–	–	5, 45
	O_{52}	6, 45	10, 46	–	–	–	–	–
	O_{53}	5, 52	6, 59	–	–	–	–	–
	O_{54}	–	–	10, 53	5, 42	–	–	–
	O_{55}	–	–	–	–	10, 48	9, 57	–

Table 6. Data for a MO-RFAJSP instance.

Machines	M_{11}	M_{12}	M_{21}	M_{22}	M_{31}	M_{32}	M_{41}
Idle energy consumption per unit	27	20	25	24	20	20	17

Table 7. Data for a MO-RFAJSP instance.

Job	I_1	I_2
Deadline	120	110

4.2 The Experimental Results

Adopt INSGA-II algorithm to solve the instance, and 6 solutions are obtained. The fitness values of the solutions are shown in Table 8.

Table 8. Solutions obtained by INSGA-II.

Index	Tardiness	Total energy consumption	Index	Tardiness	Total energy consumption
1	8	25155	4	25	22327
2	16	24987	5	29	21527
3	19	22443	6	35	21306

Figure 2 shows the Gantt chart of solution 1, and Table 9 shows the detailed result of solution 1. In Table 9, I represents the index of job, O represents the operation, the four items in *Result* represents the index of workstation, the index of machine, the start time and the end time of the operation, respectively.

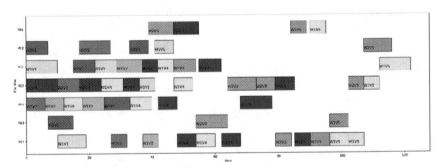

Fig. 2. The Gantt chart of solution 1.

The INSGA-II is compared with the classic algorithm or variant of classic algorithm, including: NSGAII [11] and MOEA/D_AS [14]. Use the instance designed in Sect. 4.1. Each algorithm runs 10 times and take the non-dominated solutions of the 10 results as the total non-dominated solutions, as shown in Fig. 3.

Table 9. The result of solution 1.

I	O	Result	I	O	Result	I	O	Result
1	O_{11}	3,1,0,10	1	O_{44}	3,1,42,47	2	O_{32}	3,1,47,54
	O_{12}	1,1,10,19		O_{45}	2,2,47,53		O_{33}	1,2,54,64
	O_{13}	3,1,22,29		O_{46}	1,1,54,60		O_{34}	2,2,64,73
	O_{14}	2,2,36,41		O_{51}	4,1,90,95		O_{35}	2,2,73,79
	O_{15}	3,2,41,47		O_{52}	1,1,96,102		O_{36}	1,1,79,84
	O_{21}	2,1,6,12		O_{53}	1,1,102,107		O_{41}	2,2,0,10
	O_{22}	2,1,18,25		O_{54}	2,2,107,112		O_{42}	2,2,17,24
	O_{23}	3,1,29,37		O_{55}	3,1,112,122		O_{43}	2,2,31,36
	O_{24}	1,1,37,42	2	O_{11}	3,2,0,7		O_{44}	3,1,37,42
	O_{31}	4,1,47,55		O_{12}	1,2,7,15		O_{45}	2,1,42,48
	O_{32}	3,1,55,62		O_{13}	3,1,15,22		O_{46}	1,1,48,54
	O_{33}	1,1,62,68		O_{14}	2,1,25,33		O_{51}	4,1,84,89
	O_{34}	2,1,68,78		O_{15}	3,2,33,39		O_{52}	1,1,90,96
	O_{35}	2,2,79,85		O_{21}	2,1,0,6		O_{53}	1,2,96,102
	O_{36}	1,1,85,90		O_{22}	2,2,10,17		O_{54}	2,2,102,107
	O_{41}	2,1,12,18		O_{23}	3,2,17,27		O_{55}	3,2,107,116
	O_{42}	2,2,24,31		O_{24}	1,1,27,32			
	O_{43}	2,1,33,40		O_{31}	4,1,39,47			

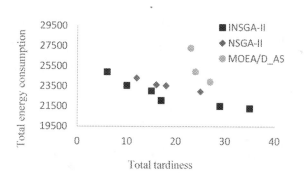

Fig. 3. Non-dominated solutions obtained with INSGA-II, NSGAII and MOEA/D_AS.

It can be observed from Fig. 3 that the non-dominated solutions obtained by the INSGA-II completely dominates the NSGAII and MOEA/D_AS. It can be concluded that the INSGA-II has obvious advantages over the others algorithms, and the INSGA-II can effectively solve the MO-RFAJSP problem.

5   Conclusions

MEMS industry is developing rapidly and it is necessary to develop a scheduling method for the scheduling of the wafer manufacturing in MEMS to improve the efficiency of the production system. First, a mathematical model is formulated as a reentrant flexible assembly job-shop scheduling problem with minimizing the total tardiness and total energy consumption. Then, an improved non-dominated sorting genetic algorithm II is proposed. The encoding and decoding method are designed according to the problem characteristics. A rule-based initialization strategy is designed to improve the quality of the initialization population. The crossover, mutation and selection operator are designed. Finally, the data of an instance is given, and numerical experiments are carried out according to the instance. A case study is carried out to show how the algorithm solve the problem. A comparison experiment is carried out to compare the proposed algorithm with the classic algorithm or its variant. The experimental results show that the algorithm can effectively solve the problem.

In the future, more constraints could be considered to get closer to real production process, such as transportation time, setting time, etc.

References

1. Tay, F.E.H., Lee, L.H., Wang, L.: Production scheduling of a MEMS manufacturing system with a wafer bonding process. J. Manuf. Syst. **21**(4), 287–301 (2002)
2. Han, Y.: Development of MEMS sensors (in Chinese). Electron. World J. **26**(01), 4–8 (2019)
3. Guo, C., Zhibin, J., Zhang, H., Li, N.: Decomposition-based classified ant colony optimization algorithm for scheduling semiconductor wafer fabrication system. Comput. Ind. Eng. **62**(1), 141–151 (2012)
4. Wu, X., Cao, Z.: An improved multi-objective evolutionary algorithm based on decomposition for solving re-entrant hybrid flow shop scheduling problem with batch processing machines. Comput. Ind. Eng. **169**, 108236 (2022)
5. Wu, X., Xie, Z.: Heterogeneous distributed flow shop scheduling and reentrant hybrid flow shop scheduling in seamless steel tube manufacturing. In: Pan, L., Pang, S., Song, T., Gong, F. (eds.) BIC-TA 2020. CCIS, vol. 1363, pp. 78–89. Springer, Singapore (2021). https://doi.org/10.1007/978-981-16-1354-8_8
6. Mittler, M., Purm, M., Gihr, O.: Set management: minimizing synchronization delays of prefabricated parts before assembly. In: 1995 Winter Simulation Conference Proceedings, pp. 829–836 (1995)
7. Dong, J., Ye, C.: Research on green job shop scheduling problem of semiconductor wafers manufacturing with learning effect (in Chinese). Oper. Res. Manage. Sci. **30**(04), 217–223 (2021)
8. Geng, K., Ye, C., Cao, L., Liu, L.: Multi-objective reentrant hybrid flowshop scheduling with machines turning on and off control strategy using improved multi-verse optimizer algorithm. Math. Probl. Eng. **2019**, 2573873 (2019)
9. Zhao, S., Han, Q., Wang, G.: Product comprehensive scheduling algorithm based on virtual component level division coding. Comput. Integr. Manuf. Syst. **21**, 2435–2445 (2015)
10. Garey, M.R., Johnson, D.S.: Computers and intractability: a guide to the theory of NP-completeness computer science. 1979338 (1979)
11. Deb, K., Pratap, A., Agarwal, S., Meyarivan, T.: A fast and elitist multiobjective genetic algorithm: NSGA-II. IEEE Trans. Evol. Comput. **6**(2), 182–197 (2002)

12. Guo, W., Lei, Q., Song, Y., Lyu, X.: A learning interactive genetic algorithm based on edge selection encoding for assembly job shop scheduling problem. Comput. Ind. Eng. **159**, 107455 (2021)
13. Weifei, G., Qi, L., Yuchuan, S., Xiangfei, L., Lei, L.I.: Integrated scheduling algorithm of complex product with no-wait constraint based on virtual component. J. Mech. Eng. **56**, 246 (2020)
14. Yan, X., Wu, X.: IMOEA/D to optimize job release problem for a reentrant hybrid flow shop. Comput. Ind. Eng. **163**, 107800 (2022)

Optimization Computing Algorithms

Multi-objective Baby Search Algorithm

Yi Liu[1], Gengsong Li[2], Wei Qin[1], Xiang Li[1], Kun Liu[1], Qiang Wang[1],
and Qibin Zheng[1(✉)]

[1] Academy of Military Science, Beijing 100091, China
zhengqibin1990@163.com
[2] National Innovation Institute of Defense Technology, Beijing 100071, China

Abstract. Multi-objective optimization problems are commonplace in real-world applications, and evolutionary algorithms are successful in solving them. Baby search algorithm is a novel evolutionary algorithm proposed recently, which has excellent ability on exploration and exploitation. However, it is designed to cater to single-objective optimization problems, but in this paper, we expand and modify it for multi-objective optimization. We introduce the boundary selection strategy to choose individuals from the Pareto archive for generating new solutions. To determine the best position of each individual we combine Pareto domination relation with random selection. Additionally, we propose an adapted Levy flight method to find promising solutions. Eleven standard multi-objective testing instances, five prevailing indicators and five state-of-art algorithms are applied to evaluate our algorithm. Experiments results demonstrate that our algorithm performs well on IGD, HV, Spread and GD measures.

Keywords: Baby Search Algorithm · Multi-objective Optimization · Evolutionary Computation · Heuristic Algorithm · Artificial Intelligence

1 Introduction

Engineering issues often require the simultaneous optimization of multiple objectives, which can be modeled as multi-objective optimization problems. Multi-objective optimization is used in various applications such as feature selection, route planning, classification rule mining, and workflow scheduling [1].

Multi-objective evolutionary algorithms (MOEAs) are methods that simulate natural phenomena or the group behavior of organisms to emerge specific behaviors to optimize complicated problems. During the last several years, several MOEAs have emerged and been employed across multiple domains. These include Multi-Objective Ant Colony Optimization (MOACO) [2], Multi-Objective Ant Lion Optimizer (MOALO) [3], Multi-Objective Evolutionary Algorithm based on Decomposition (MOEA/D) [4], and Non-Dominated Sorting Genetic Algorithm III (NSGA-III) [5].

Most MOEAs are originally single-objective optimization methods, and then they are modified and extended to handle multi-objective optimization problems, for example, MOACO, MOALO, multi-objective particle swarm optimization (MOPSO) and so on

Y. Tan et al. (Eds.): ICSI 2023, LNCS 13968, pp. 259–270, 2023.
https://doi.org/10.1007/978-3-031-36622-2_21

[6]. While some other algorithms are developed originally for tackling multi-objective optimization issues, such as MOEA/D.

Baby Search algorithm (BSA) is a novel meta-heuristic algorithm which is inspired by the behavior of infants looking for interesting things [7, 8]. BSA has a tempting skill to balance its searching performance between exploitation and exploration, which is demonstrated by comparing it with some representatives of evolutionary algorithms under general benchmarks. However, BSA is developed for dealing with single-objective optimization problems, it is not suitable for handling multi-objective optimization issues. Hence, we propose a modified version of BSA named multi-objective baby search algorithm (MOBSA) to deal with it. A detailed description of MOBSA is given, and exhaustive experiments are implemented to demonstrate its superior performance in this paper.

2 Related Works

Researchers have been investigating MOEAs for numerous years, aiming to enhance their performance and apply them to practical optimization and engineering problems.

There exist two ways to improve the performance of MOEAs, one is to solve the problems of searching special Pareto front, and another is to improve their universal optimizing performance. Fang et al. [9] develop a method, utilizing environmental selection by using local information in the objective space, that deletes worst individuals to acquire the representative solutions of latent search directions and resolve issues with irregular Pareto front. For the same objective, Liu et al. [10] introduce an innovative exploration versus exploitation model to determine suitable weights for environmental selection based on the decomposition-based approach. Moreover, Pan et al. [11] adopt a mating strategy that is inspired by manifold learning to cluster individuals and encourage diversity maintenance through mating reproduction among individuals within the same manifold.

In order to tackle the sparse Pareto solutions issues, Tian et al. [12] develop a population initialization strategy and genetic operators regarding the sparse nature of the Pareto optimal solutions to guarantee their sparsity. Other relevant researches also prove that taking the characteristics of Pareto solutions into consideration to introduce sophisticated methods can effectively improve performance [13, 14].

Many studies attempt to promote the searching ability of MOEAs by using various techniques. Zhao et al. [15] combine Nash equilibrium strategy with self-organizing mapping neural network to prevent MOPSO from trapping into local optima and choose promising leading particles. Wu et al. [16] propose an evolutionary state estimation mechanism, archive maintenance strategy and two types of leaders simultaneously to enhance solutions quality of MOPSO. Khalil et al. [17] puts up with two archive updating strategies depending on the nearest neighbor approach to make MOPSO have a fast convergence speed and splendid diversity. Gao et al. [18] employs a tempting multi-operator search strategy that obtains the manifold structure of Pareto optimal solution set and Pareto optimal front to reach a tradeoff between exploration and exploitation. Falcón-Cardona et al. [19] take advantage of algorithms based on quality indicators and uses them simultaneously by a cooperative scheme.

In addition to improving the performance of MOEAs, a lot of researches apply MOEAs in solving real-world engineering problems. For dealing with dynamic multi-objective optimization problems, Zhang et al. [20] combine self-adaptive precision controllable mutation operator, simulated isotropic magnetic particles niching strategy, and nondominated solutions guided immigration method to gain a high-quality Pareto front. While Liang et al. [21] propose a multi-objective algorithm based on decision variable classification, it adopts two different crossover operators in static optimization and reinitializes decision variables through maintenance, prediction, and diversity methods.

The constrained multi-objective optimization problems are very common in the real world, and attract plenty of researchers to resolve them. Yuan et al. [22] develop an indicator to evaluate the ability of individuals to find good Pareto solutions, then they introduce a constraint handling method base on this indicator to further enhance the diversity of the population. Ming et al. [23] employ two collaborative and complementary populations, apply a self-adaptive penalty function and a feasibility-oriented approach to handle solutions in them respectively. Liu et al. [24] draw on the idea of ensemble learning and come up with an indicator-based framework which unions indicator-based MOEAs with constraint-handling ways.

A great amount of research model real world problems as multi-objective optimization problems, and utilize MOEAs to resolve them. Tahir et al. [25] regard classification rule mining as a multi-objective optimization problem, develop a model based on classification accuracy and misclassification ratio objectives, and take NSGA-II and MOEA/D to solve it. Liu et al. [2] propose a multi-objective feature selection method to tackle imbalanced data classification, F1 and Gmean are taken as evaluation indicators, and multi-objective ant colony optimization is applied to deal with it. Liu et al. [26] combine stacked autoencoder with a multi-objective evolutionary algorithm to identify polarimetric synthetic aperture radar image. Wang et al. [27] introduce a computationally efficient multi-objective optimization approach to resolve computational costs and insufficient diversity matters in investigating the robustness of complex networks. Gustavo et al. [28] adopt a multi-objective method with a novel local search way to overcome uncertainty indicators problems for designing control systems.

We can find out that MOEAs have achieved many successes in various domains, and we need to investigate them persistently to solve more real-world problems. BSA is a new excellent meta-heuristic algorithm which simulates human babies searching for interesting things, it has two variants at present, i.e., original single-objective continuous and discrete optimization versions. Experiment results have demonstrated the powerful ability of BSA. We aim to modify it to be fit for multi-objective optimization issues to extend it to more application domains. It is evident that MOEAs have achieved significant success in different domains. Therefore, it is imperative to continuously investigate their application to real-world problems. The promising metaheuristic algorithm BSA mimics the behavior of infants seeking novel objects and has two current variants: the original single-objective continuous optimization and discrete optimization versions. Experimental results have shown the superior performance of BSA. We aim to modify it to be better suited for solving multi-objective optimization problems, thereby expanding its application to more domains.

3　MOBSA

3.1　BSA

In this part, we simply describe the details of BSA for helping readers understand MOBSA more easily. Babies are in the early stage of human development, they are inherently curious about the world and may be interested in everything that is visible and take them for fun. Inspired by the process of babies searching for interesting things, we came up with a novel fantabulous meta-heuristic algorithm called BSA [7].

BSA has three significant components, i.e. Levy flight, search model and mutation operation. Levy flight stand for a pattern of random walks that simulate heavy tail distribution which is consistent with the rules of many lives including human beings. The probability density function of Levy flights can be described by Eq. (1)

$$P(t) \sim l^{-\gamma} \tag{1}$$

where l indicates the number of steps, γ is a random value, and $1 < \gamma \leq 3$.

We model the searching process of babies in BSA as follows: i) when a baby plays with a toy reaches the time threshold TD, the baby lets it go and i–i) tries to find another toy with probability P_1 or i–ii) gets close to the baby who is holding the most interesting toy with probability $1 - P_1$; ii) if the time of toy is not up to TD, the baby will ii–i) approaches the baby who is playing with the most interesting toy with a high probability P_2 or ii–ii) finds another toy with probability $1 - P_2$. Generally speaking, $P_2 > P_1$ in most cases as it will improve the convergence ability.

The situations i–i) and ii–ii) can be described as in Eq. (2)

$$B_i^{t+1} = B_i^t + Levy_i^t + Rand(x) \times (B_{i,b} - B_i^t) \tag{2}$$

where B_i^t represents the position of the ith baby at the tth iteration. $Levy_i^t$ is the value generated by the Levy flight. $B_{i,b}$ denotes the best position of the ith baby so far. $Rand(x)$ is a random number whose range is between 0 and 1.

The situations i–ii) and ii–i) represented by formal expressions are shown in Eq. (3)

$$B_i^{t+1} = B_i^t + Rand(x) \times (B_b - B_i^t) \tag{3}$$

where B_b stands for the best position of all babies so far. The detailed description of BSA can be found in reference [7].

3.2　Modifies of BSA

To make BSA capable of solving multi-objective optimization problems, we need to modify it in three ways, namely by handling the Pareto Archive to store the Pareto-optimal solutions for each iteration; updating the individual best position, as new solutions might not dominate previous best solutions; and effectively generating Levy flights to produce new solutions.

There are many approaches for maintaining Pareto archive, which can obtain high quality Pareto solutions. We draw on the environment selection method of SPEA2 [29] in BSA as it is fast and has a fine effectiveness, being used and validated in various studies. We develop a unique way named boundary selection strategy to choose Pareto solution to generate new individuals. Boundary selection strategy works as follows: for each optimization objective, it finds out the individuals of its maximum or minimum values in Pareto archive, and we call them far solutions; we assign equal weights to normal Pareto solutions and larger weights to far solutions, then we adopt roulette wheel selection strategy to choose Pareto individuals based on their weights to generate new offspring. The assigned weights are produced by Eq. (4)

$$W(p) = \begin{cases} 1 + \sqrt{\sum_{k=1}^{N} d_{pk}}, p \in F \\ 1, \quad otherwise \end{cases} \tag{4}$$

where $W(p)$ represents the weights of solution p in Pareto archive, d_{pq} denotes the Euclidean distance between p and another Pareto individual k, N is the number of Pareto solutions in archive, F is the set of far solutions. Using this approach, we can preserve the boundary information in the objective space, thus allowing us to achieve our desired balance between convergence and diversity.

In BSA, the baby replaces its best position to date based on the fitness value to produce a better solution. However, this does not work in multi-objective optimization problems. We introduce a novel approach that uses the Pareto domination relation and random selection together. Specifically, we compare the newly generated solution with the current best solution of the baby using the Pareto domination relation. If the new solution dominates the current best solution, then we set the new solution as the best position. Otherwise, we keep the current best solution. If the two solutions do not dominate each other, we randomly choose either one as the best solution. In particular, if the current best solution dominates the new solution, we keep its time variable t; otherwise, we set it to 1.

Furthermore, as multi-objective optimization problems are more difficult than single optimization problems, we develop an adapted Levy flights method to produce promising Pareto individuals more effectively. We use Levy flight for each baby at every iteration, and dynamically control the range of values generated by Levy flights through the parameters q and w. q is used to normalize the produced values, it is a non-zero positive number, so the values range from $-q$ to q. w is taken to update q by setting $q = q * w$ at each iteration, the range of w is set between 0 to 1 as the search space gradually shrinks.

3.3 Description of MOBSA

In this part, we describe the procedure of proposed MOBSA, its pseudo code is shown in Algorithm 1.

Algorithm 1 The pseudo code of MOBSA

Input: Population size N, number of Pareto solutions in archive M, max iterations IT, Mutation Rate R, other parameters t, T, P_1, P_2, q, w

Output: Pareto archive

1. Initialize babies' positions and evaluate their fitness values
2. Update the Pareto archive and the best position of each baby
3. WHILE (current iteration < IT)
4.   Apply boundary selection strategy to choose a solution from the Pareto archive as global solution $B_{i,b}$
5.   FOR each baby
6.   Generate Levy flights by q and w
7.   Compare t of the baby with threshold T to decide whether use P_1 or P_2
8.   Use equation (2) or (3) to produce a new baby, depends on the comparison result between the random value and P_1 or P_2
9.   Perform mutation operation on new baby based on R, and obtain its fitness value
10.   Update its best position and time variable t as described in section 3.2
11.   END FOR
12.   Update the Pareto archive
13. END WHILE

We first define all necessary parameters in MOBSA. Line 1 initializes each solution and evaluate its fitness by using objective function. Line 2 utilizes the environment selection method to update the Pareto archive and determine the optimal position for each solution, as elaborated in Sect. 3.2. The main body of MOBSA consists of lines 3 to 13. Line 4 uses the boundary selection approach to opt for the global solution that generates new solutions. Line 6 leverages the Levy flight approach for every solution to search open space. Line 7 makes a comparison between t and T to determine whether to apply P_1 or P_2 in obtaining a new position. Line 8 specifies the technique utilized to create new solutions. Line 9 introduces a mutation operation on new solutions to enhance the search capability and obtain fitness value. Line 10 upgrades the best position and time t of the current baby following the approach explained in Sect. 3.2. Finally, line 12 replaces the solutions in the Pareto archive as line 2 does.

4   Experiments and Results

In this section, we compare the performance of MOBSA with five state-of-art multi-objective evolutionary algorithms, i.e., NSGA-II [30], MOEA/D [31], MOPSO [6], MOGWO [32], and multi-objective ant lion optimizer (MOALO) [33]. Five frequently used measures are applied to obtain fair and comprehensive results, i.e., inverted generational distance (IGD), hypervolume (HV), Spacing, Spread and generational distance (GD). IGD and HV are the comprehensive indicators, Spacing reflects the uniformity of the solutions distribution, Spread evaluates the diversity degree of individuals, and GD measures the convergence of results.

We adopt Windows 10 operating system, Matlab 2020a, Intel i5-7300HQ, 16 GB ram as test platform. ZDT1, ZDT2, ZDT3, ZDT6 and DTLZ1-DTLZ7 are used as testing functions, ZDT problems have two objectives, and DTLZ problems have three objectives [3], their characteristics are shown in Table 1.

Table 1. Testing multi-objective optimization instances.

Instance Name	Obj Num	Dimensions	Searching Space
ZDT1	2	30	[0, 1]
ZDT2	2	30	[0, 1]
ZDT3	2	30	[0, 1]
ZDT6	2	10	[0, 1]
DTLZ1	3	7	[0, 1]
DTLZ2	3	12	[0, 1]
DTLZ3	3	12	[0, 1]
DTLZ4	3	12	[0, 1]
DTLZ5	3	12	[0, 1]
DTLZ6	3	12	[0, 1]
DTLZ7	3	22	[0, 1]

Proper parameter values are defined for all algorithms and presented in Table 2. The methods are independently executed 40 times to attain results, which are the average values of each indicator. The outcomes for each algorithm are shown in Tables 3, 4, 5, 6 and 7.

Table 2. Parameters of five algorithms.

Algorithm Name	Parameter Values
NSGA-II	Tournament selection method, number of iterations 300, population size 100, crossover rate 0.7, mutation rate 0.02
MOEA/D	Weighted sum approach, number of iterations 300, population size 100, number of sub-problems 20
MOPSO	Adaptive grid 30, number of iterations 300, population size 100, inertia weight 0.5, $c_1 = 1$, $c_2 = 2$
MOGWO	Number of iterations 300, grid inflation parameter 0.1, number of grids per each dimension 10, leader selection pressure parameter 4, extra repository member selection pressure 2
MOALO	Number of iterations 300, population size 100, cardinality of Pareto archive 100
MOBSA	Number of iterations 300, population size 100, cardinality of Pareto archive 100, $t = 3$, $T = 5$, $P_1 = 0.6$, $P_2 = 0.4$, $R = 0.2$, w = 0.99

Table 3. IGD results on testing functions.

Name	NSGA-II	MOEA/D	MOPSO	MOGWO	MOALO	MOBSA
ZDT1	2.2836E−01	8.7384E−01	1.0023E−02	2.9967E−02	3.7564E−01	**8.4336E−03**
ZDT2	1.8930E−01	2.1499E+00	3.1580E−01	5.4941E−01	1.6251E−01	**8.0221E−03**
ZDT3	1.7626E−01	4.9139E−01	**2.1126E−02**	3.5417E−02	1.7883E−01	2.5273E−02
ZDT6	**1.0121E−02**	4.9495E+00	3.0948E−02	3.2461E−02	2.2605E−01	3.0082E−02
DTLZ1	1.2111E+00	1.1073E+01	2.2368E+01	1.2496E+01	9.3504E+00	**3.7101E−02**
DTLZ2	3.8914E−01	2.0744E−01	9.0134E−02	4.2655E−01	2.8580E−01	**6.2634E−03**
DTLZ3	1.7455E+01	1.1072E+02	1.9112E+02	1.8767E+02	1.2411E+02	**6.2379E−03**
DTLZ4	3.0255E−01	3.8618E−01	8.2930E−02	1.1759E−01	3.4544E−01	**3.4820E−02**
DTLZ5	2.3488E−01	5.4453E−02	2.1164E−02	2.7825E−02	1.0678E−01	**2.0080E−02**
DTLZ6	5.2223E+00	2.7372E−01	9.7200E−03	**9.6000E−03**	4.9608E−01	1.1946E−01
DTLZ7	1.0000E+00	5.6149E+00	**1.0702E−01**	8.7489E−01	1.4058E+00	2.2074E+00

Table 4. HV results on testing functions.

Name	NSGA-II	MOEA/D	MOPSO	MOGWO	MOALO	MOBSA
ZDT1	5.4512E−01	7.1246E−02	7.1071E−01	6.9563E−01	4.7107E−01	**7.1257E−01**
ZDT2	2.4692E−01	0.0000E+00	2.9080E−01	1.2408E−01	2.6369E−01	**7.1300E−01**
ZDT3	6.7909E−01	3.7313E−01	5.8900E−01	5.9287E−01	**6.8720E−01**	5.6271E−01
ZDT6	3.6259E−01	0.0000E+00	3.7444E−01	3.7593E−01	2.5752E−01	**4.1881E−01**
DTLZ1	1.7162E−02	0.0000E+00	0.0000E+00	0.0000E+00	8.6828E−03	**4.1765E−01**
DTLZ2	1.9911E−01	4.6801E−01	4.8853E−01	1.9131E−01	2.2963E−01	**5.9736E−01**
DTLZ3	0.0000E+00	0.0000E+00	0.0000E+00	0.0000E+00	6.4114E−03	**5.9738E−01**
DTLZ4	2.3593E−01	9.8555E−02	4.5631E−01	4.5328E−01	9.8601E−02	**4.8658E−01**
DTLZ5	7.6985E−02	1.7601E−01	1.8480E−01	1.7081E−01	8.7192E−02	**3.8515E−01**
DTLZ6	0.0000E+00	1.1067E−01	1.9749E−01	1.9740E−01	5.7237E−02	**3.5820E−01**
DTLZ7	1.1233E−01	0.0000E+00	**2.6095E−01**	1.5452E−01	1.0610E−01	1.0709E−02

Upon evaluating the testing functions, it can be established that MOBSA offers the most favorable outcomes on the IGD index, except for ZDT3, ZDT6, DTLZ6, and DTLZ7. Additionally, MOBSA outperforms other compared methods on the HV index, obtaining nine better values, except for ZDT3 and DTLZ7. Conversely, the MOEA/D method performs better than others on the Spacing measure with five minimum values, while MOBSA obtains four favorable values. However, MOBSA surpasses all other methods on the Spread index, winning seven out of eleven testing functions. Finally, MOBSA stands out as superior to other approaches in the GD index.

We will now delve deeper into analyzing the situations and reasoning behind the above statistical results. Based on the data, we can tentatively conclude that MOBSA has superior comprehensive ability compared to other methods, given that it obtains most

Table 5. Spacing results on testing functions.

Name	NSGA-II	MOEA/D	MOPSO	MOGWO	MOALO	MOBSA
ZDT1	1.0604E−02	**3.7416E−05**	1.1113E−02	1.6790E−02	1.9164E−02	2.6916E−03
ZDT2	8.2000E−03	**0**	9.0000E−03	1.9957E−03	1.8004E−02	2.6871E−03
ZDT3	1.5508E−02	5.4142E−03	1.1568E−02	1.5903E−02	3.3096E−02	**2.5349E−03**
ZDT6	1.6534E−02	**0**	2.1708E−02	2.6401E−02	1.1121E−01	2.6570E−03
DTLZ1	1.1613E−01	4.4369E−01	3.2445E+00	3.8306E+00	4.4219E+00	**2.6367E−03**
DTLZ2	2.8702E−02	4.3259E−02	5.8043E−02	1.8293E−02	9.8424E−02	**2.9854E−03**
DTLZ3	1.8806E+00	4.4288E+00	2.0620E+01	3.4540E+01	3.1378E+01	**3.0076E−03**
DTLZ4	**3.3887E−02**	6.1783E−02	6.3557E−02	5.7002E−02	9.9183E−02	4.3199E−02
DTLZ5	1.5089E−02	**4.8755E−03**	1.7341E−02	2.0189E−02	5.4226E−02	8.7069E−02
DTLZ6	4.3250E−01	2.4089E−02	1.3375E−02	**1.3300E−02**	3.2172E−01	1.0163E−01
DTLZ7	3.4719E−02	**0**	6.7197E−02	4.8362E−02	1.0397E−02	1.3278E+00

Table 6. Spread indicator results on testing functions.

Name	NSGA-II	MOEA/D	MOPSO	MOGWO	MOALO	MOBSA
ZDT1	8.7458E−01	1.0002E+00	7.2010E−01	1.1922E+00	1.1722E+00	**1.5784E−01**
ZDT2	8.9162E−01	1.0000E+00	8.2400E−01	1.0199E+00	9.1293E−01	**1.5937E−01**
ZDT3	9.0362E−01	1.0270E+00	7.2802E−01	9.8848E−01	1.3266E+00	**1.9565E−01**
ZDT6	1.1662E+00	1.0000E+00	9.3997E−01	1.0676E+00	1.5907E+00	**2.2148E−01**
DTLZ1	9.6511E−01	1.0827E+00	5.6354E−01	5.5936E−01	1.4620E+00	**2.1733E−01**
DTLZ2	9.8727E−01	1.6948E+00	3.9828E−01	7.9978E−01	1.3924E+00	**1.1315E−01**
DTLZ3	1.0318E+00	1.0482E+00	6.6710E−01	7.9630E−01	1.4105E+00	**1.1509E−01**
DTLZ4	1.0890E+00	1.1711E+00	**5.1287E−01**	5.6815E−01	1.3750E+00	5.7102E−01
DTLZ5	1.0545E+00	1.4961E+00	**5.9593E−01**	1.1468E+00	1.5114E+00	1.0375E+00
DTLZ6	**8.6947E−01**	1.4129E+00	8.8878E−01	8.8400E−01	1.5467E+00	1.0741E+00
DTLZ7	8.3243E−01	1.0000E+00	**4.9602E−01**	8.8235E−01	1.0343E+00	1.3228E+00

of the best values on IGD and HV. Additionally, MOBSA acquires better convergence and diversity solutions than other approaches, as it performed well on both the GD and Spread indexes. However, MOBSA underperforms MOEA/D on Spacing, implying that further modifications are needed to improve MOBSA's exploration ability to generate new solutions.

Overall, our experiments suggest that our modifications to extend BSA to the multi-objective optimization domain are feasible. The MOBSA algorithm effectively solves multi-objective optimization problems by generating high-quality Pareto solutions, as demonstrated by our results.

Table 7. GD indicator results on testing functions.

Name	NSGA-II	MOEA/D	MOPSO	MOGWO	MOALO	MOBSA
ZDT1	4.6529E−03	2.3613E−02	5.2458E−04	**3.3628E−04**	3.7271E−04	4.5418E−04
ZDT2	4.0832E−03	1.8970E−01	3.1708E−02	**9.6237E−06**	1.2867E−02	4.2505E−04
ZDT3	4.7883E−03	1.5352E−02	1.5519E−03	1.0178E−03	1.3738E−03	**3.7670E−04**
ZDT6	**9.6796E−05**	4.1487E−01	3.5019E−02	3.4795E−02	7.0189E−02	3.2991E−04
DTLZ1	1.5506E−01	1.5186E+00	4.7394E+00	5.7507E+00	5.0606E+00	**3.2463E−04**
DTLZ2	6.1334E−04	1.1941E−03	7.2596E−03	3.0313E−03	1.4363E−02	**4.2017E−04**
DTLZ3	2.6582E+00	1.8837E+01	2.8950E+01	4.4674E+01	2.7348E+01	**4.1069E−04**
DTLZ4	**1.4020E−03**	4.7222E−03	1.2760E−02	6.4703E−03	4.8197E−02	2.7974E−02
DTLZ5	4.0477E−05	**4.6818E−05**	2.8874E−03	5.5496E−04	1.5839E−02	5.4522E−02
DTLZ6	6.6579E−01	**2.7483E−07**	4.6657E−06	4.5795E−06	1.8968E−01	5.2180E−02
DTLZ7	1.7538E−02	4.3794E−01	4.1835E−03	1.0678E−02	**2.9768E−04**	1.6810E+00

5 Conclusions

BSA is a novel powerful evolutionary algorithm which is good at tackling single objective optimization issues. We introduce a multi-objective version based on BSA which is called MOBSA in this paper to handle multi-objective optimization problems. We adopt the effective method used in SPEA2 to update our Pareto archive. Additionally, we employ a boundary selection strategy to select individuals from the archive to produce new offspring. Our approach replaces the best position of every individual by combining Pareto domination relation with random selection. To search for favorable solutions in a wide space, we adopt the adapted Levy flight method. We compared our MOBSA algorithm with five outstanding MOEAs by using eleven testing functions and five indicators. Our experiment results indicate that our method outperforms others on IGD, HV, Spread, and GD measures but has slightly weaker performance on the Spacing index.

Our future work will be conducted in two aspects. One is to continue to improve the optimizing performance of MOBSA, more specifically, we will focus on improving its population diversity and distribution by elaborating new strategies for offspring generation; the other is to apply MOBSA in real-world engineering optimization problems.

References

1. Sharma, S., Kumar, V.: A comprehensive review on multi-objective optimization techniques: past, present and future. Arch. Comput. Methods Eng. **29**(7), 5605–5633 (2022). https://doi.org/10.1007/s11831-022-09778-9
2. Liu, Y., Wang, Y., Ren, X., Zhou, H., Diao, X.: A classification method based on feature selection for imbalanced data. IEEE Access **7**, 81794–81807 (2019)
3. Liu, Y., Qin, W., Zhang, J., Li, M., Zheng, Q., Wang, J.: Multi-objective ant lion optimizer based on time weight. IEICE Trans. Inf. Syst. **E104-D**(6), 901–904 (2021)

4. Eduardo, F., Nelson, R.V., Laura, C.R., Claudia, G.G., Carlos, A.C.: Preference incorporation in MOEA/D using an outranking approach with imprecise model parameters. Swarm Evol. Comput. **72**, 101097 (2022)
5. Peng, W., lin, J., Zhang, J., Chen, L.: A bi-objective hierarchical program scheduling problem and its solution based on NSGA-III. Ann. Oper. Res. **308**(1–2), 389–414 (2021). https://doi.org/10.1007/s10479-021-04106-z
6. Coello, C.A.C., Lechuga, M.S.: MOPSO: a proposal for multiple objective particle swarm optimization. In: 2002 Congress on Evolutionary Computation, Honolulu, USA, pp. 1051–1056. IEEE (2002)
7. Liu, Y., Li, M., Zheng, Q., Qin, W., Wang, J.: Baby search algorithm. In: 2021 4th International Conference on Advanced Electronic Materials, Computers and Software Engineering, Changsha, China, pp. 502–508. IEEE (2021)
8. Liu, Y., Zheng, Q., Li, G., Zhang, J., Ren, X., Qin, W.: Discrete baby search algorithm for combinatorial optimization problems. In: 2022 3rd International Conference on Big Data, Artificial Intelligence and Internet of Things Engineering, Xi'an, China, pp. 595–599. IEEE (2022)
9. Fang, W., Zhang, L., Yang, S., Sun, J., Wu, X.: A multiobjective evolutionary algorithm based on objective-space localization selection. IEEE Trans. Cybern. **49**(7), 2732–2743 (2019)
10. Liu, Y., Hu, Y., Zhu, N., Li, K., Zou, J., Li, M.: A decomposition-based multiobjective evolutionary algorithm with weights updated adaptively. Inf. Sci. **572**, 343–377 (2021)
11. Pan, L., Li, L., Cheng, R., He, C., Kay, C.T.: Manifold learning-inspired mating restriction for evolutionary multiobjective optimization with complicated pareto sets. IEEE Trans. Cybern. **51**(6), 3325–3337 (2021)
12. Tian, Y., Zhang, X., Wang, C., Jin, Y.: An evolutionary algorithm for large-scale sparse multiobjective optimization problems. IEEE Trans. Evol. Comput. **24**(2), 380–393 (2020)
13. Long, J., Liu, J., Mei, J.: Combining global and local information for offspring generation in evolutionary multiobjective optimization. IEEE Access **9**, 127471–127483 (2021)
14. Liu, T., Tan, L., Li, X., Song, S.: Incremental learning-inspired mating restriction strategy for evolutionary multiobjective optimization. Appl. Soft Comput. **127**, 109430 (2022)
15. Zhao, C., Guo, D.: Particle swarm optimization algorithm with self-organizing mapping for Nash equilibrium strategy in application of multiobjective optimization. IEEE Trans. Neural Netw. Learn. Syst. **32**(11), 5179–5193 (2021)
16. Wu, B., Hu, W., Hu, J., Gary, G.Y.: Adaptive multiobjective particle swarm optimization based on evolutionary state estimation. IEEE Trans. Cybern. **51**(7), 3738–3751 (2021)
17. Khalil, A., Du, W.: A fast multi-objective particle swarm optimization algorithm based on a new archive updating mechanism. IEEE Access **8**, 124734–124754 (2020)
18. Gao, X., Liu, T., Tan, L., Song, S.: Multioperator search strategy for evolutionary multiobjective optimization. Swarm Evol. Comput. **71**, 101073 (2022)
19. Falcón-Cardona, J.G., Ishibuchi, H., Coello, C.A.C., Emmerich, M.: On the effect of the cooperation of indicator-based multiobjective evolutionary algorithms. IEEE Trans. Evol. Comput. **25**(4), 681–695 (2021)
20. Zhang, K., Shen, C., Liu, X., Gary, G.Y.: Multiobjective evolution strategy for dynamic multiobjective optimization. IEEE Trans. Evol. Comput. **24**(5), 974–988 (2020)
21. Liang, Z., Wu, T., Ma, X., Zhu, Z., Yang, S.: A dynamic multiobjective evolutionary algorithm based on decision variable classification. IEEE Trans. Cybern. **52**(3), 1602–1615 (2020)
22. Yuan, J., Liu, H., Ong, Y., He, Z.: Indicator-based evolutionary algorithm for solving constrained multiobjective optimization problems. IEEE Trans. Evol. Comput. **26**(2), 379–391 (2022)
23. Ming, M., Trivedi, A., Wang, R., Srnivasan, D., Zhang, T.: A dual-population-based evolutionary algorithm for constrained multiobjective optimization. IEEE Trans. Evol. Comput. **25**(4), 739–753 (2021)

24. Liu, Z., Wang, Y., Wang, B.: Indicator-based constrained multiobjective evolutionary algorithms. IEEE Trans. Syst. Man Cybern. Syst. **51**(9), 5414–5426 (2021)
25. Tahir, S., Humar, K.Ö.: Classification rule mining based on Pareto-based Multiobjective Optimization. Appl. Soft Comput. **127**, 109321 (2022)
26. Liu, G., Li, Y., Jiao, L., Chen, Y., Shang, R.: Multiobjective evolutionary algorithm assisted stacked autoencoder for PolSAR image classification. Swarm Evol. Comput. **60**, 100794 (2021)
27. Wang, S., Liu, J., Jin, Y.: A computationally efficient evolutionary algorithm for multiobjective network robustness optimization. IEEE Trans. Evol. Comput. **25**(3), 419–432 (2021)
28. Gustavo, A.P.M., Lucas, B.M., Filipe, M.B., Bruno, H.G.B., Marco, H.T., Valdir, G.J.: Robust path-following control design of heavy vehicles based on multiobjective evolutionary optimization. Expert Syst. Appl. **192**, 116304 (2022)
29. Zitzler, E., Laumanns, M., Thiele, L.: SPEA2: improving the strength Pareto evolutioanry algorithm. Swiss Federal Institute of Technology Zurich, Zurich, Switzerland (2001)
30. Deb, K., Pratap, A., Agarwal, S., Meyarivan, T.: A fast and elitist multiobjective genetic algorithm: NSGA-II. IEEE Trans. Evol. Comput. **6**(2), 182–197 (2002)
31. Zhang, Q., Li, H.: MOEA/D: a multiobjective evolutionary algorithm based on decomposition. IEEE Trans. Evol. Comput. **11**(6), 712–731 (2007)
32. Seyedali, M., Shahrzad, S., Seyed, M.M., dos Leandro, S.C.: Multiobjective grey wolf optimizer: a novel algorithm for multi-criterion optimization. Expert Syst. Appl. **47**, 106–119 (2016)
33. Mirjalili, S., Jangir, P., Saremi, S.: Multi-objective ant lion optimizer: a multi-objective optimization algorithm for solving engineering problems. Appl. Intell. **46**(1), 79–95 (2016). https://doi.org/10.1007/s10489-016-0825-8

A Memetic Algorithm for Solving the Robust Influence Problem on Directed Networks

Zhaoxi Ou, Shuai Wang$^{(\boxtimes)}$, and Shun Cai

School of Intelligent Systems Engineering, Sun Yat-sen University, Shenzhen 518107, China
{ouzhx5,caish28}@mail2.sysu.edu.cn, wangsh368@mail.sysu.edu.cn

Abstract. The influence maximization problem has been intensively emphasized in recent studies, and the robust influence maximization problem is aimed at studying the information spreading ability of seeds under external attacks which is a current research hotspot. Elaborate diffusion models and seed determination methods are developed in previous studies. As a general form of networked systems, the plain undirected networks are thoroughly investigated. However, as indicated by related researches, directed networks show significance in daily life. Existing models and determination methods omit the effect of directness of networks, which may cause barriers in applications. Therefore, in order to study the influence robustness of directed networks, a metric, R_{SD}, is proposed to evaluate the robustness of influence pertaining to certain seeds in networks with directness. Then, a Memetic algorithm is devised to calculate the robust influence ability of seeds under intentional attacks, named MA-RIM$_D$. The algorithm considers networks with directed links, and several problem-directed operators are maintained. Experimental results on several networks demonstrate that the algorithm exhibits a competitive convergence compared with existing approaches, and superior optimized results can be attained.

Keywords: Influence maximization problem · robustness · directed network · Memetic algorithm

1 Introduction

Complex networks widely exist in nature and human society, which describe the relationship between the individuals such as social networks and electricity supply system. The networks may exhibit features like scale-free, small world, and random distribution, providing convenience to solve problems like data excavation and topological construction. This research topic has drawn increasing attention from scholars, and plenty of related problems are developed and investigated in recent decades including the influence maximization, robustness, community allocation.

© The Author(s), under exclusive license to Springer Nature Switzerland AG 2023
Y. Tan et al. (Eds.): ICSI 2023, LNCS 13968, pp. 271–283, 2023.
https://doi.org/10.1007/978-3-031-36622-2_22

The influence maximization problem (IMP) [4] is aimed at determining a set of influential seed set from networks. The application of this problem can be detected in viral marketing, virus transmission or rumor control. The IMP in social networks has been greatly emphasized in previous studies [10], and proved to be NP-hard with considerable complexity. Two factors are included in this problem, i.e., how to objectively evaluate the influence of a given seed set and how to select the seed set with maximum influence ability in the network. There are several diffusion models to depict the diffusion process including the independent cascaded (IC) model [10], the weighted cascaded (WC) model [4], and the linear threshold (LT) model [11], which are commonly employed in practice. Guided by a specific measure, the performance of seeds can be numerically evaluated [11]. Therefore, the seed determination task can be tackled as an optimization problem, and several approaches are optional to solve the problem like the greedy algorithm [12], heuristic-based methods [10], and evolutionary algorithms [9].

The transmission process depends on the integrity of networked systems. In practice, networks are exposed to disturbances and perturbances due to the complex environment. Sabotages directly cause impact on the network structure, and generate interference towards the attached information diffusion process. Related studies indicate that changes on parameters of the diffusion process may disturb the diffusion results [1,3]. Meanwhile, several pilot studies reveal that the diffusion process is expected to show tolerance against structural failures [9]. Considering potential structural damages, the seed determination process can be modeled as the robust influence maximization (RIM) problem.

At present, there has been a growing interest in the RIM problem. However, most studies only consider undirected networks, and the developed approaches can tackle the diffusion process without directness. Previous studies indicate that the directness is significant in practice [14,15,17,19], many maneuvers cannot omit the directional property. Under this context, how to evaluate the robustness of seeds in directed networks and how to detect robust and influential seeds considering the directness, such problems still remain to be addressed.

The contribution of this work can be summarized from following two aspects. First, the directness of networks is considered to solve the RIM problem which enriches the research scope of this problem to deal with more complex application scenarios. Second, based on the algorithm design of directed networks, the underlying structural information in the network is mined and the competitive results are obtained.

The rest of this paper is organized as follows: Sect. 2 describes the related work of robust influence maximization problem. Section 3 shows the designs of R_{SD} and MA-RIM$_D$. The experimental results networks are reported in Sect. 4. Finally, conclusions and discussions are shown in Sect. 5.

2   Related Work

Generally, a social network configuration can be described as a graph $G = \{V, E\}$, where $V = \{1, 2, 3, \ldots, N\}$ is N individuals in G; $E = \{e_{ij} | i, j \in V, i \neq j\}$

denotes M connection paths between individuals, and the direction information is maintained. This section describes the background knowledge of the robust influence maximization problem and directed networks.

2.1   The Influence Maximization Problem

The influence maximizing problem is aimed at selecting K nodes from the network to form a seed set S, where $S = \{s_1, s_2, \ldots, s_K\}$. The influence range generated by S is denoted as $\sigma(S)$. In the influence propagation process, the diffusive behavior of nodes is significant to complete the spreading task. Several diffusion models are defined to simulate the propagation process attached on nodes like the IC mode, the WC mode and the LT model. Nodes in the network may exhibit two status, i.e., active or inactive. The active nodes have already received the information and can work as seed to spread the influence, and the inactive ones are not touched in the diffusion process. Generally, inactive nodes can transform the state into active, but not vice versa.

Taking the IC model as an example, the diffusion process is defined as, only nodes in S are activated in the initial situation. In each diffusion step t, active seeds are attempted to activate their adjacent neighbors at a probability of p, and those successfully-activated nodes are preserved in a temporary set S_{temp}. In this way, the active node set is updated as $S_{t+1} = S_t \cup S_{temp}$, and if S_{temp} is empty (i.e., no node is activated in a time step), the diffusion process gets terminated. The number of the nodes activated is denoted as $\sigma(S)$, which reflects the influence ability of S in this network. The difference between WC model and IC model is that the activation probability p in IC model is constant while that of WC model is decided by the weight information between nodes. The LT model refers to the influence diffusion process considering the accumulation on nodes, where each node has a pre-defined threshold for activation and it can be activated until accepting enough influence in previous procedures. Considering that the IC model exhibits a distinct application value [7,11,12,23], this model is employed in following experiments.

The evaluation of $\sigma(S)$ can be accomplished via the Monte Carlo process, which is time-consuming and may not be applicable towards large-scale networks. A rapid approximation method considering the 2-hop neighborhood (the neighborhood of the neighborhood) has been developed to reach a fast evaluation [11]. The neighboring information is exploited to estimate the influential range of a certain seeds, which is defined as follows,

$$\hat{\sigma}(S) = \sum_{s \in S} \hat{\sigma}\{s\} - \left(\sum_{s \in S} \sum_{c \in C_s \cap S} p(s,c)(\sigma_c^1 - p(c,s)) \right) - \chi \tag{1}$$

where C_s is 1-hop neighbors (direct neighbors) of seeds, $p(c,s)$ is the probability of activation from active nodes (s) to inactive nodes (c), and σ_c^1 is the 1-hop neighbors of c. The first term evaluates the influence range of the seeds in S and their 1-hop neighbors; the second term deducts the overlapping influence in 1-hop of the seed set; the third term χ denotes the overlapping influence from activated nodes to the original seeds, which is as follows,

$$\chi = \sum_{s \in S} \sum_{c \in C_s \cap S} \sum_{d \in C_c \cap C_s \cap S} p(c, s)p(c, d) \tag{2}$$

The rationality of this rapid estimation method has been validated in [11], and the influence range of seed sets in undirected network can be assessed. In addition, this method exhibits the advantage of low calculational budget, and the evaluation on large-scale networks is reachable.

For directed networks, the diffusion process is restricted by the directness information attached on links, and the corresponding evaluation process requires adjustments to fit the spreading behavior.

2.2   The Robustness Evaluation of Networked Systems

The robustness of the networks is a problem considering the potential disturbances arising in the complicated practical environment. From the attackers' point of view, the destruction process can be divided into node-based attacks [16] and link-based attacks [21]. Node-based attacks concentrate on the removal of nodes to cause damage, where the connection attached on a specific node are removed accordingly to isolate this node. Link-based attacks refers to deleting links between nodal members, and the connectivity of this network gets impaired. For attackers, a limited budget needs to be considered in the destruction process, either on nodes or links. The components with significant value are thus in priority in the removal process. As in the social network, users with the higher degree nodes are likely to possess more neighbors, and may generate larger influence in the spread process. To conduct an efficient attack, the removal process is expected to be malicious, and a possible manner is to remove nodes or links with distinct degree in precedence [8,16,22,27].

In order to promote the robustness performance of networks, rational evaluation factors are necessary, which has been emphasized in recent studies. For example, the robustness measure R was proposed by Schneider et al. in [16] to utilize the change of the largest connected cluster in the destruction process. This measure provides a numerical criterion to record the connectivity change under node-based attacks. Further, Zeng et al. extended R considering the link-based attacks [27].

Employing the robustness evaluation theory, the robustness of information diffusion process is also valuable. In [22], a robust influence measure R_S is proposed to accumulate the influence range of seed set when moving nodes. A normalized result is calculated after conducting a certain number of attacks. This measure manages to evaluate the robust influence ability of seeds in undirected networks, but the impact of directness has not been touched upon.

Given such measure, the determination of robust and influential seeds can be modeled as an optimization problem. Several algorithms have been developed to work as solvers. For instance, an evolutionary algorithm is developed in [22], where the structural information is exploited to provide potential knowledge in the determination process. A multi-factorial evolutionary approach is developed in [24] to cater to the case when multiple damage scenarios happen.

2.3 The Robust Evaluation of Directed Networks

The aforementioned studies mainly concentrate on the structural information of plain networks. Yet, the directness of network has been verified to be crucial in practice [14, 15, 17, 19]. The connectivity of directed networks is closely correlated with their functionalities, but the directness should be considered in the evaluation process. Previous studies introduce the controllability theory into the robustness evaluation of directed networks, which combines the structural integrity and directed connections to comprehensively reflect the network's functionality. The minimum number of driver vertices (N_D) reflects the difficulty of controlling a directed network [25], which may work as a criterion to evaluate the performance. Employing this approach, a robustness measure of directed networks was proposed in [19], namely the controllable robustness (CR), to access the resistance of directed networks against node-based attacks. Details are as follows,

$$CR = \frac{1}{N} \sum_{q=1/N} \frac{s(q)}{N_D(q)} \tag{3}$$

where N is the number of all nodes in the network, $s(q)$ is the fraction of largest connected cluster in the network and $N_D(q)$ is the minimum number of control routes after removing $q \times N$ nodes, respectively. $\frac{1}{N}$ is the normalized factor. $s(q)$ and $N_D(q)$ depends on the network structure where $s(q)$ evaluates the integrity and $N_D(q)$ reflects the controllability. Guided by such measure, the robust optimization of directed network is achievable, and several attempts [20] have been made to successfully promote the invulnerability of directed networks.

However, seldom has an attempt to consider the robust influence maximization problem on directed networks. Considering the diffusion process, how to evaluate the robustness seeds against structural failures, and how to find considerable seed sets incorporated with directness, these questions still remain to be thoroughly investigated.

3   Algorithm Design

Memetic algorithm is a novel optimization approach which introduces local search based on evolutionary algorithm, to deal with the problems of slow convergence and difficult to find high-precision solutions. The Memetic algorithm is widely applied to solve the influence maximization problem in recent researches [22]. In this work, a Memetic algorithm is devised to solve the robust influence maximization problem on directed networks, termed MA-RIM$_D$. The included crossover and mutation operators consider the feature of this problem; in addition, the local search considers both global and neighboring searches to accelerate the convergence. Details of the performance evaluation metric and the algorithm are depicted in this section.

3.1 R_{SD}

Considering the generality, the IC model is adopted to simulate the information propagation process. Meanwhile, the malicious attack is employed to cause disturbances towards the diffusion process, and the degree centrality is used to detected the removal target. The controllable theory [19] has been introduced to the evaluation process, and the influence range of the given seed set S is recoded. A metric R_{SD} is designed utilizing such mechanism, and the features of directed network are included, as,

$$R_{SD} = \frac{1}{q \times N} \sum_{q=1/N} \frac{\sigma_S(q)}{N_D(q)} \tag{4}$$

where N is the total number of nodes in a network, $\sigma_S(q)$ are the influence range of S, and the minimum number of control routes after removing $q \times N$ nodes. $\frac{1}{q \times N}$ is the normalized factor to guarantee the comparisons on different networks comparable.

The minimum number of control routes $N_D(q)$ is considered in both R_{SD} and the measure CR in Eq. 3 to reflect the impact of directed connections in the network. A smaller $N_D(q)$ indicates that it tends to be simple to control the entire network; meanwhile, the influence generated from an important node may get spread in a wider range. Comparing with CR, R_{SD} focuses on the influence ability of seeds when the network is under attack. This measure reflects the loss of connectivity caused by node-based attacks, and records the change of influential range in the destruction process. A numerical measure is thus constructed to evaluate the robustness of seeds on directed networks.

3.2 MA-RIM$_D$

The target of solving the robust influence maximization is to select the appropriate seed set with a stable influential ability. The problem can be regarded as an optimization problem with discrete decision variables, and the gene for each candidate is a seed set $S = \{s_1, s_2, ..., s_K\}$, representing the chromosomes in MA-RIM$_D$. Guided by robust influence metric of seed set R_{SD}, MA-RIM$_D$ is designed to solve this optimization problem considering features of directed networks.

The Crossover and Mutation Operator. In genetic algorithms, crossover and mutation operations are conventional processes for exploring genetic materials and enrich the diversity of genes in the population. The crossover operator intends to generate new chromosomes by randomly swapping pieces of parental chromosomes. In this work, each chromosome of individual is a discrete sequence of a length K and single point crossover [18] is employed in this operator. The process is as follows, at a probability of p_c, for two individuals $i = \{x_1, x_2, ...x_K\}$ and $j = \{y_1, y_2, ...y_K\}$ are randomly selected from the current population. An integer *posi* is stochastically determined in the interval of $[1, K-1]$, and the gene segments around *posi* are exchanged between i and j to generate two new individuals $i' = \{x_1, x_2, ...x_{posi}, ...y_{K-1}, y_K\}$ and $j' = \{y_1, y_2, ...y_{posi}, ..., x_{K-1}, x_K\}$. If there are repetitive nodes in i' and j', the repetitive ones should be replaced by randomly chosen nodes. The two generated individuals completing crossover are preserved into the population. In the following iteration, another two individuals are randomly selected from the remaining individuals to crossover until all individuals conducting the crossover process.

The mutation operator is aimed at generating minor changes on the current individuals. Normal mutation techniques include the 1-point mutation, the uniform mutation and the Gaussian mutation [5,13]. The 1-point mutation is utilized in this work. For each chromosome $S_i = \{s_1, s_2, ..., s_i\}$, the operator is executed at a probability of p_m, whose process is as follows, an integer $posi(1 \leq posi \leq K)$ is randomly generated and the corresponding gene is replaced by a randomly chosen node m. Note that no repetitive nodes are allowed in the procedure. In this manner, the mutated chromosome is presented as $S'_i = \{s_1, s_2, ..., s_{posi-1}, m, s_{posi+1}, ..., s_K\}$.

The Local Search Operator. Generally, nodes with a larger degree tend to possess more connected neighbors; while for directed networks, nodes with a larger out-degree may exhibit better potential influence range, and these key nodes are prioritized in the destruction process. Once attacks happen, the connections between such trunk nodes and neighbors are prone to be disconnected in priority. The expected influence of seeds may be affected and decreased. Note that nodes with moderate importance may possess better influential ability under this circumstance. Considering links of these nodes are not crucial and can survive in a longer time under attacks. This conclusion may provide references for the design of the local search operator.

This operator should consider structural information pertaining to nodes such as the neighboring members in the 2-hop range. In the early generations of the search process, nodes with higher out-degree can be preferentially selected into the seed set, drawing a fast improvement on the influence propagation. However, with the evolutionary procedure conducted, the local search operator should reduce the emphasis on such nodes to avoid falling into local optimal and restraining the optimal ability. In addition, we can manage to select some candidates just randomly into the seed set to improve the diversity.

In order to keep the genetic diversity of the population and reduce the computational complexity, the local search operator is only conducted on superior chromosomes with better R_{SD}. The operation is divided into the global search, the neighboring search and the random search, where these three parts are conducted at a probability of p_l independently. In the global search, a randomly chosen candidate from the top 2% nodes in the network is determined, where nodes are ranked considering the out-degree in the entire network. A randomly chosen seed from the chromosome S is attempted to be replaced on the condition that R_{SD} gets increased. In the neighboring search, for each seed in S, the 1-hop and 2-hop neighbors are selected and preserved in a temporary set S_{nei}, where no repetitive nodes are considered. Candidates in S_{nei} attempts to replace this seed, and the corresponding R_{SD} value is calculated. The seed set with the highest R_{SD} is compared with the original S, and S is updated if a better R_{SD} can be reached. R_{SD} The random search includes random selection of nodes to replace a stochastic seed under the condition that R_{SD} is increased.

Different from the crossover and mutation operator relying on randomness, the local search operator is guided by R_{SD} to monotonously promote candidates' performances. This operator directed increases the level of fitness of the whole population.

The Selection Operator. The selection operator intends to select those superior individuals into the child population, and keep the competitiveness of the search process. There are several methods for generating child population in evolutionary algorithms including the ranking selection and the roulette selection. The ranking selection directly chooses the first Θ chromosomes into the next generation, which may lead to be trapped in local optima. The roulette wheel selection tends to consider a proportional selection on both superior and inferior individuals, and thus retaining the diversity of the population. To keep the diversity of the population and the convergence ability, the selection operator is designed via combining both the two selection methods. In detail, all individuals in the current population are ranked in a decreasing order according to their R_{SD} values. The first half of the child population is determined by the ranking selection, and those superior ones are preserved to keep the competitiveness. The second half is determined by the roulette operator, where individuals with diversified R_{SD} can be selected to guarantee the diversity.

In summary, procedures of MA-RIM$_D$ are summarized in Algorithm 1.

Algorithm 1: MA-RIM$_D$

Input:
Θ: Population size; K:nodes subset size;
G_0: Initial network; $Maxgen$:maxmum times of generation;
p_c:the crossover probability;
p_m:the mutation probability;
p_l:the local search probability;
Output: S^*:optimal seed set;
$g \leftarrow 0$ and $P^g \leftarrow$ Initialization(G_0);
while $g \leq Maxgen$ **do**
  $g \leftarrow g + 1$;
  $P_t^g \leftarrow$ Crossover Operation(P^g);
  $P^g \leftarrow$ Mutation Operation(P^g);$P_t^g \leftarrow$ Mutation Operation(P_t^g);
  $P^g \leftarrow$ Local Search Operation(P^g);$P_t^g \leftarrow$ Local Search Operation(P_t^g);
  $P^{g+1} \leftarrow$ Selection Operation(P^g, P_t^g);
end
S^* Selected from P^g;
return S^*;

---

4   Experimental Results

In the section, the performance of MA-RIM$_D$ is validated on networks, including scale-free (BA) networks [2], random (ER) networks [6] and small-world (WS) networks [26]. In the experiment, the parameters are configured as follows, the activation probability of the IC model p as 0.01, the maximal genetic iteration $Maxgen$ as 100, the population size Θ as 50, the crossover probability p_c as 0.5, the mutation probability p_m as 0.1, and the local search probability p_l as 0.7. Meanwhile, several existing approaches are implemented to provide comparison including the plain genetic algorithm (GA), the canonical Memetic algorithm (MA) and MA-RIM [22].

The numerical optimization results on BA, ER and WS networks with average degree k=4 and size of network N=200 are reported in Table 1. Each algorithm was taken 10 executions in three networks where K=5,10,20, and the table shows the best results and average results for each test case. As it can be seen in the table, MA-RIM$_D$ reaches the best numerical result of obtained seeds. In detail, GA, MA, MA-RIM and MA-RIM$_D$ obtain similar performances when K is set as 5. Considering it tends to be easy to determine influential seeds when K is small, the tested algorithm show effectiveness in this scenario. MA-RIM$_D$ exhibits superiority over other approaches when K gets increased like 10 and 20. Compared with GA that relies on randomness-based searches, MA-RIM$_D$ includes problem-directed search strategy and tends to possess better results at a given computational budget. Focusing on the search strategies, MA and MA-RIM only consider undirected networks; whilst MA-RIM$_D$ maintains a local search strategy towards networks with directedness. The utilization of structural features contributes to a favorable search ability of MA-RIM$_D$.

To further analyze the optimization ability of MA-RIM$_D$, the convergence of four evolutionary-based methods (GA, MA, MA-RIM and MA-RIM$_D$) is analyzed. As a representative, the optimal results of 10 executions for each case in the aforementioned experiments are presented, whose form are the dynamic change process of R_{SD} from generation 1 to generation 100, and other parameters are consistent with the above experiments. Taking BA network and K=10 as an example, the convergence curves of the four algorithms are drawn in Fig. 1. These curves do not show the same initial point caused by randomness-based initialization strategies. Meanwhile, GA without problem-directed search strategy has a slow convergence process for over fifty genetic iterations, whose optimization results are also inferior against other approaches. MA shows a good convergence in the early genetic process, but tends to obtain less effectiveness results as the genetic process proceeding. This phenomenon demonstrates the contribution of problem-orientated local search procedure. Comparatively, MA-RIM and MA-RIM$_D$ maintain targeted search process and consider the structural information, the obtained curves get a fast convergence. The local search strategies of MA-RIM$_D$ are efficient towards directed networks, whose curve performs better in Fig. 1. The convergence curves in ER and WS networks are also presented in Fig. 2. Although the initial trend of GA, MA and MA-RIM is slightly different from that is presented in BA networks, the curves of MA-RIM$_D$ still has great

Table 1. The numerical optimization results(R_{SD}) on several approaches. The best result of each test case is highlighted.

(a) The best result of 10 executions for each case

	K	GA	MA	MA-RIM	MA-RIMD
	5	5.8895	6.0024	6.0140	**6.0156**
BA	10	11.7583	11.8354	11.8795	**11.9045**
	20	23.1831	23.2747	23.3422	**23.4885**
	5	5.8058	5.8172	5.8284	**5.8394**
ER	10	11.5536	11.5656	11.6096	**11.6231**
	20	22.8597	22.9597	23.0060	**23.1885**
	5	7.5721	7.5866	7.6019	**7.6023**
WS	10	15.1143	15.1292	15.1596	**15.1739**
	20	30.0327	30.1073	30.1386	**30.1991**

(b) The average result of 10 executions per algorithm and standard Deviation (Average ± Standard Deviation)

	K	GA	MA	MA-RIM	MA-RIMD
	5	5.8202±0.054	5.9231±0.046	6.0038±0.0129	**6.0146±0.0021**
BA	10	11.6556±0.1342	11.7124±0.1245	11.8238±0.0671	**11.8954±0.0243**
	20	23.0135±0.1596	23.19490.0689	23.3084±0.0328	**23.4313±0.0374**
	5	5.7608±0.038	5.7941±0.0224	5.8170±0.0134	**5.8245±0.0119**
ER	10	11.5227±0.0319	11.5429±0.0277	11.5881±0.0117	**11.6089±0.0138**
	20	22.8167±0.0401	22.9055±0.0512	22.9603±0.0366	**23.1307±0.0328**
	5	7.5569±0.0149	7.5715±0.0135	7.5873±0.0137	**7.5943±0.0041**
WS	10	15.0251±0.0754	15.0681±0.0523	15.1145±0.0345	**15.1650±0.0079**
	20	29.8711±0.3894	30.0029±0.0742	30.0633±0.0558	**30.1415±0.0376**

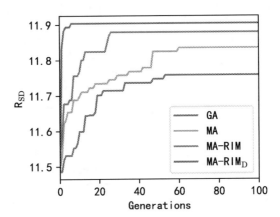

Fig. 1. The convergence curves of several evolutionary-based approaches on BA network. The best results of 10 executions per algorithm are presented as the form of dynamic change process, whose optimized results are presented in Table 1(a).

advantages in convergence speed and optimization results. This result also verifies the effectiveness of MA-RIM$_D$ in handling the robust influence maximization problem on directed networks.

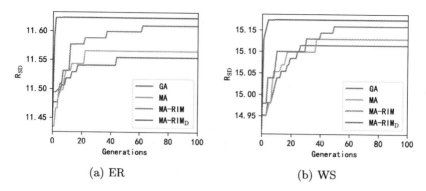

Fig. 2. The convergence curves of several evolutionary-based approaches on ER and WS networks.

Considering the characteristics of directed networks, challenges of determining robust and influential seeds lie in the directedness information. Under this context, links connected with dense connection blocks in undirected networks may possess more complicated variants in directed ones. Therefore, the search strategy in MA-RIM only aims at the undirected networks, and cannot distinguish the directions of the neighboring nodes. The search efficiency is thus less competitive when tackling directed networks, Whilst the search strategy of MA-RIM_D is guided by the problem-specific robustness evaluation factor and shows preference when selecting potentially influential nodes in the search process. In addition, the unique random search of MA-RIM_D also improves optimality to better explore the whole solution space, and the quality of obtained solutions is thus guaranteed. This strategy may lead to an increased computational budget, but an improved performance has been demonstrated in the experiment. Also, the local search process shows preference towards superior individuals, which promotes the search ability and contributes to a fast convergence. Equipped with such operators, MA-RIM_D shows better performance over MA-RIM at a given computational budget.

5   Conclusions

The robust influence maximization problem has been intensively emphasized in previous studies. In this work, a case study on directed networks is given. Combining the feature of such networks and node-based attacks, a robust influence measure R_{SD} is designed numerical evaluate the performance of seeds on directed networks. Guided by this measure, MA-RIM_D is devised to find robust and influential seeds R_{SD} against node-based malicious attacks. This algorithm includes several problem-directed genetic operators, and the local search operator exploits the structural information to speed up the convergence. It is experimentally demonstrated that MA-RIM_D shows competitive performance over existing approaches, and seeds with both considerable influential ability and tolerance against attacks are obtained. Considering the complexity and diversity of practical networks, more diffusion model can be considered to cater to diversified diffusion tasks. Besides, only the node-based attack is considered in this work, and other components like links and multiple attacks are also under risk. To further investigate the robust influence problem on directed networks, investigation on other diffusion models and different damages types are worthy in future studies.

Acknowledgements. This work was supported in part by Guangdong Basic and Applied Basic Research Foundation under Grant 2021A1515110543, and in part by National Natural Science Foundation of China under Grant 62203477.

References

1. Albert, R., Barabási, A.L.: Statistical mechanics of complex networks. Rev. Mod. Phys. **74**(1), 47 (2002)
2. Barabási, A.L., Albert, R.: Emergence of scaling in random networks. Science **286**(5439), 509–512 (1999)
3. Chen, W., Lin, T., Tan, Z., Zhao, M., Zhou, X.: Robust influence maximization. In: Proceedings of the 22nd ACM SIGKDD International Conference on Knowledge Discovery and Data Mining, pp. 795–804 (2016)
4. Chen, W., Wang, Y., Yang, S.: Efficient influence maximization in social networks. In: Proceedings of the 15th ACM SIGKDD International Conference on Knowledge Discovery and Data Mining, pp. 199–208 (2009)
5. Deep, K., Thakur, M.: A new mutation operator for real coded genetic algorithms. Appl. Math. Comput. **193**(1), 211–230 (2007)
6. Erdős, P., Rényi, A., et al.: On the evolution of random graphs. Publ. Math. Inst. Hung. Acad. Sci **5**(1), 17–60 (1960)
7. Gong, M., Yan, J., Shen, B., Ma, L., Cai, Q.: Influence maximization in social networks based on discrete particle swarm optimization. Inf. Sci. **367**, 600–614 (2016)
8. Holme, P., Kim, B.J., Yoon, C.N., Han, S.K.: Attack vulnerability of complex networks. Phys. Rev. E **65**(5), 056109 (2002)
9. Huang, D., Tan, X., Chen, N., Fan, Z.: A memetic algorithm for solving the robust influence maximization problem on complex networks against structural failures. Sensors **22**(6), 2191 (2022)
10. Kempe, D., Kleinberg, J., Tardos, É.: Maximizing the spread of influence through a social network. In: Proceedings of the Ninth ACM SIGKDD International Conference on Knowledge Discovery and Data Mining, pp. 137–146 (2003)
11. Lee, J.R., Chung, C.W.: A fast approximation for influence maximization in large social networks. In: Proceedings of the 23rd International Conference on World Wide Web, pp. 1157–1162 (2014)
12. Leskovec, J., Krause, A., Guestrin, C., Faloutsos, C., VanBriesen, J., Glance, N.: Cost-effective outbreak detection in networks. In: Proceedings of the 13th ACM SIGKDD International Conference on Knowledge Discovery and Data Mining, pp. 420–429 (2007)
13. Lim, S.M., Sultan, A.B.M., Sulaiman, M.N., Mustapha, A., Leong, K.Y.: Crossover and mutation operators of genetic algorithms. Int. J. Mach. Learn. Comput. **7**(1), 9–12 (2017)

14. Liu, X., Stanley, H.E., Gao, J.: Breakdown of interdependent directed networks. Proc. Natl. Acad. Sci. **113**(5), 1138–1143 (2016)
15. Nepusz, T., Vicsek, T.: Controlling edge dynamics in complex networks. Nat. Phys. **8**(7), 568–573 (2012)
16. Schneider, C.M., Moreira, A.A., Andrade, J.S., Jr., Havlin, S., Herrmann, H.J.: Mitigation of malicious attacks on networks. Proc. Natl. Acad. Sci. **108**(10), 3838–3841 (2011)
17. Son, S.W., Kim, B.J., Hong, H., Jeong, H.: Dynamics and directionality in complex networks. Phys. Rev. Lett. **103**(22), 228702 (2009)
18. Umbarkar, A.J., Sheth, P.D.: Crossover operators in genetic algorithms: a review. ICTACT J. Soft Comput. **6**(1) (2015)
19. Wang, B., Gao, L., Gao, Y., Deng, Y.: Maintain the structural controllability under malicious attacks on directed networks. Europhys. Lett. **101**(5), 58003 (2013)
20. Wang, S., Liu, J.: A multi-objective evolutionary algorithm for promoting the emergence of cooperation and controllable robustness on directed networks. IEEE Trans. Netw. Sci. Eng. **5**(2), 92–100 (2017)
21. Wang, S., Liu, J.: Constructing robust community structure against edge-based attacks. IEEE Syst. J. **13**(1), 582–592 (2018)
22. Wang, S., Liu, J.: A memetic algorithm for solving the robust influence maximization problem towards network structural perturbances. Chin. J. Comput. **44**(6), 1153–1167 (2021)
23. Wang, S., Liu, J., Jin, Y.: Finding influential nodes in multiplex networks using a memetic algorithm. IEEE Trans. Cybern. **51**(2), 900–912 (2019)
24. Wang, S., Tan, X.: Determining seeds with robust influential ability from multi-layer networks: a multi-factorial evolutionary approach. Knowl.-Based Syst. **246**, 108697 (2022)
25. Wang, W.X., Ni, X., Lai, Y.C., Grebogi, C.: Optimizing controllability of complex networks by minimum structural perturbations. Phys. Rev. E **85**(2), 026115 (2012)
26. Watts, D.J., Strogatz, S.H.: Collective dynamics of 'small-world' networks. Nature **393**(6684), 440–442 (1998)
27. Zeng, A., Liu, W.: Enhancing network robustness against malicious attacks. Phys. Rev. E **85**(6), 066130 (2012)

Adaptive Artificial Bee Colony Algorithm Considering Colony's Memory

Jiacheng Li[1(✉)], Masato Noto[1], and Yang Zhang[2]

[1] Department of Applied Systems and Mathematics, Kanagawa University,
Yokohama, Japan
{lijiacheng,noto}@kanagawa-u.ac.jp
[2] Department of Computer Science, Kanagawa University, Yokohama, Japan
zhangyang@kanagawa-u.ac.jp

Abstract. The artificial bee colony (ABC) algorithm is a swarm intelligence optimization algorithm inspired by the foraging behavior of honey bee colonies, which means to gradually explore and find food-rich nectar sources through the behavioral patterns of three different bees foraging between hives. We propose an adaptive ABC (AABC) algorithm to accelerate the convergence rate by considering a colony's memory in the employed-bee phase of the algorithm, since some bees will select and explore the best nectar source, and limiting the initial search range of the swarm to search for better honey sources faster. This is done by using tent chaotic search and random backward learning in the scout-bee phase to reselect the nectar source and increase the possibility of jumping out of local optimal solution. We conducted 30 numerical experiments to verify the effectiveness of the proposed adaptive ABC algorithm.

Keywords: Artificial bee colony algorithm · memory · random backward learning · tent chaotic mapping

1 Introduction

Swarm-intelligence-optimization algorithms [1] are inspired by the group behavior of insects, animals, and birds in nature and leverage the information exchange and collaboration among groups to find the optimal solution through simple and limited interactions among individuals. Commonly used swarm-intelligence-optimization algorithms include the particle swarm optimization (PSO) algorithm inspired by the predatory behavior of bird flocks in nature [2,3], sparrow-search algorithm inspired by the foraging and anti-predatory behavior of sparrows [4], and whale-optimization algorithm (WOA) that imitates the predatory behavior of whales in nature [5,6], etc. Such algorithms are more robust, stable, and adaptive compared with simplex optimization algorithms.

The artificial bee colony (ABC) algorithm [7] is a metaheuristic algorithm that mimics the foraging of bee colonies for optimizing numerical problems. The ABC algorithm has become a hot research topic research because of its strong robustness, no requirement of priori information, few parameters to be adjusted,

Y. Tan et al. (Eds.): ICSI 2023, LNCS 13968, pp. 284–296, 2023.
https://doi.org/10.1007/978-3-031-36622-2_23

and computational convenience. Zhang and Lee [8] proposed the router-oriented ABC algorithm that has a more diverse and enhanced search compared with original ABC algorithm. This algorithm also incorporates useful information from a router. Brajevic et al. [9] put forward a collaborative hybrid algorithm that is based on fireflies and multi-strategy artificial bee colonies by using a novel multi-strategy ABC algorithm for local search. Kiran [10] suggested a stigmergic behavior-based update rule for onlooker bees of a binary artificial bee colony algorithm and an extension of the xor-based update rule to improve the local-search capability and convergence properties of a binary artificial bee colony algorithm. Thirugnanasambandam et al. [11] provided an interim solution search strategy in the employed-bee and onlooker-bee phases of the ABC algorithm by considering the advantages of local, neighbor, and iterative optimal solutions and adjusting the search radius of the new candidate solution in the scout-bee phase.

We proposed an adaptive ABC algorithm that uses tent chaotic search and random backward learning to optimize the reselection of a nectar source in the scout-bee phase to accelerate convergency rate and increase the possibility of jumping out of local optimal solution. We conducted multiple numerical experiments to evaluate the proposed method by comparing it with the original ABC algorithm, PSO algorithm, and WOA algorithm.

2 ABC Algorithm

The ABC algorithm [12] consists of a nectar source and three types of bees, i.e., the employed bees, onlooker bees, and scout bees, that differ in their behaviors in a colony. The notation $x_i(i = 1, 2, \ldots, SN)$ denotes the i-th nectar, where SN is the population size. The d-th dimension of $x_i(i = 1, 2, \ldots, SN)$ is $x_{id}(d = 1, 2, \ldots, n)$, where n the number of dimensions.

An employed bee randomly searches for nectar sources at a nearby nectar source x_k and uses a greedy strategy to select the better between the new source v_{id} and original one x_{id} as the current nectar source as follows

$$v_{id} = x_{id} + a\left(x_{id} - x_{kd}\right) \tag{1}$$

where a is a random number in the range of $\begin{bmatrix} -1 & 1 \end{bmatrix}$.

Onlooker bees summarize the information of the current nectar source where the employed bee is after the latter has searched for nectar, uses the roulette-selection strategy (Eq. (2)) to select the nectar source. The position is updated using Eq. (3). Before any selection is made using the roulette strategy, the objective function should be transformed into the fitness function through Eq. (4).

$$p_i = \frac{fit\left(x_i\right)}{\sum_i^{SN} fit\left(x_i\right)} \tag{2}$$

$$v_{id} = x_{id} + a\left(x_{id} - x_{md}\right) \tag{3}$$

$$fit\left(x_i\right) = \begin{cases} \dfrac{1}{1 + f\left(x_i\right)} & f\left(x_i\right) \geq 0 \\ 1 + abs\left(f\left(x_i\right)\right) & f\left(x_i\right) < 0 \end{cases} \tag{4}$$

In Eq. (2), p_i is the probability of the i-th employed bee corresponding to the objective function, and $fit(x_i)$ is the fitness function corresponding to the i-th employed bee.

In Eq. (3), the m-th nectar source x_m is obtained with the roulette-selection strategy.

In Eq. (4), $f(x_i)$ is the objective function corresponding to the i-th employed bee.

When the employed bee reaches the "standard of abandonment" in terms of the number of searches, a better result can be achieved by abandoning the current nectar source and enabling the scout bees to search for one in accordance with Eq. (5). Therefore, the limitation of a local optimal solution will be addressed.

$$x_{id} = l_{id} + \text{rand}(0,1) * (u_{id} - l_{id}) \tag{5}$$

In Eq. (5), l_{id} and u_{id} are the lower and upper limits of x_{id}, respectively.

3   Adaptive ABC Algorithm Considering Colony's Memory

There are many improvement strategies for artificial bee colony algorithms, mainly focusing on the honey source selection method and the behavior of three types of bees [13–15]. In this paper, the ABC algorithm is improved from the two aspects of the employed bees and scout bees. A new search behavior of the employed bees is added, and using tent chaotic search and random backward learning to increase the possibility of reconnaissance bees searching for better honey sources and jumping out of local optimal solution.

3.1   Employed-Bee Phase

The employed bee has memorized the existing best nectar source before searching for a new one. Li [16] use the improved Inver-over Operation to improve the overly single search method in the reconnaissance bee stage. In this article, some employed bees decide to explore new sources at a nearby nectar source x_k, while others choose to do so at the existing best source x_{best}. Equation (6) denotes the overall position change.

$$v_{id} = \begin{cases} x_{id} + c\,(x_{id} - x_{kd}) & \text{if rand } < \beta \\ x_{id} + c\,(x_{id} - x_{best}) & \text{else} \end{cases} \tag{6}$$

The value range of the behavior probability coefficient β is [0 1]. The larger the β, the more the employed bee tends to go to the vicinity of the optimal solution, otherwise the more inclined it is to go to the vicinity of the random solution, is generally from [0.6 0.8]. c is the search step and calculated as

$$c = \left(1 - \frac{e^{(1-iter/iter_{\max})} - 1}{2(e-1)}\right) * a \tag{7}$$

where *iter* is the current iteration number, and $iter_{max}$ is the maximum iteration number. As the number of iterations reaches $iter_{max}$, the range of the search step's possible values enlarges from $\begin{bmatrix} -0.5 & 0.5 \end{bmatrix}$ to $\begin{bmatrix} -1 & 1 \end{bmatrix}$.

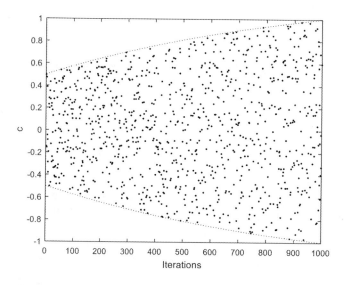

Fig. 1. Changes of the search step with number of iterations

Compared with the previously used search step falling in a fixed range, the strategy with a small search step at the beginning can avoid slow convergence of the algorithm due to a large search range, while an increase in the search step in later iterations of the algorithm can help better search the global nectar source. Changes in the search step with the number of iterations are illustrated in Fig. 1.

3.2 Scout-Bee Phase

When the number of employed bees failing to find a better solution continuously reach the "standard of abandonment", the scout bees will forsake the current nectar source and explore a new one. The original ABC algorithm uses Eq. (5) to search for a new source, which does not guarantee that the scout bees will find a better one and may even fail. Therefore, our adaptive ABC algorithm uses tent chaotic search and improved backward learning to reselect a nectar source.

Tent mapping is widely used for its uniform distribution function and the chaotic sequences it generates, which are favorable for statistics. Tent mapping is defined as

$$g\left(y_i\right) = \begin{cases} \dfrac{y_{i-1}}{\alpha} & 0 \leq y_{i-1} < \alpha \\ \dfrac{1 - y_{i-1}}{1 - \alpha} & \alpha \leq y_{i-1} \leq 1 \end{cases} \tag{8}$$

where α is a tent-mapping parameter and the value range of α is $[0\ 1]$. The scout bees explore a new nectar source x_i in accordance with Eq. (5). A new location v_i is generated using Eq. (9) in accordance with tent mapping. The greedy strategy is used for selecting a better as the new source from v_{id} and x_{id}.

$$v_{id} = \begin{cases} \dfrac{x_{id}}{\alpha\left(u_{id} + l_{id}\right)} & l_{id} \le x_{id} < \alpha\left(u_{id} + l_{id}\right) \\ \dfrac{u_{id} - x_{id}}{u_{id} - \alpha\left(u_{id} + l_{id}\right)} & \alpha\left(u_{id} + l_{id}\right) \le x_{id} \le u_{id} \end{cases} \tag{9}$$

The backward learning strategy compares the backward solution with the existing one, which is more likely to be closer to the global optimal solution than the random initial value. The main idea is to use the opposite value of the initial solution at the same time to obtain a better solution; thus, the next-generation solution can be closer to the optimal one more quickly since the initial value in problem solving is obtained from experience or simply by a random guess. General backward learning can only produce a fixed backward solution, while a new backward solution can be generated using Eq. (10) after randomness is introduced.

$$h\left(y_i\right) = u_m + l_m - y_i + \text{rand}(0, 1) * \min\left(abs\left(l_m - y_i\right), abs\left(u_m - y_i\right)\right) \tag{10}$$

where l_m and u_m are the lower and upper limits of y_i, respectively. Figure 2 shows the distribution of the random-backward-solution in the range of $\begin{bmatrix} -10 & 10 \end{bmatrix}$ with the increase of the number of iterations. Figure 3 shows the distribution statistics of 1000 random-backward-solutions. The results of the improved backward learning are random values near the original directional solution, and 1,000 results with the initial value of 3 from $\begin{bmatrix} 0 & 10 \end{bmatrix}$ are illustrated in Fig. 2 and 3. Figure 2 shows that random-backward-solutions are on both sides of the original backward solution. The closer the random results to the original backward solution are, the denser they are. It is more likely for them to be selected; thus, bees can search near the original backward solution efficiently and obtain a better result. Figure 3 clearly shows that the value of the backward solution falls between $\begin{bmatrix} 4 & 10 \end{bmatrix}$, and 7 appears more frequently when the original backward solution is closer.

The scout bees explore an x_i in accordance with Eq. (5), and a v_i is generated using Eq. (11) in accordance with random backward learning. The better source is selected as a new source.

$$v_{id} = u_{id} + l_{id} - x_{id} + \text{rand}(0, 1) * \min\left(abs\left(u_{id} - x_{id}\right), abs\left(l_{id} - x_{id}\right)\right) \tag{11}$$

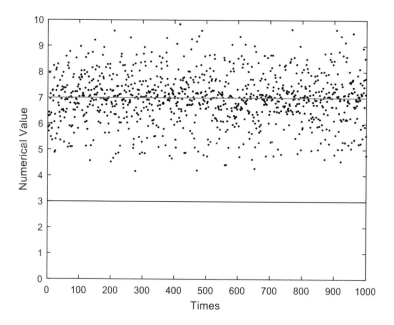

Fig. 2. Results of random-backward-solution

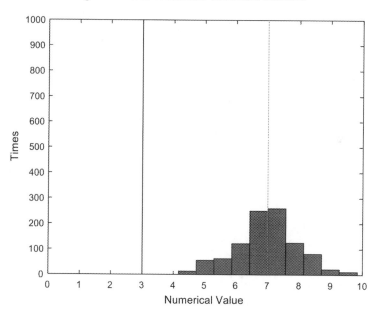

Fig. 3. Distribution of random-backward-solution values

In summary, Eqs. (5) and (12) are the position-update equations of the scout bees. The better source between x_{id} and v_{id} is selected as a new source on the basis of the greedy strategy.

$$v_{id} = \begin{cases} \begin{aligned} &u_{id} + l_{id} - x_{id} + \text{rand}(0,1) * \\ &\min\left(abs\left(u_{id} - x_{id}\right), abs\left(l_{id} - x_{id}\right)\right) \end{aligned} & \text{if rand} < 0.5 \\ \dfrac{x_{id}}{\alpha\left(u_{id} + l_{id}\right)} & \begin{aligned} &\text{if rand} \geq 0.5 \text{ and} \\ &l_{id} \leq x_{id} < \alpha\left(u_{id} + l_{id}\right) \end{aligned} \\ \dfrac{u_{id} - x_{id}}{u_{id} - \alpha\left(u_{id} + l_{id}\right)} & \begin{aligned} &\text{if rand} \geq 0.5 \text{ and} \\ &\alpha\left(u_{id} + l_{id}\right) \leq x_{id} \leq u_{id} \end{aligned} \end{cases} \tag{12}$$

4   Comparison of Test Functions

To verify the authenticity and accuracy of AABC algorithm, we conducted 30 experiments to evaluate its performance by comparing it with four other algorithms (ABC, TABC [17] (Two modified versions of artificial bee colony algorithm), PSO, and WOA) in solving 13 test functions in 100 dimensions. We used CEC2019 TEST SUITES, and the experimental environment was Intel (R) Core (TM) i7-9750H CPU @ 2.60 GHz, Windows 10 Pro, MATLAB, 2022b.

The population sizes of the AABC, TABC, ABC, PSO, and WOA algorithms were all set to 50, and their number of iterations to 3,000. Parameters β and α of the MABC algorithm were 0.7 and 0.5 respectively; the WOA algorithm's spiral parameter b was 1.

The function $f_1 - f_7$ are unimodal reference functions, and the specific expression is as follows. The function f_1 and f_7 with $D = 2$ are shown in Fig. 4 and 5 for examples.

$$f_1 = \sum_{i=1}^{n} x_i^2, x_i \in \left[-100 \quad 100\right]$$

$$f_2 = \sum_{i=1}^{n} |x_i| + \prod_{i=1}^{n} |x_i|, x_i \in \left[-10 \quad 10\right]$$

$$f_3 = \sum_{i=1}^{n} \left(\sum_{j=1}^{j} x_j\right), x_j \in \left[-100 \quad 100\right]$$

$$f_4 = \max_{i} \left\{|x_i|, 1 \leq i \leq n\right\}, x_i \in \left[-100 \quad 100\right]$$

$$f_5 = \sum_{i=1}^{n-1} \left[100\left(x_{i+1} - x_1^2\right)^2 + \left(x_i - 1\right)^2\right], x_i \in \left[-30 \quad 30\right]$$

$$f_6 = \sum_{i=1}^{n} \left([x_i + 0.5]\right)^2, x_i \in \left[-100 \quad 100\right]$$

$$f_7 = \sum_{i=1}^{n} i x_i^4 + \text{random}[0,1], x_i \in \left[-1.28 \quad 1.28\right]$$

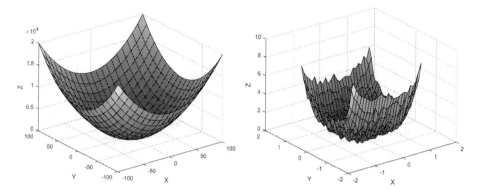

Fig. 4. The function f_1 **Fig. 5.** The function f_7

The function f_8–f_{11} are multimodal benchmark functions, and the specific expression is as follows. The function f_8 and f_9 with $D = 2$ are shown in Fig. 6 and 7 for examples.

$$f_8 = \sum_{i=1}^{n} -x_i^2 \sin\left(\sqrt{|x_i|}\right), x_i \in \begin{bmatrix} -300 & 300 \end{bmatrix}$$

$$f_9 = \sum_{i=1}^{n} \left[x_i^2 - 10\cos(2\pi x_i) + 10\right], x_i \in \begin{bmatrix} -5.12 & 5.12 \end{bmatrix}$$

$$f_{10} = -20\exp\left(-0.2\sqrt{\sum_{i=1}^{n} x_i^2}\right) - \exp\left(\frac{1}{n}\sum_{i=1}^{n} \cos(2\pi x_i)\right) + 20 + e, x_i \in \begin{bmatrix} -32 & 32 \end{bmatrix}$$

$$f_{11} = \frac{1}{4000}\sum_{i=1}^{n} x_i^2 - \prod_{i=1}^{n}\left(\frac{x_i}{\sqrt{i}}\right) + 1, x_i \in \begin{bmatrix} -600 & 600 \end{bmatrix}$$

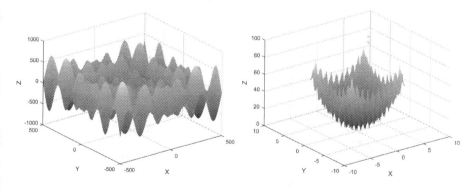

Fig. 6. The function f_8 **Fig. 7.** The function f_9

The experimental results are listed in Table 1, with the optimal values in bold. In 100 dimensions, AABC algorithm solved the optimal average optimal solution of f_1, f_3, f_5, f_6, f_7, f_9 and f_{11}, WOA algorithm solved the optimal average optimal solution of f_2, f_8 and f_{10}, WOA algorithm solved the optimal average optimal solution of f_4, and ABC algorithm results are inferior to other algorithms. The results of AABC algorithm are better than those of TABC algorithm and ABC algorithm. The average values of these experiments are basically closer to the optimal values of the corresponding test functions, and the corresponding variance was small, which proves that AABC algorithm has higher optimization accuracy and strong robustness. Therefore, AABC algorithm outperformed the three other algorithms in terms of the optimization accuracy of various test functions. From the iterative process diagram of unimodal reference functions and multimodal benchmark functions in Fig. 8, we can also see that the AABC algorithm converges faster and the result accuracy is lower.

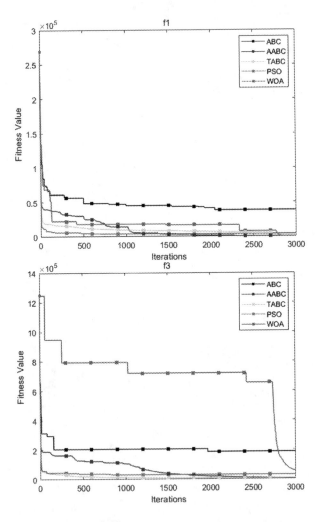

Fig. 8. Iterative process of f_1, f_3, f_8, f_{11}

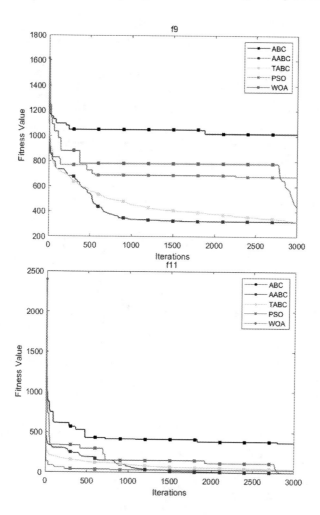

Fig. 8. (*continued*)

Table 1. CEC2019 TEST SUITES (D = 100)

Function	Index	ABC	AABC	TABC	PSO	WOA
f_1	mean	3.70E+04	**9.83E−01**	5.68E+03	5.04E+00	3.20E+03
	best	3.35E+04	**2.03E−01**	3.89E+03	4.16E+00	3.02E+03
	Std.	3.16E+03	1.30E+00	1.43E+03	**6.08E−01**	2.59E+02
f_2	mean	1.15E+02	2.64E+00	3.51E+01	**1.26E−01**	4.61E+01
	best	1.06E+02	4.19E−01	2.93E+01	**8.37E−02**	4.19E+01
	Std.	5.67E+00	3.41E+00	5.80E+00	**2.93E−02**	2.97E+00
f_3	mean	1.79E+05	**1.21E+04**	1.39E+04	5.28E+04	3.08E+04
	best	1.43E+05	**7.91E+03**	8.65E+03	4.85E+04	2.13E+04
	Std.	2.54E+04	5.69E+03	5.46E+03	**4.28E+03**	7.05E+03
f_4	mean	6.35E+01	2.82E+01	3.31E+01	4.06E+01	**1.98E+01**
	best	6.12E+01	2.44E+01	2.94E+01	2.09E+01	**1.84E+01**
	Std.	1.78E+00	3.10E+00	2.96E+00	2.44E+01	**1.60E+00**
f_5	mean	7.65E+07	1.81E+03	1.43E+06	2.41E+03	2.48E+05
	best	6.53E+07	7.96E+02	6.77E+05	**6.08E+02**	2.38E+05
	Std.	1.11E+07	**1.09E+03**	7.98E+05	1.22E+03	1.27E+04
f_6	mean	3.90E+04	**5.42E−01**	6.00E+03	1.48E+01	3.23E+03
	best	3.61E+04	**1.30E−01**	3.04E+03	1.24E+01	2.79E+03
	Std.	2.87E+03	**4.95E−01**	1.94E+03	2.44E+00	2.80E+02
f_7	mean	8.97E+01	**8.07E−02**	4.37E+00	4.04E−01	6.97E−01
	best	7.67E+01	**6.46E−02**	2.91E+00	2.78E−01	5.88E−01
	Std.	1.10E+01	**1.14E−02**	1.28E+00	1.08E−01	6.21E−02
f_8	mean	−8.04E+03	−2.17E+04	−3.62E+04	−3.89E+04	−1.58E+04
	best	−8.51E+03	−2.35E+04	−3.70E+04	−4.10E+04	−1.87E+04
	Std.	3.78E+02	2.27E+03	**9.07E+02**	1.91E+03	2.04E+03
f_9	mean	9.82E+02	**2.13E+02**	2.62E+02	4.67E+02	6.44E+02
	best	9.71E+02	**1.34E+02**	2.09E+02	3.74E+02	5.51E+02
	Std.	9.75E+00	6.07E+01	**5.05E+01**	7.69E+01	5.63E+01
f_{10}	mean	1.63E+01	1.46E+01	1.35E+01	**7.89E+00**	7.89E+00
	best	1.57E+01	1.39E+01	1.24E+01	**6.08E+00**	7.60E+00
	Std.	4.61E−01	5.37E−01	8.55E−01	2.21E+00	**2.44E−01**
f_{11}	mean	3.28E+02	**2.31E−01**	5.25E+01	7.07E−01	3.08E+01
	best	2.89E+02	**9.15E−02**	3.41E+01	4.94E−01	2.69E+01
	Std.	2.28E+01	2.07E−01	1.84E+01	**1.57E−01**	2.75E+00

5   Conclusions

We proposed an adaptive ABC algorithm considering a colony's memory to accelerate the convergence rate and mitigate the limitation of a local optimal solution throughout searches. Compared with the original ABC algorithm, the adaptive ABC algorithm has some bees in the employed-bee phase that will choose the current optimal honey source location to search, thus speeding up the search speed; And in Scout-bee phase, adopts a hybrid strategy of tent chaotic search and random reverse learning to improve the energy of the algorithm to jump out of the local optimal solution. Numerical experiments were conducted on benchmark problems of several single peaked and multi-peaked test functions in CEC2019 TEST SUITES to verify the effectiveness of our algorithm. The experimental results indicate that the adaptive AABC algorithm outperformed the original ABC algorithm and TABC algorithm as well as the PSO and WOA algorithms in single peaked and multi-peaked functions in terms of search performance.

References

1. Bonabeau, E., Dorigo, M., Theraulaz, G., Theraulaz, G.: Swarm Intelligence: From Natural to Artificial Systems, no. 1. Oxford University Press, USA (1999)
2. Kennedy, J.: Particle swarm optimization. In: Sammut, C., Webb, G.I. (eds.) Encyclopedia of the Sciences of Learning, pp. 760–766. Springer, Heidelberg (2011). https://doi.org/10.1007/978-0-387-30164-8_630
3. Shi, Y., Eberhart, R.: A modified particle swarm optimizer. In: 1998 IEEE International Conference on Evolutionary Computation Proceedings. IEEE World Congress on Computational Intelligence (Cat. No. 98TH8360), pp. 69–73. IEEE (1998)
4. Xue, J., Shen, B.: A novel swarm intelligence optimization approach: sparrow search algorithm. Syst. Sci. Control Eng. **8**(1), 22–34 (2020)
5. Mirjalili, S., Lewis, A.: The whale optimization algorithm. Adv. Eng. Softw. **95**, 51–67 (2016)
6. Abd El Aziz, M., Ewees, A.A., Hassanien, A.E.: Whale optimization algorithm and moth-flame optimization for multilevel thresholding image segmentation. Expert Syst. Appl. **83**, 242–256 (2017)
7. Karaboga, D., et al.: An idea based on honey bee swarm for numerical optimization. Technical report, Technical report-tr06, Erciyes university (2005)
8. Zhang, S., Lee, C.K.: An improved artificial bee colony algorithm for the capacitated vehicle routing problem. In: 2015 IEEE International Conference on Systems, Man, and Cybernetics, Hong Kong, China, pp. 2124–2128. IEEE (2016)
9. Brajević, I., Stanimirović, P.S., Li, S., Cao, X.: A hybrid firefly and multi-strategy artificial bee colony algorithm. Int. J. Comput. Intell. Syst. **13**(1), 810–821 (2020)
10. Kiran, M.S.: A binary artificial bee colony algorithm and its performance assessment. Expert Syst. Appl. **175**, 114817 (2021)
11. Thirugnanasambandam, K., Rajeswari, M., Bhattacharyya, D., Kim, J.: Directed artificial bee colony algorithm with revamped search strategy to solve global numerical optimization problems. Autom. Softw. Eng. **29**, 1–31 (2022)

12. Karaboga, D., Basturk, B.: A powerful and efficient algorithm for numerical function optimization: artificial bee colony (ABC) algorithm. J. Global Optim. **39**(3), 459–471 (2007)
13. Hadidi, A., Azad, S.K., Azad, S.K.: Structural optimization using artificial bee colony algorithm. In: 2nd International Conference on Engineering Optimization, pp. 6–9 (2010)
14. El-Abd, M.: Generalized opposition-based artificial bee colony algorithm. In: 2012 IEEE Congress on Evolutionary Computation, pp. 1–4. IEEE (2012)
15. Zhang, D.L., Ying-Gan, T., Xin-Ping, G.: Optimum design of fractional order PID controller for an AVR system using an improved artificial bee colony algorithm. Acta Automatica Sinica **40**(5), 973–979 (2014)
16. Li, W.H., Li, W.J., Yang, Y., Liao, H.Q., Li, J.L., Zheng, X.P.: Artificial bee colony algorithm for traveling salesman problem. In: Advanced Materials Research, vol. 314–316, pp. 2191–2196. Trans Tech Publication (2011)
17. Alizadegan, A., Asady, B., Ahmadpour, M.: Two modified versions of artificial bee colony algorithm. Appl. Math. Comput. **225**, 601–609 (2013)

An Improved Seagull Algorithm for Numerical Optimization Problem

Waqas Haider Bangyal[1](✉), Rabia Shakir[2], Najeeb Ur Rehman[3], Adnan Ashraf[4], and Jamil Ahmad[5]

[1] Department of Computer, Kohsar University, Murree 47150, Pakistan
waqas.bangyal@kum.edu.pk
[2] Department of Computer Science, FUUAST, Islamabad, Pakistan
[3] Department of Computer Science, University of Gujrat, Gujrat, Pakistan
[4] Beijing Institute of Technology (BIT), Beijing 100081, China
[5] Department of Computer Science, Hazara University Manshera, Mansehra, KPK, Pakistan

Abstract. In Artificial Intelligence, numerical optimization is an instantly rising research domain. Swarm Intelligence (SI) and Evolutionary Algorithm (EA) are widely used to answer the problems where the optimal solution is required. Inspired by Seagull's natural behavior, the Seagull Optimization Algorithm (SOA) is a meta-heuristic, swarm-based intelligent search method. SOA algorithm is a population-based intelligent stochastic search procedure that inherited the manner of seagulls to seek food. In SOA, population initialization is crucial for making rapid progress in a d-dimensional search space. In order to address the issue of premature convergence, this research presents a new variation called the Adaptive Seagull Optimization Algorithm (ASOA). Second, a variety of starting methods have been suggested as ways to enhance seagulls' propensity for exploratory activity. To improve the diversity and convergence factors, instead of applying the random distribution for the initialization of the population, Qusai-random sequences are used. This paper reveals the state-of-the-art population initialization, and a new SOA variant is introduced using adaptive mutation strategies to prevent local optima. To simulate and validate the results of ASOA and initialization techniques, 8 different benchmark test functions are applied; some are uni-modal, and some are multimodal. The simulation results depict that proposed variant ASOA provides superior results.

Keywords: Pseudo-RNG · Metaheuristic · Seagull · Optimization · ASOA

1 Introduction

In the field of Computer Science, Evolutionary Computing (EC) has been one of the most exciting recent areas of study. EC originates with Darwin's theory of natural selection, which states that environmental changes cause gradual but cumulative changes in every member of a species [1]. Each person needs to be proficient in completing the environmental and reproductive processes in order to thrive in this planet. EC refers to a computing paradigm in which global optimization is attained through the application of

Y. Tan et al. (Eds.): ICSI 2023, LNCS 13968, pp. 297–308, 2023.
https://doi.org/10.1007/978-3-031-36622-2_24

population-based algorithms. An appropriate definition of this group of meta-heuristic population-based algorithms would be "the family of algorithms that draws inspiration from the process of biological evolution" [2]. Evolutionary algorithms (EA) and swarm intelligence (SI) are the two main subfields within EC (SI).

Swarm Intelligence (SI) is used to demonstrate the algorithm inspired by the similar actions of different groups of animals and insects like bird flocking, fish schooling, and ant colonies. These types of SI systems are usually made up of simple agents of the population interacting with each other and the environment [3].

Swarm Intelligence does not apply only to robotics systems; after this novel invention, SI has been employed in every domain of daily life, either in engineering or medicine. Moreover, in research, Swarm algorithms are used repeatedly and efficiently. There has been a lot of success with these algorithms in the context of global optimization, which has many applications in the scientific community [4]. There are many other soft computing techniques on which many researchers are working on ACO, PSO [5], BCO, HS, Bat Algorithm (BA), SOA, etc. Further discussion will be related to Seagull Optimization Algorithm [6].

Dhiman and Kumar presented a novel bio-inspired method that they called the Seagull Optimization Algorithm (SOA) in 2018 for the purpose of solving difficult computational issues [7]. Seagulls are sea birds that live all over the world. They eat fishes, insects, reptiles, earthworms, and amphibians. These are smart birds and live in colonies. The algorithm's structure is straightforward and intimates the behaviour of seagulls. An essential aspect of Seagull's behaviour is their migration and attacking behaviour [8].

Population-based algorithms emphasize exploration and exploration processes in the search area, each and every optimization algorithm has a duty to keep a healthy equilibrium between the processes of exploration and exploitation [7].

SOA has many applications, e.g. data classification, clustering, feature selection, image processing, machine learning application, Artificial Neural Networks classification, and function optimization. It has been proved to be the more effective algorithm for solving global optimization problems [9–14].

Seagull Optimization Algorithm is a process that copies the behaviour of the seagull's life. It is a population-based sophisticated randomized method of searching encouraged by the natural behaviour of swarms that search for food. In the SOA algorithm, population initialization is critical [15]. In the basic Seagull Optimization Algorithm, Seagulls were initialized with random number generators. Usage of a random number generator does not provide efficient population generation [16]. Nature-inspired meta-heuristic algorithms are used to solve real-world global optimization problems [17–19]. These nature-inspired algorithms are frequently stuck in local minima and do not provide adequate results. To overcome this problem, a new variant is proposed named Adaptive Seagull Optimization Algorithm; instead of using a random distribution for population initialization in SOA, we have used Quasi-random sequencing and Pseudo-random numbers generators that are more beneficial for improving diversity and convergence factors.

The standard Seagull Optimization Algorithm (SOA) has been recognized as a robust optimization technique for low-dimensional functions and various applications. Despite such progress, as the dimensions of the problem increase, its performance may worsen. These dimensionality-related concerns affect nearly all other algorithms as well. However, several potential solutions will be investigated in this paper. Exploration and exploitation are distinctive properties of optimization algorithms because they influence their efficiency. The ability of an algorithm to discover a better solution in unexplored search space is known as exploration (or global search). This also assists the algorithm in getting out from local minima. On the other hand, exploitation (or local search) refers to the algorithm's capacity to improve a current solution. Better exploitation capabilities lead to a better solution, convergence speed, and higher accuracy.

The rest of the text takes shape subsequently; for example, Sect. 3 details the research procedures for the various proposed adjustments. The simulation results are described in Sect. 4. Section 5 serves to illustrate the study's final findings.

2 Literature Review

Approaches based on swarms can be broken down into three categories: evolutionary-based, physics-based, and swarm-based algorithms. To overcome the local search ability [20] have introduced Hybrid SOA in which SOA and TOE are combined because SOA has a good exploration search ability and TEO has a robust local search ability. In order to solve the problem of feature selection, the researcher employs three different hybrid approaches that are based on the seagull optimization algorithm (SOA) and thermal exchange optimization (TEO). The hybrid approaches proposed in this paper are to choose the algorithm by the iterating process for the position update. Secondly, iterate the SOA algorithm and add the TEO algorithm position update formula. Thirdly, the heat exchange formula that is part of the TEO algorithm is included into the SOA algorithm in order to enhance its capacity for exploitation. Twenty benchmark datasets from the UCI repository were used to test the proposed methodology, and the results were compared to those of three hybrid optimization methods. The standards of the two evaluation approaches are average size and classification accuracy. After comparing the results, the SOA-TEO3 algorithm was reported to be the best for enhancing the exploitation process and minimizing CPU computational time. The final findings also demonstrated that the proposed algorithm could balance the exploration and exploitation processes [20].

An expanded version of SOA [21] has been introduced, and it has been given the name Multi-objective Optimization Algorithm to reflect its application to situations with many objectives (MOSOA). The non-dominated Pareto optimal solution is accumulated in a dynamic archive, according to this approach, has been presented in the study as a new notion. The behavior of attacking and migrating seagulls is imitated in the roulette wheel selection procedure so that the best possible archived results can be chosen.

Multi-objective optimization using an evolutionary twist is the focus of the Evolutionary Multi-Objective Seagull Optimization Algorithm (EMSOA), a spinoff of the seagull optimization algorithm (SOA) (EMoSOA). Because of the use of a changing archive id, this tactic can store previously computed Non-dominated Pareto solutions in a cache for later use, a grid mechanism, leader selection, and genetic operators. Both theoretical and practical engineering problems (including 24 standard benchmark functions) are used to verify the effectiveness of the suggested method. Data shows that our suggested technique outperforms commonly-used alternatives [22].

The paper has introduced a new variant of the seagull optimization approach for PEMFC stack estimating the parameters. The Lévy flight phenomena is utilized by the novel strategy to attain higher convergence rates. Using BCS 500-W and NedStack, two empirical PEMFC modeling methodologies, we examine the squared error between observed values and the theoretically predicted one [23]. The research presents an ideal model for proton exchange membrane fuel cells (PEMFC). For the simulation of the results, BCS500W and NedStack were selected as real-world case studies. The suggested model is compared to the outcomes of numerous different optimization techniques, such as BSOA, SOA, GA, PSO, WCO, etc. The outcomes of the algorithm showed that the new, improved bio-inspired BSOA technique showed the dynamic behaviour of PEMFC. With the proposed method, PEMFC has efficient performance and appropriate output current [24, 25].

3 Methodology

SOA can easily get stuck in the local optimum because the Seagull could not converge to the optimal placement of the best seagull and does not provide a better convergence rate. A considerable amplitude is needed in the early iteration phase to enhance their proficiency and reach the best state of SOA. This will provide an improved global search function and low amplitude in the last iterations to ensure a better convergence rate. For this, we have used an adaptive mutation strategy. An Adaptive mutation is a technique that can ensure the dynamic implementation of these two initialization strategies that guarantee the balance between exploitation and exploration process with that it enhances the accuracy and the searching ability of the algorithm.

An adaptive mutation technique is developed to improve Seagull Optimization Algorithm's global searching capabilities and convergence speed, as mentioned in 错误!未找到引用源。. Every Seagull fly to its prior best Seagull, Pbest and global best Seagull, Gbest. Particles of Pbest and Gbest become smaller as the number of generations increases. As a resultant value of Pbest and Gbest decreases. While searching for the best solution, when the best seagull is stuck in local minima, all the seagulls presented in Swarm will swiftly converge to local optima. In basic version of PSO, every Seagull flies towards the best Seagulls Pbest and Gbest in the group of swarms. It is defined as dynamically modifying the mutation size regarding the current search area.

$$gb_j(t+1) = gb_j(t) + [b_j(t) - a_j(t)] * rand() \qquad (1)$$

$$a_j(t) = Min(x_i(t)), b_j(t) = Max(x_i(t)) \qquad (2)$$

$$i = 1, 2, ..., PopSize, j = 1, 2, ..., D$$

here in this equations, gb_j represents the jth generation of the best global particle, $a_j(t)$ and $b_j(t)$ represent the minimum and maximum values of the jth dimension in the current search space, rand() is a random number between 0 and 1, and t = 1, 2,..., represents the generations.

In the proposed work, we discussed the Qusai random Sequence technique and Pseudo-random Number Generator initialization technique with SOA and Adaptive SOA. PRNG are deterministic; it approximates the properties of random numbers. A PRNG value start state is a seed state. On the other hand, QRS try to cover the feasible search space in an optimized way. Common QRS include Nierder-reiter, Sobol, Halton, Faure, and Torus. Our numerical results are conducted using QRS techniques like Sobol, Halton, Hammersley, and Torus (Fig. 1).

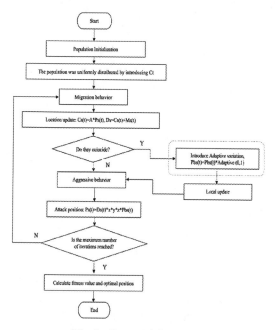

Fig. 1. Proposed Approach

4   Results and Discussion

Inside this section, detailed results of our proposed Adaptive Seagull Optimization algorithm and different initialization techniques are computed and compared with basic version of Seagull Optimization Algorithm. Twenty different standard objective functions are used to validate the proposed strategies. In the first modification Seagull Optimization Algorithm is initialized with different initialization techniques of Qusai-Random Sequence and Pseudo-random Number generators. Secondly, these initialization techniques are implemented on different medical datasets. Along with that with Basic SOA and Adaptive SOA.

4.1 Parameter Setting

Eight benchmark functions are used individually with 30 population size to validate the proposed variant. For any given function, there exists a d-dimensional search space in which the optimal value can be found. Each function is run on 10, 20, 30 dimensions, and the mean value is calculated after 10 runs. 1000, 2000, 3000 iterations are used, respectively. The suggested SOA's parameter settings are displayed in Table 1.

Table 1. Experimental setting of SOA's parameters.

Parameter	Value		
Dimensions	10	20	30
Search Space	[100, −100]		
Iterations	1000	2000	3000
Pop size	40		
Runs	10		

4.2 Result Discussion

The Table 3 displays the results of simulations comparing the traditional SOA with the proposed modifications. The optimal value of standard SOA is shown as a mean value and standard deviations after extensive empirical study has been conducted on each function across a number of dimensions (10–30) and iterations. This information is provided after each function has been experimentally studied (1000–3000).

Seagull Optimization Algorithm is used for solving complex problems. In our proposed work, we have used different initialization strategies with respect to Seagull Optimization Algorithm (SOA). Different initialization techniques are performed to enhance the performance, accuracy, and convergence speed and improve the global search of SOA. The impact of SOA on twenty-benchmark functions was tested with four initialization techniques: SOA-Sobol, SOA-Halton, SOA- Hammersley and SOA-Torus. Among all of the initialization techniques, SOA-Torus has given better performance and convergence rate.

We compare the convergence performance of Rand base adaptive seagull optimization algorithm, Sobol-based adaptive seagull optimization algorithm, Hamersley-based adaptive seagull optimization algorithm, Halton-based seagull optimization algorithm, and Torus-based seagull optimization in Figs. 2, 3, 4, 5, 6, 7, 8 and 9 present the mean best fitness value of the 10 runs for each test function. For graph comparison, we selected 1000, 2000, and 3000 iterations with respect to 10, 20, and 30 dimensions. ASOA-Torus is performing well on 10, 20 and 30 dimensions compared to the Existing Algorithm. ASOA-Torus can find the optimal result easily except F6, F10 and F11 on 10 Dim and except F5, F6, F10, and F11 on 20 dimensions and except F5, F6, F8, F11, F13, and F15 on 30 dimensions. Other existing algorithms show bad convergence on specific functions.

Results for the nine standard test functions are summarized in Table 2, where it can be seen that the proposed SOA versions likely perform better. The function F-08 does not converge to zero in any dimension, but the proposed modifications make this possible. Table 3 shows that the effectiveness of the proposed variations of SOA improves as the dimensions of each function grow larger.

The unique function outcomes with respect to the iterations, dimensions, and related optimal values are depicted in each convergence graph, which also serves as a graphical representation of Table 3. For this convergence, the optimal values are plotted along the y-axis, while the Dimension-Iterations are shown along the x-axis.

4.3   Characteristics of Benchmark Functions

Table 2. Characteristics of benchmark functions.

Function ID	Function Name	Objective	Search Space	Min Value
F-01	Sphere	$Minf(x) = \sum_{i=1}^{D} x_i^2$	$-100 \leq x_i \leq 100$	0
F-02	Schwefel's Problem 2.22	$Minf(x) = \sum_{i=1}^{D} |x_i| + \prod_{i=1}^{D} |x_i|$	$-10 \leq x_i \leq 10$	0
F-03	Schwefel's Problem 1.2	$Minf(x) = \sum_{i=1}^{D} \left(\sum_{j=1}^{i} x_j \right)^2$	$-100 \leq x_i \leq 100$	0
F-04	Schwefel's Problem 2.21	$Minf(x) = \max_{1 \leq i \leq D} |x_i|$	$-100 \leq x_i \leq 100$	0
F-05	Generalized Rosenbrock's	$Minf(x) = \sum_{i=1}^{d} \left[100\left(x_{i+1} - x_i^2\right)^2 + (x_i - 1)^2 \right]$	$-30 \leq x_i \leq 30$	0
F-06	Step	$Minf(x) = \sum_{i=1}^{D} |x_i + 0.5|^2$	$-100 \leq x_i \leq 100$	0
F-07	Quartic	$Minf(x) = \sum_{i=1}^{d} ix_i^4 + rand[0, 1)$	$-1.28 \leq x_i \leq 1.28$	0
F-08	Generalized Schwefel's 2.26	$Minf(x) = \sum_{i=1}^{D} |x_i|^{(i+1)}$	$-500 \leq x_i \leq 500$	0

4.4 Simulation Results of Benchmark Functions

Table 3. SOA and its proposed variants simulation results.

Function Id	Dim	Iterations	ASOA	ASOA-Sobol	ASOA-Hammersley	ASOA-Halton	ASOA-Torus
F-01	10	1000	8.14E-89	2.96E-88	6.89E-88	5.43E-87	**3.72E-91**
	20	2000	1.05E-74	1.70E-76	1.20E-73	1.42E-75	**4.45E-77**
	30	3000	6.27E-73	1.80E-72	**1.27E-73**	4.31E-72	5.29E-73
F-02	10	1000	4.47E-55	1.78E-54	4.26E-56	3.45E-58	**8.78E-59**
	20	2000	7.60E-54	3.21E-54	4.41E-54	9.03E-54	**3.78E-55**
	30	3000	7.81E-51	1.02E-54	6.03E-54	3.52E-54	**9.51E-55**
F-03	10	1000	5.15E-21	5.99E-23	8.30E-23	5.93E-22	**7.28E-24**
	20	2000	2.27E-10	1.75E-10	1.62E-09	9.97E-06	**8.58E-11**
	30	3000	3.03E-08	2.19E-04	2.89E-04	2.60E-04	**7.85E-10**
F-04	10	1000	6.93E-22	2.79E-23	**9.42E-24**	9.47E-23	1.05E-22
	20	2000	5.32E-15	1.27E-16	2.15E-15	8.82E-16	**1.19E-17**
	30	3000	5.80E-13	1.15E-13	3.36E-14	7.80E-14	**3.27E-15**
F-05	10	1000	5.05E + 01	6.14E + 00	**5.87E + 00**	3.77E + 01	2.73E + 01
	20	2000	**2.51E + 01**	4.66E + 01	3.94E + 01	2.72E + 01	3.23E + 01
	30	3000	4.31E + 01	4.09E + 01	2.70E + 01	**2.63E + 01**	3.63E + 01
F-06	10	1000	**2.16E-04**	2.16E-04	2.28E-04	2.32E-04	2.51E-04
	20	2000	9.18E-04	**6.80E-04**	9.26E-04	8.70E-04	8.48E-04
	30	3000	2.07E-03	2.17E-03	1.99E-03	**1.76E-03**	1.92E-03
F-07	10	1000	1.48E-03	1.66E-03	1.54E-03	1.49E-03	**1.09E-03**
	20	2000	8.65E-03	4.82E-03	1.91E-03	2.49E-03	**1.32E-03**
	30	3000	2.83E-03	3.27E-03	3.10E-03	2.96E-03	**2.79E-03**
F-08	10	1000	−1.84E + 04	−2.56E + 04	**−3.30E + 04**	2.01E + 04	4.05E + 04
	20	2000	−4.83E + 04	**−8.21E + 04**	−7.85E + 04	5.90E + 04	6.18E + 04
	30	3000	−8.53E + 04	−5.12E + 04	**−1.05E + 05**	8.18E + 04	7.42E + 04

4.5 Characteristics of Benchmark Functions

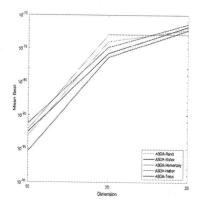

Fig. 2. Convergence Graph of F-01

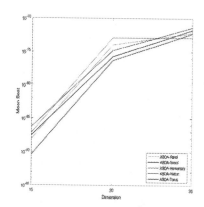

Fig. 3. Convergence Graph of F-02

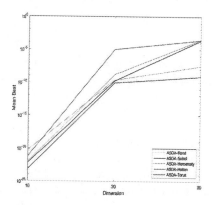

Fig. 4. Convergence Graph of F-03

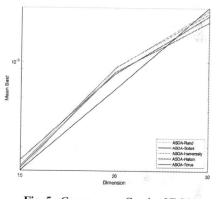

Fig. 5. Convergence Graph of F-04

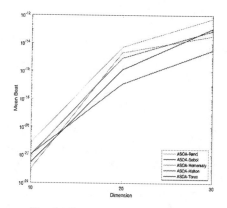

Fig. 6. Convergence Graph of F-05

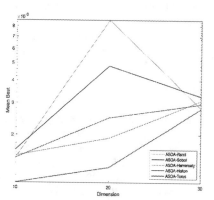

Fig. 7. Convergence Graph of F-06

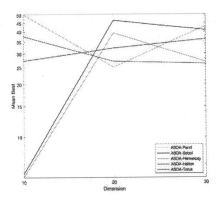

Fig. 8. Convergence Graph of F-07

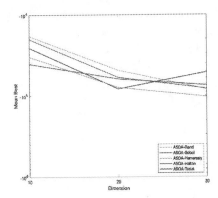

Fig. 9. Convergence Graph of F-08

5 Conclusion

The main objective of the research is to simulate the SOA, which is extensively used to solve real-world nonlinear computationally intensive issues in various fields. The researcher has recommended numerous assessment variants to improve SOA's performance, although SOA still requires a thorough examination to enhance the results. This paper has introduced a new adaptive mutation strategy and different initialization procedures to tackle the local minimum issue of premature convergence and achieve the best results. This paper attempted to provide an overview of various initiation procedures QRS based and examine each one separately. Along with the proposed a new variant of SOA named as Adaptive SOA (ASOA). In ASOA, adaptive mutation strategy is used and the mutative operator will reduce the searching area with the help of the mutation equation, allowing the algorithm to escape from the local minima and search for the optimal solution and search in a greater area. Among all the initialization techniques with ASOA, torus-based ASOA has given better results in terms of efficiency measures on most of the Optimization Functions. It is expected that this investigation simulation-based survey will draw more attention to these issues and that major research will help to replicate fundamental insight. As it explores the whole search space and provides us the optimum result. It explores the whole search space and finds solutions until it reaches for the best solution. Nowadays soft computing is an emerging field in computer science and needs specific attention for more improved results.

References

1. Jabeen, H., Jalil, Z., Baig, A.R.: Opposition based initialization in particle swarm optimization (O-PSO), p. 2047 (2009). https://doi.org/10.1145/1570256.1570274
2. Jiang, H., Yang, Y., Ping, W., Dong, Y.: A novel hybrid classification method based on the opposition-based seagull optimization algorithm. IEEE Access **8**, 100778–100790 (2020). https://doi.org/10.1109/ACCESS.2020.2997791
3. Bangyal, W.H., Rauf, H.T., Batool, H., Bangyal, S.A., Ahmed, J., Pervaiz, S.: An improved particle swarm optimization algorithm with disturbance term. Int. J. Adv. Comput. Sci. Appl. **10**(3), 100–108 (2019). https://doi.org/10.1007/11816102_11

4. Bangyal, W.H., et al.: New modified controlled bat algorithm for numerical optimization problem. Comput. Mater. Contin. **70**(2), 2241–2259 (2022). https://doi.org/10.32604/CMC.2022.017789

5. Pervaiz, S., Ul-Qayyum, Z., Bangyal, W.H., Gao, L., Ahmad, J.: A systematic literature review on particle swarm optimization techniques for medical diseases detection. Comput. Math. Methods Med. **2021** (2021). https://doi.org/10.1155/2021/5990999

6. Li, J., Qin, D.: The mutation seagull algorithm optimizes the speech emotion recognition of BP neural network. In: ACM International Conference Proceeding Series, pp. 160–164 (2021)

7. Dhiman, G., Kumar, V.: Seagull optimization algorithm: theory and its applications for large-scale industrial engineering problems. Knowl.-Based Syst. **165**, 169–196 (2019)

8. Chen, X., Li, Y., Zhang, Y., Ye, X., Xiong, X., Zhang, F.: A novel hybrid model based on an improved seagull optimization algorithm for short-term wind speed forecasting. Processes **9**(2), 1–21 (2021)

9. Ashraf, A., Almazroi, A.A., Bangyal, W.H., Alqarni, M.A.: Particle swarm optimization with new initializing technique to solve global optimization problems. Intell. Autom. Soft Comput. **31**(1), 191–206 (2022)

10. Gan, C., Cao, W., Wu, M., Chen, X.: A new bat algorithm based on iterative local search and stochastic inertia weight. Expert Syst. Appl. **104**, 202–212 (2018). https://doi.org/10.1016/j.eswa.2018.03.015

11. Ashraf, A., et al. Training of artificial neural network using new initialization approach of particle swarm optimization for data classification. In: 2020 International Conference on Emerging Trends in Smart Technologies (ICETST), IEEE (2020)

12. Bangyal, W., Ahmad, J., Abbas, Q.J.I.J.O.E.T.: Analysis of learning rate using CPN algorithm for hand written character recognition application. Int. J. Eng. Technol. **5**(2), 187 (2013)

13. Bangyal, W.H., et al.: Constructing domain ontology for Alzheimer disease using deep learning based approach. Electronics **11**(12), 1890 (2022)

14. Bangyal, W., et al.: Recognition of off-line isolated handwritten character using counter propagation network **5**(2), 227 (2013)

15. Junaid, M., Bangyal, W.H., Ahmad, J.: A novel bat algorithm using sobol sequence for the initialization of population. In: IEEE 23rd International Multitopic Conference, INMIC 2020, November 2020. https://doi.org/10.1109/INMIC50486.2020.9318127

16. Ji, X., Pan, Y., Jia, G., Fang, W.: A neural network-based prediction model in water monitoring networks. Water Sci. Technol. Water Supply **21**(5), 2347–2356 (2021)

17. Bangyal, W.H., Nisar, K., Ibrahim, A.A.B.A., Haque, M.R., Rodrigues, J.J.P.C., Rawat, D.B.: Comparative analysis of low discrepancy sequence-based initialization approaches using population-based algorithms for solving the global optimization problems. Appl. Sci. **11**(16), 7591 (2021)

18. Ul Hassan, N., et al.: Improved opposition-based particle swarm optimization algorithm for global optimization. Symmetry **13**(12), 2280 (2021)

19. Abbas, Q., Bangyal, W.H., Ahmad, J.: The impact of training iterations on ANN applications using BPNN algorithm. Int. J. Future Comput. Commun. **2**(6), 567 (2013)

20. Jia, H., Xing, Z., Song, W.: A new hybrid seagull optimization algorithm for feature selection. IEEE Access **7**, 49614–49631 (2019)

21. Dhiman, G., et al.: MOSOA: a new multi-objective seagull optimization algorithm. Expert Syst. Appl. **167**, 114150 (2020)

22. Dhiman, G., et al.: EMoSOA: a new evolutionary multi-objective seagull optimization algorithm for global optimization. Int. J. Mach. Learn. Cybern. **12**(2), 571–596 (2020). https://doi.org/10.1007/s13042-020-01189-1

23. Bangyal, W.H., Malik, Z.A., Saleem, I., Rehman, N.U.: An analysis of initialization techniques of particle swarm optimization algorithm for global optimization. In: The 4th International Conference on Innovative Computing, ICIC 2021 (2021). https://doi.org/10.1109/ICIC53490. 2021.9692931

24. Cao, Y., Li, Y., Zhang, G., Jermsittiparsert, K., Razmjooy, N.: Experimental modeling of PEM fuel cells using a new improved seagull optimization algorithm. Energy Rep. 5, 1616–1625 (2019). https://doi.org/10.1016/j.egyr.2019.11.013

25. Bangyal, W.H., Hameed, A., Alosaimi, W., Alyami, H.: A new initialization approach in particle swarm optimization for global optimization problems. Comput. Intell. Neurosci. 2021, 1–17 (2021)

A Modified Gaining-Sharing Knowledge Algorithm Based on Dual-Population and Multi-operators for Unconstrained Optimization

Haoran Ma, Jiahao Zhang, Wenhong Wei, Wanyou Cheng, and Qunfeng Liu[✉]

School of Computer Science and Technology, Dongguan University of Technology, Dongguan 523808, China
liuqf@dgut.edu.cn

Abstract. Evolutionary Algorithms (EAs) have a great power to solve complex optimization problems. In recent years, Gaining-Sharing Knowledge algorithm (GSK) is one of the novel nature-inspired algorithms to solve real life optimization problems. This paper introduces a modified Gaining-Sharing Knowledge algorithm based on dual-population framework and multi-operators (mGSK-DPMO). Meanwhile, a hybrid parameter adaptive technique is implemented in order to enhance its performance. The numerical result shows that the presented algorithm (mGSK-DPMO) is found to be highly competitive and significantly superior to other state-of-the-art algorithms including CEC2020's top 3 algorithms in CEC2020 benchmark problems.

Keywords: Evolutionary computation · Nature-inspired algorithms · Global optimization · Parameter adaptation technique

1 Introduction

Optimization is essential for many disciplines and real-life applications [3]. The optimization problems we are facing with are becoming more complex and difficult, which increases the need of developing advanced optimization algorithms correspondingly. Among all the approaches of optimization proposed by researchers in the past decades, great importance has been attached to the evolutionary algorithms (EAs) for its success of tackling optimization problems.

Evolutionary algorithms are often metaheuristics or nature-inspired algorithms based on population, which adopt operations inspired by nature phenomenons and biological behaviors including mutation, crossover and selection. These operations are applied on a set of individuals, i.e., solutions from the population space. In the beginning of the evolutionary algorithm, all the individuals

Supported in part by National Natural Science Fundation of China (No. 61773119), and in part by Guangdong Universities' Special Projects in Key Fields of Natural Science (No. 2019KZDZX1005).

are randomly generated within the search range of the problems to be solved. After the initialization, the individuals will be evolved by operations above to find the best so far solution during iterations. At the end of the iteration, the best individual with the best function value will be considered as the final solution for the optimization problem.

Over the past few decades, modifications of evolutionary algorithms in population and operators have been proved to be capable of enhancing the performance of EAs [4,8,9,12–14]. Generally, the whole population will be divided into several parts, and each sub-population will implement different mutation strategies respectively, which is so-called multi-operators mechanism. This mechanism takes advantage of different operators' characteristics in order to tackle complex problems. Hence, multi-operators mechanism in EAs usually perform much better than single operator mechanism in EAs.

Based on the above observation, a new algorithm is presented in this paper called modified Gaining-Sharing Knowledge algorithm based on dual-population framework and multi-operators strategies (mGSK-DPMO). mGSK-DPMO uses a dual-population framework with multi-operators mechanism. Also a hybrid parameter adaptation approach is proposed to enhance its performance. The results are compared under CEC competition ranking score metric and performance profile.

The structure of the paper is as follows. Section 2 briefly reviewed the basic Gaining-Sharing Knowledge algorithm. In Sect. 3, the approaches of mGSK-DPMO are presented. The numerical experiments and comparisons with other state-of-the-art algorithms are shown in Sect. 4. Conclusion will be presented in Sect. 5.

2   Review of Basic GSK Algorithm

Gaining-Sharing Knowledge optimization algorithm (GSK) is based on the philosophy of gaining and sharing knowledge during the human life span [11]. It has two main phases, junior phase and senior phase. The gaining and sharing dimensions for each individual using both junior and senior phases are calculated and updated in each iteration using a non-linear equation in [10]. The initial population is randomly generated. The set of potential NP solutions, i.e., individuals are vectors with length D, where D are the disciplines of a person to master.

During junior phase, all individuals will gain knowledge from the closest individuals. The nearest better (x_{i-1}) and worse (x_{i+1}) individuals are for gaining knowledge. A random individual (x_r) different from these three individuals for sharing knowledge. During senior phase, the population will be divided into three categories: best individuals x_{p-best}, middle individuals x_m and worst individuals $x_{p-worst}$. Each individual uses two random chosen vectors selected from the top and bottom $100 * p\%$ of the individuals respectively in the current population for gaining knowledge, while choosing another random vector from middle individuals for sharing knowledge.

Both phases use two parameters. The k_f for knowledge factor is for controlling the amount of gained and shared knowledge. The k_r for knowledge ratio is the ratio between the current and obtained experience.

3   Proposed Algorithm

In this section, detailed approaches of the proposed algorithm (mGSK-DPMO) will be explained.

3.1   Dual-Population and Multi-operators Gaining-Sharing Scheme

In the basic GSK algorithm, there exists only one population, i.e., each individual is gaining and sharing information in a fully connected network. From the population topology perspective, this structure may owns more probability to be trapped in the local optima in difficult tasks. Hence, to mimic the human learning characteristics, we use multi-operators as strategies of gaining-sharing schemes of the populations based on dual-population framework. In the proposed scheme, the whole population are separated into n_{pop} sub-populations. Here, we set $n_{pop} = 2$ and denote them as pop_1, pop_2, respectively. Each population has its own gaining-sharing scheme for junior and senior phase, representing different groups of people has different leaning strategies. For pop_1 and pop_2, the gaining-sharing scheme are in the following.

$-$. $pop1$:

$$junior : x_i^{new} = x_i^{old} + k_f^{jun} * \left[(x_{r1} - x_{r2}) + (x_{r3} - x_i^{old}) \right] \tag{1}$$

$$senior : x_i^{new} = x_i^{old} + k_f^{sen} * \left[(x_{\phi best} - x_{m-worst}) + (x_m - x_{r3}) \right] \tag{2}$$

$-$ $pop2$:

$$junior : x_i^{new} = x_i^{old} + k_f^{jun} * \left[(x_{r2} - x_{r1}) + (x_i^{old} - x_{r3}) \right] \tag{3}$$

$$senior : x_i^{new} = x_i^{old} + k_f^{sen} * (x_{\phi worst} - x_{\phi best}) \tag{4}$$

where x_i^{new}, x_i^{old} are the newly generated gaining-sharing individual and current individual respectively. $r_1 \neq r_2 \neq r_3$ are randomly generated integers, x_{r1}, x_{r2}, x_{r3} are randomly selected from the whole population, $x_{\phi best}, x_{m-worst}$ and $x_{\phi worst}$ are chosen from the best $\phi\%$ individuals of the whole population, the middle and worst $(100 - 2\phi)\%$ individuals and the worst $\phi\%$ individuals, respectively. k_f^{jun} and k_f^{sen} use different adaptive parameter settings.

In the proposed scheme, for the junior cases, individuals do not have much experiences for knowledge learning, hence they prefer to learn from random sources of information. For the senior cases, individuals have its independent thoughts, so they would prefer to gain and share based on their judgment. In pop_1, junior phase obtains difference of random individuals as their gaining source and let another random individual be the sharing source, while senior phase obtains $\phi best$ and middle-worst individuals as gaining source, while middle and random individuals be the sharing source. In pop_2, junior phase adopt the contrary scheme of pop_1 junior phase, and the senior phase obtain $\phi best$ and $\phi worst$ as the only gaining source. This setting would allow individuals have a better chance to jump out of the local optima by learning with not only the better individuals but also the worsen individuals.

3.2   Hybrid Parameter Adaptation Setting

From the experimental results, it could be found out that the performance of GSK algorithm is very sensitive to its four control parameters, i.e., knowledge factor k_f, knowledge ratio k_r, knowledge rate K and population size NP. Thus, a suitable parameter setting and adaptation scheme are crucial for GSK algorithm. In mGSK-DPMO, a hybrid parameter adaptation setting is proposed in the following to improve the performance.

Adaptive Settings for k_f and k_r. In the proposed algorithm, the parameter adaptation used for k_f and k_r is a hybrid version of the proposed techniques in [11] and [6]. In this adaptation setting, an individual who has both junior phase and senior phase is associated with two sets of parameters, i.e., k_f^{jun}, k_r^{jun} for junior phase and k_f^{sen}, k_r^{sen} for senior phase. These two sets of parameters are cross controlled by adaptation schemes, namely hybrid adaptive settings.

For the k_f^{jun} and k_r^{sen}, they are set similarly to a modified success-historical-based parameter scheme proposed by [6]. In this setting, the k_f^{jun} and k_r^{sen} are set in a probabilistic distribution for the initial part of iteration. A historical memory which has H elements for both k_f^{jun} and k_r^{sen} are respectively denoted as $\mu_{k_f^{jun}}$ and $\mu_{k_r^{sen}}$. The memory will be a maintained pool for adapting the value of k_f^{jun} and k_r^{sen} during iterations.

Additionally, the weighted Lehmer mean $mean_w$:

$$mean_w(S) = \frac{\sum_{k=1}^{\|S\|} w_k S_k^2}{\sum_{k=1}^{\|S\|} w_k S_k} \tag{5}$$

where the weight is modified as:

$$w_k = log(\frac{\|S\| + 0.5}{k}) \tag{6}$$

The change for Lehmer mean is suitable for the proposed algorithm because the information of the original improvement calculation for w_k is already considered in the new ranking scheme, which will be mentioned in the corresponding subsection.

For the k_f^{sen} and k_r^{jun}, they are set as the AGSK adaptation scheme for k_f and k_r. The adaptation process is handled with parameter setting pools and probability parameter K_{wp} for selecting the pool elements [11]. The pools contain pairs of k_f^{sen} and k_r^{jun} as candidate parameter settings, which will be assigned for each individual. For the first 10% of total function evaluations, the probability for each pair of settings are fixed. The adaptation of the probability parameter will be taken into setting after evaluating 10% of the overall function evaluations.

Considering that the k_f, k_r have different adaptation in different gaining-sharing schemes, instead of using pairs for each individuals in [11], k_f^{sen} and k_r^{jun} are only considered as parameters for a part of the gaining-sharing scheme. This hybrid adaptation parameter setting reflects a human learning period in a more specific way. For the junior phase, a person will have a larger chance to have a more randomized learning scales and a more fixed learning will. Likewise, a person in senior phase will have better chance to have randomized learning wills and fixed learning scales. In addition, considering that senior phase has obtained more judgment to the environment, k_f^{sen} is further set using following formulas:

$$k_f^{sen} = \begin{cases} 0.7 * k_f^{sen}, & nfes < 0.2 * MaxFES \\ 0.8 * k_f^{sen}, & 0.2 * MaxFES \le nfes \le 0.4 * MaxFES \\ 1.1 * k_f^{sen}, & otherwise \end{cases} \qquad (7)$$

Adaptation for Knowledge Rate K. In GSK, the heterogeneous nature of populations in nature is taken into consideration for the knowledge rate K. It controls the dimensions changes of junior to senior phases. To reflect that in the human life span a person will be probable to obtain a faster pave to a well-learned individual, a modified adaptation scheme is proposed as follows:

$$K = \begin{cases} K_{init}, & rand < (N_{fes}/MaxFES) \\ 4 * K_{init}, & otherwise \end{cases} \qquad (8)$$

Population Size Reduction. To improve the performance of mGSK-DPMO, Non-Linear Population Size Reduction(NLPSR) mechanism is used, which was also used in AGSK.

New Ranking Scheme. In mGSK-DPMO, we use a new ranking scheme which involves the fitness function value in current generation while the improvement of the fitness function value is also considered. In particular, in the first generation, we still use the original ranking scheme only based on fitness value, since there is no previous generation. From the second generation, the new ranking scheme is implemented. Firstly, arrange all individuals in ascending order according to their fitness function value, noted as ranking scheme 1 (RS1), and the rank of each candidate in RS1 noted as $rank_1$. Secondly, calculate the improvement of the fitness function value of each individual, and arrange them in descending order, noted as ranking scheme 2 (RS2). The rank of each candidate in RS2 is noted as $rank_2$. Finally, the rank of each candidate is the convex combination of $rank_1$ and $rank_2$ as a score, the smaller the score, the better:

$$score = \omega * rank_1 + (1 - \omega) * rank_2 \qquad (9)$$

The score array is sorted in ascending order, and the new ranking scheme is set up eventually. To balance the contributions between $rank_1$ and $rank_2$, we set a weight ω with boundaries of 0.1 and 0.9, it will increase linearly with the number of iterations:

$$\omega = 0.1 + (0.9 - 0.1) * N_{fes}/MaxFES; \qquad (10)$$

4   Experimental Results and Analysis

To test the performance of the proposed mGSK-DPMO algorithm, CEC2020 benchmark suite [15] that represented in CEC2020 competition on bound constrained numerical optimization problems is adopted. This benchmark absorbed uni-modal function, basic functions, hybrid functions and composition functions in 5, 10, 15 and 20 dimensions in a search range of $[-100, 100]^D$ comprehensively. Thus, finding the global optimum in this benchmark is considered as a challenging task. More details about the benchmark could be found in [15]. The proposed mGSK-DPMO algorithm is coded in MATLAB R2021b and run on a PC with 3.30 GHz Ryzen 5 3400GE processor, 16 GB RAM, Windows 10. The parameters of other competitive algorithms are obtained from relevant articles.

4.1   Parameter Settings of the Algorithm

The values of the parameters for mGSK-DPMO are set as follows.

1. *Population size:* $N_{min} = 4 + floor(3 * log(D))$, for 5D, $N_{init} = 250$, for 10D, 15D, 20D, $N_{init} = 240 * D$.
2. *Control parameters:* $H = 20 * D$, $S_{kf}^0 = 0.5$, $S_{kr}^0 = 0.5$, k_f^{sen} pools are set as [0.1 1.0 0.5 1.0] and [−0.15 −0.05 −0.05 −0.15], k_r^{jun} pool is set as [0.2 0.1 0.9 0.9], $Kw_{P,init} = [0.85, 0.05, 0.05, 0.05]$, $K_{init} = 10$, $c = 0.05$.

For the comparison with other algorithms, the parameters are set to their initial settings.

4.2   Comparison with Other Algorithms

In this subsection, we compare mGSK-DPMO with other state-of-the-art algorithms as follows: **ELSHADE-SPACMA** [5], 3rd in CEC2018 competition. **EBOwithCMAR** [7], 1st in CEC2017 competition. **HSES** [16], 1st in CEC2018 competition. **IMODE** [12], **AGSK** [11], **j2020** [1], which are top 3 in CEC2020 competition.

Results Based on CEC2020 Metric. Table 2 and Table 1 display the comparison results between mGSK-DPMO and other algorithms in CEC2020 benchmark, we now start to analyze the results in detail.

In Table 1, EBOwithCMAR, ELSHADE-SPACMA and HSES were compared with mGSK-DPMO. All algorithms are able to obtain the global optimum solution in function F1 as F1 is the easiest problem in CEC2020 benchmark. In 5D and 15D, the performance of mGSK-DPMO is better than other algorithms in six out of eight problems and six out of ten problems, respectively. In 10D and 20D, mGSK-DPMO is better than other algorithms in seven out of ten problems. Table 3 present the computed values of Score1 (SE), Score2 (RS), and Total Score (S) obtained from each algorithm. We can see that mGSK-DPMO is the best algorithm according to CEC2020 scoring metric.

In Table 2, IMODE, AGSK and j2020 are compared with mGSK-DPMO. In 5D, mGSK-DPMO is very competitive with IMODE, the winner in CEC2020 competition, and better in 10D, 15D and 20D. Especially in function F10, mGSK-DPMO performs the best in other 3 algorithms despite that it failed to obtain the global optimum solution, which proves that our mutation strategies can help algorithm jump out of the local optimal solution to a certain extend. Also, for function F9, mGSK-DPMO is potential to obtain the global optimum solution through 30 runs independently in 20D, while other algorithms don't. Table 4 displays the final score and rank based on CEC2020 metric, we can see that mGSK-DPMO is significantly better than IMODE, AGSK and j2020, again, which are the top 3 algorithms in CEC2020 competition.

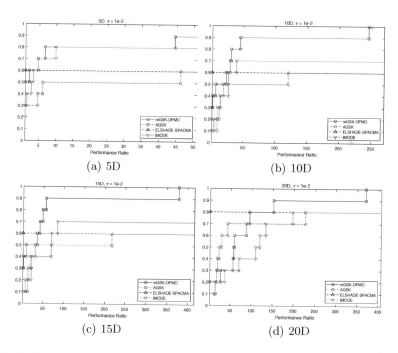

Fig. 1. Performance profiles of mGSK-DPMO, AGSK, ELSHADE-SPACMA and IMODE based on evaluation history results for 5D, 10D, 15D and 20D

Table 1. Compare mGSK-DPMO with EBOwithCMAR, ELSHADE-SPACMA and HSES on CEC2020 benchmark in 5,10,15,20D.

Prob	mGSK-DPMO			ELSHADE-SPACMA			EBOwithCMAR			HSES		
	Best	Mean	Std.	Best	Mean	Std.	Best	Mean	Std.	Best	Mean	Std.
Dimension = 5												
F1	0.00E+00	0.00E+00	0.00E+00	0.00E+00	0.00E+00	0.00E+00	0.00E+00	0.00E+00	0.00E+00	0.00E+00	0.00E+00	0.00E+00
F2	5.56E-01	8.72E-01	6.55E-00	2.16E-04	4.06E-01	1.20E+00	0.00E+00	3.82E-01	1.22E-00	0.00E+00	3.99E-01	5.69E-01
F3	3.62E-02	1.82E+00	1.33E-00	5.15E-00	5.26E-00	1.78E-01	5.15E-00	5.16E-00	6.17E-02	5.15E-00	5.43E-00	2.38E-01
F4	4.62E-04	1.05E-01	9.08E-02	0.00E+00	6.67E-04	2.54E-03	9.88E-03	5.28E-02	2.09E-02	1.05E-01	2.80E-01	1.18E-01
F5	0.00E+00	0.00E+00	0.00E+00	0.00E+00	4.16E-02	1.58E-01	0.00E+00	1.41E-01	3.30E-01	9.95E-01	3.26E-00	1.71E-00
F6												
F7												
F8	0.00E+00	0.00E+00	0.00E+00	0.00E+00	3.35E-01	1.83E-01	0.00E+00	7.55E-01	2.54E-01	0.00E+00	5.68E-01	4.16E-01
F9	0.00E+00	0.00E+00	0.00E+00	0.00E+00	1.00E-02	2.63E-01	0.00E+00	9.00E-01	3.05E-01	1.00E-02	1.36E-02	7.57E-01
F10	0.00E+00	3.00E+00	5.35E-01	3.00E-02	3.46E-02	8.66E-01	3.00E-02	3.46E-01	8.65E-00	3.47E-02	3.47E-01	1.11E-02
Dimension = 10												
F1	0.00E+00	1.33E-01	0.00E+00	0.00E+00	0.00E+00	0.00E+00	0.00E+00	0.00E+00	0.00E+00	0.00E+00	0.00E+00	0.00E+00
F2	8.03E-01	1.33E-01	8.10E-00	1.05E-04	4.23E+00	2.91E+00	1.87E-01	5.48E-00	4.42E-00	3.48E-00	7.84E-00	3.34E-00
F3	1.61E+00	8.17E-00	3.45E-00	1.10E-01	1.30E-01	1.95E-00	1.04E-01	1.04E-01	2.35E-02	6.54E-01	1.14E-01	7.16E-01
F4	9.94E-03	2.00E-01	1.14E-01	0.00E+00	5.63E-02	6.95E-02	9.86E-02	1.27E-01	2.44E-02	2.09E-01	9.77E-01	2.02E-01
F5	5.51E-03	1.45E-00	1.26E-00	0.00E+00	1.95E-00	4.79E-00	2.08E-01	8.36E-00	1.16E-01	2.09E-01	4.69E-01	1.34E-02
F6	1.75E-01	4.35E-01	1.37E-01	1.77E-01	4.46E-01	1.75E-01	6.82E-03	1.91E-01	1.29E-01	2.65E-01	1.65E-01	4.09E-01
F7	1.37E-03	1.37E-02	1.52E-02	3.01E-01	3.01E-01	3.40E-01	7.54E-06	1.43E-01	1.66E-01	4.74E-03	1.39E-01	1.11E-02
F8	0.00E+00	1.39E+00	4.39E-00	1.00E-02	1.00E-02	6.34E-02	1.70E-01	9.72E-01	1.52E-01	1.00E-02	1.00E-02	0.00E+00
F9	0.00E+00	6.67E-00	2.54E-01	1.00E-02	2.64E-02	1.09E-02	2.32E-01	1.54E-02	1.01E-02	3.27E-02	3.29E-02	6.29E-01
F10	0.00E+00	9.67E-01	1.83E-01	3.98E-01	4.15E-02	2.24E-01	3.98E-02	4.12E-02	2.12E-01	4.44E-02	4.46E-02	1.00E+00
Dimension = 15												
F1	0.00E+00	1.19E-01	0.00E+00	0.00E+00	0.00E+00	0.00E+00	0.00E+00	1.47E-01	0.00E+00	0.00E+00	0.00E+00	0.00E+00
F2	5.74E-01	1.19E-01	9.32E-00	3.05E-01	6.04E+00	7.20E-00	1.25E-01	1.47E-01	2.99E-01	1.19E-02	2.29E-01	1.47E-02
F3	4.12E-01	4.62E-01	5.78E-00	1.56E-01	1.60E-01	7.29E-01	1.56E-01	1.56E-01	9.46E-12	1.62E-01	1.76E-01	1.00E-00
F4	2.29E-01	5.47E-01	1.72E-01	1.56E-01	2.86E-01	3.22E-01	1.48E-01	2.06E+00	2.26E-02	9.28E-01	1.45E-01	2.91E-00
F5	1.12E-01	3.13E-01	9.00E-01	3.12E-01	7.90E-00	2.17E-01	3.12E-01	4.73E-01	6.00E-01	6.02E-01	4.55E-01	5.67E-01
F6	3.79E-01	2.49E-01	2.39E-01	7.37E-02	7.93E-01	1.62E-00	6.38E-02	3.14E-01	1.53E-01	6.28E-01	1.61E-01	1.08E-01
F7	3.60E-01	7.66E-01	2.29E-01	1.58E-01	4.75E-01	2.19E-01	4.64E-02	4.65E-00	2.16E-01	6.21E-01	1.82E-01	1.17E-01
F8	0.00E+00	1.66E+01	1.29E-01	1.00E-02	1.00E-02	0.00E+00	1.00E-02	1.00E-02	8.30E-14	1.00E-02	1.00E-02	0.00E+00
F9	0.00E+00	7.33E+01	4.50E-01	3.90E-01	3.90E-02	1.13E-00	1.00E+00	2.74E-02	1.37E-02	3.86E-02	3.88E-02	1.74E-00
F10	1.00E-02	3.00E+02	1.44E-02	4.00E-02	4.00E-02	0.00E+00	4.00E-02	4.00E-02	0.00E+00	4.00E-02	4.00E-02	0.00E+00
Dimension = 20												
F1	0.00E+00	2.42E-02	0.00E+00	0.00E+00	0.00E+00	0.00E+00	0.00E+00	0.00E+00	0.00E+00	0.00E+00	1.98E-01	0.00E+00
F2	9.37E-02	2.42E-02	2.25E-00	6.30E-02	1.63E-00	1.11E-00	6.25E-02	1.27E+00	7.90E-01	1.96E-01	1.98E-01	1.68E-02
F3	0.00E+00	6.97E+00	9.66E-00	2.04E-01	2.08E-01	6.00E-01	2.04E-01	2.04E-01	2.39E-01	2.12E-01	2.45E-01	3.80E-00
F4	2.59E-01	6.93E-01	1.47E-01	2.56E-01	5.15E-01	4.96E-01	1.97E-01	2.96E-01	3.83E-02	1.31E-01	1.97E-01	3.88E-01
F5	1.81E-00	8.91E-02	4.64E-01	1.22E-01	3.28E-01	5.73E-01	1.20E-01	2.96E-01	6.89E-02	2.09E-02	2.16E-02	1.22E-02
F6	9.28E-02	2.18E-01	6.66E-02	1.22E-01	4.09E-01	1.33E-01	5.30E-02	3.65E-01	2.65E-01	4.72E-01	1.33E-01	3.61E-01
F7	4.14E-01	1.17E+00	6.95E-01	6.21E-01	7.03E-01	1.07E-01	2.21E-01	9.73E-00	2.21E-01	6.62E-01	2.40E-01	3.70E-01
F8	6.45E-01	8.38E-01	1.04E-01	1.00E-02	1.00E-02	0.00E+00	1.00E+00	1.00E-02	2.23E-13	1.00E-02	1.00E-02	2.27E-13
F9	0.00E+00	8.82E+01	3.11E-01	4.02E-02	4.10E-02	5.48E-00	1.00E-02	3.65E-02	8.77E-01	3.87E-02	3.95E-02	2.98E+00
F10	3.99E-02	3.99E+02	2.39E-13	4.14E-02	4.14E-02	2.98E-02	3.99E-02	4.13E-02	3.44E-01	4.14E-02	4.29E-02	2.79E-01

Table 2. Compare mGSK-DPMO with IMODE, AGSK and j2020 on CEC2020 benchmark on 5,10,15,20D.

Prob	mGSK-DPMO			IMODE			AGSK			j2020		
	Best	Mean	Std	Best	Mean	Std	Best	Mean	Std	Best	Mean	Std
Dimension = 5												
F1	0.00E+00	0.00E+00	0.00E+00	0.00E+00	0.00E+00	0.00E+00	0.00E+00	0.00E+00	0.00E+00	0.00E+00	0.00E+00	0.00E+00
F2	5.56E-01	8.72E-01	6.55E-00	0.00E+00	8.33E-02	8.89E-02	6.14E-01	1.64E-01	2.58E-01	1.91E-04	3.23E-00	3.74E-00
F3	3.62E-02	1.82E+00	1.33E-00	5.15E-01	5.15E-00	0.00E+00	4.38E-07	2.87E-00	2.05E-00	0.00E+00	3.42E-00	2.33E-00
F4	4.62E-04	1.05E-01	9.08E-02	0.00E+00	0.00E+00	0.00E+00	1.67E-03	1.11E-01	6.05E-02	0.00E+00	7.68E-02	6.40E-02
F5	0.00E+00	0.00E+00	0.00E+00	0.00E+00	0.00E+00	0.00E+00	0.00E+00	0.00E+00	0.00E+00	0.00E+00	1.37E-01	2.86E-01
F6	-	-										
F7	-	-										
F8	0.00E+00	0.00E+00	0.00E+00	0.00E+00	0.00E+00	0.00E+00	0.00E+00	0.00E+00	0.00E+00	6.28E-01		2.39E-00
F9	0.00E+00	0.00E+00	0.00E+00	0.00E+00	0.00E+00	0.00E+00	0.00E+00	3.33E-01	4.79E-01	0.00E+00	2.05E-01	3.75E-01
F10	0.00E+00	3.00E+00	5.35E+01	0.00E+00	2.44E-02	1.36E-02	0.00E+00	2.25E-01	1.32E-02	0.00E+00	1.26E-02	9.03E-01
Dimension = 10												
F1	0.00E+00	0.00E+00	0.00E+00	0.00E+00	0.00E+00	0.00E+00	0.00E+00	0.00E+00	0.00E+00	0.00E+00	0.00E+00	0.00E+00
F2	8.03E-01	1.33E-01	8.10E+00	1.25E-01	4.20E-00	3.70E-00	4.09E-00	2.84E-01	3.21E-01	0.00E+00	6.79E-01	1.16E+00
F3	1.61E-00	8.17E-00	3.45E-00	1.07E-01	1.21E-01	7.83E-01	6.12E-01	9.93E-00	4.26E-00	0.00E+00	8.06E+00	3.88E+00
F4	9.94E-03	2.00E-01	1.14E-01	0.00E+00	0.00E+00	0.00E+00	1.94E-03	5.83E-02	3.11E-02	0.00E+00	1.09E-01	9.04E-02
F5	5.51E-03	1.45E-00	1.26E-00	4.03E-06	3.88E-01	3.83E-01	0.00E+00	3.18E-01	3.06E-01	0.00E+00	3.02E-01	3.13E-01
F6	1.75E-01	4.35E-01	1.37E-01	2.67E-02	9.15E-02	5.08E-02	2.20E-02	1.55E-01	1.17E-01	2.91E-02	4.78E-01	2.49E-01
F7	1.37E-03	1.37E-02	1.52E-02	1.41E-05	8.54E-04	1.10E-03	2.20E-03	1.54E-03	1.71E-03	3.10E-07	6.73E-02	1.25E-01
F8	0.00E+00	1.39E+00	4.39E-00	0.00E+00	2.72E-00	7.46E-00	0.00E+00	1.80E-01	2.38E-01	0.00E+00	1.54E-00	4.00E+00
F9	0.00E+00	6.67E+00	2.54E+01	0.00E+00	4.10E+00	4.46E-01	0.00E+00	7.63E-01	4.29E-01	0.00E+00	8.00E-01	4.07E-01
F10	0.00E+00	9.67E-01	1.83E-01	3.98E-02	3.98E-02	0.00E+00	1.00E-02	2.98E-01	1.43E-02	1.00E-02	1.40E-02	8.12E-01
Dimension = 15												
F1	0.00E+00	0.00E+00	0.00E+00	0.00E+00	0.00E+00	0.00E+00	0.00E+00	0.00E+00	0.00E+00	0.00E+00	0.00E+00	0.00E+00
F2	5.74E-01	1.19E-01	9.32E+00	1.25E-01	3.14E+00	3.22E-00	3.12E-00	1.85E-01	1.46E+01	0.00E+00	5.72E-02	4.32E-02
F3	4.12E-01	4.62E+00	5.78E-00	1.56E-01	1.61E-01	3.12E-01	0.00E+00	1.42E-01	4.27E-01	0.00E+00	6.78E-00	7.82E-00
F4	2.29E-01	5.47E-01	1.72E-01	0.00E+00	0.00E+00	0.00E+00	4.74E-02	1.42E-01	5.71E-02	0.00E+00	1.99E-01	7.47E-02
F5	1.12E-01	3.13E-01	9.00E-00	1.15E-00	7.79E-00	3.66E+00	3.12E-01	6.25E+00	4.32E-00	0.00E+00	7.58E-00	7.69E-00
F6	3.79E-01	2.49E-00	2.39E-00	2.81E-01	6.92E-01	2.52E-01	1.72E-01	4.02E-01	2.23E-01	1.65E-03	8.45E-01	2.09E-00
F7	3.60E-01	7.66E-01	2.29E-01	1.28E-01	5.30E-01	2.23E-01	1.09E-02	2.47E-01	2.00E-01	6.81E-02	9.83E-01	2.03E-00
F8	0.00E+00	1.66E-01	1.29E-01	0.00E+00	4.18E-00	9.61E+00	1.09E-02	6.85E-01	3.83E-01	1.00E-02	9.49E-00	2.74E-01
F9	0.00E+00	7.33E+00	4.50E-01	0.00E+00	9.33E-01	2.54E-01	9.67E-01	9.67E-01	1.83E+01	1.00E-02	1.23E-02	5.68E-01
F10	1.00E-02	3.00E-02	1.44E-02	4.00E-02	4.00E-02	0.00E+00	4.00E-02	4.00E-00	0.00E+00	1.00E-02	3.90E-02	5.48E-01
Dimension = 20												
F1	0.00E+00	0.00E+00	0.00E+00	0.00E+00	0.00E+00	0.00E+00	0.00E+00	0.00E+00	0.00E+00	0.00E+00	0.00E+00	0.00E+00
F2	9.37E-02	2.42E-01	2.25E-00	3.12E-01	5.13E-01	7.12E-01	9.88E-02	9.68E-01	1.23E-00	0.00E+00	2.60E-02	2.47E-02
F3	0.00E+00	6.97E+00	9.66E-00	2.04E-01	2.05E-01	1.25E-01	2.04E-01	2.04E-01	7.23E-15	0.00E+00	1.44E-01	9.29E-00
F4	2.59E-01	6.93E-01	1.47E-01	0.00E+00	0.00E+00	0.00E+00	6.83E-02	1.45E-01	5.47E-02	2.98E-02	1.80E-01	7.84E-02
F5	1.81E+00	8.91E-01	4.64E-01	2.61E-01	1.09E-01	4.33E+00	2.20E-00	4.50E-01	3.67E-01	3.12E-01	7.78E-01	5.75E-01
F6	9.28E-02	2.18E-01	6.68E-02	1.76E-01	3.02E-01	8.17E-02	8.54E-02	1.68E-01	4.45E-02	6.84E-02	1.91E-01	1.01E-01
F7	4.11E-01	1.17E-02	6.95E-01	2.38E-01	5.24E-01	1.64E-01	1.83E-01	6.81E-01	9.09E-01	1.95E-02	1.98E-00	4.02E-00
F8	6.48E-01	8.38E-01	1.04E-01	3.05E-01	8.40E+01	1.89E-01	7.46E-01	9.92E-01	4.63E+00	0.00E+00	9.27E-01	2.21E-01
F9	0.00E+00	8.82E-01	3.11E-01	1.34E-04	9.67E-01	1.83E-01	1.00E-02	1.00E-00	1.00E-02	1.00E-02	3.39E-02	1.28E-02
F10	3.99E-02	3.99E+02	2.39E-13	3.99E-01	4.00E-02	6.18E-01	3.99E-02	3.99E-02	1.59E-02	3.99E-02	3.99E-02	4.02E-02

Table 3. Results in CEC Score Metric with other state-of-the-art algorithms

Algorithm	R	S1	S2	S
mGSK-DPMO	1	50.00	50.00	100.00
EBOwithCMAR	2	25.41	38.84	64.25
ELSHADE-SPACMA	3	24.54	33.81	58.35
HSES	4	17.46	23.79	41.25

Table 4. Results in CEC Score Metric with CEC2020 state-of-the-art algorithms

Algorithm	R	S1	S2	S
mGSK-DPMO	1	50.00	47.01	97.01
IMODE	2	41.46	50.00	91.46
AGSK	3	38.35	45.33	83.67
j2020	4	36.28	44.19	80.46

Results Based on Performance Profile. Performance profile is a tool for analyzing benchmark experiments results [2]. Specifically, it is used to compare the performance of several solvers using several test problem and a comparison goal. Basically, the higher the curve in the profile, the better the performance for the solver. Figure 1 displays the performance curve of mGSK-DPMO, IMODE, AGSK and ELSHADE-SPACMA with $\tau = 1e-2$. It is clear that mGSK-DPMO can solve 90% problems in 5D while almost solve 100% problems in other dimensions comparing with other algorithms, and it also occupied the first place among other performance curves.

5   Conclusion

In this paper, based on the idea of Gaining-Sharing Knowledge algorithm, a new algorithm called mGSK-DPMO is presented. The difference from the original GSK or AGSK is that a dual-population strategy is used to enhance the population diversity. To improve the performance of mGSK-DPMO, a hybrid parameter adaptation mechanism is adopted. The CEC2020 benchmark is used to test the performance of mGSK-DPMO and the result outperformed other state-of-the-art algorithms including the top 3 algorithms in CEC2020 competition. For future work, research efforts will be directed to further enhance the performance of mGSK-DPMO to solve more complicated and challenging problems, and modify mGSK-DPMO to handle multi-objective optimization problems.

Acknowledgements. This work was supported in part by National Natural Science Fundation of China (No. 61773119), and in part by Guangdong Universities' Special Projects in Key Fields of Natural Science (No. 2019KZDZX1005).

References

1. Brest, J., Maučec, M.S., Bošković, B.: Differential evolution algorithm for single objective bound-constrained optimization: algorithm J2020. In: 2020 IEEE Congress on Evolutionary Computation (CEC), pp. 1–8. IEEE (2020)
2. Dolan, E.D., Moré, J.J.: Benchmarking optimization software with performance profiles. Math. Program. **91**(2), 201–213 (2002)
3. Du, D.Z., Pardalos, P.M., Wu, W.: History of Optimization. Springer, Boston (2009)
4. Elsayed, S., Hamza, N., Sarker, R.: Testing united multi-operator evolutionary algorithms-II on single objective optimization problems. In: 2016 IEEE Congress on Evolutionary Computation (CEC), pp. 2966–2973. IEEE (2016)
5. Hadi, A.A., Mohamed, A.W., Jambi, K.M.: Single-objective real-parameter optimization: enhanced LSHADE-SPACMA algorithm. In: Yalaoui, F., Amodeo, L., Talbi, E.-G. (eds.) Heuristics for Optimization and Learning. SCI, vol. 906, pp. 103–121. Springer, Cham (2021). https://doi.org/10.1007/978-3-030-58930-1_7
6. Kumar, A., Biswas, P.P., Suganthan, P.N.: Differential evolution with orthogonal array-based initialization and a novel selection strategy. Swarm Evol. Comput. **68**, 101010 (2022)
7. Kumar, A., Misra, R.K., Singh, D.: Improving the local search capability of effective butterfly optimizer using covariance matrix adapted retreat phase. In: 2017 IEEE Congress on Evolutionary Computation (CEC), pp. 1835–1842. IEEE (2017)
8. Liu, Q., Wei, W., Yuan, H., Zhan, Z.H., Li, Y.: Topology selection for particle swarm optimization. Inf. Sci. **363**, 154–173 (2016)
9. Mallipeddi, R., Suganthan, P.N., Pan, Q.K., Tasgetiren, M.F.: Differential evolution algorithm with ensemble of parameters and mutation strategies. Appl. Soft Comput. **11**(2), 1679–1696 (2011)
10. Mohamed, A.W., Hadi, A.A., Mohamed, A.K.: Gaining-sharing knowledge based algorithm for solving optimization problems: a novel nature-inspired algorithm. Int. J. Mach. Learn. Cybern. **11**(7), 1501–1529 (2020)
11. Mohamed, A.W., Hadi, A.A., Mohamed, A.K., Awad, N.H.: Evaluating the performance of adaptive gainingsharing knowledge based algorithm on CEC 2020 benchmark problems. In: 2020 IEEE Congress on Evolutionary Computation (CEC), pp. 1–8. IEEE (2020)
12. Sallam, K.M., Elsayed, S.M., Chakrabortty, R.K., Ryan, M.J.: Improved multi-operator differential evolution algorithm for solving unconstrained problems. In: 2020 IEEE Congress on Evolutionary Computation (CEC), pp. 1–8. IEEE (2020)
13. Sun, G., Cai, Y., Wang, T., Tian, H., Wang, C., Chen, Y.: Differential evolution with individual-dependent topology adaptation. Inf. Sci. **450**, 1–38 (2018)
14. Wu, G., Shen, X., Li, H., Chen, H., Lin, A., Suganthan, P.N.: Ensemble of differential evolution variants. Inf. Sci. **423**, 172–186 (2018)
15. Yue, C., et al.: Problem definitions and evaluation criteria for the CEC 2020 special session and competition on single objective bound constrained numerical optimization. Computational Intelligence Laboratory, Zhengzhou University, Zhengzhou, China, Technical report 201911 (2019)
16. Zhang, G., Shi, Y.: Hybrid sampling evolution strategy for solving single objective bound constrained problems. In: 2018 IEEE Congress on Evolutionary Computation (CEC), pp. 1–7. IEEE (2018)

Neural Network Search and Large-Scale Optimization

GEFWA: Gradient-Enhanced Fireworks Algorithm for Optimizing Convolutional Neural Networks

Maiyue Chen[1,2] and Ying Tan[1,2,3(✉)]

[1] School of Intelligence Science and Technology, Peking University, Beijing, China
mychen@pku.edu.cn
[2] Key Laboratory of Machine Perceptron (MOE), Peking University, Beijing, China
ytan@pku.edu.cn
[3] Institute for Artificial Intelligence, Peking University, Beijing, China

Abstract. The efficacy of evolutionary and swarm intelligence-based black-box optimization algorithms in machine learning has increased their usage, but concerns have been raised about their low sample efficiency owing to their reliance on sampling. Consequently, improving the sample efficiency of conventional black-box optimization algorithms while retaining their strengths is crucial. To this end, we propose a new algorithm called Gradient Enhanced Fireworks Algorithm (GEFWA) that incorporates first-order gradient information into the population-based fireworks algorithm (FWA). We enhance the explosion operator with the gradient-enhanced explosion (GEE) and take advantage of attraction-based cooperation (ABC) for firework collaboration. Experimental results illustrate that GEFWA outperforms traditional first-order stochastic gradient descent-based optimization methods such as Adm and SGD when it comes to optimizing convolutional neural networks. These results demonstrate the potential of integrating gradient information into the FWA framework for addressing large-scale machine learning problems.

Keywords: Fireworks algorithm · Deep learning · Convolutional neural network · Swarm intelligence

1 Introduction

In recent years, deep learning has emerged as a powerful tool for solving a wide range of problems in fields such as computer vision [12], natural language processing [8], speech recognition [19], and robotics [5]. This success can be attributed to the ability of deep learning models to learn complex representations of data, which in turn enables them to make accurate predictions and generate realistic output. However, the ability to learn these complex representations comes at a cost. Training these models often involves optimizing millions of parameters to find the optimal values for the given task. This is where optimization techniques play a crucial role.

© The Author(s), under exclusive license to Springer Nature Switzerland AG 2023
Y. Tan et al. (Eds.): ICSI 2023, LNCS 13968, pp. 323–333, 2023.
https://doi.org/10.1007/978-3-031-36622-2_26

Optimization is the process of finding the set of parameters that minimize a loss function, which measures the difference between the predictions of the model and the true output. In deep learning, optimization algorithms are used to update the parameters of a model to minimize the loss function. The goal of optimization is to find the best set of parameters that can accurately predict the output for a given input. Optimization is crucial in modern deep learning because the size and complexity of deep learning models make it impossible to optimize them manually.

Optimization algorithms are used in deep learning for a variety of tasks including training neural networks, tuning hyperparameters, and adjusting the learning rate. Many gradient-based optimization methods have been proposed in recent years, such as stochastic gradient descent [1,3], Adagrad [4], Adam [10] and AdamW [15]. These optimization methods have been shown to improve the performance of deep learning models significantly.

However, current gradient-based methods face the problem of falling into local optimas or saddle points, hindering their ability to find the global optimum. This is especially problematic in deep learning, where the objective function is often non-convex and high-dimensional. To solve this problem, in this work, we propose a novel Gradient-Enhanced Fireworks Algorithm (GEFWA) to fuse the gradient information into the FWA framework. The proposed GEFWA algorithm is a hybrid algorithm that combines the advantages of both gradient-based and black-box optimizing methods. It utilizes the gradient information to guide the search process via the Gradient-Enhanced Explosion (GEE) operator and the Attraction-Based Cooperation (ABC) to help the fireworks learn from each other.

The rest of the paper is organized as follows. Section 2 provides a brief overview of fireworks algorithm and related works. Section 3 describes our proposed method in detail. Section 4 presents the experimental setup and results. Finally, Sect. 5 concludes the paper.

2   Background

2.1   Fireworks Algorithm

The Fireworks Algorithm (FWA) was first introduced as a swarm intelligence algorithm in [25]. The algorithm takes inspiration from real-world fireworks explosions in the night sky. By following a local explosion-based search behavior and promoting collaborative or competitive cooperation between fireworks, FWA exhibits a rich set of behaviors that can tackle diverse problem structures. Various FWA variants have also been proposed to tackle specific challenges. For example, Li et al. proposed that fireworks could inspect other fireworks' performance to estimate their future potential and introduced LoT-FWA in [13] to solve multimodal optimization problems. In [27], MC-FWA was proposed to handle problems with hierarchical problem structures by controlling firework cooperation at different levels. To achieve maximum search space coverage, Li et al. proposed to utilize multiple fireworks to partition the search space in a non-overlapping way in [14]. Overall, the FWA algorithm is essentially a very expressive algorithm framework that can be instantiated into different instances to solve a large variety of problems.

2.2   Related Works

In the field of optimization, many problems, such as those encountered in deep learning, involve non-convex and high-dimensional objective functions which make finding the global minimum challenging. Stochastic gradient descent (SGD) is an iterative algorithm that can optimize models by utilizing noisy gradients to update their parameters [1,3]. However, SGD can slow down near critical but suboptimal values such as local minima and saddle points [17]. Conversely, neuroevolution is a family of techniques that use evolutionary algorithms to train neural networks using Darwinian principles such as selection, mutation, and crossover. It has several advantages over SGD, including the ability to handle high-dimensional and non-convex search spaces and find globally optimal or near-optimal solutions. Recently, neuroevolution has been successfully applied to reinforcement learning (RL) problems where only imperfect gradient information is available. Several neuroevolution-based methods, such as ES [20] and GA [24], have shown to be competitive alternatives to SGD-based methods for deep RL. Additionally, there exist several variants of neuroevolution for supervised and unsupervised learning tasks, such as NEAT [23], HyperNEAT [22], and CoDeepNEAT [16].

Since both SGD and neuroevolution have unique drawbacks and benefits, there have been attempts to combine them to acquire the advantages of both. One such algorithm is PGD, which utilizes the mutation operator in GA to help the optimizer escape saddle points [7]. EGD builds on this idea and incorporates both the mutation and population nature of EAs. It alternates between SGD steps and EA steps to supplement each other [26]. In RL, researchers are also exploring hybrid algorithms based on GA and ES, which have shown to achieve better results than both EA and deep RL algorithms. The hybrid algorithms include Khadka's algorithm [9], Bodnar's algorithm [2], and Pourchot's algorithm [18].

3   Methods

Figure 1 depicts the general flow of GEFWA in an intuitive way. The overall procedure is the same as the classical FWA framework where each firework conducts independent explosion-based search and cooperates with other fireworks via the cooperation strategy. Subsequent sections will provide comprehensive descriptions of its fundamental elements.

3.1   Gradient-Enhanced Explosion (GEE)

The core idea of GEE is to enhance the explosion operator in FWA with gradient information. To be specific, in each generation for a specific firework network F, k

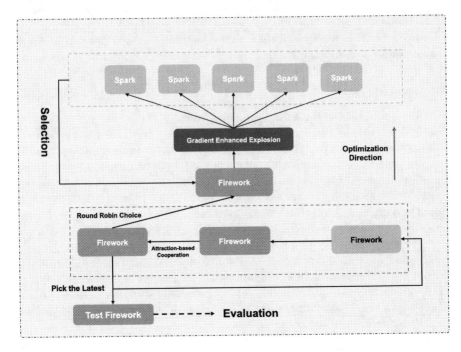

Fig. 1. This figure shows the overall framework of GEFWA.

sparks will be initilized with the same parameters as F and we randomly split the training set into a spark training set and a fitness evaluation validation set. Later, all sparks will be trained with gradient descent for n steps to acquire the first-order information of the objective function. Next, we flatten the trained sparks to a vector form and calculate the difference from F and produce a collection of difference vectors. Then, these vectors are used to build a diagonal covariance matrix to capture the local curvature of the objective function. Finally, we use the covariance matrix to sample k spark networks from a multivariate normal distribution.

For the above procedure, the classic cross-entropy loss is used to train and calculate the fitness of the spark networks:

$$L(y, \hat{y}) = -\frac{1}{N} \sum_{i=1}^{N} [y_i \log(\hat{y}_i) + (1 - y_i) \log(1 - \hat{y}_i)]. \tag{1}$$

where y is the ground truth label of the image, $\hat{y}$ is the predicted label, and N is the total number of images in the dataset. Algorithm 1 describes the above procedure in detail.

3.2   Attraction-Based Cooperation (ABC)

Inter-firework cooperation is en essential mechanism for all FWA variants. In this work, we propose a very simple yet effective cooperation strategy called

---

Algorithm 1: Gradient-Enhanced Explosion (GEE)

Input: Firework networks F, number of sparks k, gradient steps n, trainning data loader $\mathcal{D}_{train}$

Output: Sparks $\{S_1, ..., S_k\}$

1 **for** $j = 1, ..., k$ **do**
2 $\quad$ **for** $t = 1, ..., n$ **do**
3 $\quad\quad$ Sample a batch of data b_t from $\mathcal{D}_{train}$
4 $\quad\quad$ Copy parameters from F to worker A_j
5 $\quad\quad$ Update A_j with b_t according to the loss function in Equation (1)
6 $\quad$ $P_j = \text{Vec}(A_j) - \text{Vec}(F)$
7 $\bar{P} = \frac{1}{m} \sum_i^m P_i$
8 Build the diagonal covariance model $D = \sum_i^m \frac{(P_i - \bar{P})^2}{m-1}$
9 **for** $i = 1, ..., m$ **do**
10 $\quad$ $\mathbf{z} \sim \mathcal{N}(0, \mathbf{I})$
11 $\quad$ $\text{Vec}(S_i) = \text{Vec}(F) + \bar{P} + \mathbf{z} * \sqrt{D}$
12 $\quad$ $S_i \leftarrow \text{Vec}(S_i)$

Attraction-Based Cooperation (ABC). The core idea is to use the last-trained firework $\hat{F}$ to attract the current firework F_i to the last-updated firework's position with a randomized attraction force. This way, the last-updated firework can be used as a reference point to guide the search of the current firework. The attraction force is calculated as follows:

$$\mathbf{z} \sim \mathcal{U}(0, 1). \tag{2}$$
$$\mathbf{AF} = \gamma * \mathbf{z} * (F_i - \hat{F}). \tag{3}$$

where the coefficient γ is a hyperparameter defining the maximum amplitude of the attraction force. The attraction force is then added to the current firework's position to update it. Equation (4) describes the above procedure in detail.

$$F_i = F_i + \mathbf{AF}. \tag{4}$$

3.3   GEFWA

With the introduction of GEE and ABC, the GEFWA algorithm is described in detail in Algorithm 2. Firstly, a set of fireworks is initialized. Then, the algorithm iteratively generates sparks using a given firework and randomly generated training and fitness dataloaders. Next, the spark networks are evaluated on a held-out validation set where their validation loss serves as fitness values. These fitness values are used to conduct the classical linear recombination-based selection proposed by Hansen et al. [6]. Periodically, fireworks conduct attraction-based cooperation to accelerate the learning process by merging information gathered from other fireworks. Additionally, the algorithm tests the performance of the last-trained firework on a test set at regular intervals.

Algorithm 2: GEFWA

Input: Evaluation interval E, Cooperation interval T
Output: The last-trained firework $\hat{F}$

1 Initialize l fireworks
2 $g = 0$
3 $i = 0$
4 **while** *(Maximum time steps not reached)* **do**
5   Regenerate new train and fitness dataloaders $\mathcal{D}_{train}$ and $\mathcal{D}_{fitness}$
6   Generate k sparks with F_p and $\mathcal{D}_{train}$ according to Algorithm 1
7   Evaluate the fitness of sparks with $\mathcal{D}_{fitness}$
8   Update F_i according to Equation (4)
9   **if** $g\%T = 0$ **then**
10   Pick the last-trained firework $\hat{F}$
11   Conduct firework cooperation according to Equation (4)
12   **if** $g\%E = 0$ **then**
13   Pick the last-trained firework $\hat{F}$ to evaluate performance on the test set
14     $g = g + 1$
15     $i = (i + 1)\%l$

---

4 Experiments

To verify the effectiveness of GEFWA, we conduct extensive experiments on the CIFAR-100 dataset [11]. The dataset has 60,000 32×32 color images grouped into 100 classes, split into 50k training and 10k testing images. We use the VGG convolutional network [21] as the backbone, which consists of convolutional and max-pooling layers, followed by fully connected layers. In our experiments, the VGG11 variant is used, which contains 9,271,780 parameters. Its use of 3×3 filters enables deeper networks with fewer parameters, making it a popular option for computer vision studies. For evaluation, the top-1 classification accuracy on the test set is used as the metric.

4.1 Comparative Study

We performed a comparative experiment to evaluate the performance of GEFWA against popular stochastic gradient descent-based optimizers, including SGD [1,3], Adam [10], and AdamW [15]. Each optimizer was run five times with different random seeds, and we reported the mean and standard deviation of the test accuracies. The results are presented in Table 1, which demonstrates that GEFWA outperforms the other optimizers, achieving a median accuracy of 51.70%. This finding is statistically significant, indicating that GEFWA is a promising optimizer for similar applications. Additionally, the standard deviation of GEFWA is lower than that of the other optimizers, indicating that GEFWA exhibits more consistent performance. Learning curves for each optimizer are presented in Fig. 2, which shows that GEFWA typically converges slower than the baseline methods but achieves significantly better generalization. The training loss for GEFWA also exhibits a slower convergence rate and larger loss than the other optimizers, which suggests that it is less prone to over-fitting.

Table 1. This table shows the comparison of the test accuracies of the VGG convolutional network of GEFWA and gradient descent-based optimizers on the CIFAR-100 dataset.

Optimizers	Median	Mean	Std	Gain
Adam	33.06%	32.63%	1.26%	58.20% ↑
AdamW	33.59%	33.12%	1.29%	55.85% ↑
SGD	31.84%	32.47%	1.79%	58.98% ↑
GEFWA (Ours)	**51.70%**	**51.62%**	**0.71%**	-

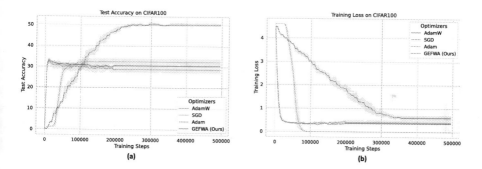

Fig. 2. This figure shows the learning curves for optimizing the VGG convolutional network of GEFWA and gradient descent-based optimizers on the CIFAR-100 dataset.

4.2 Ablation

To verify the effectiveness of the two proposed mechanisms GEE and ABC, we also conducted two ablative experiments.

We initiated the experiment by assigning the number of fireworks to 1 while the number of sparks was set to 4 in order to eliminate the potential impact

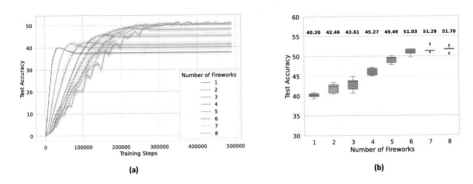

Fig. 3. This figure shows the learning curves for optimizing the VGG convolutional network of GEFWA with different settings of the number of fireworks.

Table 2. This table shows the comparison of the test accuracies of the VGG convolutional network of GEFWA with different settings of the number of fireworks. For all variants, the number of sparks is set to 4.

Optimizer	Fireworks	Median	Mean	Std
GEFWA	1	40.20%	40.19%	0.60%
GEFWA	2	42.46%	42.01%	1.33%
GEFWA	3	43.61%	42.97%	1.71%
GEFWA	4	45.27%	45.94%	1.04%
GEFWA	5	49.49%	49.08%	0.97%
GEFWA	6	51.03%	50.95%	0.85%
GEFWA	7	51.29%	51.56%	0.79%
GEFWA	8	**51.70%**	**51.62%**	**0.71%**

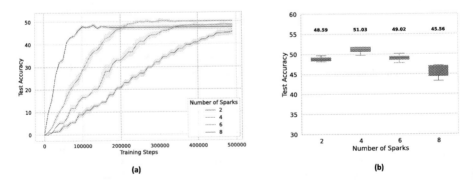

Fig. 4. This figure shows the learning curves for optimizing the VGG convolutional network of GEFWA with different settings of the number of sparks.

Table 3. This table shows the comparison of the test accuracies of the VGG convolutional network of GEFWA with different settings of the number of sparks. For all variants, the number of sparks is set to 6.

Optimizer	Sparks	Median	Mean	Std
GEFWA	2	48.59%	48.78%	**0.64%**
GEFWA	4	**51.03%**	**50.95%**	0.85%
GEFWA	6	49.02%	48.97%	0.88%
GEFWA	8	45.56%	45.49%	1.67%

of inter-firework cooperation. This allowed us to evaluate the effectiveness of the GEE mechanism. As displayed in the first row of Table 2, the GEE-only variation of GEFWA model yielded a mean top-1 accuracy of 40.19%, which is considerably superior to the performance of the AdamW model (mean: 33.12%, std: 1.29%) that is regarded as the best baseline method. Thus, these findings validate the proposed GEE mechanism's efficiency. Further experiments were conducted, considering the number of sparks between 2 to 8, with a 2-step increment. Additional results and graphs are shown as Table 3 and Fig. 4. The graph of Fig. 4(b) substantiates that GEFWA model's mean top-1 accuracy did not increase uniformly with respect to the increase in number of sparks. Surprisingly, the model yielded the best performance when the number of sparks was set to 4. Additionally, reducing the number of sparks led to a faster convergence rate, but at the expense of compromising the model's generalization.

Next, the number of sparks is fixed to 4 and the number of fireworks is varied from 1 to 8. The final mean accuracies are summarized in Table 2 and Fig. 3 shows the learning curves and box plots of different settings. The results shows that increasing the number of fireworks leads to a gradual increase in median and mean. Specifically, the model yields an average top-1 accuracy of 40.19% when the number of fireworks is set to 1. However, as the number of fireworks increases, the performance of the optimizer improves, and the model achieves the highest top-1 accuracy of 51.70% when the number of fireworks is set to 8. This result suggests that the attraction-based cooperation is indeed helpful in improving the performance of the optimizer during training.

In conclusion, the ablative study indicates that both the GEE and ABC are essential for the performance of GEFWA.

5   Conclusion

In this paper, a novel optimization method called the Gradient-Enhanced Fireworks Algorithm (GEFWA) is proposed. The purpose of the GEFWA algorithm is to improve the FWA framework by integrating gradient information. By incorporating the Gradient-Enhanced Explosion (GEE) operator and the Attraction-Based Cooperation (ABC) method, the algorithm can use the gradient information to guide its search process and allow each firework to learn from one another. The results of our experiments show that the GEFWA algorithm outperforms gradient-based baseline methods in terms of generalization ability. Our ablative studies confirm the effectiveness of both the GEE and ABC methods. Moving forward, we plan to extend the proposed GEFWA algorithm to other optimization problems in deep learning and explore additional methods to fuse gradient information into the FWA framework.

Acknowledgement. This work is supported by the National Natural Science Foundation of China (Grant No. 62250037, 62076010 and 62276008), and partially supported by Science and Technology Innovation 2030-"New Generation Artificial Intelligence" Major Project (Grant Nos.: 2018AAA0102301).

References

1. Amari, S.I.: Backpropagation and stochastic gradient descent method. Neurocomputing **5**(4–5), 185–196 (1993)
2. Bodnar, C., Day, B., Lió, P.: Proximal distilled evolutionary reinforcement learning. In: Proceedings of the AAAI Conference on Artificial Intelligence, vol. 34, no. 04, pp. 3283–3290 (2020)
3. Bottou, L.: Large-scale machine learning with stochastic gradient descent. In: Lechevallier, Y., Saporta, G. (eds.) COMPSTAT 2010, pp. 177–186. Springer, Heidelberg (2010). https://doi.org/10.1007/978-3-7908-2604-3_16
4. Duchi, J., Hazan, E., Singer, Y.: Adaptive subgradient methods for online learning and stochastic optimization. J. Mach. Learn. Res. **12**(7) (2011)
5. Haarnoja, T., Zhou, A., Abbeel, P., Levine, S.: Soft actor-critic: off-policy maximum entropy deep reinforcement learning with a stochastic actor. In: Proceedings of the 35th International Conference on Machine Learning, pp. 1861–1870. PMLR (2018)
6. Hansen, N.: The CMA Evolution Strategy: A Tutorial. arXiv:1604.00772 (2016)
7. Jin, C., Ge, R., Netrapalli, P., Kakade, S.M., Jordan, M.I.: How to escape saddle points efficiently. In: International Conference on Machine Learning, pp. 1724–1732. PMLR (2017)
8. Kenton, J.D.M.W.C., Toutanova, L.K.: BERT: pre-training of deep bidirectional transformers for language understanding. In: Proceedings of NAACL-HLT, pp. 4171–4186 (2019)
9. Khadka, S., Tumer, K.: Evolution-guided policy gradient in reinforcement learning. In: Advances in Neural Information Processing Systems, vol. 31. Curran Associates, Inc. (2018)
10. Kingma, D.P., Ba, J.: Adam: a method for stochastic optimization. arXiv preprint arXiv:1412.6980 (2014)
11. Krizhevsky, A., Hinton, G.: Learning multiple layers of features from tiny images (2009)
12. Krizhevsky, A., Sutskever, I., Hinton, G.E.: ImageNet classification with deep convolutional neural networks. In: Advances in Neural Information Processing Systems, vol. 25. Curran Associates, Inc. (2012)
13. Li, J., Tan, Y.: Loser-out tournament-based fireworks algorithm for multimodal function optimization. IEEE Trans. Evol. Comput. **22**(5), 679–691 (2018)
14. Li, Y., Tan, Y.: Hierarchical collaborated fireworks algorithm. Electronics **11**(6), 948 (2022)

15. Loshchilov, I., Hutter, F.: Decoupled weight decay regularization. In: International Conference on Learning Representations (2022)
16. Miikkulainen, R., et al.: Evolving deep neural networks. In: Artificial Intelligence in the Age of Neural Networks and Brain Computing, pp. 293–312. Elsevier (2019)
17. Nesterov, Y.: Introductory Lectures on Convex Optimization: A Basic Course, 1st edn. Springer, New York (2014). https://doi.org/10.1007/978-1-4419-8853-9
18. Pourchot, A., Perrin, N., Sigaud, O.: Importance mixing: improving sample reuse in evolutionary policy search methods. arXiv:1808.05832 (2018)
19. Radford, A., Kim, J.W., Xu, T., Brockman, G., McLeavey, C., Sutskever, I.: Robust speech recognition via large-scale weak supervision. arXiv preprint arXiv:2212.04356 (2022)
20. Salimans, T., Ho, J., Chen, X., Sidor, S., Sutskever, I.: Evolution Strategies as a Scalable Alternative to Reinforcement Learning (2017)
21. Simonyan, K., Zisserman, A.: Very deep convolutional networks for large-scale image recognition. arXiv preprint arXiv:1409.1556 (2014)
22. Stanley, K.O., D'Ambrosio, D.B., Gauci, J.: A hypercube-based encoding for evolving large-scale neural networks. Artif. Life 15(2), 185–212 (2009)
23. Stanley, K.O., Miikkulainen, R.: Evolving neural networks through augmenting topologies. Evol. Comput. 10(2), 99–127 (2002)
24. Such, F.P., Madhavan, V., Conti, E., Lehman, J., Stanley, K.O., Clune, J.: Deep Neuroevolution: Genetic Algorithms Are a Competitive Alternative for Training Deep Neural Networks for Reinforcement Learning. arXiv:1712.06567 (2018)
25. Tan, Y., Zhu, Y.: Fireworks algorithm for optimization. In: Tan, Y., Shi, Y., Tan, K.C. (eds.) ICSI 2010. LNCS, vol. 6145, pp. 355–364. Springer, Heidelberg (2010). https://doi.org/10.1007/978-3-642-13495-1_44
26. Xue, K., Qian, C., Xu, L., Fei, X.: Evolutionary gradient descent for non-convex optimization. In: Twenty-Ninth International Joint Conference on Artificial Intelligence, vol. 3, pp. 3221–3227 (2021)
27. Li, Y., Tan, Y.: Enhancing fireworks algorithm in local adaptation and global collaboration. In: Tan, Y., Shi, Y. (eds.) ICSI 2021. LNCS, vol. 12689, pp. 451–465. Springer, Cham (2021). https://doi.org/10.1007/978-3-030-78743-1_41

Neural Architecture Search Based on Improved Brain Storm Optimization Algorithm

Xiaojie An[1], Lianbo Ma[1,2($\boxtimes$)], Yuee Zhou[1], Nan Li[1], Tiejun Xing[3($\boxtimes$)], Yingyou Wen[4], Chang Liu[5], and Haibo Shi[5]

[1] College of Software, Northeastern University, Shenyang 110819, China
malb@swc.neu.edu.cn
[2] Key Laboratory of Smart Manufacturing in Energy Chemical Process, Ministry of Education, East China University of Science and Technology, Shanghai, China
[3] Neusoft Institute of Intelligent Medical Research, Shenyang 110167, China
xingtj@neusoft.com
[4] School of Computer Science and Engineering, Northeastern University, Shenyang 110167, China
[5] Digital Factory Department, Shenyang Institute of Automation Chinese Academy of Sciences, Shenyang 110016, China

Abstract. The performance of deep neural networks (DNNs) often depends on the design of their architectures. But designing a DNN with good performance is a difficult and knowledge-intensive process. In this paper, we propose a neural architecture search method based on improved brain storm optimization (BSO) algorithm to efficiently deal with image classification tasks. BSO successfully transposes the human brainstorming process to design of optimization algorithms, which typically uses grouping, substitution, and creation operators to generate as many solutions as possible to approach the global optimization of the problem generation by generation. However, the BSO algorithm using clustering methods for grouping increases the computational burden, so we use the BSO algorithm based on simple grouping methods to solve the optimal architecture of the neural architecture search (NAS). We also redesigned the search space and designed an efficient encoding strategy for each individual.

Keywords: Neural architecture search · Brain storm optimization algorithm · Simple grouping method

1 Introduction

The great achievement of deep neural networks (DNNs) is often determined by the design of DNN architectures [1–5]. This is demonstrated by VGG [6], ResNet [2], and DenseNet [3], they have good performance and take an important step. Typically, these DNN architectures are designed manually with rich professional knowledge [2, 3, 6]. Therefore, more and more researchers are discovering their incredible potential. Unfortunately, due to their lack of expertise in designing DNN architectures, the architectures that are designed often have poor performance. This problem is solved by neural

Y. Tan et al. (Eds.): ICSI 2023, LNCS 13968, pp. 334–344, 2023.
https://doi.org/10.1007/978-3-031-36622-2_27

architecture search (NAS), which automatically designs architectures without human intervention. NAS [7, 8] aims to solve the problem of finding architectures with good performance, using optimization algorithms to automatically find DNN architectures that meet people's expectations.

According to the search paradigms, most of NAS algorithms widely divided into reinforcement learning (RL)-based [9], gradient-based [10], and evolutionary computing-based (EC) [11, 12] (called ENAS). Particularly, the ENAS algorithms usually require less computational budget than RL-based algorithms. ENAS methods are automatic and produce appropriate architectures without any human intervention compared with gradient-based algorithms [13, 14]. ENAS uses EC techniques to solve NAS problems. Specifically, EC includes genetic algorithm (GA) [15], genetic programming (GP) [16], differential evolution (DE) [17], particle swarm optimization (PSO) [18], brain storm optimization (BSO) algorithm [19], and artificial bee colony (ABC) algorithm [20–22], they have good global search capabilities and are used to solve optimization problems by using the method of population evolution [23–25].

As the search space for NAS continues to expand, the choice of search methods as well as their design becomes more important. Swarm intelligence has a stronger global search capability. BSO is usually used as swarm intelligence algorithm that effectively combines local and global search capabilities in order to search the optimal solution to the proposed problem, i.e. the optimal architecture [26, 27].

BSO widely used to solve many machine learning problems recently [26, 28, 29]. In the brainstorming process, the brainstorming starts with the first round of ideas, and after each round of ideas is presented, the best ideas are selected. The ideas in the next round are expanded on the basis of the ideas presented and finally get the solution of the problem. This process is to find the best architecture in a search space [30]. BSO usually uses grouping, replacement, and creation to generate more ideas to solve global optimization problems generations. Nevertheless, the traditional BSO uses clustering methods in grouping operations, which increase the computational burden of the algorithm, we use an improved method [31] in grouping operations to make the algorithm easy to implement and reduce the computational burden.

For search space, since the previous search space designed by ResNet blocks (RB) and DenseNet blocks (DB) achieves better performance, we design the search space based on them as well as the pooling layer [14]. We add more information representations such as the number of ResNet Units, DenseNet Units, and the number and type of convolutional kernels in their structures, so that we can search for a better architecture. At the same time, the improved BSO algorithm (SBSO) can optimize the architectures.

In this paper, we use the improved BSO in the search strategy of NAS to optimize its architecture. Accordingly, the main contributions of this work include:

1) We use BSO algorithm in the search strategy of NAS to find better individuals as the best solution. Specifically, BSO is an optimization algorithm, which has populations, divides the populations, and updates the positions of the solutions. Eventually finds the best solution. Then this optimal solution can be used as the optimal architecture in neural architecture search.

2) We apply the improved BSO algorithm to the search strategy of neural architecture search using the simple grouping method instead of the traditional clustering method

in the grouping operation to reduce the time overhead when searching and save computing resources.

3) We use the combination of ResNet blocks (RB) and DenseNet blocks (DB) to build search space and design an efficient encoding strategy. Specifically, information about the number and type of convolutional kernels is added, and more information is incorporated into the encoding method.

The other parts of this paper are structured as below: Sect. 2 introduces BSO and ENAS. Section 3 introduces the encoding strategy and improvement of BSO, and describes the solution at length. Section 4 represents the experimental datasets and parameter settings, and analyzes the results. We summarize the conclusion in Sect. 5.

2   Related Work

2.1   Brain Storm Optimization (BSO) Algorithm

BSO algorithm refers to gathering some people to brainstorm problems that are hard to handle by a single person, so as to generate inspiration for solving problems. It is more important in this process to continuously generate new ideas and expand existing ones in order to find solutions to problems rather than whether or not the ideas are valid at the time. The algorithm achieves good results in optimizing the test function. This new algorithm regards each solution as a data point, and finds the best solution to the problem by clustering the data points.

In the BSO algorithm, there are three basic operators: solution set clustering, new solution generation and new solution selection operators [32]. The steps are as follows:

Step 1. Initialization: Stochastically produce n individuals as the initial population and estimate the fitness values of n individuals.

Step 2. Cluster: Applying k-means to cluster the initial population into m clusters, the optimal individual is called the cluster center.

Step 3. Replace: Stochastically select a cluster center and replace it with a stochastically generated individual.

Step 4. Process: Select any one or two clusters, and select any individual or cluster center from the selected clusters, generating a new idea based on this;

Generate new solutions according to Eq. (1) and (2):

$$x_{new}^i = x_{old}^i + \xi(t) \times rand() \tag{1}$$

$$\xi(t) = \log sig((0.5 \times T - t)/s) \times rand() \tag{2}$$

where i represents the dimension; $rand()$ generates a uniformly distributed random number between $[0, 1)$; x_{old}^i is formed by a solution or a combination of two solutions in the current set of solutions; the parameter T is the maximum iterations; t is the current iteration; c could change the gradient of the step function $\xi(t)$ of the logsig() function. The function logsig() is defined as follows:

$$\log sig(a) = \frac{1}{1 + \exp(a)} \tag{3}$$

Step 5. Select: Compare new solutions with the original solutions, leaving the good solutions.

Step 6. Repeat steps 2 to 5 until the maximum number of iterations is terminated. Calculate the function values of n individuals.

2.2 Evolutionary Neural Architecture Search

Evolutionary algorithm is a search algorithm that solves optimization. The neural architecture search method based on evolutionary algorithms includes:

Step 1. First initialize population, i.e. initialize the population of n individuals in the constructed search space.

Step 2. Develop an encoding strategy to construct individuals in search space.

Step 3. Each individual is evaluated for fitness.

Step 4. After initialization, enter the population evolution stage and stochastically select individuals from the initialized population as parents.

Step 5. Implement evolutionary search in the established search space, and evaluate the fitness of generated individuals.

Step 6. Based on the optimization strategy, the retained individuals are selected and added to the population until the conditions are met and the optimal individual is returned.

The above summarizes the process of the ENAS, including the encoding of individuals, the population optimization strategy, and the method of evaluation.

3 Proposed Method

In Sect. 3, the framework of the method is discussed at length, as well as the main components. The method will be called SBSO-NAS in the later sections.

3.1 Algorithm Overview

Algorithm 1 illustrates the components of SBOS-NAS. The population N is first initialized and each individual is encoded using the proposed coding strategy (line 1). Each individual is then evaluated (row 2). Then, all individuals were involved in the process of an improved (using a simple grouping approach) BSO with maximum generation T (lines 3–6). Finally, the best individual is selected based on their fitness values, which is decoded as the best architecture output (line 7). In the evolution of the population participating in the improved BSO (SBOS), for each subset of groupings, new solutions are generated using the SBSO method; and new subsets are generated using the SBSO method. The "encoding strategy" and "the improved BSO" phases are documented in Sects. 3.2 and 3.3, respectively.

Algorithm 1 SBSO-NAS.

    Input: The population size N, the *max_ generation* number T;
    Output: Best architecture;

1   **Initialization**: $P_0 \leftarrow$ Initializing the population N using the proposed encoding strategy;

2   **Evaluation**: Get training datasets and evaluate the fitness of each individual in P_0;

3 **while** *current_ generation ≠ max_ generation* **do**

4   **Search the best solution**: generate new solutions by SBSO;

5   *current_ generation*++;

6 **end**

7 Estimate the accuracy of the best architecture on the test dataset;

3.2 Encoding Strategy

For the search space, we need to design it to contain as many architectures as possible. In this paper, we build it based on RBs, DBs and pooling layers. This approach has been proven to be effective [14], so we build DNNs based on it. Each individual represents an architecture and the solution to the problem, and the final task is to search for the optimal architecture using evolutionary methods.

The initialized population contains n individuals. For each individual, if you want to obtain the information they contain, you must develop an encoding strategy for them. In this paper, a new encoding strategy is designed that can better characterize the information of individuals. For the RBs used, we initialize the filter size of convolution layers to 3 × 3, which is also used for the convolution layers in the DBs. We set it to 2 × 2 in pooling layers [14]. For this purpose, the unknown parameters for the RBs are set to the size of the input and output, the filter size F1 of the convolutional layers, and the number of convolution kernels num1. The same for DBs, respectively F2, num2, with the extra k. The unknown parameters for the pooling layers are set to their type only, i.e. the maximum or average pooling type. The proposed encoding strategy is based on the above types and their position in the architecture.

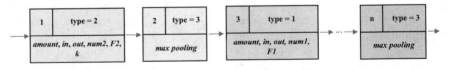

Fig. 1. Diagram of the encoding strategy.

Figure 1 shows the encoding strategy of n units of DNN. The upper left corner indicates the position of different units in DNN. When the type is 1, 2, or 3, the units are RBU, DBU, or PU. The length of each individual is variable [14].

3.3 The Improved BSO

The k-means method was used in the grouping operator of BSO [34] and to group n ideas into m clusters. However, the k-means make the implementation of the BSO algorithm more complicated and computationally intensive. Since evolution contains uncertainties in its process, BSO implements the grouping operator at each generation, so there is no need to use a precise method such as k-means to categorize ideas into different groups [31]. As the population, a set of N solutions ($A_i = [a_{i1}, a_{i2}, \ldots, a_{iDim}]$, where $1 \leq i \leq N$, the dimension of the problem is Dim) are generated at random during the initialization. We use simple grouping method to implement the grouping operation in SBSO, as follows:

Step 1. In the current iteration, choose S different architectures at random and use these S different architectures as centers, these S centers are denoted as $C_j (1 \leq j \leq S)$.

Step 2. For the solution $A_i (1 \leq i \leq N)$ in the current iteration, calculate the distance between it and each center c_j:

$$dis_{ij} = \|A_i, C_j\| = \sqrt{\left(\sum_{d=1}^{Dim} \left(a_{id} - c_{jd}\right)^2\right)/Dim} \tag{4}$$

Step 3. Compare these S distances and assign A_i to the center closest to it, center S remains the same.

Step 4. Return to step 2 and continue to calculate the distance between the next solution and each center c_j and assigning it until each solution A_i has been assigned.

4 Experiments and Result

We design and run experiments on the widely used CIFAR10 and CIFAR100 datasets to test the performance of SBSO-NAS, and then compared it to some other competitors.

4.1 Datasets and Parameter Settings

Typically we compare the classification performance of DNNs. The currently widely used datasets are CIFAR10 and CIFAR100. Therefore, we also chose to conduct comparative experiments on these two datasets.

As the name implies, cifar10 contains 10 classes and cifar100 contains 100 classes. They are two datasets used to identify some natural things like horses, tanks and aircraft. Each class of CIFAR10 has 6000 images, while each class of CIFAR100 has 600 images.

In this paper, we extract the results of other methods in their papers, and the extracted experimental data has experimented with each algorithm on the same GPU resources [41]. Samples are randomly selected from 10 categories of CIFAR10 and CIFAR100, with 100 samples in each category.

We follow the rules of setting values for all parameters in this paper can reduce the difficulty for researchers who want to look for the best architecture. The parameter details are shown in Table 1.

We set the maximum number of fitness evaluations to 10 and the threshold to 0.6, which is based on existing knowledge and experience [26, 35].

Table 1. Parameter settings

N	maximal_generation	m	P_r	$P_{replace}$	P_{one}	P_{one_center}	P_{two_center}
20	20	2	0.005	0.2	0.8	0.4	0.5

We allocate the training, validation, and test set of the datasets in the ratio of 6:2:2. Finally, we verify the proposed method on the test dataset, using the classification error rates as a criterion for comparison.

Furthermore, according to the structure of DenseNet, the k in DB can be 12, 20 and 40, and the convolutional layers in DB is 5 and 10. The maximum number of the units in this paper are both set to 4. The number of DBs and RBs is 3–10 respectively [14].

These settings are referenced in [14]. The experimental algorithm proposed in this paper models on NVIDIA GeForce GTX 3090 and executes on two GPU cards.

4.2 Result and Discussion

We evaluate the classification error, the parameters, and the computational resources of the method in our experiments. Using "GPU Days" as a metric for computational complexity.

Table 2. Compared the proposed algorithm with other competitors on CIFAR10 and CIFAR100 datasets

	CIFAR10	CIFAR100	# of Parameters	GPU Days
DenseNet [3]	5.24	24.42	1.0M	–
ResNe [2]	6.43	25.16	1.7M	–
ResNet [2]	7.93	27.82	10.2M	–
VGG [6]	6.66	28.05	20.04M	–
Genetic CNN [36]	7.1	29.05	–	17
EAS [37]	4.23	–	23.4M	10
Large-scale Evolution [38]	5.4	–	5.4M	2750
NAS [7]	6.01	–	2.5M	22400
OFA-random [39]	6.2	–	7.7M	–
MF-KD [40]	6.07	28.00	–	–
SBOS-NAS	4.2	20.2	5.3M	35

Table 2 displays the results of our method and others in this paper. The first line indicates the selected datasets and parameters, the second to fifth lines show the manually designed algorithms; the sixth to seventh rows are the semi-automatically designed algorithms; the eighth to ninth rows are the automatically designed algorithms. The last row is our algorithm.

From Table 2, SBOS-NAS performs better than the peer competitors that are manually designed on CIFAR10. The classification error of SBOS-NAS is about 1.0% lower than DenseNet, 2.2% lower than ResNet (the first one) and VGG, and 3.7% lower than ResNet. On CIFAR100, the classification error of SBOS-NAS was significantly lower than VGG and ResNet, and slightly lower than DenseNet, ResNet (the first one).

Compared with semi-automatic peer competitors, the performance of SBOS-NAS on CIFAR10 and CIFAR100 is significantly better than Genetic CNN, and better than EAS on CIFAR10.

Compared with the automatic methods, SBOS-NAS performs best in classification error and the number of GPU days consumed on CIFAR10 and CIFAR100. Comparison results make known our method acquires outstanding performance in automatic peer competitors.

4.3 Efficiency of SBSO

For improved BSO (SBSO), it reduces search time, accelerates search speed, and uses fewer computing resources.

Specifically, the traditional k-means clustering method is more accurate, but in the evolutionary process, we need to preserve the diversity of species and do not need such accuracy; and save computational time, so that the evolutionary process is more efficient.

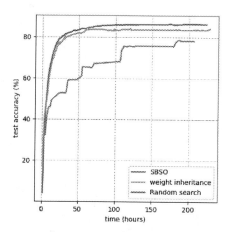

Fig. 2. Comparison of SBSO and traditional search strategies.

Secondly, compared to other search strategies, the method using random search is longer and far less precise than the evolutionary method using SBSO; for the method using weight inheritance (without evolution), its search time is not much different from SBSO, but SBSO also saves some time, and more importantly, the precision of the evolutionary method using SBOS is higher than that using weight inheritance (without evolution), as shown in Fig. 2.

5 Conclusion

Based on the SBSO algorithm, we develop a DNN architecture method that can automatically generate the best DNN architecture.

Firstly, based on ResNet blocks, DenseNet blocks, and pooling layer, we design an novel encoding strategy. Based on the previous excellent design, this paper adds information about the number of ResNet blocks and DenseNet blocks, the number and type of their convolutional kernels to help find the optimal architecture.

Secondly for the BSO, k-means is used in the grouping operation. However, k-means make the implementation of the BSO method complicated and involves extra computation. Therefore, SBSO algorithm is proposed, which is implemented using the simple grouping method in the grouping operation and uses it to search for the best architecture. SBSO significantly enhances the performance of BSO and makes the algorithm simple and implement easily.

Finally we verify the method on the datasets and compare with four manually designed architectures, two architectures designed in a semi-automatic way and four architectures designed entirely automatically. Experimental results show that the proposed method achieves good performance.

Acknowledgments. This work was supported in part by the Fundamental Research Funds for the Central Universities No. N2117005, the Joint Funds of the Natural Science Foundation of Liaoning Province und Grant 2021-KF-11-01 and the Fundamental Research Funds for the Central Universities.

References

1. Li, N., Ma, L., Yu, G., et al.: Survey on evolutionary deep learning: principles, algorithms, applications and open issues. arXiv preprint arXiv:2208.10658 (2022)
2. He, K., Zhang, X., Ren, S., Sun, J.: Deep residual learning for image recognition. IEEE (2016)
3. Huang, G., Liu, Z., Van Der Maaten, L., Weinberger, K.Q.: Densely connected convolutional networks. In: Proceedings of the IEEE Conference on Computer Vision and Pattern Recognition, pp. 4700–4708 (2017)
4. Devlin, J., Chang, M.-W., Lee, K., Toutanova, K.: BERT: pre-training of deep bidirectional transformers for language understanding. arXiv preprint arXiv:1810.04805 (2018)
5. Zhang, Y., Chan, W., Jaitly, N.: Very deep convolutional networks for end-to-end speech recognition. In: 2017 IEEE International Conference on Acoustics, Speech and Signal Processing (ICASSP), pp. 4845–4849. IEEE (2017)
6. Simonyan, K., Zisserman, A.: Very deep convolutional networks for large-scale image recognition. arXiv preprint arXiv:1409.1556 (2014)
7. Zoph, B., Le, Q.V.: Neural architecture search with reinforcement learning. arXiv preprint arXiv:1611.01578 (2016)
8. Ma, L., Li, N., Yu, G., et al.: How to simplify search: classification-wise Pareto evolution for one-shot neural architecture search. arXiv preprint arXiv:2109.07582 (2021)
9. Kaelbling, L.P., Littman, M.L., Moore, A.W.: Reinforcement learning: a survey. J. Artif. Intell. Res. **4**, 237–285 (1996)
10. Liu, H., Simonyan, K., Yang, Y.: Darts: differentiable architecture search. arXiv preprint arXiv:1806.09055 (2018)

11. Bäck, T., Fogel, D.B., Michalewicz, Z.: Handbook of evolutionary computation. Release **97**(1), B1 (1997)
12. Ma, L., Wang, X., Huang, M., et al.: A novel evolutionary root system growth algorithm for solving multi-objective optimization problems. Appl. Soft Comput. **57**, 379–398 (2017)
13. Sun, Y., Xue, B., Zhang, M., Yen, G.G.: Evolving deep convolutional neural networks for image classifification. IEEE Trans. Evol. Comput. **24**(2), 394–407 (2020)
14. Sun, Y., Xue, B., Zhang, M., Yen, G.G.: Completely automated CNN architecture design based on 9 blocks. IEEE Trans. Neural Netw. Learn. Syst. **31**(4), 1242–1254 (2020)
15. Mitchell, M.: An Introduction to Genetic Algorithms. MIT Press, Cambridge (1998)
16. Banzhaf, W., Nordin, P., Keller, R.E., Francone, F.D.: Genetic Programming: An Introduction: On the Automatic Evolution of Computer Programs and its Applications. Morgan Kaufmann Publishers Inc. (1998)
17. Das, S., Suganthan, P.N.: Differential evolution: a survey of the state-of-the-art. IEEE Trans. Evol. Comput. **15**(1), 4–31 (2011)
18. Kennedy, J., Eberhart, R.: Particle swarm optimization. In: Proceedings of ICNN 1995 International Conference on Neural Networks, vol. 4, pp. 1942–1948. IEEE (1995)
19. Tuba, E., Strumberger, I.: Classification and feature selection method for medical datasets by brain storm optimization algorithm and support vector machine. Procedia Comput. Sci. **162**, 307–315 (2019)
20. Ma, L., Wang, X., Huang, M., Lin, Z., Tian, L., Chen, H.: Two-level master-slave RFID networks planning via hybrid multiobjective artificial bee colony optimizer. IEEE Trans. Syst. Man Cybern.: Syst. **49**(5), 861–880 (2019). https://doi.org/10.1109/TSMC.2017.272 3483
21. Ma, L., Hu, K., Zhu, Y., et al.: Discrete and continuous optimization based on hierarchical artificial bee colony optimizer. J. Appl. Math. **2014**, 1–20 (2014). https://doi.org/10.1155/2014/402616
22. Chen, H., Zhu, Y., Hu, K., et al.: Bacterial colony foraging algorithm: combining chemotaxis, cell-to-cell communication, and self-adaptive strategy. Inf. Sci. **273**, 73–100 (2014)
23. Ma, L., Li, N., Guo, Y., et al.: Learning to optimize: reference vector reinforcement learning adaption to constrained many-objective optimization of industrial copper burdening system. IEEE Trans. Cybern. **52**(12), 12698–12711 (2022). https://doi.org/10.1109/TCYB.2021.308 6501
24. Zhang, B., Wang, X., Ma, L., Huang, M.: Optimal controller placement problem in internet-oriented software defined network. In: 2016 International Conference on Cyber-Enabled Distributed Computing and Knowledge Discovery (CyberC), Chengdu, China, pp. 481–488 (2016). https://doi.org/10.1109/CyberC.2016.98
25. Sun, Y., Yen, G.G., Yi, Z.: IGD indicator-based evolutionary algorithm for many-objective optimization problems. IEEE Trans. Evol. Comput. **23**(2), 173–187 (2018)
26. Xue, Y., Zhao, Y.: Structure and weights search for classification with feature selection based on brain storm optimization algorithm. Appl. Intell. **52**(5), 5857–5866 (2022)
27. Zeng, R., Su, M., Yu, R., Wang, X.: CD2: fine-grained 3D mesh reconstruction with twice chamfer distance. ACM Trans. Multimed. Comput. Commun. Appl. Just Accepted (2023). https://doi.org/10.1145/3582694
28. Cheng, S., Zhang, M., Ma, L., Lu, H., Wang, R., Shi, Y.: Brain storm optimization algorithm for solving knowledge spillover problems. Neural Comput. Appl. **35**, 12247–12260 (2021). https://doi.org/10.1007/s00521-020-05674-0
29. Ma, L., Cheng, S., Shi, Y.: Enhancing learning efficiency of brain storm optimization via orthogonal learning design. IEEE Trans. Syst. Man Cybern.: Syst. **51**(11), 6723–6742 (2020)
30. Tran, B., Xue, B., Zhang, M.: A new representation in PSO for discretization-based feature selection. IEEE Trans. Cybern. **48**(6), 1733–1746 (2017)

31. Zhan, Z., Zhang, J., Shi, Y., et al.: A modified brain storm optimization. In: 2012 IEEE Congress on Evolutionary Computation, pp. 1–8. IEEE (2012)
32. Cheng, S.,Shi, Y., Qin, Q., et al.: Solution clustering analysis in brain storm optimization algorithm. In: Proceedings of the 2013 IEEE Symposium on Swarm Intelligence (SIS 2013), Singapore, pp. 111–118. IEEE (2013)
33. Sun, Y., Sun, X., Fang, Y., et al.: A novel training protocol for performance predictors of evolutionary neural architecture search algorithms. IEEE Trans. Evol. Comput. **25**(3), 524–536 (2021)
34. Shi, Y.: Brain storm optimization algorithm. In: Tan, Y., Shi, Y., Chai, Yi., Wang, G. (eds.) Advances in Swarm Intelligence, pp. 303–309. Springerg, Heidelberg (2011). https://doi.org/10.1007/978-3-642-21515-5_36
35. Xue, B., Zhang, M., Browne, W.N.: Particle swarm optimization for feature selection in classification: novel initialisation and updating mechanisms. Appl. Soft Comput. J. **18**, 261–276 (2014)
36. Xie, L., Yuille, A.: Genetic CNN. In: Proceedings of the IEEE International Conference on ComputerVision (ICCV), Venice, Italy, pp. 1388–1397 (2017)
37. Cai, H., Chen, T., Zhang, W., Yu, Y., Wang, J.: Efficient architecturesearch by network transformation. In: Proceedings of the 32nd AAAI Conference on Artificial Intelligence, New Orleans, LA, USA, pp. 1–8 (2018)
38. Real, E., et al.: Large-scale evolution of image classifiers. In: Proceedings of the 34thInternational Conference on Machine Learning, pp. 2902–2911 (2017)
39. Cai, H., Gan, C., Wang, T., Zhang, Z., Han, S.: Once for all: train one network and specialize it for efficient deployment. In: International Conference on Learning Representations (2020)
40. Trofimov, I., Klyuchnikov, N., Salnikov, M., Filippov, A., Burnaev, E.: Multi-fidelity neural architecture search with knowledge distillation. IEEE Access. https://doi.org/10.1109/ACCESS.2023.3234810
41. Xie, X., Liu, Y., Sun, Y., et al.: Benchenas: a benchmarking platform for evolutionary neural architecture search. IEEE Trans. Evol. Comput. **26**(6), 1473–1485 (2022)

Beam-GBAS for Multidimensional Knapsack Problems

Jianjun Cao[1,2], Chumei Gu[1,2,3]($\boxtimes$) (iD), Baowei Wang[3], Zhen Yuan[1,2], Nianfeng Weng[1,2], and Yuxin Xu[1,2,3]

[1] The Sixty-Third Research Institute, National University of Defense Technology, Nanjing, China
{caojj,wengnf}@nudt.edu.cn, m15261820030@163.com
[2] Laboratory for Big Data and Decision, National University of Defense Technology, Changsha, China
[3] School of Computer Science, Nanjing University of Information Science and Technology, Nanjing, China

Abstract. In order to improve the performance of Graph-Based Ant System (GBAS) to solve Multidimensional Knapsack Problem (MKP), a method combining GBAS with Beam Search (Beam-GBAS) is proposed. GBAS transforms the unordered information of the problem into ordered influence on ants. Beam search is used to increase the number of candidate paths obtained by GBAS and the optimal path will be obtained by comparing. The experimental results show that the proposed method not only improves the performance of searching the optimal solution, but also improves the stability compared with GBAS.

Keywords: Graph-Based Ant System · Beam Search · Multidimensional Knapsack Problem · subset problem

1 Introduction

Ant colony optimization (ACO) is a swarm intelligence algorithm developed by natural evolution. Its search process adopts distributed computing and it can get positive feedback results [1]. The basic idea of ACO is to use walking path of ant to represent one feasible solution, and all paths constitute the solution space of the optimal problem [2].

The solving process of ordered combinatorial optimization problems such as Travelling Salesman Problem (TSP) is similar to the behavior of ants looking for the shortest path [3]. TSP problem is one of the earliest problems solved by ACO and subsequent improvements of ACO are also based on ordered combinatorial optimization problems [4]. In order to find better paths in the TSP, some literatures combined beam search method with ACO [5]. For example, Manuel et al. [6] proposed the use of stochastic sampling as a useful alternative to bounding information of Beam-ACO algorithm. They chose TSP with time windows as a case study and the results shown the improved Beam-ACO algorithm found a better solution to TSP. Hajewski et al. [7] proposed a greedy variant of Beam-ACO that used a greedy path selection heuristic. This approach reduced the internal state of Beam-ACO, making it simpler to parallelize. Experiments demonstrated that greedy Beam-ACO was faster than traditional Beam-ACO in time and less dependent on hyperparameter settings. In order to solve the problem of large calculation of Beam-ACO, Hajewski et al. [8] proposed a parallel version of Beam-ACO. This method run both the ant search and beam evaluation in parallel. Experiments verified that parallel Beam-ACO is faster than the most traditional ACO and Beam-ACO. Blum et al. [9] adapted a previous published Beam-ACO from the classical longest common subsequence problem to the repetition-free longest common subsequence problem. Compared to the best ones of the heuristics from the literature, the results showed that Beam-ACO generally outperforms these heuristics, often even by a large margin. F. Simoes et al. [10] focusd on the development of a new hybridization between beam search and the population-based ant colony optimization algorithm. The experiments showed all algorithms achieving exceptional performance on a hard benchmark problem. Caldeira et al. [11] proposed the use of hybrid algorithm Beam-ACO to implement the same management concept. The results showed that the distributed optimization paradigm can still be applied on supply chains. In order to solve the problem of assembly lines for the manufacturing of products, Blum et al. [12] proposed a Beam-ACO which is an algorithm that results from hybridizing ant colony optimization with beam search. The experimental results showed that this algorithm is a state-of-the-art meta-heuristic for this problem. In order to solve the problem of assembly line balancing problems are concerned with the distribution of work required to assemble a product in mass or series production among a set of work stations on an assembly line, Blum et al. [13] proposed an extended version of Beam-ACO published in [12]. Beam-ACO is a hybrid method. The experimental showed that this algorithm is able to find 128 new best solutions in 269 cases.

The solution of unordered combinatorial optimization problems such as knapsack problem (KP) is a set independent of the order of elements, which is inconsistent with the natural behavior of ants seeking the shortest foraging path. This kind of problem brings new challenge to ACO. Ren et al. [14] proposed a hybrid algorithm that integrated ACO with Lagrangian relaxation for solving NP-hard and strongly constrained MKP. This algorithm took ACO as the basic framework and defined a novel utility index for MKP based on LR dual information. Experiments shown that the algorithm is of strong robustness and efficiency. Yang et al. [15] proposed a greedy binary lion swarm optimization algorithm according to MKP. That paper introduced the idea of greedy algorithm on the basic of lion group algorithm and binary coding was used. The experiments shown that the algorithm had good stability, convergence and global search ability. An effective way to solve unordered combinatorial optimization problems is to transform the problem into walking process of ants on the directed graph. Specifically, Cao et al. [16] proposed a Graph-Based Ant System (GBAS) for Multidimensional Knapsack Problem. For the Multidimensional Knapsack Problem, the construction graph and the equivalent path were defined, and the pheromone updating strategy based on the equivalent path was proposed, which transformed the disorder information of the problem into the ordered influence on ants.

In order to improve the ability of GBAS to find the optimal path, this paper introduces Beam-ACO of TSP into GBAS, and proposes Beam-GBAS to solve the subset problem. This method combines GBAS with Beam Search.

2 Multidimensional Knapsack Problem

Knapsack problem is a classical NP-hard subset problem, which can be used as a standard example to study subset problems. The mathematical model of Multidimensional Knapsack Problem (MKP) is shown as follows.

$$
\max \sum_{i=1}^{n} p_i x_i
$$
$$
\text{s.t.} \sum_{i=1}^{n} r_{di} x_i \leq c_d, d = 1, 2, ..., C \tag{1}
$$
$$
x_i \in \{0, 1\}, i = 1, 2, ..., n
$$

In Eq. (1), n is the number of items in each knapsack, p_i is the value of the i-th item, x_i is the solution vector and the solution of x_i is one or zero, where one means to select this item and zero means not to select this item. R_{di} is the occupation amount of the corresponding constraint, c_d is the d-th constraint vector, the number of constraint vectors c is the dimension of the problem. The c-dimensional knapsack problem can be summarized as: under the condition that c constraints are satisfied, a group of items (subset) are selected from n items (original set) to maximize the value of all the selected items.

3 Beam-GBAS

3.1 GBAS for MKP

In order to solve MKP, Graph-Based Ant System (GBAS) was used in literature [11]. GBAS and structure graph's digraph of MKP were introduced, which was shown in Fig. 1.

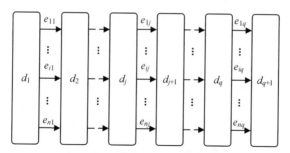

Fig. 1. Structure graph's digraph of MKP

In Fig. 1, n is the cardinality of the alternative set, q is the maximum possible cardinality of the obtained subset, node $d_k(k = 1, 2, ..., q + 1)$, $e_{ij} = <d_j, d_{j+1}>, j = 1, 2, ..., q$ represents the i-th element of the alternative set and is associated with the conditions related to the i-th element. In MKP, the edges of the digraph are candidate items and are associated with the constraint conditions. The path maps to a selected item subset and the value of objective function.

At time t, the number of ants at node d_1 is set as A, and under the constraint condition, each ant independently selects one edge to the next node according to the pheromone and heuristic factor on the edge. At time t, the probability of the a-th ant transferring from d_j to d_{j+1} through line segment e_{ij} is shown as follows:

$$P_{ij}^a(t) = \begin{cases} \dfrac{[\tau_{ij}(t-1)]^\alpha \eta_i^\beta}{\sum\limits_{e_{bj} \notin tabu_a} [\tau_{bj}(t-1)]^\alpha \eta_b^\beta}, & e_{ij} \notin tabu_a \\ 0, & others \end{cases} \tag{2}$$

In Eq. (2), $\tau_{ij}(t)$ is the amount of pheromone on the edge e_{ij} at time $t(t = 0, 1, 2,...)$, and the initial time $\tau_{ij}(0) = D$ (D is a constant). The heuristic factor η_i indicates the expected degree of selecting the i-th element, and $\alpha(\alpha \in R)$ and $\beta(\beta \in R)$ are the importance of pheromone amount and the heuristic factor respectively. The edge passed by the a-th ant is recorded with a tabu table $tabu_a$.

For MKP in this paper, the heuristic factor is shown as follows:

$$\eta_i = \frac{p_i}{\max\limits_{d=1}^{c}\left(\frac{rdi}{cd}\right)}, \quad i = 1, 2,...,n \tag{3}$$

Equation (3) shows that it is desirable to select items with high value and little impact on constraints.

According to the definition 2 and theorem 1 in literature [11], in the pheromone update phase, when the pheromone of a path is enhanced, the pheromone of the equivalent paths also need to be enhanced, that is, the pheromone update strategy based on the equivalent path enhancement.

After one iteration, the pheromone on the path $\Gamma(tabu^t)$ is enhanced at time t, the pheromone is updated according to the following updating equation:

$$\tau_{ij}(t) = \begin{cases} (1 - \rho)\tau_{ij}(t - 1) + \dfrac{\Phi'(tabu^t)}{Q}, & e_{ij} \in \Gamma(tabu^t) \\ (1 - \rho)\tau_{ij}(t - 1), & others \end{cases} \tag{4}$$

In Eq. (4), ρ is the pheromone volatilization coefficient, $\Phi'(tabu^t)/Q$ is the pheromone increment formula, $\Phi'(tabu^t)$ is the objective function value of the pheromone enhanced path and Q is a constant.

3.2 Beam-GBAS for MKP

In order to improve the ability of GBAS which is used to solve MKP, beam search will be used to select more paths in GBAS. As a heuristic graph search algorithm, the basic idea of beam search is similar to branch and bound [17]. However, in order to reduce the space and time occupied by search, compared with branching all nodes in branch and bound, beam search will delete some nodes with poor quality during the depth expansion of each layer [18]. The schematic diagram of beam search with beam width of 2 in MKP is shown in Fig. 2 and the specific process is shown as follows.

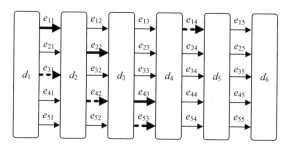

Fig. 2. Schematic diagram of beam search with beam width of 2

(1) First, for example, the total number of items is 5, the beam width is set as 2.
(2) Second, starting from node d_1, five items can be chosen in the first layer.
(3) Third, according to the path selection probability in Eq. (2) and roulette method [8]. The first probability value greater than the random number which is generated by roulette method is selected and the item number is returned, then two items are selected separately as beam nodes.
(4) Fourth, the two beam nodes (item e_{11} and e_{31} in Fig. 2) are used as the search points to continue branching and the beam nodes based on GBAS according to Eq. (2) and phenomenon update strategy in Eq. (4) are selected of this layer. And so on until the growth stopping condition is met.
(5) After beam search, the value is calculated and compared, and two feasible paths (beam width) will be obtained (That is, the two paths represented by solid arrows and dotted arrows in Fig. 2), and the optimal path will be chosen after comparison of the two paths. It should be noted that the selection of items in the two paths does not interfere with each other.

Compared with GBAS method, Beam-GBAS is more effective because more paths can be achieved and the optimal path will be chosen. For one ant in GBAS, a path is constructed item by item. But for one ant in Beam-GBAS, when the beam width of beam search is set to $k(k \in N^*)$, the ant will select k items for each partial path it constructs. Finally, k better paths are selected. To sum up, compared with GBAS, Beam-GBAS will explore much larger variety paths, which is why Beam-GBAS is more effective than GBAS in finding the optimal solution.

4 Experiments

The experiments are run on a computer with i7–4770 3.40GHz 4 core processor and 24 GB, and the development environment is MATLAB R2017a.

Experimental data are from the website: http://elib.zib.de/pub/Packages/mp-testdata/ip/sac94-suite/index/index.html.

4.1 Experiments of GBAS and Beam-GBAS

In order to test the performance of proposed method for MKP, the following two algorithms are compared: the GBAS method in literature [16], Beam-GBAS method for MKP improved by beam search algorithm and beam width k is set to 3. The following two examples are used for the experiment.

Example 1: pb2.dat-knapsacks: 4, objects: 35, the optimal solution (maximum value) provided is 3186. The initialized parameters are: $\tau_{ij}(0) = 100$, $\alpha = 1.2$, $\beta = 0.2$, $\rho = 0.1$, $Q = 200$, $A = 20$.

Example 2: weish14.dat-knapsacks: 5, objects: 60, the optimal solution provided is 6954. The initialized parameters are: $\tau_{ij}(0) = 100$, $\alpha = 1$, $\beta = 0.2$, $\rho = 0.2$, $Q = 300$, $A = 20$.

Example 3: flei.dat-knapsacks: 10, objects: 20, the optimal solution provided is 2139. The initialized parameters are: $\tau_{ij}(0) = 100$, $\alpha = 1.2$, $\beta = 0.2$, $\rho = 0.1$, $Q = 200$, $A = 20$.

Example 4: weing6.dat-knapsacks: 2, objects: 28, the optimal solution provided is 130623. The initialized parameters are: $\tau_{ij}(0) = 100$, $\alpha = 1$, $\beta = 0.2$, $\rho = 0.1$, $Q = 200$, $A = 20$.

The number of iterations is set to 200, and the mean value and standard deviation test results of the solution obtained by running 20 times are shown in Fig. 3(a)–(h).

Then take 40, 60, 80, 100, 120, 140, 160, 180 and 200 iterations respectively, the number of running times is still set to 20. The number of solutions found by GBAS and Beam-GBAS are shown in Table 1.

In Fig. 3, the mean value of Beam-GBAS is basically greater than GBAS, and the standard deviation is basically less than GBAS, showing a better ability to find better solutions than GBAS, and its solutions have higher stability. In Table 1, the number of optimal solutions provided by Beam-GBAS is basically more or equal to that of GBAS. It can be seen that increasing the number of candidate path in Beam-GBAS can improve the performance of finding better solutions.

4.2 Experiments on the Influence of Beam Width

The number of candidate paths in Beam-GBAS is dynamically adjusted depending on the beam width, and there are multiple structured paths in the generated candidate paths. As an important parameter, beam width affects the performance of Beam-GBAS algorithm, so it is necessary to verity the impact of beam width on Beam-GBAS through experiments.

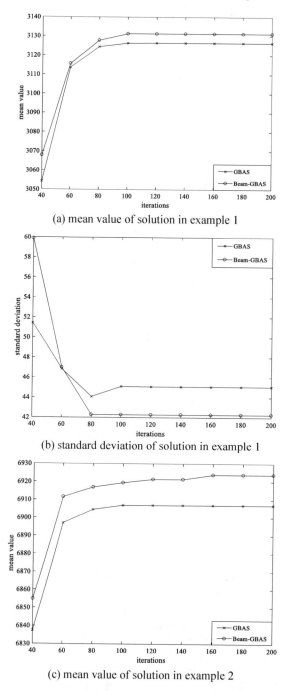

(a) mean value of solution in example 1

(b) standard deviation of solution in example 1

(c) mean value of solution in example 2

Fig. 3. Experimental results of GBAS and Beam-GBAS for MKP

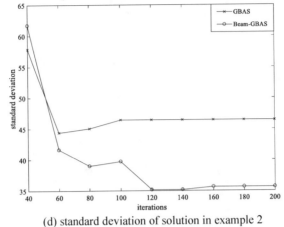

(d) standard deviation of solution in example 2

Fig. 3. (*continued*)

Table 1. Number of times to find the optimal solution provided

Iterations	Example 1		Example 2	
	GBAS	Beam-GBAS	GBAS	Beam-GBAS
40	0	0	0	**3**
60	2	2	0	**2**
80	2	2	2	**2**
100	2	**3**	3	**4**
120	2	**4**	3	3
140	3	**4**	1	**4**
160	1	1	3	**5**
180	0	**2**	0	**2**
200	0	**1**	3	3

The parameter settings of example 1 and example 2 are the same as Sect. 4.1. The number of iterations is set to 200 and the first 120 times are shown in Fig. 4, and influence of different k (k = 1 to 7) on the mean value and standard deviation test results of the solution obtained by running 20 times are shown in Fig. 4(a)–(d).

In Fig. 4, with the increase of k value, the mean value increases or remains unchanged, and the standard deviation decreases or remains unchanged. This shows that to a certain extent, combing Beam Search with GBAS can increase the number of candidate paths, cut off some paths with poor output quality, and the optimization efficiency is higher.

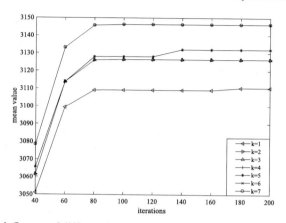

(a) influence of different k on mean value of solution in example 1

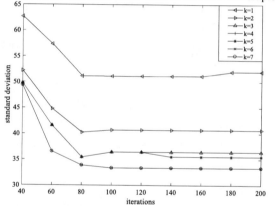

(b) influence of different k on standard deviation of solution in example 1

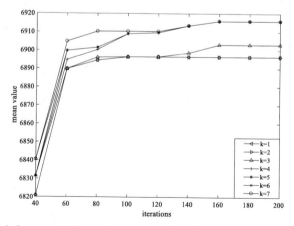

(c) influence of different k on mean value of solution in example 2

Fig. 4. Influence of different k

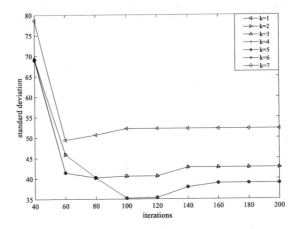

(d) influence of different k on standard deviation of solution in example 2

Fig. 4. (*continued*)

5   Conclusions

In order to improve the performance of GBAS to solve MKP, this paper focuses on introducing beam search method into GBAS. The main achievements are as follows:

(1) A mathematical model with the maximum value of the selected items as evaluation criterion is established, and Beam-GBAS method with stronger search ability is proposed.
(2) Beam-GBAS algorithm is used to solve the model, the excellent search ability and stability of beam search are used to optimize GBAS. The candidate paths obtained by Beam-GBAS is expanded and after comparing the value of the candidate paths, the optimal path can be obtained. The experimental results show that compared with GBAS, Beam-GBAS can solve MKP more accurately, obtain a better path scheme, and improve the stability.

The research described in this paper realizes the optimization of GBAS by using beam search method, taking into account the impact of beam width on the algorithm, but does not consider the problem of time consumption while increasing the beam width. Therefore, how to balance the size of beam width and the amount of time spent will be a point of following research.

References

1. Yang, L., Li, K., Zhang, W., et al.: An improved chaotic ACO clustering algorithm. In: 20th International Conference on High Performance Computing and Communications, pp. 1642–1649. IEEE (2018)
2. Manuel, L., Christian, B., Dhananjay, T., et al.: Beam-ACO based on stochastic sampling for makespan optimization concerning the TSP with time windows. In: Proceedings of Conference on LNCS, pp. 97–108 (2009)

3. Liu, R., Li, L., Zhao, Y., et al.: Beam-ACO for the lock chamber arrangement. In: 3rd International Conference on Computer and Communication Engineering Technology, pp. 186–189. IEEE (2020)

4. Dhananjay, T., Christian, B., Bernd, M., et al.: Hybridizing beam-ACO with constraint programming for single machine job scheduling. In: Proceedings of Conference on LNCS, pp. 30–44 (2009)

5. Sun, H., Han, J.: Virtual resequencing in automobile paint shops based on beam search-based algorithm. Comput. Integr. Manuf. Syst. **22**(8), 1900–1906 (2016)

6. Manuel, L., Christian, B.: Beam-ACO based on stochastic sampling: a case study on the TSP with time windows. In: Proceedings of the Conference on LNCS, pp. 59–73 (2009)

7. Hajewski, J., Oliveira, S., Stewart, D.E., et al.: gBeam-ACO: a greedy and faster variant of beam-ACO. In: Proceedings of the 20th Conference on GECCO, pp. 1434–1440 (2020)

8. Hajewski, J., Oliveira, S., Stewart, D.E, et al.: Exploring trade-offs in parallel beam-ACO. In: Proceedings of the 11th Annual Computing and Communication Workshop and Conference (CCWC), pp. 1525–1534 (2021)

9. Blum, C., Blesa, M.J., Calvo, B., et al.: Beam-ACO for therepetition-free longest common subsequence problem. In: Proceedings of the Conference on EA, pp. 79–90 (2014)

10. Simoes, L.F., Izzo, D., Haasdijk, E., et al.: Multi-rendezvousspacecraft trajectory optimization with Beam P-ACO. In: Proceedings of the Conference on EvoCOP, pp. 141–156 (2017)

11. Caldeira, J., Azevedo, R., Silva, C.A., et al.: Beam-ACO distributedoptimization applied to supply-chain management. In: Proceedings of the Conference on IFSA, pp. 799-809 (2007)

12. Blum, C., Bautista, J., Pereira, J.: An extended Beam-ACO approachto the time and space constrained simple assembly line balancing problem. In: Proceedings of the Conference on EvoCOP, pp. 85–96 (2008)

13. Blum, C., Bautista, J., Pereira, J.: Beam-ACO appliedto assembly line balancing. In: Proceedings of the Conference on ANTS, pp. 96-107 (2006)

14. Ren, Z., Zhao, S., Huang, S., et al.: Hybrid optimization algorithm of ant colony optimization and Lagrangian relaxation for solving multidimensional knapsack problem. Control Decis. **31**(7), 1178–1184 (2016)

15. Yang, Y., Liu, J., Zhou, Y.: Greedy binary lion swarm optimization algorithm for solving multidimensional knapsack problem. J. Comput. Appl. **40**(5), 1291–1294 (2020)

16. Cao, J., Zhang, P., Wang, Y., et al.: Graph-based ant system for subset problems. J. Syst. Simul. **20**(22), 6146–6150 (2008)

17. Yan, Q., Lu, J., Jiang, W., et al.: Path optimization of stacker in compact storage system with dual-port layout. J. Shanghai Jiao Tong Univ. **56**(7), 858–867 (2022)

18. Wu, K., Zhou, X., Li, Z., et al.: Path selection for Chinese knowledge base question answering. J. Chin. Inf. Process. **35**(9), 113–122 (2021)

Brain Storm Optimization Integrated with Cooperative Coevolution for Large-Scale Constrained Optimization

Yuetong Sun[1], Peilan Xu[1(✉)], Ziyu Zhang[1], Tao Zhu[2], and Wenjian Luo[3]

[1] School of Artificial Intelligence/School of Future Technology, Nanjing University of Information Science and Technology, Nanjing 210044, Jiangsu, China
{202283460028,xpl,202283460036}@nuist.edu.cn
[2] School of Computer Science, University of South China, Hengyang 421001, Hunan, China
tzhu@usc.edu.cn
[3] School of Computer Science and Technology, Harbin Institute of Technology, Shenzhen 518000, Guangdong, China
luowenjian@hit.edu.cn

Abstract. Large-scale constrained optimization problems (LSCOPs) are challenging to solve because of the high dimensionality and constraint limitations. Although cooperative coevolution (CC) has been applied to LSCOPs, more efficient optimizers that could be adapted to CC are still required. In this paper, we propose ConBSO, a variant of the brain storm optimization (BSO) designed for constrained optimization. Then, ConBSO is integrated into constraint-objective cooperative coevolution (COCC), denoted as COCC-ConBSO. To evaluate the performance of COCC-ConBSO, we test it on the benchmark suite with 12 LSCOPs and compared it to several algorithms, including two algorithms based on the COCC framework and three state-of-the-art large-scale constrained optimization algorithms. Experimental results demonstrate the adaptability of ConBSO to COCC and highlight the competitiveness of COCC-ConBSO in solving LSCOPs.

Keywords: Large-scale constrained optimization · Cooperative coevolution · Brain storm optimization

1 Introduction

In the real world, many optimization problems involve a large number of decision variables and are subject to constraints, such as satellite design [13], electric aircraft design [10], and power system problems [17]. These problems are known as large-scale constrained optimization problems (LSCOPs). The explosion of

This work is partly supported by the Startup Foundation for Introducing Talent of NUIST (No. 2022r121) and National Natural Science Foundation of China (No. 62006110).

parameter dimensions leads to the curse of dimensionality, which significantly deteriorates the performance of many optimization algorithms [15], while constraint limitations may partition the search space into multiple disconnected feasible regions [14]. Therefore, the development of efficient optimization algorithms to search for the feasible optimal solution of LSCOPs is a critical research topic [12,18].

Swarm intelligence (SI) [6] is a popular class of meta-heuristic optimization algorithms based on population search, widely employed to tackle LSCOPs. Depending on whether variable interactions are analyzed or not, SI for LSCOPs could be classified into two categories [18]. The first is to design efficient optimization strategies to improve the performance of small-scale constrained optimization algorithms for solving LSCOPs, such as offspring generation [11] and surrogate modeling [23]. The second is cooperative coevolution (CC) [21,29], which employs the idea of divide-and-conquer and has shown great promise in solving LSCOPs. Many strategies have been proposed for improving the solving ability of CC for LSCOPs, such as the problem decomposition strategies [1,3,25], computational resource allocation strategies [31], and optimization strategies [2,19,20]

Brain storm optimization (BSO) [26] is a type of SI that simulates the brainstorming behavior of human societies when faced with challenging problems. The main steps of BSO include the clustering and mutation of solutions. The multi-cluster search feature of BSO has a unique advantage in coping with the discontinuous feasible regions arising from constraint limitations, and there have been a few studies employing BSO to solve constrained optimization problems [5,8]. Moreover, the cooperative search viewpoint adopted by BSO is promisingly suitable for integration with the CC framework [9].

Motivated by the advantages of both BSO and CC, this paper proposes an optimization algorithm, called COCC-ConBSO, for solving LSCOPs. First, a variant of BSO for constrained optimization, called ConBSO, is proposed. In the intra-cluster mutation, subpopulations explore the feasible or optimal regions, respectively, while in the inter-cluster mutation, the worse subpopulations learn from the better ones. Then, the proposed ConBSO is embedded into the constraint-objective cooperative coevolution (COCC) [31], and the search regions of different subpopulations are associated with the subproblems to be optimized. COCC-ConBSO is tested on the large-scale constrained optimization benchmark proposed in Ref. [31], and experimental results show that the proposed COCC-ConBSO is competitive with two algorithms based on the COCC framework, and three large-scale constraint optimization algorithms, εCCPSOd [20], εCCPSOdg2 [19], and DELoCoS-DG2 [2].

The remainder of this paper is organized as follows. Section 2 reviews related work. The proposed COCC-ConBSO is presented in Sect. 3. Section 4 describes the experimental design and results. Section 5 concludes this paper.

2   Related Work

In this section, we first introduce the constraint-objective cooperative coevolution (COCC) framework developed for constrained optimization. Then, we describe the bairn storm optimization (BSO) and one of its improvements called BSO20.

2.1   Constraint-Objective Cooperative Coevolution

COCC framework was developed to address LSCOPs [31]. The fundamental idea of COCC is to allocate computational resources to components that have a significant impact on objective values or constraint violations. COCC comprises three primary components: the decomposer, the selector, and the optimizer.

The decomposer in COCC adopts recursive differential grouping [27] to decompose the objective function and constraint functions separately, obtaining two sets of subproblems. The selector and optimizer are then cyclically invoked until the termination condition is met.

The selector in COCC first selects a set from the sets of objective subproblems and constraint subproblems, based on the constraint handling techniques used, such as the superiority of the feasible solution (SF) [7], ε-constraint (EC) [28], and stochastic ranking (SR) [24]. Then, a subproblem is selected based on its contribution to the objective value or its correction to constraint violations.

The optimizer in COCC uses a small-scale constrained optimization algorithm to optimize the selected subproblems. Finally, the representative solutions of subproblems are assembled to obtain the final solution.

2.2   Bairn Storm Optimization

BSO was proposed by Shi [26], drawing inspiration from the creative problem-solving brainstorming process used by humans. BSO comprises three main steps: clustering, mutation, and selection of solutions, which are executed in each iteration following population initialization. In the original BSO, these steps are executed in the following order.

- **Clustering**: The original BSO adopts the well-known k-means algorithm to divide the population into clusters, where each cluster center is randomly initialized with probability p_{init}.
- **Mutation**: The original BSO generates new offspring $\mathbf{x}'_i$ using Formula (1), where $\mathbf{y}_i$ represents the base individual, $\xi(t)$ is calculated according to Formula (2), and t and T represent the current and maximum number of iterations, respectively. Furthermore, $\mathcal{N}(0,1)$ and $U(0,1)$ represent random numbers following standard normal and uniform distributions, respectively.

$$x'_{i,d} = y_{i,d} + \xi(t) \cdot \mathcal{N}(0,1), \tag{1}$$

$$\xi(t) = \text{logsig}\left(\frac{0.5 \times T - t}{20}\right) \cdot U(0,1). \tag{2}$$

For intra-cluster mutation, the original BSO randomly selects an individual from the cluster as the base individual $\mathbf{y}_i$, with probability p_{one_best} of being the cluster center. For inter-cluster mutation, the original BSO selects two individuals from two different random clusters to generate the base individual $\mathbf{y}_i$, with probability p_{two_best} of being the cluster centers.

- **Selection**: The offspring $\mathbf{x}'_i$ is compared with parent $\mathbf{x}_i$, and the better one joins the next generation population.

BSO20 [30] is one of the improvements of BSO. First, BSO20 employs a more efficient hybrid clustering algorithm that combines the nearest-better clustering (NBC) [22] to cluster better individuals with the random grouping strategy (RGS) [4] to cluster the remaining poor individuals. During the evolutionary process, the proportion of individuals clustered by NBC is dynamically adjusted. Secondly, BSO20 uses intra- and inter-cluster mutations. For intra-cluster mutation, an individual is randomly selected to learn from the leader to generate the base individual $\mathbf{y}_i$. For inter-cluster mutation, a randomly selected individual shares information with individuals in the other two clusters.

3   Proposed Algorithm

To address the challenges of LSCOPs, which include the large decision space and constraint limitations, this paper proposes COCC-ConBSO. The proposed algorithm comprises two primary components: a variant of BSO designed for constraint optimization, named ConBSO, and the integration of ConBSO into COCC. In the following sections, we provide a detailed explanation of ConBSO and how it is incorporated into COCC to solve LSCOPs.

3.1   Brain Storm Optimization for Constrained Optimization

In constrained optimization, the feasible region of a given problem is typically partitioned by constraints into multiple disjoint regions. These regions represent the set of all feasible solutions, among which the optimization algorithm is expected to locate the optimal solution. To achieve this, an optimization algorithm should simultaneously explore multiple feasible regions while balancing the exploration of feasible regions with the exploitation of optimal regions during the evolutionary process.

This paper presents a variant of BSO named constrained BSO (ConBSO) for constrained optimization. ConBSO utilizes a solution clustering strategy to divide the population into multiple clusters, enabling each cluster to search different regions simultaneously. The intra-cluster mutation operator explores both feasible and optimal regions within each cluster, while the inter-cluster mutation operator facilitates the exchange of information about feasible and optimal regions across different clusters. Further details on ConBSO are provided in the following subsections.

Clustering Strategy of ConBSO. ConBSO is designed to simultaneously search for feasible and optimal regions in constrained optimization. ConBSO divides clusters into two categories using the hybrid clustering strategy proposed in BSO20 [30], which combines NBC and RGS. Specifically, the better individuals are clustered using NBC, while RGS is used to cluster the remaining individuals. The resulting clusters are labeled as $\Theta_1 = \{C_1, \ldots, C_{k_n}\}$ and $\Theta_2 = \{C_{k_n+1}, \ldots, C_k\}$, respectively, and two types of clusters are allocated to explore feasible or optimal regions during the evolutionary process.

In contrast to BSO20, ConBSO sets the number k_r of clusters generated by RGS to a constant value, because it seeks to balance the search for feasible and optimal regions throughout the optimization process. Additionally, ConBSO uses NBC-minsize [16] to generate clusters without limiting the number of clusters generated.

The clustering strategy of ConBSO is outlined in Algorithm 1. The algorithm starts by sorting the population P and determining the size s_n of the subpopulation clustered by NBC, which is computed as $s_n = \mathrm{NP} - k_r s_r$, where s_r is a constant indicating the size of clusters generated by RGS and NP is the population size.

Next, the population is divided into two subpopulations, P_n and P_r, and NBC-minsize is used to cluster the better subpopulation P_n, while RGS is used to cluster the subpopulation P_r. The resulting clusters are recorded in Λ to keep track of the cluster assignment of each individual in the population.

Algorithm 1. Clustering strategy of ConBSO

1: $P_n = \{\mathbf{x}_1, \ldots, \mathbf{x}_{s_n}\}$;
2: $P_r = \{\mathbf{x}_{s_n+1}, \ldots, \mathbf{x}_{\mathrm{NP}}\}$;
3: Cluster subpopulation P_n by NBC-minsize to form Θ_1;
4: Shuffle P_r;
5: **for** $i \leftarrow k_n + 1$ **to** k **do**
6:     Form i-th cluster with the top s_r individuals in P_r;
7:     Remove the top s_r individuals from P_r;
8: **end for**
9: Record the cluster to which each individual belongs in Λ;

Mutation and Selection Strategy of ConBSO. The mutation strategy of ConBSO includes both intra-cluster and inter-cluster mutation. The intra-cluster mutation operator is designed to search either the feasible or optimal regions independently, depending on the boolean variable μ_i. The clusters are focused on exploring the feasible regions when μ_i is true, otherwise exploiting the optimal regions. Moreover, the inter-cluster mutation operator enables the inferior cluster to learn from the superior cluster.

Algorithm 2 provides the pseudocode of the mutation and selection strategy in ConBSO, where $\mu = \{\mu_1, \ldots, \mu_{\mathrm{NP}}\}$ is a boolean vector to be input, all true or all false.

Algorithm 2. Mutation and selection strategy of ConBSO

1: **for** $i \leftarrow 1$ **to** NP **do**
2:     $C_i \leftarrow \Lambda(\mathbf{x}_i)$;
3:     **if** $C_i \in \Theta_1$ **or** $r < p_{one}$ **then**
4:         Generate a base individual $\mathbf{y}_i$ by intra-cluster mutation;
5:         **if** $C_i \in \Theta_2$ **then**
6:             $\mu_i \leftarrow \neg\mu_i$;
7:         **end if**
8:     **else**
9:         Generate a base individual $\mathbf{y}_i$ by inter-cluster mutation;
10:     **end if**
11:     Add Gaussian random numbers to generate a new individual $\mathbf{x}'_i$;
12:     **if** μ_i **or** $\phi(\mathbf{x}^*) = 0$ **then**
13:         **if** $f(\mathbf{x}'_i) \leq f(\mathbf{x}_i)$ **then**
14:             Update individual $\mathbf{x}_i$ by $\mathbf{x}'_i$;
15:         **end if**
16:     **else**
17:         **if** $\phi(\mathbf{x}'_i) \leq \phi(\mathbf{x}_i)$ **then**
18:             Update individual $\mathbf{x}_i$ by $\mathbf{x}'_i$;
19:         **end if**
20:     **end if**
21: **end for**

The specifics are described as follows. For an individual $\mathbf{x}_i$, the algorithm retrieves the corresponding cluster C_j from set Λ. If the cluster C_j is produced by NBC, the intra-cluster mutation is exclusively performed. Moreover, the intra-cluster mutation is executed with a probability of p_{one} for individuals clustered by RGS.

The base individual $\mathbf{y}_i$ is generated based on the modified BSO algorithm [33],

$$\mathbf{y}_i = \mathbf{x}_{r_1} + \mathbf{r} \cdot (\mathbf{x}_{r_2} - \mathbf{x}_{r_3}). \tag{3}$$

where $\mathbf{r}$ is a random vector with D dimensions that follows a uniform distribution within the range of $[0,1]$. In the intra-cluster mutation, $\mathbf{x}_{r_1}$, $\mathbf{x}_{r_2} \neq \mathbf{x}_{r_3}$ are individuals randomly chosen from cluster C_i. In the inter-cluster mutation, $\mathbf{x}_{r_1}$ is randomly picked from cluster C_i, whereas $\mathbf{x}_{r_2}$ and $\mathbf{x}_{r_3}$ are randomly selected from the two distinct clusters created by NBC.

Subsequently, a new individual $\mathbf{x}'_i$ is generated using Formula (1), in which the function $\xi(t)$ is described by Formula (4).

$$\xi(t) = \begin{cases} \text{logsig}\left(\dfrac{0.5 \times T - t}{20}\right) \cdot U(0,1), & \text{if } r > p_r, \\ \eta \cdot U(0,1), & \text{otherwise.} \end{cases} \tag{4}$$

where η is a uniformly distributed random number in the interval $[0, 0.1]$.

Finally, a comparison is made between the new individual $\mathbf{x}'_i$ and the original individual $\mathbf{x}_i$. The individuals are compared based on their objective value when

μ_i is true, whereas those are compared based on their constraint violations when μ_i is false.

3.2   ConBSO Integrated with COCC

COCC considers both the decomposition of the objective and constraint functions of the LSCOPs. Before selecting a subproblem, it is decided whether to optimize the objective or constraint subproblem sets depending on the constraint handling technique used. Utilizing this idea, ConBSO could be well integrated with COCC. As an example of SF as the constraint handling technique, The pseudocode of COCC-ConBSO is given in Algorithm 3, where the Boolean vector μ is defined in lines 5 and 8.

Algorithm 3. COCC-ConBSO

1: Initialize algorithm parameters and population P;
2: $\{G_{obj}, G_\phi\} = \text{Decomposer}(P)$;
3: **while** Termination conditions are not met **do**
4:     **if** $\phi(\mathbf{x}^*) = 0$ **then**
5:         $\mu = \text{True}(1, \text{NP})$;
6:         Select a subproblem from group G_{obj};
7:     **else**
8:         $\mu = \text{False}(1, \text{NP})$;
9:         Select a subproblem from group G_ϕ;
10:    **end if**
11:    Optimize selected subproblem employing ConBSO;
12: **end while**

---

4   Experiments

This section adopts a large-scale constrained optimization benchmark to test the performance of the proposed COCC-ConBSO. First, the benchmark and the experimental design are presented. Then, the comparison results between COCC-ConBSO and several large-scale constrained optimization algorithms are given.

4.1   Benchmark and Experimental Design

The performance of COCC-ConBSO algorithm is tested on the benchmark suite consisting of 12 LSCOPs proposed in Ref. [31]. To ensure the accuracy of the results, each test function independently runs 25 times with the maximum number of fitness evaluations MaxFEs $= 3 \times 10^6$. Upon termination of each independent run, the objective values and constraint violations of the optimal individual found by the algorithm are recorded. The mean objective value *mean*,

mean constraint violation vio, and the success rate sr are calculated over multiple independent runs. Here, the success rate is the percentage of successful runs to the total number of runs, and a successful run is defined as one where the algorithm finds at least one feasible solution in a single independent run.

The experiments of this paper adopt several compared algorithms, including two algorithms based on the COCC framework [31], COCC-DE, and COCC-BSO20. The latter algorithm utilizes BSO20 [4] as the optimizer and ε-constraint [28] as the constraint handling technique. The constraint handling technique used in the COCC framework is the superiority of the feasible solution [7]. Additionally, three state-of-the-art large-scale constrained optimization algorithms, εCCPSOd [20], εCCPSOdg2 [19], and DELoCoS-DG2 [2], are also used as compared algorithms. The symbols $A_i^k(sr)$, $A_i^k(vio)$, and $A_i^k(mean)$ denote the success rate, the mean constraint violation, and the mean objective value achieved by the i-th comparison algorithm on the k-th problem, respectively, and the detailed comparison rules are as follows.

- For the performance of two algorithms A_i^k and A_j^k in the k-th function, if $A_i^k(sr) \neq A_j^k(sr)$, the algorithm $A = \arg\min\{A_i^k(sr), A_j^k(sr)\}$ performs better on the k-th problem, otherwise compare their vio;
- If $A_i^k(sr) = A_j^k(sr)$ and $A_i^k(vio) \neq A_j^k(vio)$, the algorithm $A = \arg\max\{A_i^k(vio), A_j^k(vio)\}$ performs better on the k-th problem, otherwise compare their $mean$;
- If $A_i^k(sr) = A_j^k(sr)$, $A_i^k(vio) = A_j^k(vio)$, and $A_i^k(mean) \neq A_i^k(mean)$, then the algorithm $A = \arg\min\{A_i^k(mean), A_j^k(mean)\}$ performs better on the k-th problem, otherwise there is no difference in performance between the comparison algorithms.

The parameters for the proposed COCC-ConBSO algorithm are specified as follows. The population size is set to 100. The number of clusters generated by the RGS is set to 2, with each cluster containing 25 individuals. The probability of intra-cluster mutation is set to 0.3. The selection probability of step size is set to 0.8. The parameter settings for the compared algorithms are following the recommendations provided in their respective literature.

4.2   Experimental Results

Table 1 presents the experimental results of COCC-ConBSO and five compared algorithms on 12 LSCOPs, with winning results highlighted in bold for each test function. The last row of the table displays symbols "$\pm$" indicating whether COCC-ConBSO is superior or inferior to the compared algorithm.

Overall, the comparison of experimental results of COCC-ConBSO and several large-scale constrained optimization algorithms in Table 1 demonstrate that COCC-ConBSO has better performance and is promising and competitive. COCC-ConBSO outperforms two algorithms based on the COCC framework, COCC-DE and COCC-BSO20 in 7 and 12 functions separately. COCC-ConBSO also performs better than three large-scale constrained optimization algorithms,

Table 1. Experimental results of COCC-ConBSO and five compared algorithms on 12 LSCOPs

No.		COCC-ConBSO	COCC-DE	COCC-BSO20	εCCPSOd	εCCPSOdg2	DELoCoS-DG2
F1	mean	**−6.6336E+02**	6.5781E+02	−4.9771E+03	−2.2458E+08	−2.9489E+08	−4.8176E+02
	vio	**0.0000E+00**	1.0633E-01	2.7204E-04	4.6471E+02	1.2989E+00	2.9314E+00
	sr	**100%**	0%	44%	0%	0%	0%
F2	mean	1.7562E+02	2.4836E+02	1.9235E+05	2.5002E+05	5.1850E+04	**3.2217E+01**
	vio	0.0000E+00	0.0000E+00	1.3564E+02	3.3280E+03	2.1036E+05	**0.0000E+00**
	sr	100%	100%	0%	0%	0%	**100%**
F3	mean	2.9522E+03	3.3503E+03	5.1411E+04	3.4495E+04	4.7472E+04	**2.8312E+03**
	vio	1.4828E-01	2.4661E-01	9.0699E+05	7.1188E+04	7.1351E+05	**0.0000E+00**
	sr	88%	76%	0%	0%	0%	**100%**
F4	mean	**5.4676E+07**	7.2987E+07	4.1752E+07	3.6368E+07	9.3264E+07	9.2180E+07
	vio	**4.2610E+05**	3.2414E+07	3.6138E+07	5.1493E+06	9.5151E+06	3.6716E+06
	sr	**0%**	0%	0%	0%	0%	0%
F5	mean	**7.7858E+05**	9.7230E+08	1.1007E+07	1.4271E+08	1.5706E+09	2.4553E+08
	vio	**0.0000E+00**	0.0000E+00	0.0000E+00	0.0000E+00	0.0000E+00	0.0000E+00
	sr	**100%**	100%	100%	100%	100%	100%
F6	mean	1.6222E+06	**1.1899E+04**	3.6669E+07	2.4066E+06	3.7654E+12	3.2548E+12
	vio	0.0000E+00	**0.0000E+00**	0.0000E+00	0.0000E+00	2.5463E+06	0.0000E+00
	sr	100%	**100%**	100%	100%	0%	100%
F7	mean	1.1635E+06	**6.4707E+05**	2.8290E+06	1.1749E+07	9.4003E+05	3.4013E+07
	vio	0.0000E+00	**0.0000E+00**	0.0000E+00	2.0218E+00	1.2314E+06	1.0737E+00
	sr	100%	**100%**	100%	0%	0%	0%
F8	mean	1.6622E+10	1.5982E+10	1.6810E+10	1.2044E+10	1.4923E+10	**1.4228E+10**
	vio	0.0000E+00	0.0000E+00	1.6364E+02	1.3891E+02	5.0778E+02	**0.0000E+00**
	sr	100%	100%	0%	12%	0%	**100%**
F9	mean	**1.6104E+06**	9.1206E+05	5.9617E+05	5.9951E+03	2.3251E+07	3.6433E+08
	vio	**2.1479E+03**	7.3593E+03	6.9991E+03	3.4410E+03	1.2419E+04	1.3549E+04
	sr	**0%**	0%	0%	0%	0%	0%
F10	mean	5.1993E+02	**6.8623E+02**	3.0400E+03	8.3632E+02	3.8874E+03	5.7272E+02
	vio	4.6716E+05	**4.6481E+05**	9.5262E+07	8.8251E+06	3.2110E+07	8.4330E+05
	sr	0%	**0%**	0%	0%	0%	0%
F11	mean	**4.0454E+06**	1.0939E+06	6.8858E+05	2.9656E+05	5.0089E+11	1.5468E+07
	vio	**1.9301E+01**	6.1086E+01	5.4292E+01	1.3915E+02	1.8357E+04	1.3820E+02
	sr	**0%**	0%	0%	0%	0%	0%
F12	mean	3.0230E+04	**1.3901E+04**	6.7201E+04	4.7925E+04	6.3712E+04	1.5679E+04
	vio	0.0000E+00	**0.0000E+00**	0.0000E+00	0.0000E+00	0.0000E+00	0.0000E+00
	sr	100%	**100%**	100%	100%	100%	100%
±			7/5	12/0	12/0	12/0	8/4

εCCPSOd, εCCPSOdg2, and DELoCoSDG2 in 12, 12, and 8 functions, respectively. Particularly, COCC-ConBSO could find feasible solutions on function F1 for each independent run, and significantly improves its optimization on function F5, outperforming the comparison algorithms by at least two orders of magnitude.

As for COCC-BSO20, one of two algorithms based on the COCC framework, the results in Table 1 demonstrate that COCC-ConBSO outperforms COCC-BSO20 significantly and wins on all functions. The experiments show that the proposed ConBSO excels in handling constrained optimization functions. By assigning distinct tasks to clusters in intra-cluster mutation and guiding inferior clusters to learn from superior clusters in inter-cluster mutation, the cooperative multi-cluster search feature could be effectively exploited. Thus, the results are better than the combination of BSO20 and ε-constraints.

As for COCC-DE, another algorithm based on the COCC framework, the comparison results with COCC-DE show that COCC-ConBSO achieves a lead in 7 functions. In particular, COCC-ConBSO has better performance in finding feasible solutions. For example, the success rate on functions F1 and F3 are 100% and 88%, respectively. The mean of constraint violation on F4, F9, and F11 is also lower than COCC-DE. Although COCC-ConBSO does not perform as well as COCC-DE on the five tested functions, the constraint violation is higher than COCC-DE only on function F10. The experimental results demonstrate the greater ability of COCC-ConBSO to find feasible solutions.

Finally, the comparison results with the three large-scale constrained optimization algorithms show that the proposed COCC-ConBSO also has a clear advantage over εCCPSOd and εCCPSOdg2 for all functions, and over DELoCoS-DG2 for 8 functions. it is worth noting that these compared algorithms use the popular PSO and DE as optimizers. the win of COCC-ConBSO shows that in large-scale constrained optimization problems, BSO is expected to be one of the competitive optimizers.

5   Conclusion

In this paper, COCC-ConBSO has been proposed for solving large-scale constrained optimization problems. First, we have proposed a variant of BSO for solving constrained optimization, ConBSO, which is characterized by assigning different search tasks to each cluster to explore the feasible or optimal domains, in the intra-cluster mutation, and learning from inferior clusters to superior clusters in the inter-cluster mutation to guide the population to converge to the feasible optimal region. Then, we have integrated ConBSO into the COCC framework. COCC-ConBSO utilizes the feature that COCC selects a subproblem to be optimized from a set of objective or constraint subproblems to determine the behavior of different clusters in ConBSO. Finally, we have tested the proposed COCC-ConBSO on a large-scale constrained optimization benchmark. Experimental results have demonstrated that the proposed algorithm is promising and competitive.

In future work, we will further improve the performance of ConBSO in terms of both clustering strategy and mutation strategy. In addition, we will try to apply COCC-ConBSO to solve more complex large-scale constrained optimization problems, such as the problem with heterogeneous coupled modules mentioned in Ref. [32].

References

1. Aguilar-Justo, A.E., Mezura-Montes, E.: Towards an improvement of variable interaction identification for large-scale constrained problems. In: Proceedings of the 2016 IEEE Congress on Evolutionary Computation (CEC), pp. 4167–4174. IEEE (2016)
2. Aguilar-Justo, A.E., Mezura-Montes, E.: A local cooperative approach to solve large-scale constrained optimization problems. Swarm Evol. Comput. **51**, 100577 (2019)
3. Blanchard, J., Beauthier, C., Carletti, T.: A cooperative co-evolutionary algorithm for solving large-scale constrained problems with interaction detection. In: Proceedings of the Genetic and Evolutionary Computation Conference, pp. 697–704 (2017)
4. Cao, Z., Shi, Y., Rong, X., Liu, B., Du, Z., Yang, B.: Random grouping brain storm optimization algorithm with a new dynamically changing step size. In: Tan, Y., Shi, Y., Buarque, F., Gelbukh, A., Das, S., Engelbrecht, A. (eds.) ICSI 2015. LNCS, vol. 9140, pp. 357–364. Springer, Cham (2015). https://doi.org/10.1007/978-3-319-20466-6_38
5. Cervantes-Castillo, A., Mezura-Montes, E.: A modified brain storm optimization algorithm with a special operator to solve constrained optimization problems. Appl. Intell. **50**(12), 4145–4161 (2020). https://doi.org/10.1007/s10489-020-01763-8
6. Chakraborty, A., Kar, A.K.: Swarm intelligence: a review of algorithms. In: Patnaik, S., Yang, X.-S., Nakamatsu, K. (eds.) Nature-Inspired Computing and Optimization. MOST, vol. 10, pp. 475–494. Springer, Cham (2017). https://doi.org/10.1007/978-3-319-50920-4_19
7. Deb, K.: An efficient constraint handling method for genetic algorithms. Comput. Methods Appl. Mech. Eng. **186**(2–4), 311–338 (2000)
8. Duan, H., Li, S., Shi, Y.: Predator-prey brain storm optimization for DC brushless motor. IEEE Trans. Magn. **49**(10), 5336–5340 (2013)
9. El-Abd, M.: Cooperative coevolution using the brain storm optimization algorithm. In: Proceedings of the 2016 IEEE Symposium Series on Computational Intelligence (SSCI), pp. 1–7. IEEE (2016)
10. Ha, T.H., Lee, K., Hwang, J.T.: Large-scale design-economics optimization of eVTOL concepts for urban air mobility. In: AIAA Scitech 2019 Forum, p. 1218 (2019)
11. He, C., Cheng, R., Tian, Y., Zhang, X., Tan, K.C., Jin, Y.: Paired offspring generation for constrained large-scale multiobjective optimization. IEEE Trans. Evol. Comput. **25**(3), 448–462 (2021)
12. Hwang, J.T., Jain, A.V., Ha, T.H.: Large-scale multidisciplinary design optimization-review and recommendations. In: AIAA Aviation 2019 Forum, p. 3106 (2019)
13. Hwang, J.T., Lee, D.Y., Cutler, J.W., Martins, J.R.R.A.: Large-scale multidisciplinary optimization of a small satellite's design and operation. J. Spacecr. Rocket. **51**(5), 1648–1663 (2014)
14. Kumar, A., Wu, G., Ali, M.Z., Mallipeddi, R., Suganthan, P.N., Das, S.: A testsuite of non-convex constrained optimization problems from the real-world and some baseline results. Swarm Evol. Comput. **56**, 100693 (2020)
15. Li, X., Yao, X.: Cooperatively coevolving particle swarms for large scale optimization. IEEE Trans. Evol. Comput. **16**(2), 210–224 (2011)

16. Lin, X., Luo, W., Xu, P.: Differential evolution for multimodal optimization with species by nearest-better clustering. IEEE Trans. Cybern. **51**(2), 970–983 (2021)
17. Mishra, S., Kumar, A., Singh, D., Misra, R.K.: Butterfly optimizer for placement and sizing of distributed generation for feeder phase balancing. In: Verma, N., Ghosh, A. (eds.) Computational Intelligence: Theories, Applications and Future Directions, pp. 519–530. Springer, Singapore (2019). https://doi.org/10.1007/978-981-13-1135-2_39
18. Omidvar, M.N., Li, X., Yao, X.: A review of population-based metaheuristics for large-scale black-box global optimization-part II. IEEE Trans. Evol. Comput. **26**(5), 823–843 (2021)
19. Peng, C., Hui, Q.: Comparison of differential grouping and random grouping methods on sCCPSO for large-scale constrained optimization. In: Proceedings of the 2016 IEEE Congress on Evolutionary Computation (CEC), pp. 2057–2063. IEEE (2016)
20. Peng, C., Hui, Q.: Epsilon-constrained CCPSO with different improvement detection techniques for large-scale constrained optimization. In: Proceedings of the 2016 49th Hawaii International Conference on System Sciences (HICSS), pp. 1711–1718. IEEE (2016)
21. Potter, M.A., De Jong, K.A.: A cooperative coevolutionary approach to function optimization. In: Davidor, Y., Schwefel, H.-P., Männer, R. (eds.) PPSN 1994. LNCS, vol. 866, pp. 249–257. Springer, Heidelberg (1994). https://doi.org/10.1007/3-540-58484-6_269
22. Preuss, M.: Niching the CMA-ES via nearest-better clustering. In: Proceedings of the 12th Annual Conference Companion on Genetic and Evolutionary Computation, pp. 1711–1718 (2010)
23. Regis, R.G.: Evolutionary programming for high-dimensional constrained expensive black-box optimization using radial basis functions. IEEE Trans. Evol. Comput. **18**(3), 326–347 (2014)
24. Runarsson, T., Yao, X.: Stochastic ranking for constrained evolutionary optimization. IEEE Trans. Evol. Comput. **4**(3), 284–294 (2000)
25. Sayed, E., Essam, D., Sarker, R., Elsayed, S.: Decomposition-based evolutionary algorithm for large scale constrained problems. Inf. Sci. **316**, 457–486 (2015)
26. Shi, Y.: Brain storm optimization algorithm. In: Tan, Y., Shi, Y., Chai, Y., Wang, G. (eds.) ICSI 2011. LNCS, vol. 6728, pp. 303–309. Springer, Heidelberg (2011). https://doi.org/10.1007/978-3-642-21515-5_36
27. Sun, Y., Kirley, M., Halgamuge, S.K.: A recursive decomposition method for large scale continuous optimization. IEEE Trans. Evol. Comput. **22**(5), 647–661 (2018)
28. Takahama, T., Sakai, S.: Constrained optimization by the ε constrained differential evolution with gradient-based mutation and feasible elites. In: Proceedings of the 2006 IEEE International Conference on Evolutionary Computation (CEC), pp. 1–8. IEEE (2006)
29. Xu, P., Luo, W., Lin, X., Chang, Y., Tang, K.: Difficulty and contribution based cooperative coevolution for large-scale optimization. IEEE Trans. Evol. Comput. (2022)
30. Xu, P., Luo, W., Lin, X., Cheng, S., Shi, Y.: BSO20: efficient brain storm optimization for real-parameter numerical optimization. Complex Intell. Syst. **7**(5), 2415–2436 (2021). https://doi.org/10.1007/s40747-021-00404-y
31. Xu, P., Luo, W., Lin, X., Zhang, J., Qiao, Y., Wang, X.: Constraint-objective cooperative coevolution for large-scale constrained optimization. ACM Trans. Evol. Learn. Optim. **1**(3), 1–26 (2021)

32. Xu, P., Luo, W., Lin, X., Zhang, J., Wang, X.: A large-scale continuous optimization benchmark suite with versatile coupled heterogeneous modules. Swarm Evol. Comput. **78**, 101280 (2023)
33. Zhan, Z.H., Zhang, J., Shi, Y.H., Liu, H.L.: A modified brain storm optimization. In: Proceedings of the 2012 IEEE Congress on Evolutionary Computation (CEC), pp. 1–8. IEEE (2012)

Multi-objective Optimization

On the Privacy Issue of Evolutionary Biparty Multiobjective Optimization

Zeneng She[1], Wenjian Luo[1,2(✉)], Yatong Chang[1], Zhen Song[1], and Yuhui Shi[3]

[1] Guangdong Provincial Key Laboratory of Novel Intelligence Technologies, School of Computer Science and Technology, Harbin Institute of Technology, Shenzhen 518055, Guangdong, China
{20s151103,20s151150,21s151097}@stu.hit.edu.cn
[2] Peng Cheng Laboratory, Shenzhen 518055, Guangdong, China
luowenjian@hit.edu.cn
[3] Department of Computer Science and Engineering, Southern University of Science and Technology, Shenzhen 518055, China
shiyh@sustech.edu.cn

Abstract. Some evolutionary algorithms have been proposed to address biparty multiobjective optimization problems (BPMOPs). However, all these algorithms are centralized algorithms which directly obtain the privacy information including objective functions from decision makers (DMs). This paper transforms the centralized algorithm OptMPNDS2 into a distributed framework for BPMOPs and focuses on the privacy issue in the framework. The framework has a server and two clients, and each client belongs to a DM. The clients keep their objective functions locally, evaluate individuals, and upload Pareto levels and crowding distances of all individuals to the server. The server performs the other operations including reproduction and selection of offspring. Experimental results show that the performance of the framework is very close to OptMPNDS2. Besides, two privacy attacks are proposed when one client is malicious. Experimental results show that the client could steal approximate Pareto optimal solutions of the other honest DM.

Keywords: Multiobjective optimization · Biparty multiobjective optimization · Evolutionary computation · Privacy

1 Introduction

In real-world applications, there are a lot of decision-making problems with multiple conflicting objectives, which are called multiobjective optimization problems (MOPs) [2,4,17]. MOPs widely exist in different fields including computer

This study is supported by the National Key R&D Program of China (Grant No. 2022YFB3102100), Shenzhen Fundamental Research Program (Grant No. JCYJ20220818102414030), the Major Key Project of PCL (Grant No. PCL2022A03, PCL2021A02, PCL2021A09), Guangdong Provincial Key Laboratory of Novel Security Intelligence Technologies (Grant No. 2022B1212010005).

Y. Tan et al. (Eds.): ICSI 2023, LNCS 13968, pp. 371–382, 2023.
https://doi.org/10.1007/978-3-031-36622-2_30

science [9,10], engineering [7,12], economics [15], etc. There are a special class of MOPs involving multiple decision makers (DMs), which are defined as multiparty multiobjective optimization problems (MPMOPs) [8] or group decision making for multiobjective problems (GDM-MOP) [19]. In an MPMOP, each DM focuses on different conflicting objectives. MPMOPs with two decision makers are called biparty multiobjective optimization problems (BPMOPs).

There is some work about designing centralized algorithms to obtain the Pareto optimal solutions of MPMOPs, and most algorithms are adapted from multiobjective evolutionary algorithms. In 2020, Liu *et al.* proposed a multiparty multiobjective evolutionary algorithm called OptMPNDS [8], based on NSGA-II [3], to solve the MPMOPs with common Pareto optimal solutions which are Pareto optimal for all DMs. OptMPNDS takes advantage of the structure of common Pareto optimal solutions and assigns a high priority to solutions which have the same Pareto level in respective sorts of all decision makers. In 2021, She *et al.* [13] defined a new sorting among individuals under multiple parties, which is used to evaluate individuals in the algorithm OptMPNDS2. In 2022, Chang *et al.* [1] embedded the party-by-party strategy into MOEA/D [21] and proposed MOEA/D-MP algorithm to solve MPMOPs. In [16], Song *et al.* considered the multiparty multiobjective knapsack problems and proposed an algorithm called SPEA2-MP to solve the problems.

So far, the study about the MPMOP mainly concentrates on the design of centralized algorithms. However, as an emerging topic in the computational intelligence field, MPMOPs still have many aspects to be further studied and there is little work concerning the privacy issue. Even if in the whole evolutionary computation community, the privacy issue also have not been paid much attention [6,18,20,24]. Consequently, this paper studies the BPMOP, which is a class of MPMOPs, from the viewpoint of privacy issue. Existing centralized algorithms need to directly obtain some sensitive data of DMs, including objective functions and individuals' objective values. To protect the privacy of DMs, in this paper, a distributed framework is constructed to perform a variant of the algorithm OptMPNDS2 in [13]. The framework uses Client/Server distributed mode. The server controls the optimization process including reproduction and selection of offspring. Two clients keep their own objective functions and evaluate individuals locally, while the evaluation fitness results including Pareto levels and crowding distances of all individuals are sent to the server to complete the performing of the evolution. Thus, in the proposed distributed framework, it is unnecessary to pass the objective functions of DMs to the server.

Even though, this framework is under the risk of privacy leakage when DMs are malicious. This paper proposes two privacy attacks when one DM is malicious to steal the privacy of the other DM in the framework. The two attacks, i.e., the lazy attack and the noise-based attack, show that the malicious client could send the counterfeit information to the server and guide the optimization direction to the Pareto optimal solutions of the honest client. And then the malicious client could obtain the Pareto optimal solutions from the server.

Overall, the contributions of this paper are described as follows.

– This paper proposes a distributed framework with a server and two clients for BPMOPs based on OptMPNDS2. The server controls the evolution of the population, and the clients of two decision makers evaluate the individuals locally, respectively.
– Privacy issue is further explored in the above framework, where one of decision makers is honest and the other is malicious. From the viewpoint of the malicious decision maker, this paper designs two methods to steal the Pareto optimal solutions of the honest decision maker, i.e., the lazy attack and the noise-based attack. Experimental results show that these two attacks could steal the privacy data by misleading the direction of the optimization.

The rest of this paper is organized as follows. First, we introduce some related concepts of the biparty multiobjective optimization problems and algorithms in Sect. 2. Next, we describe a distributed evolutionary framework in Sect. 3. Then, Sect. 4 briefly discusses the privacy issue and Sect. 5 gives two attack methods. Finally, Sect. 6 draws a brief conclusion.

2   Related Work

2.1   Multiparty Multiobjective Optimization Problems

A multiparty multiobjective optimization problems (MPMOP) is a multiobjective optimization problems (MOPs) with multiple parties, where each party stands for a decision maker (DM) and each DM cares different objectives. The related concepts about MPMOPs are based on MOPs but different from MOPs. For convenience, an MOP is defined as a minimization problem [5]:

$$Minimize \quad F(x) = (f_1(x), f_2(x), \ldots, f_m(x)),$$
$$Subject \quad to \begin{cases} h_i(x) = 0, & i = 1, \ldots, n_p, \\ g_j(x) \leq 0, & j = 1, \ldots, n_q, \\ x \in [x_{min}, x_{max}], \end{cases} \tag{1}$$

where F denotes the m-dimensional vector of objective functions which should be minimized, in which f_i denotes the i-th objective function. $h_i(x) = 0$ represents the i-th equality constraint whose number is n_p; $g_j(x) \leq 0$ represents the j-th inequality constraint whose number is n_q. x stands for a decision vector, whose lower bound is x_{min} and upper bound is x_{max}.

Dominance is a partial relation in the decision space [25]. Given two decision vectors x and y, it is said that x Pareto dominates y, denoting as $x \prec y$ [11], if and only if $f_i(x) \leq f_i(y), \forall i \in \{1, \cdots, m\}$ and $f_j(x) < f_j(y), \exists j \in \{1, \cdots, m\}$. A decision vector x is defined to be a Pareto optimal solution if and only if no solution dominates x. Pareto optimal set (PS) is a set of all the Pareto optimal solutions. Pareto optimal front (PF) is a set of corresponding objective vectors of PS, which is formally defined as PF $= \{(f_1(x), f_2(x), \cdots, f_n(x)) | x \in$ PS$\}$.

There are some parallel concepts of MPMOPs differing from MOPs, because an MPMOP involves multiple DMs, where each DM focuses on different objectives and at least one DM has at least two objectives. In Formula (1), $f_i(x)$ denotes one objective of the solution x in an MOP. But, MPMOPs make $f_i(x)$ be a vector function, which represents all objectives of the i-th party. Specifically, $f_i(x) = (f_{i1}(x), f_{i2}(x), \ldots, f_{ij_i}(x))$, and j_i denotes the number of objectives of the i-th party. The goal of solving an MPMOP is to find the solutions close to the PFs of all parties as much as possible. When involving two decision makers, the MPMOP is called the biparty multiobjective optimization problem (BPMOP).

2.2 OptMPNDS2

OptMPNDS2 by She *et al.* [13] was proposed to solve MPMOPs, which is an improved version of OptMPNDS [8] and the latter is modified from the classic multiobjective evolutionary algorithm NSGA-II. Similar to NSGA-II [3], OptMPNDS2 first initializes and evaluates the population, then repeats reproducing offspring and selection next generation until the condition is satisfied. Particularly, it uses the multiparty Pareto dominance to substitute the Pareto dominance to evaluate individuals in the process of selection.

In each generation of OptMPNDS2, the process of evaluating individuals is shown as follows. To obtain Pareto levels of individuals in each party, it applies the non-dominated sorting to each party, respectively. After obtaining multiple Pareto levels of all parties, OptMPNDS2 takes all the levels of each individual as its "objectives" to perform non-dominated sorting again.

3 The Distributed Framework

3.1 Framework

OptMPNDS2 directly gets objective functions of DMs and performs the evolutionary process to obtain the final solutions [13]. To protect the private objective functions, the natural idea is to build a distributed system, where a server controls the evolution process, and objectives are calculated in the local devices which DMs control. It is sufficient that DMs calculate the Pareto levels of individuals in local devices and then pass the Pareto levels to the server to direct the optimization. Thus, it is unnecessary to let the server know the detailed objective functions. Therefore, we construct a distributed optimization framework for biparty multiobjective optimization problems as in Fig. 1.

The framework involves two decision makers. In this framework, a central server is set up to control the evolutionary process, and two clients are used for two parties to evaluate individuals, respectively. The central server receives the evaluated values of the individuals from two clients, including the Pareto levels and crowding distances, and then sorts the individuals. So the central server do not know the specific objective values of individuals at both DMs. Similarly, the decision makers could not know objective functions of each other since the

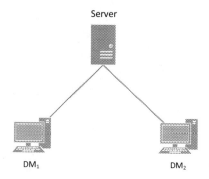

Fig. 1. The architecture of the framework

clients only receive the decision vectors of all individuals from central server and send back both the Pareto levels and the crowding distances.

This framework is a distributed version of OptMPNDS2 [13]. The main procedure of the server in the framework is depicted in Algorithm 1, and explained as follows.

(1) A population P_0 whose size is N is initialized randomly.
(2) Generation number t is initialized as 0. Offspring Q_t is set as $\emptyset$.
(3) The loop from steps 3 to 19 shows the process of one generation. Current population P_t and its offspring Q_t are gathered into R_t. The server obtains the Pareto levels L and global crowding distance D of individuals from two clients in turn. After that, the server uses the non-dominated sorting in NSGA-II to sort the individuals into different ranks $\mathcal{F}$, where Pareto levels L are viewed as objectives. Then, the crowding distance D is used to sort individuals in the same rank. Subsequently, the next generation population is obtained from the sorted $\mathcal{F}$. Specifically, generation number t is updated and N individuals are selected from the $\mathcal{F}$ by order as the next generation P_t. Next, the offspring Q_t is generated from P_t by both crossover and mutation operators, and the loop is repeated until the terminated condition is satisfied.
(4) Return the final biparty optimal solutions BOS.

The sorting strategy is the same as OptMPNDS2, where the Pareto levels of each individual from both DMs are assigned to the individual as objectives and then individuals are sorted according the new objectives. Differently, the server runs the sorting after receiving the Pareto levels and crowding distances from clients. For decreasing the communication cost, the crowding distance at each party is calculated in the whole population instead of the same rank.

The main procedure of each client is shown as Algorithm 2, which indirectly provides the fitness information to the central server. First, the client receives decision vectors from the server and evaluate the individuals. Then, it applies NonDominatedSorting($\cdot$) to sort individuals into different Pareto levels and calculates the modified crowding distances according to the objectives. Finally, the client sends the Pareto levels and crowding distance of each individual to the

---

Algorithm 1. Server

Require: N (the population size)
Ensure: BOS (the biparty optimal solutions)
1: Initialize population P_0 with size N
2: $t = 0, Q_t = \emptyset, R_t = \emptyset$
3: **while** Maximum fitness evaluations FE is not reached **do**
4:     $R_t = P_t \cup Q_t$
5:     $\mathcal{L} = \emptyset, \mathcal{D} = \emptyset$
6:     **for** $i \in \{1, 2\}$ **do**
7:         Receive Pareto levels from the i-th client and store in $\mathcal{L}(:, i)$
8:         Receive crowding distances from the i-th client and store in cd_i
9:         $\mathcal{D} = \mathcal{D} + cd_i$
10:     **end for**
11:     $\mathcal{F} = \text{NonDominatedSorting}(R_t, \mathcal{L})$
12:     Sort $\mathcal{F}$ by crowding distance $\mathcal{D}$ for each rank
13:     $t = t + 1, i = 1, P_t = \emptyset$
14:     **while** $|P_t \cup \mathcal{F}_i| \leq N$ **do**
15:         $P_t = P_t \cup \mathcal{F}_i$
16:         $i = i + 1$
17:     **end while**
18:     $P_t = P_t \cup \mathcal{F}_i(1 : N - |P_t|)$
19:     Generate the offspring Q_t of P_t
20: **end while**
21: $BOS \leftarrow$ biparty optimal solutions in P_t

Algorithm 2. The Client of DM_i

Require: R_t (individuals), $f_i(x)$ (the objective functions of DM_i)
Ensure: L (the Pareto optimal levels), cd (crowding distances)
1: Receive R_t from server
2: $L = \text{NonDominatedSorting}(R_t, f_i(x))$
3: Calculate crowding distance cd
4: Send L and cd to the server

---

server. Here, NonDominatedSorting($\cdot$) is a function using the Pareto dominance relationship to sort all the individuals in the population, which is the same as that in [23].

3.2   Experiments

This paper chooses a set of biparty distance minimization problems to test the performance of the distributed framework. The test problems are from MPDMP1 to MPDMP8 [14], which involve two parties. All these algorithms are run 30 times independently on these problems to compare the average results. For each run, the population size is set to 200; FE (the maximum fitness evaluations) is set to 80000. The source code of this framework is available at https://github.com/MiLabHITSZ/2023SheFramework.

The inverted generational distance (IGD) [22] is a comprehensive performance metric, which evaluates both convergence and distribution performance of the solutions. The smaller the value is, the better the performance of the algorithm is. IGD is calculated as follows:

$$IGD(P^*, P) = \frac{\sum_{v \in P^*} d(v, P)}{|P^*|}, \tag{2}$$

where P^* represents the true Pareto optimal solutions and P represents the final solutions that algorithms obtained. $|P^*|$ is the size of the set P^*. $d(v, P)$ is the minimum distance from the solution v to the set P, which is the distance from v to the nearest solution in set P. To measure solutions of MPMOPs/BPMOPs, the slight modification of $d(v, P)$ is used, which is defined as follows [8].

$$d(v, P) = \min_{s \in P} \sum_{i=1}^{M} \sqrt{(v_{i1} - s_{i1})^2 + \cdots + (v_{ij_i} - s_{ij_i})^2}, \tag{3}$$

where M represents the number of parties and is equal to 2 in these test problems; v_{ij} and s_{ij} denote the j-th objective of the i-th party, which belong to solutions v and s respectively; j_i means the number of objectives of the i-th party. For convenience, the modified IGD is called MPIGD.

Table 1. MPIGD values of the distributed framework and OptMPNDS2

Problem	Distributed framework	OptMPNDS2	ΔMPIGD
MPDMP1	1.1506e+00(6.9022e-01)	**5.4518e-02(2.9469e-02)**	1.0961e+00
MPDMP2	1.3624e+00(6.4256e-01)	**6.4854e-02(3.4374e-02)**	1.2975e+00
MPDMP3	2.6536e-01(5.2710e-02)	**2.1066e-01(4.7955e-02)**	5.4700e-02
MPDMP4	2.3734e+00(1.3832e+00)	**9.5788e-04(2.1092e-03)**	2.3724e+00
MPDMP5	4.5086e+00(2.9318e-01)	**2.9676e+00(9.1629e-02)**	1.5410e+00
MPDMP6	1.6495e+00(2.4139e-01)	**1.3157e-01(8.9143e-03)**	1.5179e+00
MPDMP7	4.5361e+00(3.6616e-01)	**2.4542e+00(1.3254e-01)**	2.0819e+00
MPDMP8	3.8490e+00(2.7221e-01)	**2.4560e+00(6.5224e-02)**	1.3930e+00

Experimental results are given in Table 1. The best result on each problem is marked in bold font, and ΔMPIGD is the difference between the MPIGD value that distributed framework obtained and the result that OptMPNDS2 obtained. The MPIGD values of the framework are not so good as the OptMPNDS2 since the crowding distance is not calculated in the same Pareto level but in the whole population. However, the ΔMPIGD values are relatively small, especial for MPDMP3.

4   Privacy Problem

In a BPMOP, each DM focuses on an MOP, where the objectives of the MOP is composed of the DM's objectives. In the above framework, the MOP of each DM is considered as the private information. Also, the Pareto Front and Pareto Set of the MOP of each DM are sensitive data. The framework does not require clients send their original information including objective functions and objective values of solutions. In some extent, it protects the privacy of decision makers because the client calculates the objectives locally.

However, the framework works under the prerequisite that both parties are honest and transfer the correct evaluation information. The malicious client could modify the Pareto levels of individuals, which would disturb the optimization. Furthermore, the malicious client may steal some sensitive information from the disturbed optimization result.

To correctly perform the optimization process and detect the malicious party, the server could check whether the information from clients is consistent and correct or not. For example, the server could check whether the dominance relations between two individuals in different generations are consistent or not.

5   Privacy Attacks

Suppose one DM in the distributed framework is malicious, and the other is honest. Since the ranks of individuals in the population is jointly determined by two DMs, the malicious client can control the optimization direction by transferring counterfeit to the server.

Theoretically, if the malicious client assigns the same Pareto level to all individuals in the population, the biparty sorting result is completely determined by the honest client. Thus, the final result is the same as the result of the multi-objective optimization problem of the honest DM. Finally, the server sends the result to each client, then the malicious client can obtain all the Pareto optimal solutions of the honest DM.

However, if the malicious client always set the same Pareto levels for all individuals, the server could detect such abnormal behaviors easily. Thus, the malicious party could set abnormal Pareto levels for some generations, but not all the time. According to this principle, in this section, two attacks are designed to steal the approximate Pareto solutions of the honest DM in this section.

5.1   Lazy Attack

All individuals are assigned 1 as their Pareto levels, which means that the evolution process is about to converge. Therefore, the malicious client can set the Pareto levels of all individuals to 1 to disguise that the population has converged. Specifically, the malicious client can correctly evaluate individuals for some early generations and then assign 1 to all individuals as their Pareto levels in later generations.

Algorithm 3. Lazy Attack

Require: R_t (individuals)
Ensure: L (the Pareto optimal levels), cd (crowding distances)
 1: $count \leftarrow 0$
 2: Receive R_t from server
 3: **if** $count < v$ **then**
 4:     $L = \text{NonDominatedSorting}(R_t, f_1(x))$
 5: **else**
 6:     Set every dimension of L to 1
 7: **end if**
 8: Calculate crowding distances cd
 9: Send L and cd to the server
10: $count \leftarrow count + 1$

As shown in Algorithm 3, the client records generation number $count$ and allocates all individuals in the first Pareto level when $count$ is greater than v (set to 100 in experiments). Here, we suppose DM_1 is the malicious party, and the corresponding MOP is $f_1(x)$. As we can see, what the malicious party need to do in this attack is very simple, so it is called the lazy attack.

5.2 Noise-Based Attack

In order to avoid continually sending the same evaluation results of all individuals, the malicious client could add some noise using the uniform distribution in some generations. The pseudocode of such a noise-based attack is shown in Algorithm 4, where randi($\cdot$) selects N numbers from a uniform distribution $U(1, N)$.

Algorithm 4. Noise-based Attack

Require: R_t (individuals)
Ensure: L (the Pareto optimal levels), cd (crowding distance)
 1: Receive the R_t from server
 2: **if** $rand < 0.8$ **then**
 3:     Set every dimension of L to 1
 4: **else**
 5:     $N = $ size of R_t
 6:     $L = \text{randi}(N, 1, N)$
 7: **end if**
 8: Calculate crowding distance cd
 9: Send L and cd to the server

5.3 Experiments

The test problems, namely biparty distance minimization problems, are the same with those in the Sect. 3. The parameters of the framework which is used to test the two attacks are the same as the normal framework. The compared algorithm is NSGA-II [3], which is used to optimize the objectives of the honest DM. The maximum fitness evaluation of NSGA-II is set to 40000, and other parameters are the same as the distributed framework. Two attacks and NSGA-II are independently run 30 times to compare their average results. The best result among these algorithms in each problem is marked in bold font.

The IGD metric is used to evaluate the performances of both two attacks. To measure the solutions that the malicious DM steals, P^* of Eq. (2) is set as the true PS of the honest DM. If the IGD value obtained by the attack is better than or the same as that of NSGA-II, the attack is successful.

Table 2. IGD values of two attacks and NSGA-II

Problem	Lazy attack	Noise-based attack	NSGA-II	ΔL	ΔN
MPDMP1	1.6395e-01(5.3206e-03)	4.8362e+00(9.3386e+00)	**1.4907e-01(3.2295e-03)**	1.4880E-02	4.6871E+00
MPDMP2	1.6541e-01(6.3810e-03)	5.0040e+00(9.7470e+00)	**1.4907e-01(3.2295e-03)**	1.6340E-02	4.8549E+00
MPDMP3	1.0201e-01(3.6073e-03)	3.1738e+00(6.3089e+00)	**9.6172e-02(3.5441e-03)**	5.8380E-03	3.0776E+00
MPDMP4	1.3487e+00(5.2941e-02)	4.7466e+00(6.4886e+00)	**1.3146e+00(4.7676e-02)**	3.4100E-02	3.4320E+00
MPDMP5	2.5936e+00(9.9380e-02)	8.9122e+00(1.2381e+01)	**2.4123e+00(9.2714e-02)**	1.8130E-01	6.4999E+00
MPDMP6	**1.5162e+00(5.5556e-02)**	5.1648e+00(7.4129e+00)	1.6451e+00(7.3060e-02)	-1.2890E-01	3.5197E+00
MPDMP7	**2.5194e+00(7.3291e-02)**	9.0968e+00(1.2958e+01)	2.7997e+00(1.5958e-01)	-2.8030E-01	6.2971E+00
MPDMP8	**2.0470e+00(6.4300e-02)**	6.5928e+00(8.5036e+00)	2.2416e+00(8.7335e-02)	-1.9460E-01	4.3512E+00

Experimental results are shown in Table 2, where ΔL and ΔN are the differences between the results by two attacks and the results by NSGA-II, respectively. Since the smaller IGD value means the better performance of an algorithm, the smaller ΔL and ΔN also mean the better performance of the attacks. The IGD values obtained by both two attacks are close to the results obtained by NSGA-II at most cases, especially the lazy attack.

The lazy attack performs quite good in MPDMP6, MPDMP7 and MPDMP8. In these three problems, the ΔL values of the lazy attack are negative. This means that the solutions obtained by the lazy attack are even better than the solutions that NSGA-II obtained. The ΔL values of the lazy attack are generally smaller than the ΔN values of the noise-based attack. In conclusion, both methods could steal some approximate Pareto optimal solutions of the honest DM, while the performance of the lazy attack is better.

6   Conclusion

In this paper, a distributed framework is proposed to optimize the BPMOPs and the privacy issue is pointed out in the framework which involve a malicious DM and an honest DM. Taking biparty distance minimization problems involving two DMs as examples, this paper illustrates that the malicious client could steal approximate optimal solutions of the honest client. In the future, we will construct a more efficient framework to perform the evolutionary procedure of solving BPMOPs, and test the privacy issues of these frameworks. We will also study on the privacy protection techniques, and propose effective and efficient defense strategies.

References

1. Chang, Y., Luo, W., Lin, X., She, Z., Shi, Y.: Multiparty multiobjective optimization by MOEA/D. In: Proceedings of the 2022 IEEE Congress on Evolutionary Computation (CEC), pp. 1–8. IEEE (2022)
2. Coello, C.C.: Evolutionary multi-objective optimization: a historical view of the field. IEEE Comput. Intell. Mag. $1(1)$, 28–36 (2006)
3. Deb, K., Pratap, A., Agarwal, S., Meyarivan, T.: A fast and elitist genetic algorithm: NSGA-II. IEEE Trans. Evol. Comput. $6(2)$, 182–197 (2002)
4. Fonseca, C.M., Fleming, P.J.: An overview of evolutionary algorithms in multiobjective optimization. Evol. Comput. $3(1)$, 1–16 (1995)
5. Geng, H., Zhang, M., Huang, L., Wang, X.: Infeasible elitists and stochastic ranking selection in constrained evolutionary multi-objective optimization. In: Wang, T.-D., et al. (eds.) SEAL 2006. LNCS, vol. 4247, pp. 336–344. Springer, Heidelberg (2006). https://doi.org/10.1007/11903697_43
6. Kong, L., Luo, W., Zhang, H., Liu, Y., Shi, Y.: Evolutionary multi-label adversarial examples: an effective black-box attack. IEEE Trans. Artif. Intell. (2022)
7. Lin, Y.H., Lin, M.D., Tsai, K.T., Deng, M.J., Ishii, H.: Multi-objective optimization design of green building envelopes and air conditioning systems for energy conservation and CO_2 emission reduction. Sustain. Urban Areas **64**, 102555 (2021)
8. Liu, W., Luo, W., Lin, X., Li, M., Yang, S.: Evolutionary approach to multiparty multiobjective optimization problems with common pareto optimal solutions. In: Proceedings of the 2020 IEEE Congress on Evolutionary Computation (CEC), pp. 1–9. IEEE (2020)
9. Lokuciejewski, P., Plazar, S., Falk, H., Marwedel, P., Thiele, L.: Multi-objective exploration of compiler optimizations for real-time systems. In: Proceedings of the 13th IEEE International Symposium on Object/Component/Service-Oriented Real-Time Distributed Computing, pp. 115–122 (2010)
10. Martínez-Álvarez, A., Cuenca-Asensi, S., Ortiz, A., Calvo-Zaragoza, J., Tejuelo, L.A.V.: Tuning compilations by multi-objective optimization: application to apache web server. Appl. Soft Comput. **29**, 461–470 (2015)
11. Miettinen, K.: Nonlinear Multiobjective Optimization, vol. 12. Springer, New York (2012). https://doi.org/10.1007/978-1-4615-5563-6
12. Rosso, F., Ciancio, V., Dell'Olmo, J., Salata, F.: Multi-objective optimization of building retrofit in the mediterranean climate by means of genetic algorithm application. Energy Build. **216**, 109945 (2020)
13. She, Z., Luo, W., Chang, Y., Lin, X., Tan, Y.: A new evolutionary approach to multiparty multiobjective optimization. In: Tan, Y., Shi, Y. (eds.) ICSI 2021. LNCS, vol. 12690, pp. 58–69. Springer, Cham (2021). https://doi.org/10.1007/978-3-030-78811-7_6
14. She, Z., Luo, W., Lin, X., Chang, Y., Shi, Y.: Evolutionary multiparty distance minimization. arXiv preprint arXiv:2207.13390 (2022)
15. Shi, W., Chen, W.N., Lin, Y., Gu, T., Kwong, S., Zhang, J.: An adaptive estimation of distribution algorithm for multipolicy insurance investment planning. IEEE Trans. Evol. Comput. **23**(1), 1–14 (2017)
16. Song, Z., Luo, W., Lin, X., She, Z., Zhang, Q.: On multiobjective knapsack problems with multiple decision makers. In: Proceedings of the 2022 IEEE Symposium Series on Computational Intelligence (SSCI), pp. 156–163. IEEE (2022)
17. Tamaki, H., Kita, H., Kobayashi, S.: Multi-objective optimization by genetic algorithms: a review. In: Proceedings of the IEEE International Conference on Evolutionary Computation, pp. 517–522. IEEE (1996)

18. Wu, C., Luo, W., Zhou, N., Xu, P., Zhu, T.: Genetic algorithm with multiple fitness functions for generating adversarial examples. In: Proceedings of the 2021 IEEE Congress on Evolutionary Computation (CEC), pp. 1792–1799. IEEE (2021)
19. Xiong, J., Tan, X., Yang, K.W., Chen, Y.W.: Fuzzy group decision making for multiobjective problems: tradeoff between consensus and robustness. J. Appl. Math. **2013** (2013)
20. Zhan, Z.H., Wu, S.H., Zhang, J.: A new evolutionary computation framework for privacy-preserving optimization. In: Proceedings of the 2021 13th International Conference on Advanced Computational Intelligence (ICACI), pp. 220–226. IEEE (2021)
21. Zhang, Q., Li, H.: MOEA/D: a multiobjective evolutionary algorithm based on decomposition. IEEE Trans. Evol. Comput. **11**(6), 712–731 (2007)
22. Zhang, Q., Zhou, A., Zhao, S., Suganthan, P.N., Liu, W., Tiwari, S., et al.: Multiobjective optimization test instances for the CEC 2009 special session and competition. University of Essex, Colchester, UK and Nanyang Technological University, Singapore, special session on performance assessment of multi-objective optimization algorithms, technical report, vol. 264, pp. 1–30 (2008)
23. Zhang, X., Tian, Y., Cheng, R., Jin, Y.: An efficient approach to nondominated sorting for evolutionary multiobjective optimization. IEEE Trans. Evol. Comput. **19**(2), 201–213 (2014)
24. Zhao, B., et al.: PRIMPSO: a privacy-preserving multiagent particle swarm optimization algorithm. IEEE Trans. Cybern. (2022)
25. Zhou, A., Qu, B.Y., Li, H., Zhao, S.Z., Suganthan, P.N., Zhang, Q.: Multiobjective evolutionary algorithms: a survey of the state of the art. Swarm Evol. Comput. **1**(1), 32–49 (2011)

Dynamic Multi-modal Multi-objective Evolutionary Optimization Algorithm Based on Decomposition

Biao Xu[1], Yang Chen[1], Ke Li[1(✉)], Zhun Fan[1,2], Dunwei Gong[3], and Lin Bao[4]

[1] College of Engineering, Shantou University, Shoutou 515041, China
ericlee@stu.edu.cn

[2] School of Energy and Mechanical Engineering, Nanchang Institute of Technology, Nanchang 330013, China

[3] College of Information Science and Technology, Qingdao University of Science and Technology, Qingdao 266100, China

[4] School of Electronics and Information, Jiangsu University of Science and Technology, Zhenjiang 212100, China

Abstract. In this paper, the independent convergent and non-convergent decision variables are firstly obtained by analyzing the contribution of decision variables to the objective function based on the existing research results of multi-objective optimization algorithms. Secondly, according to their characteristics, the multi-population is employed, so that the population can search the corresponding multiple Pareto optimal solution set in each individual environment. Then, when the problem changes, two more targeted response strategies are proposed for different types of decision variables and their effects on the objective function. As the environment changes, the algorithm can ensure the rapid convergence of the population in the objective space, while maintaining the diversity of the population in the decision space and the objective space. Therefore, the proposed algorithm has the ability of quickly respond to the change of the problem and maintain the diversity of the solution set.

Keywords: dynamic optimization · multi-objective optimization · multi-modal optimization · decomposition

1 Introduction

Multi-Modal Multi-Objective Optimization (MMMO) is a kind of multi-objective optimization problem in which there are several different optimal solutions in the optimization problem [1]. These optimal solutions may correspond to different objectives or combinations of objectives, leading to the emergence of multi-modal multi-objective optimization problems. The following is a general form of multi-modal multi-objective optimization:

$$\begin{cases} \min f(x) = [f_1(x), f_2(x), \ldots, fm(x)] \\ \text{s.t.} \ g_i(x) \leqslant 0, i = 1, 2, \ldots, k \\ \quad hj(x) = 0, j = 1, 2, \ldots, p \end{cases} \tag{1}$$

where m is the number of objectives to be optimized, $x = (x1, x2, \ldots, x_n)$ is the n-dimensional decision variable, $g_i(x) \leqslant 0 (i = 1, 2, \ldots, k)$ is the k inequality constraint, and $h_j(x) = 0 (j = 1, 2, \ldots, p)$ is the p equality constraint. The space R^n that satisfies both the above equations and inequality constraints is called the decision space; the space R^m that maps the decision space through the objective function is called the objective space

There are more complex multi-objective optimization problems in production life, which often need to take into account multi-modality and dynamics, and are called dynamic multi-modal multi-objective optimization problems. Similar to dynamic multi-objective optimization problems, dynamic multi-modal multi-objective optimization problems also suffer from uncertainty and variability of the external environment. In order to adapt to the changes and uncertainties of the external environment, the dynamic multi-modal multi-objective optimization problem needs to consider the changes between multiple time steps and the interrelationships between multiple objective functions. Due to its inherent multi-modal nature, the dynamic multi-modal multi-objective optimization problem also needs to find multiple optimal solutions to cope with the changing demands. Solving dynamic multi-modal multi-objective optimization problems is a complex task that requires consideration of the interrelationships among multiple objective functions, changes among multiple time steps, and the possible existence of multiple optimal solutions.

In order to better solve the dynamic multi-modal multi-objective optimization problem, this paper proposes a dynamic multi-modal multi-objective evolutionary optimization method (DMMMOEA-CP&DM) based on convergent solution set prediction and distributed solution set preservation by considering dynamic changes based on the multi-modal multi-objective optimization problem solved in the literature [1]. The algorithm follows the two preservation sets designed in DMMMOEA-C&D [1], the convergence preservation set and the diversity preservation set. The structure diagram of the algorithm MMMOEA-C&D is shown in Fig. 1.

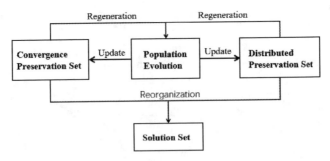

Fig. 1. The structure of the algorithm MMMOEA-C&D

2 Proposed Algorithm

2.1 Basic Framework

First, this paper improves the basis of MMMMOEA-C&D in the literature [1] by using the existing research results of multi-objective optimization algorithms to obtain independently convergent and non-convergent decision variables from the perspective of the contribution of decision variables to the objective function, and uses multiple populations to optimize them separately according to their respective characteristics, so that the populations can search for the corresponding multiple Pareto optimal in each individual environment solution set in each individual environment. Second, when changes in the problem are detected, two more targeted response strategies are proposed for different types of decision variables and their effects on the objective function. When the environment changes, the algorithm can ensure the rapid convergence of the population in the objective space while maintaining the diversity of the population in the decision space and the objective space to achieve the objectives of rapid response to the problem change and maintaining the diversity of the solution set, and the main algorithmic framework is as follows.

Algorithm 1.1: A framework for dynamic multi-modal multi-objective optimization algorithms

Input: Evolutionary populations P; Population size N; Convergence Preservation Set A_C^0; Diversity Preservation Set A_D^0; Maximum number of iterations per environment $MaxGen_;$.

Output: Final Solution Set FS

1: $t = 0$;

2: Performing decision variable analysis, independent convergent decision variables are obtained X_{IC};

3: Initializing the population P;

4: The convergence and diversity preservation sets $A_C(t)$ and $A_D(t)$; are updated, respectively;

5: **while** Termination conditions not met **do**

6:     Implement environmental change detection strategies;

7:     **if** Problem Change **then**

8:         $t = t + 1$;

9:         Prediction strategies for solutions corresponding to convergent decision variables;

10:        Distributional retention strategies for solutions corresponding to multiple and mixed decision variables;

11:        Update $A_C(t)$ and $A_D(t)$;

12:        **for** $g = 1$ to MaxGen **do**

13:            Perform selection, crossover, and mutation operations on individuals in $A_C(t)$ and $A_D(t)$;

14:            Update $A_C(t)$ and $A_D(t)$;

15:        **end for**

16:        The final solution $FS(t)$ under environment t is obtained by reorganizing the preservation sets $A_C(t)$ and $A_D(t)$;

17:    **end if**

18: **end while**

2.2 Detection of Problematic Changes

In this paper, we focus on one of the scenarios, i.e., the scenario where the multi-modal properties of the problem are assumed to be the same as those of the original problem after the change of environment, and the multi-modal problem remains the same with the same number of modes, and the change of the problem only causes a translation of the modes.

In view of this, the method proposed by Deb [6] is still used to detect whether the problem has changed. A certain number of individuals from the distributional preservation set A_D (t) are selected for re-evaluation, and the problem is considered to have changed whenever the fitness value of one of the objective functions or constraints has changed.

Considering that the solution goal of a multi-modal problem is to find the multiple Pareto optimal solution set of this type of problem, and a large number of multiple Pareto optimal solution sets are kept in the convergence preservation set, the objective function values of these multiple Pareto optimal solution sets may be duplicated. If individuals are selected from this solution set, it may cause double calculations and affect the accuracy of environmental change detection. A similar situation exists for the individuals in the final solution set. In contrast, the solutions in the diversity preservation set are mainly designed to increase the diversity of the Pareto optimal solution set with different fitness values. Therefore, only individuals from the distribution preservation set $A_D(t)$ are selected in the selection of individuals.

3   Experiments

3.1 Benchmark Optimization Problem

The study of dynamic multi-modal multi-objective optimization problems has not yet started, and no relevant benchmark optimization problems and mathematical models based on real problems have been found. Some test functions for multi-modal multi-objective optimization problems are given in the literature [10, 11]. Considering that they are too simple to distinguish the performance of different algorithms, Liu [1] et al. integrated a set of multi-modal multi-objective benchmark optimization problems (CEC 2013 [12], CEC 2015 [13]) and multi-objective benchmark optimization problems (CEC 2009 [14], DTLZ [15]) on the basis of the most popular multi-modal benchmark optimization problems (CEC 2009 [14], DTLZ [15]). The objective functions and the dimensionality of the decision variables of the given benchmark optimization problems can be adjusted by the user.

In this section, we propose a set of dynamic multi-modal multi-objective optimization problems, DMMMOP1 to DMMMOP3, based on the test functions proposed in [1], which contain the multi-modal characteristics of the original problems and have problem-dependent PF(t)/PS(t).

3.2 Parameter Setting

Since the method in this paper utilizes the MMMOEA-C&D and MDP strategies, the relevant parameters involved remain basically the same as those in the original literature.

The difference is that the number of function evaluations of the algorithm after each environment is significantly reduced. This is because, in dynamic optimization problems, the algorithm is required to obtain converged and well-distributed solutions in a short time, unlike the MMMOEA-C&D algorithm, which can give a sufficient number of function evaluations for convergence. It is more reasonable to reduce the number of evaluations of the algorithm to check the superiority of the algorithm. For all test functions DMMMOP1-DMMMOP3, the magnitude of the environmental variation $n_t=10$ and the frequencies $\tau_t=30$ and $\tau_t=50$, respectively, are used to investigate the effect of the variation frequency on the algorithm. The various algorithms track 20 environmental changes uniformly. The details are shown in the following table.

Table 1. Parameter settings

Problem	M	n	k_A	k_B	c	d	N
DMMMOP1	2	3	1	1	\	\	100
DMMMOP2	2	3	1	1	\	\	100
DMMMOP3	2	2	0	1	\	3	300

In Table 1, c and d represent all values of c and d in this optimization problem, respectively, N is the population size of all algorithms, and NoE (Number of Evaluations) is the maximum number of individual evaluations of the algorithm given each change in the environment.

3.3 Convergence and Distribution Analysis of the Algorithm in the Target Space

Table 2 shows the values of MHV for $n_t = 10$ and τ_t equal to 30 and 50 for each comparison algorithm. The larger the value of τ_t, the greater the number of iterations of the algorithm, and the greater the probability that the algorithm will find a good solution. Based on this, the following conclusions can be drawn from the analysis of Tables 2.

First, for the problem DMMMOP1, the proposed method DMMMOEACP&DM achieves the maximum MHV for τ_t of 30 and 50, and there is a significant difference with the MHV values obtained by the algorithms MMMMOEA-C&D and DMMMOEAB. For the solution of the DMMMOP2 problem, the optimal MHV value is obtained by the algorithm DMMMOEA-MDP, which is significantly larger than that obtained by the algorithm MMMMOEA-C&D. However, there is no significant difference with the MHV values obtained by other algorithms. Therefore, the performance of the algorithm MMMOEA-C&D is poor in solving the problems DMMMOP1 and DMMMOP2, and the convergence and distribution of the obtained solutions are not satisfactory.

Second, for problems DMMMOP3-DMMMOP3, there is no significant difference between the proposed method and the compared algorithms, i.e., the convergence and distribution of all algorithms are considered to be the same.

Finally, for solving problem DMMMOP1, the algorithms obtained significantly better results for $\tau_t = 50$ than for $\tau_t = 30$, indicating that the algorithms obtained better

Table 2. Mean and standard deviation of MHV for each algorithm

Problems	$(n_t,$ $\tau_t)$	MMMOEA-C\&D	DMMMOE–A	DMMMOEA–B	DMMMOEA-MDP	DMMMOEA-CP\&DM
DMMMOP1	(10, 30)	0.8341*†	0.8991†	0.7703*†	0.8958†	0.9004†
		0.0824	0.0334	0.0334	0.0305	0.0104
	(10, 50)	0.8973*†	0.9237†	0.9053*†	0.9235†	0.9256†
		0.0516	0.0049	0.0049	0.0043	0.0037
DMMMOP2	(10, 30)	0.5916*	0.6301	0.6296	0.6314	0.6303
		0.0236	0.0127	0.0127	0.0117	0.0098
	(10, 50)	0.5982*	0.6329	0.6328	0.6333	0.6240
		0.0319	0.0116	0.0116	0.0064	0.0274
DMMMOP3	(10, 30)	2.8284	2.8283	2.8287	2.8274	2.8275
		1.6263	1.6263	1.6263	1.6258	1.6257
	(10, 50)	2.8268	2.8267	2.8270	2.8261	2.8260
		1.6254	1.6253	1.6253	1.6250	1.6249

convergence and distribution for solving DMMMOP1 for $\tau_t = 50$. However, the value of τ_t has no significant effect on the obtained results when dealing with other problems. Thus, for $\tau_t = 30$, the algorithms do not achieve optimal convergence and distribution in solving DMMMOP1.However, they have largely converged and remained stable in solving other problems.

4   Conclusions

This paper conducts a preliminary study of dynamic multi-objective optimization problems with multi-modal characteristics. The experimental results show that the Pareto optimal solution set obtained by the proposed method not only exhibits good distribution and diversity in the decision space, but also is a multiple Pareto optimal solution set in the decision space. In contrast, the comparison algorithm can only guarantee the distribution and convergence of the obtained solutions in the decision space, but cannot find the multiple optimal solutions for multi-modal problems.

Acknowledgements. This work was jointly supported by National Key R&D Program of China (2021ZD0111502), National Natural Science Foundation of China (51907112, U2066212), Jiangxi"Double Thousand Plan" Project (JXSQ20210019), Natural Science Foundation of Guangdong Province of China (2021A1515011709), Scientific Research Staring Foundation of Shantou University (NTF20009). The funding body have played a role in the purchase of experimental equipment and expert consultation.

References

1. Liu, Y., Yen, G.G., Gong, D.: A multi-modal multi-objective evolutionary algorithm using two-archive and recombination strategies. IEEE Trans. Evol. Comput. **23**(4), 660–674 (2018)
2. Li, X.: Niching without niching parameters: particle swarm optimization using a ring topology. IEEE Trans. Evol. Comput. **14**(1), 150–169 (2010)
3. Min, L., Wen-Hua, Z.: Memory enhanced dynamic multi-objective evolutionary algorithm based on decomposition. J. Softw. **24**(7), 1571–1588 (2013)
4. Hatzakis I, Wallace, D.: Dynamic multi-objective optimization with evolutionary algorithms. In: Proceedings of the 8 th Annual Conference on Genetic and Evolutionary Computation. pp. 1201–1208, ACM, New York, USA (2006)
5. Yue, C., Qu, B., Liang, J.: A multi-objective particle swarm optimizer using ring topology for solving multi-modal multi-objective problems. IEEE Trans. Evol. Comput. **22**(5), 805–817 (2018)
6. Deb, K., Udaya, B.R.N., Karthik, S.: Dynamic multi-objective optimization and decision-making using modified NSGA-II: a case study on hydro-thermal power scheduling. In:Evolutionary Multi-Criterion Optimization, pp. 803–817 (2007)
7. Zhou, A., Jin, Y., Zhang, Q.: A population prediction strategy for evolutionary dynamic multi-objective optimization. IEEE Trans. Cybern. **44**(1), 40–53 (2014)
8. Zhou, A., Jin, Y., Zhang, Q., Sendhoff, B., Tsang, E.: Prediction-based population re-initialization for evolutionary dynamic multi-objective optimization. In: InternationalConference on Evolutionary Multi-Criterion Optimization, pp. 832–846 (2007)
9. Rong, M., Gong, D., Zhang, Y., Jin, Y., Pedrycz, W.: Multidirectional prediction approach for dynamic multi-objective optimization problems. IEEE Trans. Cybern. **49**(9), 3362–3374 (2019)
10. Deb, K., Tiwari, S.: Omni-optimizer: a generic evolutionary algorithm for single and multi-objective optimization. Eur. J. Oper. Res. **185**(3), 1062–1087 (2008)
11. Liang, J.J., Yue, C.T., Qu, B.Y. Multi-modal multi-objective optimization: a preliminarystudy. In: IEEE Congress on Evolutionary Computation, pp. 2454–2461 (2016)
12. Li, X., Engelbrecht, A., Epitropakis, M.: Benchmark functions for CEC 2013 special sessionand competition on niching methods for multi-modal function optimization. Technical report, RMIT University, Australia (2013)
13. Qu, B.Y., Liang, J.J., Wang, Z.Y., Chen, Q., Suganthan, P.N.: Novel benchmark functions for continuous multi-modal optimization with comparative results. Swarm Evol. Comput. **26**, 23–34 (2016)
14. Zhang, Q., Zhou, A., Zhao, S., Suganthan, P.N., Liu, W., Tiwari, S.: Multi-objective optimization test instances for the CEC 2009 special session and competition. Technicalreport, University of Essex, Colchester, UK and Nanyang technological University, Singapore (2008)
15. Deb, K., Thiele, L., Laumanns, M., Zitzler, E. Scalable Test Problems for EvolutionaryMulti-objective Optimization. In: Abraham, A., Jain, L., Goldberg, R. (eds.) Evolutionary Multiobjective Optimization. Advanced Information and Knowledge Processing, pp. 105–145. Springer, London (2001). https://doi.org/10.1007/1-84628-137-7_6

A WGAN-Based Generative Strategy in Evolutionary Multitasking for Multi-objective Optimization

Tianwei Zhou[1,2], Xizhang Yao[1,2], Guanghui Yue[3], and Ben Niu[1,2,4(✉)]

[1] College of Management, Shenzhen University, Shenzhen 518060, People's Republic of China
drniuben@gmail.com
[2] Great Bay Area International Institute for Innovation, Shenzhen University, Shenzhen 518060, People's Republic of China
[3] School of Biomedical Engineering, Health Science Center, Shenzhen University, Shenzhen 518060, People's Republic of China
[4] Institute of Big Data Intelligent Management and Decision, Shenzhen University, Shenzhen 518060, People's Republic of China

Abstract. Multitasking for multi-objective optimization (MTMO) is one of the most important issues in evolutionary computation. The information exchange mechanism among inter-tasks is the key factor in enhancing the algorithm. Evolutionary multitasking algorithm based on generative strategies (EMT-GS), the current mainstream algorithm, employs a generative adversarial network (GAN) to acquire, propagate and exploit knowledge among tasks, yet GAN suffers from a series of intractable drawbacks, such as training difficulties and mode collapse, etc. To address the issues listed above and achieve better performance, this paper proposes a new algorithm named MTMO-WGAN, which leverages Wasserstein GAN(WGAN) with weight clipping and gradient penalty as the generative strategies to deal with MTMO problems, respectively. Based on the MTMOO benchmark problems, MTMO-WGAN outperforms EMT-GS in the bulk of tasks and has great potential for advancement in the future, which unlocks possibilities for the application of deep generative models in the field of MTMO.

Keywords: Evolutionary multitasking · multi-objective optimization · generative adversarial networks · Wasserstein GAN

1 Introduction

Evolutionary algorithms (EAs) are heuristic algorithms with substantial search capabilities. Inspired by biological evolution in nature [1, 2], EAs are frequently employed to solve challenging multi-objective optimization problems (MOPs). Since real-world problems are interconnected, the knowledge extracted from one task can be transferred to another similar one. In light of this, the evolutionary multitasking (EMT) paradigm [3] was proposed by replicating the capacity of parallel processing in the human brain. The first EMT algorithm, known as MO-MFEA [4], was subsequently put forward to tackle several MOPs. Multi-objective EMT algorithms are superior to typical ones in solving myriad interrelated MOPs by transferring knowledge among tasks.

© The Author(s), under exclusive license to Springer Nature Switzerland AG 2023
Y. Tan et al. (Eds.): ICSI 2023, LNCS 13968, pp. 390–400, 2023.
https://doi.org/10.1007/978-3-031-36622-2_32

The knowledge transfer between inter-tasks is the cornerstone of the multi-objective EMT method. Many initiatives have been made to increase the effectiveness of information transmission in several aspects, including knowledge acquisition [5-11], knowledge transfer [12-18], and knowledge utilization [15-20]. Despite the significant advancements of multi-objective EMT algorithms, it remains challenging to deliver information of a high caliber. In the existing literature, the majority of corresponding algorithms usually choose one or more individuals from the source task population to transfer knowledge. However, the quality of transferred knowledge cannot be guaranteed, especially in the situation where the chosen ones get stuck in local optima or suffer from inadequate diversity.

To address the aforementioned issues, EMT-GS [21] introduced a generative adversarial network (GAN) to deal with multitasking for multi-objective optimization (MTMO) problems, ingeniously integrating the deep generative model into the MTMO domain. Compared with other models, EMT-GS obtains SOTA performances on numerous benchmarks, yet suffers from a series of intractable drawbacks, such as training difficulties and mode collapse, etc.

To cope with the problems of GAN, Wasserstein GAN (WGAN) introduces the Wasserstein distance, which can theoretically solve the gradient vanishing problem due to its superior smoothing properties relative to the loss function used by GAN. It has been demonstrated that WGAN could increase learning stability, eliminate issues such as mode collapse, and offer meaningful learning curves helpful for debugging and hyperparameter selection. Therefore, WGAN is a potential method to deal with MTMO problems. In this paper, based on the above motivations, we go beyond previous backbones and propose the MTMO-WGAN as a new paradigm for MTMO problems. Empowered by WGAN (Wasserstein GAN), the model is enabled to enhance the efficiency of knowledge sharing. Concretely, MTMO-WGAN leverages the Wasserstein distance to produce a value function that has better theoretical properties than EMT-GS.

The contributions of this paper are outlined below in brief.

- A new algorithm is proposed to tackle the MTMO problem based on the WGAN generation strategy. The drawbacks of GAN are overcome by WGAN, which can estimate the potential population distribution of tasks and facilitate knowledge transfer among various tasks more proficiently.
- The two enforcement strategies for WGAN are implemented and compared in the MTMO problem. Furthermore, it is demonstrated that the modifications in the deep generation model are also applicable to knowledge sharing between MTMO problems, which amplifies the application of deep generative models in the MTMO domain.

The rest of this paper is organized as follows. Section 2 reviews the related work. Section 3 illustrates the proposed MTMO-WGAN. Section 4 presents the experimental results. Finally, Sect. 5 concludes this paper and discusses some future directions.

2 Related Work

The information exchange mechanism among inter-tasks plays a fundamental role in MTMO problems. Nonetheless, the differences in search spaces between two different tasks may lead to deficient knowledge transfer. To deal with this problem, one feasible method is to limit the likelihood of unfavorable information transfer by mapping the search spaces of various tasks to each other and then transferring the mapped individuals as knowledge.

Incipiently, to map randomly picked individuals from one task to another, MFEA-GHS [8], MOMFEA-SADE [9], and EMT-LTR [6] employed the genetic transform method, subspace alignment technique, and a procedure of learning task associations, respectively. Afterward, EMTEGT [22] was designed with a denoising autoencoder to discharge the mapping strategy. Subsequently, for the individual identification of each task, EMTIL [7] constituted a naive Bayesian classifier based on incremental learning and implemented an enhanced upper and lower bound mapping approach. Thereafter, EMT/ET [10] adopted a similar mapping method for every task to map a randomly chosen individual from the non-dominated individuals and their closest neighbors. Thereupon, a mapping strategy based on a K-nearest-neighbor classifier was speculated by MMOTK [23] utilizing an analogous mapping mechanism.

Notwithstanding, the efficiency of knowledge transfer in the preceding approach may be sensitive to the nominated individuals since a poorly phrased individual may diminish the model's performance. The implementation of distributed estimation may mitigate this issue. EMT-PD [11], for instance, advocates extracting knowledge via the maximum likelihood estimation of the probability models. This strategy integrates the search patterns of various models and exhibits a more considerable level of robustness. In contrast, EMT-PD involves a predefined data distribution, which enhances the model's instability. Generative adversarial networks (GANs), in antithesis, do not require predefined data distribution and are capable of directly distinguishing the transition information from source task to target task. EMT-GS [21] leveraged GANs into the MTMO problem and yielded SOTA results on the comprehensive range of test datasets.

Based on the above analysis, knowledge sharing among multiple tasks turns indispensable in MTMO problems. Traditional methods usually select some individuals from the source task as knowledge to transfer to the target ones, which could lead to instability of the paradigm owing to its sensitiveness to the selected individuals. Nonetheless, the GANs implemented by the EMT-GS employ the paradigm of distribution estimation and could eliminate the above issues. Correspondingly, EMT-GS outperforms others fundamentally. Nonetheless, there exists much room for improvement in EMT-GS owing to the intrinsic drawbacks of GANs.

3 Proposed MTMO-WGAN

3.1 GANs and WGANs

A potent class of generative models known as Generative Adversarial Networks (GANs) [24] portrays generative modeling as a contest between two rival networks, including a generator and a discriminator. A source of noise is mapped to the input space by the generator network. The discriminator network is compelled to be able to clearly distinguish between a generated sample and a sample of actual data. The discriminator is collimated for the generator's tricks. Essentially, the minimax objective is the match between the generator G and the discriminator D.

$$\min_{G} \max_{D} V(D, G) = \mathop{\mathbb{E}}_{x \in P_r} [\log D(x)] + \mathop{\mathbb{E}}_{\tilde{x} \in P_g} [1 - \log D(\tilde{x}))] \tag{1}$$

where P_r is the data distribution and P_g is the model distribution implicitly defined by $\tilde{x} = G(z), z \in p(z)$ (the input z to the generator is sampled from one simple noise distribution p, such as the uniform distribution or a spherical Gaussian distribution, the generated data set $\tilde{x}$ obeys the distribution $G(z)$).

Nevertheless, It is proven that the Kullback-Leibler (KL) divergences that GANs routinely minimize may not be continuous, leading to training difficulty and mode collapse, etc. [25]. As an alternative, the Earth-Mover distance $W(q, p)$, also known as the Wasserstein-1 distance, was adopted to take the place of the KL divergence. $W(q, p)$ is technically defined as the minimum cost of transporting mass to shift the distribution q into the distribution p instead, where the cost is mass times transport distance. Under reasonable assumptions, $W(q, p)$ is continuous and practically differentiable. The Wasserstein GAN (WGAN) [25] value function is constructed using the Kantorovich-Rubinstein duality [26] to obtain

$$\min_{G} \max_{D \in \tilde{D}} V(D, G) = \mathop{\mathbb{E}}_{x \in P_r} [D(x)] + \mathop{\mathbb{E}}_{\tilde{x} \in P_g} [D(\tilde{x}))] \tag{2}$$

where $\tilde{D}$ is the set of 1-Lipschitz functions.

To impose the Lipschitz stipulation, the methodology of clipping the weights [25] of the discriminator to fit in a compact space $[-c, c]$ was propounded. A subset of the k-Lipschitz functions for some k, which rely on c and the critic topology, makes up the set of functions satisfying this restraint.

Despite the advancement of alleviating the dilemmas of GAN, the discriminator trained with weight clipping overlooks higher moments of the data distribution. To overcome the bottleneck, a soft variant of the restriction with a penalty on the gradient norm, named WGAN-GP [27], was introduced and the corresponding objective is

$$L = \mathop{\mathbb{E}}_{\tilde{x} \sim \mathbb{P}_g} [D(\tilde{x})] - \mathop{\mathbb{E}}_{x \sim \mathbb{P}_r} [D(x)] + \lambda \mathop{\mathbb{E}}_{\hat{x} \sim \mathbb{P}_{\hat{x}}} [(\nabla_{\hat{x}} D(\hat{x})_2 - 1)^2]. \tag{3}$$

where $P_{\hat{x}}$ is implicitly emphasized to sample uniformly along straight lines between pairs of points sampled from the data distribution P_r and the generator distribution P_g, and λ is the penalty coefficient.

3.2 EMT-GS

EMT-GS [21], a multi-objective EMT algorithm manufactured on GANs and Inertial Differential Evolution (IDE), was envisioned to facilitate the implementation of knowledge transfer and offspring generation. The approach includes a soft sharing mechanism to induce intermediate individuals as the knowledge funneled from one job to another while habitually training a GAN model for each source-target task pair. The IDE technique is implemented to incentivize the progression of convergence of the algorithm and to emit offspring in a favorable path toward the Pareto Front (PF).

On top of that, several additional characteristics were included in [4] to facilitate for comparison of individuals as EMT evolved:

1) Skill Factor: The skill factor τ_i of p_i denotes the task, among all tasks, on which the individual performs the best.
2) Factorial Cost: The factorial cost of p_i denotes the objective value(s) of p_i on the τ_i-th task. The factorial costs of p_i on the other tasks are assumed to be infinity.

3.3 MTMO-WGAN

Considering the aforementioned motivations, we eclipse the proceeding backbones and put forward the MTMO-WGAN with the same general framework as EMT-GS. Concerning the specific implementation, MTMO-WGAN incorporates two tactics known as WGAN with weight clipping (WGAN-WC) and WGAN with gradient penalty (WGAN-GP).

The mechanism of ENT-GS and MTMO-WGAN is defined in Algorithm 1. At the initial stage, the individual population P of size N is randomly initialized. Subpopulations of size $n = N/k$ are created from the N individuals uniformly. A corresponding skill factor is allocated to each subpopulation. After initialization, the individuals in P are assessed employing factorial costs. An archive P' and an offspring population C are also initialized to P' and $\emptyset$ (empty set), respectively (lines 1–3). Throughout evolution, in an interval of t steps, a GAN model, including original GANs and WGANs, is trained using GANTrain () (specified in Algorithm 2) for each pair of source-target tasks, engaging the subpopulations of the source and target tasks as real and fake data, accordingly. Each task serves as both a knowledge source and a knowledge target for every other task (lines 5–11). During offspring generation, two individuals are arbitrarily chosen from the incumbent population P each time (line 13). If the two predetermined individuals exhibit various skill factors, two offspring are manufactured (lines 15–18) based on knowledge generation operation KG ($\cdot$) (specified in Algorithm 3) and IDE ($\cdot$) operation (specified in Algorithm 4). Otherwise, two individuals are generated with IDE ($\cdot$) operation (lines 20–21) directly. The offspring generation procedure is implemented repeatedly until N offspring are assembled. Subsequently, the factorial cost of each individual in C is appraised in line 25. The population P is conserved into P' (line 26) and a new population is procured based on the environmental selection of the union of P and C (line 27). It needs to be noted that Algorithm 2, Algorithm 3, and Algorithm 4 are detailed in the paper of EMT-GS [21].

Algorithm 1 Procedure of EMT-GS / MTMO-WGAN

Input: k (number of tasks), N (population size), t (training interval of GANs), s (batch size), T (maximum number of iterations for training GANs)

Output: a series of nondominated solutions

1: $P \leftarrow$ population initialization in Φ and evaluation;

2: $P' \leftarrow P$

3: $C \leftarrow \varnothing$

4: **while** *the termination condition is not met* **do**

5:      **if** *generation* $\% \, t == 0$ **then** /* GANs training */

6:          **for** $i, j = 1$ **to** k **do**

7:              **if** $j \neq i$ **then**

8:                  $\Pi_{i,j} \leftarrow$ GANTrain (i, j, s, P, T) (Algorithm 2);

9:              **end if**

10:         **end for**

11:     **end if**

12:     **while** $\| C \| < N$ **do** /* offspring generation */

13:         $p_i, p_j \leftarrow$ randomly select two individuals from P ;

14:         **if** $\tau_i \neq \tau_j$ **then** /* inter-task */

15:             $q \leftarrow$ KG $(p_i, \Pi_{\tau_j, \tau_i}, \Pi_{\tau_i, \tau_j})$ (Algorithm 3);

16:             $c1 \leftarrow$ IDE (q, τ_i, P, P') (Algorithm 4);

17:             $q \leftarrow$ KG $(p_i, \Pi_{\tau_j, \tau_i}, \Pi_{\tau_i, \tau_j})$ (Algorithm 3);

18:             $c2 \leftarrow$ IDE (q, τ_j, P, P') (Algorithm 4);

19:         **else** /* intra-task */

20:             $c1 \leftarrow$ IDE (p_i, τ_i, P, P') (Algorithm 4);

21:             $c2 \leftarrow$ IDE (p_j, τ_j, P, P') (Algorithm 4);

22:         **end if**

23:         $C \leftarrow C \cup c1 \cup c2$

24:     **end while**

25:     Evaluate the factorial cost of each individual in C;

26:     $P' \leftarrow P$;

27:     $P \leftarrow$ environmental selection on $C \cup P$;

28:     $C \leftarrow \varnothing$;

29: **end while**

4 Experiment

4.1 Experimental Design

Test Problems and Baseline. The classic MTMOO benchmark problems are applied to assess the performance of MTMO-WGAN. Seven sets of problems are chosen, and each contains two MOPs. It stands to reason that EMT-GS is adopted as the baseline algorithm for comparison with MTMO-WGAN.

Parameter Settings. The generator and discriminator in the GAN model are simply designed to employ three-layer and two-layer fully connected neural networks, respectively. Empirically, a training interval of 10 is engaged for GANs. The discriminator and generator utilize learning rates of 0.0002 and 0.0003, respectively. The maximum number of iterations $T = 200$ and the batch size $s = 10$ are selected to train the GANs and WGANs.

The most prominent hyper-parameters for WGAN-WC and WGAN-GP are c and λ, respectively. Each has seven defined distinct values, with $c \in [0.0001, 0.0005, 0.001, 0.003, 0.005, 0.01, 0.1]$ and $\lambda \in [0.1, 0.5, 1, 3, 5, 10, 20]$.

Metric. An important issue in MOP is the performance comparison of different algorithms. In the case of MOP evolutionary algorithms, the final result could be regarded as an approximation of the corresponding PF and thus denoted as an approximation set. Therefore the distance between the generated overcome and PF becomes a paramount indicator of the algorithm evaluation. In light of this, two metrics of performance assessment about PF are selected, which are listed below.

Inverted Generational Distance (IGD) [28] is used in this paper to evaluate the performance of the algorithms. Particularly, IGD is formulated as follows:

$$IGD = \frac{\sum_{x \in P^*} dist(x^*, A)}{|P^*|} \tag{4}$$

where P^* represents a predefined set of reference points on *PF*. $dist(x^*, A)$ stands for the shortest Euclidean distance between the reference point x^* in the goal space and the non-dominated solution set A obtained by the algorithm. *IGD* possesses the capability of quantifying both the convergence and diversity of A. A lower *IGD* score signifies a greater functionality of the algorithm.

Another criterion applied to assess the performance of the algorithms is the Mean and Standard Score (MSS) [29].

$$MSS = \frac{1}{k} \sum_{i=1}^{k} \frac{I_i - \mu_i}{\sigma_i} \tag{5}$$

where k is the number of tasks in a test problem, μ_i and σ_i illustrate the average and standard deviation of the *IGD* value garnered by all compared algorithms on the i-th task over all iterations, respectively. $I_i(i = 1, ...k)$ symbolizes the *IGD* score for the i-th task that an algorithm has fulfilled in a single iteration. A reduced *MSS* value denotes a superior overall performance of an algorithm on a test scenario.

Implementation Details. MTMO-WGAN was implemented in MATLAB R2022a and run on a Windows PC with an Intel Core i7-9750H CPU at 2.60 GHz and 16 GB RAM.

4.2 Experimental Results

The average *IGD* values over the classic MTMOO benchmark problems for the EMT-GS, MTMO-WGAN-WC, and MTMO-WGAN-GP during 30 independent runs are shown in Table 1. The finest results are emphasized in bold and red, while the second-best ones are highlighted in blue. The marks "+," "−," and "≈" signify that the relevant results are, superior to, inferior to, and comparable to those of EMT-GS distinctively.

Table 1. AVERAGED IGD VALUE OF EMT-GS, MTMO-WGAN-WC, AND MTMO-WGAN-GP ON THE MTMOO FOR 30 INDEPENDENT RUNS

Problem	Task	EMT-GS	MTMO-WGAN-WC						
			c = 0.0001	c = 0.0005	c = 0.001	c = 0.003	c = 0.005	c = 0.01	c = 0.1
CIHS	T1	4.99E-03	4.30E-03(+)	4.41E-03(+)	4.27E-03(+)	4.36E-03(+)	4.37E-03(+)	4.16E-03(+)	**4.11E-03(+)**
	T2	4.14E-03	4.15E-03(-)	4.14E-03(≈)	4.05E-03(+)	4.19E-03(-)	4.14E-03(≈)	4.11E-03(+)	**4.04E-03(+)**
CIMS	T1	4.90E-03	4.42E-03(+)	4.22E-03(+)	4.36E-03(+)	4.22E-03(+)	4.19E-03(+)	4.20E-03(+)	4.27E-03(+)
	T2	3.98E-03	3.91E-03(+)	3.92E-03(+)	3.92E-03(+)	3.91E-03(+)	3.90E-03(+)	4.03E-03(-)	3.90E-03(+)
CILS	T1	**3.75E+01**	6.47E+01(-)	5.38E+01(-)	4.58E+01(-)	5.18E+01(-)	8.76E+01(-)	7.56E+01(-)	6.38E+01(-)
	T2	4.09E-03	4.07E-03(+)	4.04E-03(+)	4.01E-03(+)	4.06E-03(+)	4.05E-03(+)	4.08E-03(+)	4.08E-03(+)
PIHS	T1	4.07E-03	4.03E-03(+)	4.07E-03(≈)	4.00E-03(+)	4.05E-03(+)	4.09E-03(-)	4.00E-03(+)	**3.97E-03(+)**
	T2	1.30E+02	8.92E+01(+)	1.03E+02(+)	1.60E+02(-)	8.83E+01(+)	9.01E+01(+)	1.46E+02(+)	1.08E+02(+)
PIMS	T1	3.96E-02	3.89E-02(+)	3.54E-02(+)	3.53E-02(+)	3.80E-02(+)	3.80E-02(+)	3.61E-02(+)	3.81E-02(+)
	T2	4.87E+02	2.21E+02(+)	**2.14E+02(+)**	2.48E+02(+)	4.13E+02(+)	5.91E+02(-)	5.67E+02(-)	5.60E+02(-)
PILS	T1	3.96E-03	3.98E-03(-)	3.92E-03(+)	3.97E-03(-)	2.03E-02(-)	8.64E-03(-)	3.94E-03(+)	3.98E-03(-)
	T2	**2.00E+01**	2.00E+01(≈)	2.00E+01(≈)	2.00E+01(≈)	2.00E+01(≈)	2.00E+01(≈)	2.00E+01(≈)	2.00E+01(≈)
NIHS	T1	1.03E+02	4.58E+01(+)	1.52E+02(-)	9.10E+01(+)	**3.34E+01(+)**	1.01E+02(+)	1.01E+02(+)	4.53E+01(+)
	T2	4.11E-03	4.05E-03(+)	4.05E-03(+)	4.08E-03(+)	4.01E-03(+)	4.09E-03(+)	4.05E-03(+)	4.04E-03(+)
+/-/≈			10/3/1	9/2/3	10/3/1	10/3/1	8/4/2	9/4/1	10/3/1

Problem	Task	EMT-GS	MTMO-WGAN-GP						
			λ = 0.1	λ = 0.5	λ = 1	λ = 3	λ = 5	λ = 10	λ = 20
CIHS	T1	4.99E-03	4.28E-03(+)	4.51E-03(+)	4.24E-03(+)	4.40E-03(+)	4.27E-03(+)	4.37E-03(+)	4.32E-03(+)
	T2	4.14E-03	4.14E-03(+)	4.18E-03(-)	4.19E-03(-)	4.13E-03(+)	4.12E-03(+)	4.17E-03(-)	4.15E-03(-)
CIMS	T1	4.90E-03	**4.15E-03(+)**	4.22E-03(+)	4.15E-03(+)	4.19E-03(+)	4.15E-03(+)	4.21E-03(+)	4.17E-03(+)
	T2	3.98E-03	3.92E-03(+)	3.95E-03(+)	3.92E-03(+)	3.86E-03(+)	3.92E-03(+)	3.90E-03(+)	**3.84E-03(+)**
CILS	T1	**3.75E+01**	6.57E+01(-)	8.22E+01(-)	7.76E+01(-)	1.39E+02(-)	1.27E+02(-)	1.04E+02(-)	9.16E+01(-)
	T2	4.09E-03	4.11E-03(-)	**3.99E-03(+)**	4.09E-03(+)	4.06E-03(+)	4.05E-03(+)	4.03E-03(+)	4.09E-03(+)
PIHS	T1	4.07E-03	4.01E-03(+)	4.06E-03(+)	4.04E-03(+)	4.10E-03(-)	4.06E-03(+)	4.00E-03(+)	4.00E-03(+)
	T2	1.30E+02	1.50E+02(-)	8.16E+01(+)	1.14E+02(+)	**7.88E+01(+)**	1.44E+02(-)	8.54E+01(+)	1.09E+02(+)
PIMS	T1	3.96E-02	**3.16E-02(+)**	3.33E-02(+)	3.66E-02(+)	3.75E-02(+)	3.71E-02(+)	3.40E-02(+)	4.03E-02(-)
	T2	4.87E+02	3.98E+02(+)	2.39E+02(+)	2.34E+02(+)	2.82E+02(+)	2.15E+02(+)	3.65E+02(+)	4.77E+02(+)
PILS	T1	3.96E-03	3.96E-03(+)	3.92E-03(+)	3.92E-03(+)	1.10E-02(-)	**3.89E-03(+)**	3.99E-03(-)	1.32E-02(-)
	T2	**2.00E+01**	2.00E+01(≈)	2.00E+01(≈)	2.00E+01(≈)	2.00E+01(≈)	2.00E+01(≈)	2.00E+01(≈)	2.00E+01(≈)
NIHS	T1	1.03E+02	9.97E+01(+)	4.73E+01(+)	1.48E+02(-)	8.14E+01(+)	8.83E+01(+)	1.57E+02(-)	4.55E+01(+)
	T2	4.11E-03	4.04E-03(+)	4.07E-03(-)	**4.00E-03(+)**	4.01E-03(+)	4.10E-03(-)	4.04E-03(+)	4.00E-03(+)
+/-/≈			10/3/1	10/3/1	10/3/1	10/3/1	8/4/2	11/2/1	9/4/1

Due to the employment of WGAN-based generation techniques, both MTMO-WGAN-WC and MTMO-WGAN-GP exhibit enhanced performance on the majority of benchmark issues contrasted to EMT-GS. Correspondingly, the latter outperforms the former fundamentally since the gradient penalty has the upper hand on weight chipping. To elucidate, the superiority of the latter over the former is detailed in two points. First and foremost, the prime element is that MTMO-WGAN-GP exhibits lower values predominantly. On top of that, the second noteworthy argument is reflected by the phenomenon that MTMO-WGAN-GP has a more even distribution of performance and turns less sensitive to the variation of hyper-parameters.

However, the WGAN-based algorithms are slightly worse than EMT-GS in some low-similarity problems. On T1 of CILS with the incredibly minimal congruity of 0.07, EMT-GS accomplishes the best performance thanks to the imperfect capacity to complement the object probability distribution, which inhibits EMT-GS from getting caught in modeling the resemblance between two tasks with little correlation to overlook the global convergence. On T2 of PILS with the extremely tiny coherence of 0.002, it is unveiled that all approaches exhibit high consistency with identical scores, which reveals that all generation strategies will collapse, facing tasks with near-zero similitude.

The MSS values for the algorithms on the MTMOO benchmark issues over 30 independent runs are displayed in Table 2. The table's best results for each algorithm are highlighted in bold and red, while the second-best results are noted in blue. The aftereffects displayed in this table are in line with the outcomes in terms of IGD. Based on the above analysis, MTMO-WGANs yield better overall performance.

Table 2. CSS VALUE OF EMT-GS, MTMO-WGAN-WC, AND MTMO-WGAN-GP ON THE MTMOO FOR 30 INDEPENDENT RUNS

Problem	EMT-GS	MTMO-WGAN-WC						
		c = 0.0001	c = 0.0005	c = 0.001	c = 0.003	c = 0.005	c = 0.01	c = 0.1
CIHS	1.59E+00	5.46E-02	1.51E-01	-1.19E+00	6.18E-01	4.41E-02	-7.96E-01	-1.70E+00
CIMS	2.32E+00	2.79E-01	-1.47E-01	2.95E-01	-1.78E-01	-4.59E-01	1.08E+00	-1.64E-01
CILS	-1.88E-01	-1.18E-01	-6.77E-01	-1.31E+00	-4.19E-01	-4.47E-02	2.11E-01	2.31E-02
PIHS	7.26E-01	-4.40E-01	3.34E-01	3.99E-01	-3.15E-01	3.00E-01	1.94E-01	-9.63E-01
PIMS	1.04E+00	-4.81E-02	-8.05E-01	-6.98E-01	4.42E-01	1.08E+00	5.91E-01	9.72E-01
PILS	-4.41E-01	-4.39E-01	-4.46E-01	-4.41E-01	1.24E+00	3.99E-02	-4.44E-01	1.17E+00
NIHS	1.03E+00	-4.78E-01	7.52E-01	4.59E-01	-1.21E+00	7.88E-01	1.20E-01	-6.74E-01
+/-/≈		10/3/1	9/2/3	10/3/1	10/3/1	8/4/2	9/4/1	10/3/1

Problem	EMT-GS	MTMO-WGAN-GP						
		λ = 0.1	λ = 0.5	λ = 1	λ = 3	λ = 5	λ = 10	λ = 20
CIHS	1.59E+00	-1.95E-01	9.21E-01	3.09E-01	2.28E-03	-3.43E-01	4.41E-01	9.13E-02
CIMS	2.32E+00	-2.74E-01	2.25E-01	-3.14E-01	-8.53E-01	-3.46E-01	-3.61E-01	-1.11E+00
CILS	-1.88E-01	5.31E-01	-1.00E+00	4.61E-01	1.07E+00	7.54E-01	-3.02E-02	7.37E-01
PIHS	7.26E-01	3.81E-01	-2.54E-01	9.69E-02	1.85E-01	8.95E-01	-9.85E-01	-5.54E-01
PIMS	1.04E+00	-9.37E-01	-1.15E+00	-4.84E-01	-1.34E-01	-4.44E-01	-5.64E-01	1.14E+00
PILS	-4.41E-01	-4.41E-01	-4.46E-01	-4.46E-01	2.79E-01	-4.49E-01	-4.38E-01	1.71E+00
NIHS	1.03E+00	2.54E-02	-2.84E-01	-1.08E-02	-6.66E-01	6.73E-01	6.97E-01	-1.21E+00
+/-/≈		10/3/1	10/3/1	10/3/1	10/3/1	8/4/2	11/2/1	9/4/1

5 Conclusion and Future Work

This paper proposed a novel algorithm namely MTMO-WGAN, which utilizes WGAN with weight clipping (WGAN-WC) and WGAN with gradient penalty (WGAN-GP) as the generative strategies respectively, to address multitasking for multi-objective optimization problems. The WGAN-based approach triumphs over the blemish of training difficulties and mode collapse emerging in GANs, and apprehends the resemblance between distinct probability distributions more perspicaciously, with better and more stable performance. MTMO-WGAN is validated on seven benchmarks of MTMOO. It is demonstrated that the MTMOO-WGAN family of models outperforms EMT-GS on the whole and exhibits great promise for advancement for future development.

Despite the advancement of MTMO-WGAN, there exists a huge challenge that the training cost of the model is still quite high, especially facing myriad tasks simultaneously. Extending MTMO-WGAN turns not only a challenging further direction but also represents the potential for the implementation and rollout of deep generative models in the domain of MTMO.

Acknowledgment. The study was supported in part by the Natural Science Foundation of China under Grant No. 62103286, No. 62001302, in part by Ministry of Education of the People's Republic of China Humanities and Social Sciences Youth Foundation under Grant No. 21YJC630181, in part by Guangdong Basic and Applied Basic Research Foundation under Grant No. 2021A1515011348, in part by Natural Science Foundation of Shenzhen under Grant No. JCYJ20190808145011259, in part by Natural Science Foundation of Guangdong Province under Grant No. 2020A1515010752, No. 2020A1515110541, in part by Guangdong Province Innovation Team under Grant 2021WCXTD002, in part by Shenzhen Science and Technology Program under Grant RCBS20200714114920379.

References

1. Back, T., Hammel, U., Schwefel, H.-P.: Evolutionary computation: comments on the history and current state. IEEE Trans. Evol. Comput. **1**, 3–17 (1997)
2. Tanabe, R., Ishibuchi, H.: A review of evolutionary multimodal multiobjective optimization. IEEE Trans. Evol. Comput. **24**, 193–200 (2019)
3. Gupta, A., Ong, Y.-S., Feng, L.: Multifactorial evolution: toward evolutionary multitasking. IEEE Trans. Evol. Comput. **20**, 343–357 (2015)
4. Gupta, A., Ong, Y.-S., Feng, L., Tan, K.C.: Multiobjective multifactorial optimization in evolutionary multitasking. IEEE Trans. Cybern. **47**, 1652–1665 (2016)
5. Wang X., Dong Z., Tang L., Zhang Q.: Multiobjective multitask optimization-neighborhood as a bridge for knowledge transfer. IEEE Trans. Evol. Comput. (2022)
6. Chen, Z., Zhou, Y., He, X., Zhang, J.: Learning task relationships in evolutionary multitasking for multiobjective continuous optimization. IEEE Trans. Cybern. **52**(6), 5278–5289 (2020)
7. Lin, J., Liu, H.-L., Xue, B., Zhang, M., Gu, F.: Multiobjective multitasking optimization based on incremental learning. IEEE Trans. Evol. Comput. **24**, 824–838 (2019)
8. Liang, Z., Zhang, J., Feng, L., Zhu, Z.: A hybrid of genetic transform and hyper-rectangle search strategies for evolutionary multi-tasking. Expert Syst. Appl. **138**, 112798 (2019)
9. Liang, Z., Dong, H., Liu, C., Liang, W., Zhu, Z.: Evolutionary multitasking for multiobjective optimization with subspace alignment and adaptive differential evolution. IEEE Trans. Cybern. **52**(4), 2096–2109 (2020)

10. Lin, J., Liu, H.-L., Tan, K.C., Gu, F.: An effective knowledge transfer approach for multiobjective multitasking optimization. IEEE Trans. Cybern. **51**, 3238–3248 (2020)

11. Liang, Z., Liang, W., Wang, Z., Ma, X., Liu, L., Zhu, Z.: Multiobjective evolutionary multitasking with two-stage adaptive knowledge transfer based on population distribution. IEEE Trans. Syst. Man Cybern. Syst. **52**(7), 4457–4469 (2021)

12. Bali, K.K., Gupta, A., Ong, Y.-S., Tan, P.S.: Cognizant multitasking in multiobjective multifactorial evolution: mo-mfea-Ii. IEEE Trans. Cybern. **51**, 1784–1796 (2020)

13. Yang, C., Ding, J., Tan, K.C., Jin, Y.: Two-stage assortative mating for multi-objective multifactorial evolutionary optimization. In: 2017 IEEE 56th Annual Conference on Decision and Control (CDC), pp. 76–81. IEEE (2017)

14. Xu, Z., Liu, X., Zhang, K., He, J.: Cultural transmission based multi-objective evolution strategy for evolutionary multitasking. Inf. Sci. **582**, 215–242 (2022)

15. Binh, H.T.T., Tuan, N.Q., Long, D.C.T.: A multi-objective multi-factorial evolutionary algorithm with reference-point-based approach. In: 2019 IEEE Congress on Evolutionary Computation (CEC), pp. 2824–2831. IEEE (2019)

16. Yao, S., Dong, Z., Wang, X., Ren, L.: A multiobjective multifactorial optimization algorithm based on decomposition and dynamic resource allocation strategy. Inf. Sci. **511**, 18–35 (2020)

17. Hashimoto, R., Urita, T., Masuyama, N., Nojima, Y., Ishibuchi, H.: Effects of local mating in inter-task crossover on the performance of decomposition-based evolutionary multiobjective multitask optimization algorithms. In: 2020 IEEE Congress on Evolutionary Computation (CEC), pp. 1–8. IEEE (2020)

18. Wei, T., Zhong, J.: Towards generalized resource allocation on evolutionary multitasking for multi-objective optimization. IEEE Comput. Intell. Mag. **16**, 20–37 (2021)

19. Chen, Y., Zhong, J., Tan, M.: A fast memetic multi-objective differential evolution for multitasking optimization. In: 2018 IEEE Congress on Evolutionary Computation (CEC), pp. 1–8. IEEE (2018)

20. Zhou, L., et al.: Toward adaptive knowledge transfer in multifactorial evolutionary computation. IEEE Trans. Cybern. **51**, 2563–2576 (2020)

21. Liang, Z., Zhu, Y., Wang, X., Li, Z., Zhu, Z.: Evolutionary multitasking for multi-objective optimization based on generative strategies. IEEE Trans. Evol. Comput. **52**(4), 2096–2109 (2022)

22. Feng, L., et al.: Evolutionary multitasking via explicit autoencoding. IEEE Trans. Cybern. **49**, 3457–3470 (2018)

23. Chen, H., Liu, H.-L., Gu, F., Tan, K.C.: A multi-objective multitask optimization algorithm using transfer rank. IEEE Trans. Evol. Comput. (2022)

24. Goodfellow, I., et al.: Generative adversarial networks. Commun. ACM **63**, 139–144 (2020)

25. Arjovsky, M., Chintala, S., Bottou, L.: Wasserstein generative adversarial networks. In: International Conference on Machine Learning, pp. 214–223. PMLR (2017)

26. Villani, C.: Optimal Transport: Old and New. Springer, Heidelberg (2009). https://doi.org/10.1007/978-3-540-71050-9

27. Gulrajani, I., Ahmed, F., Arjovsky, M., Dumoulin, V., Courville, A.C.: Improved training of wasserstein GANs. In: Advances in Neural Information Processing Systems, vol. 30 (2017)

28. Zitzler, E., Thiele, L., Laumanns, M., Fonseca, C.M., Da Fonseca, V.G.: Performance assessment of multiobjective optimizers: an analysis and review. IEEE Trans. Evol. Comput. **7**, 117–132 (2003)

29. Yuan, Y., et al.: Evolutionary Multitasking for Multiobjective Continuous Optimization: Benchmark Problems, Performance Metrics and Baseline Results. arXiv preprint arXiv:1706.02766 (2017)

A Two-Stage Evolutionary Algorithm with Repair Strategy for Heat Component-Constrained Layout Optimization

Ke Shi, Yu Zhang(ID), Xinyue Li(ID), and Wang Hu(✉)(ID)

School of Computer Science and Engineering, University of Electronic Science and Technology of China, Chengdu, Sichuan, China
huwang@uestc.edu.cn

Abstract. The heat component-constrained layout optimization (HCLO) problem is difficult to be solved due to its many constraints and large search space. In this paper, a novel two-stage evolutionary algorithm (TEA/HCLO) is proposed to improve the effectiveness in solving the HCLO. The HCLO problem is decomposed into two subproblems and optimized in two stages to obtain a satisfying layout. In the first stage, a novel environmental selection strategy using the individual density in objective and decision space cooperatively is proposed to reduce the search space. In the second stage, a repair strategy based on the novel neighborhood search is applied to repair the infeasible solutions obtained in the first stage. The comparative experiments show that the TEA/HCLO algorithm has achieved the best performance in five HCLOs compared with some competitive algorithms.

Keywords: evolutionary algorithm · repair strategy · constraint optimization · heat component · layout optimization

1 Introduction

The high concentration of electronic components in an integrated control system leads to increase the heat generation of the system [1]. Electronic components will get unacceptable harm if the heat cannot be released in time. So, the heat pipe is required to dissipate heat from the integrated control system [2]. However, the heat pipe can exert the best effect only when the constraint conditions are satisfied well. For example, each component which needs heat dissipation should contact the heat pipes, the components connected to each heat pipe are limited by the heat dissipation capacity, and so on. As a result, the component layout in the heat system is a hard optimization problem with many variables and constraints [3]. The large number of decision variables and the complex layout constraints in the HCLO problems are the challenges for the traditional optimization methods.

The evolutionary algorithm [4], inspired from the natural behaviors, can deal with constraints problems better since it has the strong ability to jump out of local optimal, resulting in the ability on searching the global optimal solution. Therefore, the using

Y. Tan et al. (Eds.): ICSI 2023, LNCS 13968, pp. 401–412, 2023.
https://doi.org/10.1007/978-3-031-36622-2_33

of evolutionary algorithm to solve the constraint optimization problems are becoming a hot research field.

Evolutionary algorithm includes genetic algorithm [5], particle swarm optimization [6], differential evolution algorithm [7], SMA [21], GBS based brain storm optimization [22], GWO-Based optimal fuzzy controllers [20] and so on. In this paper, a two-stage strategy is proposed for solving the heat component-constrained layout optimization (HCLO) problem. This problem is firstly decomposed into two relatively independent subproblems to reduce search space, and then a two-stage strategy is used to handle this problem. In the first stage, a novel environmental selection strategy which utilizes the individual density in objective and decision space cooperatively (ES-OD) to reduce the searching space in order to purse the best solution. In the second stage, a repair strategy based on the novel neighborhood search is applied to repair infeasible solutions obtained in the first stage. All the components are summarized into the new algorithm called TEA/HCLO to solve the constraint optimization problems better.

Most used environmental selection strategies of evolutionary algorithm are the roulette wheel selection strategy [8] and tournament selection [9] strategy. An algorithm with the roulette wheel selection strategy has a good convergence in the early stage, but poor in the middle and later stage because of the similar objective of the population. Therefore, the roulette wheel selection has no difference from random selection which makes it difficult to select the good solutions. As for the algorithm with tournament selection, an individual with good objective may be discarded in sometimes in the early and middle stage because of the strong randomness of tournament selection. To overcome those shortcomings of the two environmental selections strategy, the ES-OD is proposed to enhance the search ability in this HCLO and avoid falling into local optimal area.

Among local search algorithms, the neighborhood search technique is a simple and effective method. The main idea of neighborhood search is to randomly generate solutions near the current solutions, and then evaluate and screen them by constraint violation value. Gaussian distribution is usually assumed according to the central limit theorem of probability theory which can better fit the unknown distribution. Therefore, a novel neighborhood search strategy based on Gaussian disturbance is used in the second stage to repair the infeasible solutions.

The rest of this paper is organized as follows. The problem description and formulation of HCLO will be described in Sect. 2. Section 3 describes the details of TEA/HCLO. In Sect. 4, the comparative experiments are carried out to verify the effectiveness of the TEA/HCLO. Finally, conclusions and future works are discussed in Sect. 5.

2   Problem Description and Formulation

In this section, the HCLO is introduced and formulated with decision variables, layout constraints, and the optimization objective.

2.1   Problem Description

A general integrated system consists of components, heat pipes, and circuit boards. The components are distributed discretely on the circuit board, and heat pipes are usually distributed parallelly and evenly on the circuit board. An example of general integrated system is shown in Fig. 1, the grey area represents the layout board, the blue rectangle represents the component, and the orange rectangle represents the heat pipe. Both components and heat pipes are restricted to limited area. Main properties of components are length, width, heating power and weight. For heat pipe properties, the main properties are width and interval between the heat pipes. All the components and heat pipes are limited in the certain board.

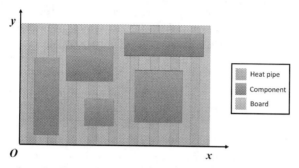

Fig. 1. The illustration of HCLO problem. The heat pipes, components and board are simplified as orange rectangles, blue rectangles and grey rectangles, respectively. Heat pipes and components are located on the board area.

2.2 Problem Formulation

With some simplifications, the HCLO problem can be transformed into a two-dimensional plane model. In this model, components and heat pipes are simplified to some rectangles. This model can be formulated as follows.

For each component, its coordinates are defined as the low left corner of each component. The coordinates of the i-th component can be represented as (x_i, y_i). Therefore, a sequence of variables $(x_1, y_1, x_2, y_2, \ldots, x_{Nc}, y_{Nc})$ can be used as decision variables to describe the layout of all components.

As described in Subsect. 2.1, the optimization objective is to make the power load of each heat pipe as similar as possible. Therefore, the goal is to minimize the variance of the power load of these heat pipes. The objective function can be constructed as Eq. (1), where N_c represents the number of components, p_j^{hp} represents the heat dissipation power of the heat pipe j and $\overline{P}$ is the average of real load power of all heat pipes in the system. The value of p_j^{hp} can be calculated by Eq. (2),

$$minf = \frac{1}{N_c}\sqrt{\sum_{j=1}^{N_c}(P_j^{hp} - \overline{P})^2}, \tag{1}$$

$$p_j^{hp} = \sum_{i=1}^{n}(P_i^c/k_i), \tag{2}$$

where P_i^c denotes the heat power of i-th components overlapping with j-th heat pipe, n denotes the number of components overlapping with j-th heat pipe and k_i denotes i-th component overlap with k_i heat pipes, which means the heat power is evenly divided into k parts.

There are four constraints should be satisfied in HCLO. First of all, for every two components, they cannot overlap. Therefore, the first constraint (G1) is that the overlap area between every two components must be not larger than zero. Secondly, the static stability constraint requires to balance the mass of whole components. In other words, the distance between theoretical centroid and actual centroid cannot be too much. Therefore, the second constraint (G2) is the static stability constraint. Thirdly, in the third constraint (G3), each component must overlap with the heat pipe to transfer heat to the heat pipe. Finally, for the last constraint (G4), the actual heat dissipation power of each heat pipe must be not larger than the maximum heat dissipation capacity.

To sum up, the HCLO can be formulated as optimization problem with one objective and four constraints in this paper, where the objective ensures the heat dissipation power of each heat pipe to be balanced to reach the maximum heat dissipation efficiency and the solutions can keep feasible after satisfying all the four constraints mentioned above. In the next section, the proposed two-stage evolutionary algorithm will be introduced to solve the HCLO problem via the problem decomposition strategy.

3   Proposed Algorithm

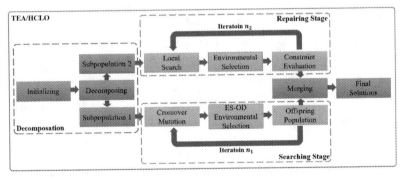

Fig. 2. Workflow of the proposed algorithm. The algorithm divides the whole population into two subpopulations. The subpopulation $\mathbf{P}^X$ will be optimized in the first stage to obtain a solution with minimal objective, while the subpopulation $\mathbf{P}^Y$ will be optimized in the second stage to repair the infeasible solution. Finally, a feasible solution with the best objective value can be obtained after the two-stage optimization.

In this section, a two-stage evolutionary algorithm based on decomposition (TEA/HCLO) will be proposed according to the analysis in Sect. 1 and 2. The position of the component can be divided into two independent parts as the vertical position and horizontal position. Therefore, the two parts can be considered to be optimized in two stages separately. The specific steps of the TEA/HCLO are shown in Fig. 2. The new proposed algorithm is divided into two stages as search stage and repair stage. The whole population is divided into two subpopulations name as subpopulation P^X and P^Y. Firstly, P^X will be optimized in the first stage to search the optimal objective. Then P^Y will be repaired to obtain the feasible solution in the second stage. Finally, a feasible solution with the best objective value can be obtained after the two-stage optimization.

3.1 Problem Decomposition and Population Encoding

Since the heat pipe is evenly and vertically placed in the board, the heating power of each heat pipe is only related with the components' horizontal position, which determines whether the components intersect with the heat pipe or not. Therefore, in the first stage, only the horizontal coordinate of the components is considered to find the optimal solution.

The decoding method of this decomposition strategy is described as follows. Firstly, the position of i-th component is represented by ordinate pair (x_i, y_i). Next, the position of all components in this problem can be encoded as a sequence of variables $p_i = (x_1, y_1, x_2, y_2, \ldots x_{Nc}, y_{Nc})$, which represents a solution of the component's layout. Then, the solution is split into two groups of sub-solutions, $p_i^x = (x_1, x_2, \ldots, x_{Nc})$ and $p_i^y = (y_1, y_2, \ldots, y_{Nc})$, which respectively correspond to the horizontal and vertical positions of components in the layout. As a result of the decomposition strategy, the whole population $\mathbf{P} = \{p_1, p_2, \ldots, p_{N_c}\}$ is divided into subpopulation $\mathbf{P}^X = \{p_1^X, p_2^X, \ldots, p_{N_c}^X\}$ and $\mathbf{P}^Y = \{p_1^Y, p_2^Y, \ldots, p_{N_c}^Y\}$. These subpopulations will be optimized respectively in the search stage and repair stage in the Subsect. 3.2 and 3.3, respectively.

3.2 Environmental Selection Based on Cooperative Density on Objective and Decision Space (ES-OD)

In the first stage of optimization, the genetic algorithm based on simulated binary crossover [10] and polynomial mutation [11] is used for optimizing subpopulation P^X to find the solution with the best objective. In this stage, only the main objective and the constraints G3 and G4 are considered.

Once the heat pipe contacts with the component, it is considered that heat power generated by the component is transferred by the heat pipe. As a result, the relative position of heat pipe and component can be adjusted within a certain range. Because different component layouts may lead to the same heating power transmission. Therefore, each objective value corresponds to a region in the decision space. The conclusion is that there is the high probability that individuals with identical objective value are in the same feasible region. Based on this analysis, an environmental selection with convergence controlling is required to maximize the search ability. Thus, in the proposed ES-OD, individuals with the identical objective value should be grouped together. Then, individual with minimum density is selected to represent the whole group of the same objective.

The specific process of the proposed ES-OD is summarized in Algorithm 1. At the beginning of the stage, the population P^X will be evaluated in Line 1. Then the population is classified according to objective value in Lines 2 and 3. The individuals with the identical objective values is assigned to the same group. After that, the density d_i of each individual p_i will be calculated in Line 7 according to the decision variable of individuals in Eq. (3).

$$d_i = \sqrt{\sum_{k=1, k \neq i}^{k=|S|} (p_i - p_k)^2},$$ (3)

where S represents a classification set, p_i and p_j are the individuals with the identical objective value in S. Then individuals in the group with minimum density will be retained, while the others will be replaced by the new randomly generated individuals in Lines 9–14.

Algorithm 1: Environmental selection based on individual density in objective and decision space (ES-OD).

Input: the combination of population and offspring $\mathbf{P'^X}$

Output: the population after environmental selection $\mathbf{P^X}$

1. Evaluate the objective of $\mathbf{P'^X}$;

2. Assign the ranks of $\mathbf{P'^X}$ according to objective value;

3. Decompose $\mathbf{P'^X}$ into $S = \{s_1, s_2, ..., s_n\}$ according to the identical objective value of individuals; // n is the number of groups

4. **For each s_j in S do**

5.   **If $|s_j| \geq n_s$ then** // n_s is the size of groups

6.   **For each p_i^X in s_j do**

7.   Calculate d_i according to Eq. (7);

8. $s = \emptyset$;

9.   **If d_i of p_i^X is the minimum then**

10. $s = p_i^X$;

11.   **Else**

12.   Generate p^X randomly;

13. $s = p^X$;

14.   **End if**

15. $s_j = s$;

16.   **End for**

17.   **End if**

18. **End for**

19. $\mathbf{P^X} = \{s_1 \cup s_2 \cup ... \cup s_n\}$; // Join all groups

20. **Return $\mathbf{P^X}$**

After the first search stage, the optimal solution p_{opt}^X satisfying constraints G3 and G4 is found, here, constraints G1 and G2 are not considered yet. Therefore, in the second stage, p^Y will be optimized further to meet the constraints G1 and G2, to be introduced in Subsect. 3.3.

3.3 Repair Strategy Based on Neighborhood Search

After the first stage of optimization, the solution p^X with the minimal objective value can be obtained under the condition that constraints G3 and G4 are satisfied while constraints G1 and G2 are not considered. Therefore, in the second stage, the subpopulation P^Y is required to adjust according to the local neighborhood search strategy to repair the infeasible solution for satisfying the constraints G1 and G2.

When the number of samples is large enough and the distribution is unknown, Gaussian disturbance [12] is usually used according to the central limit theorem of probability theory. Therefore, Gaussian disturbance is considered to generate new population according to the i-th individual with the minimal constraint violation c_i in order to satisfy the constraints G1 and G2, where c_i is the sum of G_1 and G_2. As a result, the components will be relocated in the vertical space by optimizing the p^Y so as to satisfy the constraints G1 and G2.

Therefore, the solutions $\{p_1^Y, p_2^Y, \ldots, p_n^Y\}$ is generated according to Gaussian disturbance at the beginning. Next, constraint violation c_i will be evaluated for Each individual p_i^Y. Then, the result with minimal constraint violation will be obtained. After that, if solution with minimum constraint violation is feasible, then break the while-loop and return the feasible solution. Otherwise, the while-loop will be continued till the second termination condition met. Finally, the best solution p_{opt}^X found in the first search stage merges with the feasible solution p_{opt}^Y found in the second stage so as to constitute the final solution p_{opt} with best objective and all satisfying constraints.

An example for the second stage optimization is shown in Fig. 3 where Subfig. 3(a) correspond to the component's layout before and after repair, respectively. It can be seen from the figure that overlapping components are relocated to avoid overlapping between components, which satisfy the constraint G1 after the second stage. The Subfig. 3(b) shows that the centroid of layout is adjusted into a feasible region after the repair stage in order to satisfy the constraint G2. Note that the horizontal position of components has not been modified in the repair stage. Therefore, this strategy repairs the infeasible solutions to meet all the constraints.

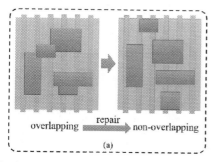 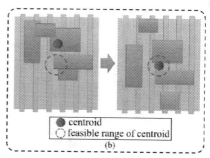

Fig. 3. The optimal layout of HCLO problem. (a) Process of repairing overlapping; (b) Process of adjusting centroid.

3.4 Proposed Algorithm TEA/HCLO

The pseudo code of the TEA/HCLO is shown in Algorithm 2. The population, $\mathbf{P} = \{\boldsymbol{p}_1, \boldsymbol{p}_2, \ldots, \boldsymbol{p}_{N_c}\}$, will be initialized in Line 1, where N_c is the population size. Then population $\mathbf{P}$ is divided into $\mathbf{P^X} = \{\boldsymbol{p}_1^X, \boldsymbol{p}_2^X, \ldots, \boldsymbol{p}_{N_c}^X\}$ and $\mathbf{P^Y} = \{\boldsymbol{p}_1^Y, \boldsymbol{p}_2^Y, \ldots, \boldsymbol{p}_{N_c}^Y\}$ in Line 2, where $\boldsymbol{p}_i^X = (x_1, x_2, \ldots, x_n)$ represents the horizon coordination x of the i-th individual. The first stage begins after decomposition. At the beginning of the first stage, offspring generated by using simulated binary crossover and polynomial mutation operators based on $\mathbf{P^X} = \{\boldsymbol{p}_1^X, \boldsymbol{p}_2^X, \ldots, \boldsymbol{p}_{N_c}^X\}$ in Line 5. Then, the current population $\mathbf{P^X}$ is combined with its offspring and is executed the ES-OD environmental selection strategy in Line7. When reaching the maximum evaluations n_1 of the first stage, the algorithm enters the second stage. Then, at the second stage, the subpopulation $\mathbf{P^Y} = \{\boldsymbol{p}_1^Y, \boldsymbol{p}_2^Y, \ldots, \boldsymbol{p}_{N_c}^Y\}$ is generated by Gaussian disturbance and optimized by using neighborhood search iteratively to satisfy the other two constraints in Lines 13–19. Finally, the feasible solution $\boldsymbol{p}_{opt}$ with optimal objective value can be obtained in Line 20.

Algorithm 2: Two-stage evolutionary algorithm

Input: the population size N_c, the number of evaluations in the first stage n_1, the number of evaluations in the second stage n_2

Output: the approximate optimal solution $\mathrm{p_{opt}}$.

1. Initialize the population $\mathbf{P} = \{\boldsymbol{p}_1, \boldsymbol{p}_2, \ldots, \boldsymbol{p}_{N_c}\}$;

2. Decomposing $\mathbf{P}$ into $\mathbf{P^X}$ and $\mathbf{P^Y}$;

3. *evaluation* = 0;

4. **While** *evaluation* $< n_1$ **do**

5.   Generate offspring $\mathbf{O^X} = \{\boldsymbol{o}_1^X, \boldsymbol{o}_2^X, \ldots, \boldsymbol{o}_{N_c}^X\}$ based on $\mathbf{P^X}$ with mutation and crossover operator;

6. $\mathbf{P'^X} = \{\mathbf{P^X} \cup \mathbf{O^X}\}$;// Concatenate the current population and offspring

7. $\mathbf{P^X} =$ES-OD $(\mathbf{P'^X})$; // Obtain the $\mathbf{P^X}$ using the Algorithms 1;

8.   *evaluation = evaluation +1;*

9. **End while**

10. Find $\boldsymbol{p}_{opt}^X$ with the minimal objective value in $\mathbf{P^X}$;

11. Randomly generate $\boldsymbol{p}^Y = (y_1, y_2, \ldots, y_n)$;

12. *evaluation* = 0;

13. **While** *evaluation* $< n_2$ **do**

14. $\boldsymbol{p}_{opt} =$*Neighborhood_search*$(\boldsymbol{p}_{opt}^X, \boldsymbol{p}^Y)$; // Obtain the $\boldsymbol{p}_{opt}$ using the Algorithms 2;

15.   **If** $\boldsymbol{p}_{opt}$ is the feasible solution **then**

16.   **Break**;

17.   **End if**

18.   *evaluation = evaluation +1;*

19. **End while**

20. **Return** $\boldsymbol{p}_{opt}$.

The computational complexity of Algorithm 1 is $O(N \times n_1)$ due to the loop structure from Lines 4–18 in Algorithm 1, where N is the population size and n_1 is the generation number of the first stage. The total computational complexity of TEA/HCLO is $O(N \times G)$, where G is the max generation number of those two stages.

4 Experimental Results and Discussions

In this section, the proposed TEA/HCLO is compared with six representative algorithms, ABC [13], CMAES [14], OFA [15], SHADE [16], CCMO [17] and CMOEA-MS [18] to verify the performance of the TEA/HCLO. The CCMO and CMOEA-MS are effective EAs of constraint multi-objective optimization, where CCMO has a coevolutionary framework and CMOEA-MS is a two-stage evolutionary algorithm, which adjusts the objective evaluation strategies during the evolutionary process to adaptively balance objective optimization and constraint satisfaction. The two algorithms are modified to the single-objective algorithm by changing the non-dominant sorting of multi-objectives to ordinary sort of single objective.

Five HCLO problems with increasing scales are used to verify the performance of the comparative algorithms. For HCLO problems, the parameters are the number of components, the number of heat pipes, and the size of board. And the detailed information about these problem parameters is presented in Table 1, where the number of components and heat pipes varies from 6 up to180, and 4 up to 60 respectively. The size of board varies from (80,50) up to (6000,6000).

Table 1. The parameter settings of problems used in the experiment.

Problem	Components	Heat pipes	Board
P1	6	4	(80,50)
P2	15	6	(1000,1000)
P3	40	16	(2000,2000)
P4	90	30	(4000,4000)
P5	180	60	(6000,6000)

The parameters in the comparative algorithms are set as suggested in their original papers. In ABC, the limitation of the number of trials for releasing a food source is set to 20. In CMAES, the population size λ is set to 10 and parent number μ is set to 1. In OFA, the optimal solution is obtained when optimization error is less than 1.0e-6. In SHADE, the population size and memory size are set to 100.

For fairness, the population size of all the compared EAs is set to 100, the maximum number of function evaluations of all the compared EAs is set to 100000, and the independent run number of each algorithm is 30. Besides, the number of function evaluations of TEA/HCLO is set to 70000 times in the first stage and 30000 times in the second stage, respectively.

On the one hand, the average number of feasible solutions after 30 independent runs for each algorithm on five HCLO problems is shown in Table 2. Among them, the TEA/HCLO can always get the highest feasibility rate. Other algorithms are not so effective because of the large constraints number in HCLO. Compared with other algorithms, the TEA/HCLO has obvious advantages in searching for the feasible solutions owe to the decomposition strategy and repair strategy. Results indicates that the two-stage strategy effectively reduces the difficulty of solving the two sub-stages by problem decomposition, therefore, less constraints are considered in each stage.

Table 2. The feasibility rate of solutions of TEA/HCLO and other six EAs on five HCLO problems.

Problem	P1	P2	P3	P4	P5
TEA/HCLO	100%	100%	100%	100%	100%
ABC	100%	100%	92%	72%	60%
OFA	100%	77%	67%	0%	0%
CMAES	0%	0%	0%	0%	0%
SHADE	100%	100%	100%	100%	0%
CCMO	100%	100%	100%	100%	80%
CMOEA_MS	100%	100%	100%	100%	47%

On the other hand, the mean and standard deviation of objective values of TEA/HCLO and other six EAs on five HCLO problem are shown in Table 3. If an algorithm score None on objective value, it means that this algorithm cannot find the feasible solutions after all 30 independent runs. It should be noted that all experimental results in this paper are adopted with the Wilcoxon rank-sum test with a significance level of 0.05 [19]. The "+", "−" and "≈" in the last row of the table indicate that the results are significantly better than, worse than, and statistically similar to the TEA/HCLO, respectively. The best average value of each HCLO problem in each row is highlighted in **bold**.

As seen from Table 3, the performance of TEA/HCLO on the HCLO problems is much better than that of the six comparative EAs overall. Among the five problems, the TEA/HCLO obtains four best objective values, which is the best algorithm with the largest number of optima. On the objective value indicator, the comprehensive performance of TEA/HCLO has achieves the best performance with the increasing scale of problems of P3, P4 and P5. Results has shown that the ES-OD strategy in the first stage has enhanced the searching ability to find the best objective.

It can be concluded that the TEA/HCLO achieved the better performance than other EAs from the comparative experimental results according to objective values. It shows that the new environmental selection strategy based on individual density proposed in Subsect. 3.2 can select more favorable solutions for the HCLO problem and repair strategy based on neighborhood search proposed in Subsect. 3.3 can be more effective in repairing infeasible solution.

Table 3. Objective values of TEA/HCLO and other six EAs on five HCLO problems.

Problem	P1	P2	P3	P4	P5	$\pm / =$
TEA/HCLO	**−6.2500e + 00 (0.00e + 00)**	**4.4367e + 01 (2.10e + 00)**	**8.7018e + 00 (3.92e + 00)**	**2.6635e + 01 (1.09e + 01)**	**3.9713e + 01 (4.14e + 01) −**	
ABC	1.2872e + 01 (1.10e + 01) −	6.2500e + 00 (0.00e + 00) −	2.7643e + 01(1.12e + 00) −	4.7992e + 01 (1.12e + 01) −	5.6044e + 02 (8.06e + 01) −	0/5/0
OFA	**6.2500e + 00 (0.00e + 00) =**	6.0245e + 02 (2.89e + 02) −	4.1432e + 02 (8.45e + 01) −	None	None	0/4/1
CMAES	None	None	None	None	None	0/5/0
SHADE	**6.2500e + 00 (0.00e + 00) =**	**2.6961e + 01 (1.12e + 00) +**	4.7992e + 01 (1.12e + 01) −	5.6044e + 02 (8.06e + 01) −	None	1/3/1
CCMO	1.6503e + 01 (2.26e + 01) −	1.0739e + 01 (8.83e + 00) −	6.7244e + 01 (3.61e + 00) −	5.4742e + 01 (5.61e + 00) −	7.1206e + 01 (4.01e + 00) −	0/5/0
CMOEA_MS	1.0739e + 01 (8.83e + 00) −	6.7244e + 00 (3.61e + 00) −	1.4742e + 01 (5.61e + 00) −	4.8797e + 01 (7.06e + 00) −	6.8596e + 01 (2.78e + 00) −	0/5/0

5 Conclusion

In this paper, a new two-stage evolutionary algorithm has proposed for solving the HCLO via decomposing the HCLO into two subproblems to be optimized in two stages separately. In the first stage, an environmental selection strategy based on individual density in decision and objective space (ES-OD) is applied to search the optimal layout efficiently. In the second stage, a repair strategy based on neighborhood search is introduced to repair the infeasible solutions. Experiments results on five different scale HCLO problems have shown that the TEA/HCLO can find the better layout and ensure the feasibility of the solutions when compared with the other six competitive algorithms.

Acknowledgements. This research work is supported by the National Natural Science Foundation of China (Grant No. 61976046) and the Spark Program of Earthquake Sciences (Grant No. XH20042).

References

1. Popovic, J., Ferreira, J.A.: Concepts for high packaging and integration efficiency. In: 35th Annual IEEE Power Electronics Specialists Conference (PESC 04), pp. 1–13. Springer, Heidelberg (2016)

2. Mansouri, N., Weasner, C., Zaghlol, A.: Characterization of a heat sink with embedded heat pipe with variable heat dissipating source placement for power electronics applications. In: 17th IEEE Intersociety Conference on Thermal and Thermomechanical Phenomena in Electronic Systems (ITherm), CA, San Diego, pp. 311–317 (2018)

3. Qiao, Z., Zhang, W., Zhu, J., Tong, G.: Layout optimization of multi-component structures under static loads and random excitations. Eng. Struct. **43**(5), 120–128 (2012)

4. Whitley, D., Rana, S., Dzubera, J., Mathias, K.E.: Evaluating evolutionary algorithms. Artif. Intell. **84**(1–2), 357–358 (1996)

5. Scrucca, L.: GA: a package for genetic algorithms in R. J. Stat. Softw. **53**, 1–37 (2013)

6. Bratton, D., Kennedy, J.: Defining a standard for particle swarm optimization. In: IEEE Swarm Intelligence Symposium (2007)

7. Storn, R., Price, K.: Differential evolution-a simple and efficient heuristic for global optimization over continuous spaces. J. Global Optim. **11**(4), 341–359 (1997)

8. Kumar, R., Jyotishree: Blending roulette wheel selection & rank selection in genetic algorithms (2012)

9. Blickle, T., Thiele, L.: A Mathematical Analysis of Tournament Selection. Morgan Kaufmann Publishers Inc., Burlington (1998)

10. Deb, K., Beyer, H.G.: Self-Adaptation in Real-Parameter Genetic Algorithms with Simulated Binary Crossover. Morgan Kaufmann Publishers Inc., Burlington (1999)

11. Hamdan, M.: A dynamic polynomial mutation for evolutionary multi-objective optimization algorithms. Int. J. Artif. Intell. Tools **20**(01), 209–219 (2011)

12. Nyawade, K.O.: Generalized inverse Gaussian distributions under different parametrizations research report. Mathematics, Number 27 (2018)

13. Karaboga, D.: An idea based on honey bee swarm for numerical optimization, Technical Report - TR06 (2005)

14. Hansen, N., Ostermeier, A.: Completely Derandomized Self-Adaptation in Evolution Strategies. MIT Press, Cambridge (2001)

15. Zhu, G.Y., Zhang, W.B.: Optimal foraging algorithm for global optimization. Appl. Soft Comput. **51**, 294–313 (2017)

16. Tanabe, R., Fukunaga, A.: Success-history based parameter adaptation for differential evolution. In: 2013 IEEE Congress on Evolutionary Computation, Cancun, Mexico, pp. 71–78. IEEE (2013)

17. Tian, Y., Zhang, T., Xiao, J., Zhang, X., Jin, Y.: A coevolutionary framework for constrained multi-objective optimization problems. IEEE Trans. Evol. Comput. **1**(25), 102–116 (2021)

18. Tian, Y., Zhang, Y., Su, Y., Zhang, X., Tan, K.C., Jin, Y.: Balancing objective optimization and constraint satisfaction in constrained evolutionary multi-objective optimization. IEEE Trans. Cybern. **9**(52), 9559–9572 (2022)

19. Derrac, J., García, S., Molina, D., Herrera, F.: A practical tutorial on the use of nonparametric statistical tests as a methodology for comparing evolutionary and swarm intelligence algorithms. Swarm Evol. Comput. **1**(1), 3–18 (2011)

20. Bojan-Dragos, C.A., et al.: GWO-based optimal tuning of type-1 and type-2 fuzzy controllers for electromagnetic actuated clutch systems. In: 4th IFAC Conference on Embedded Systems, Computational Intelligence and Telematics in Control (CESCIT), Valenciennes, France, pp. 189–194 (2021)

21. Precup, R.E., David, R.C., Roman, R.C., Petriu, E.M., Szedlak-Stinean, A.I.: Slime mould algorithm-based tuning of cost-effective fuzzy controllers for servo systems. Int. J. Comput. Intell. Syst. **1**(14), 1042–1052 (2021)

22. Cai, Z.H., Gao, S.C., Yang, X., Yang, G., Cheng, S., Shi, Y.H.: Alternate search pattern-based brain storm optimization. Knowl.-Based Syst. **238**, 107896 (2022)

A Large-Scale Multi-objective Brain Storm Optimization Algorithm Based on Direction Vectors and Variance Analysis

Yang Liu[1], Tiejun Xing[3(✉)], Yuee Zhou[1], Nan Li[1], Lianbo Ma[1,2(✉)], Yingyou Wen[4], Chang Liu[5], and Haibo Shi[5]

[1] College of Software, Northeastern University, Shenyang 110819, China
`malb@swc.neu.edu.cn`
[2] Key Laboratory of Smart Manufacturing in Energy Chemical Process, Ministry of Education, East China University of Science and Technology, Shanghai, China
[3] Neusoft Institute of Intelligent Medical Research, Shenyang 110167, China
`xingtj@neusoft.com`
[4] School of Computer Science and Engineering Northeastern University, Shenyang 110167, China
[5] Digital Factory Department, Shenyang Institute of Automation Chinese Academy of Sciences, Shenyang 110016, China

Abstract. Large-scale multi-objective optimization problems (LSMOPs) can lead to the conventional reproduction operator being inefficient for searching. Therefore, we propose a large-scale multi-objective brain storm optimization algorithm based on direction vectors and variance analysis (LMOBSO-DV) to enhance the efficiency of tackling LSMOPs. Specifically, we adopt brain storm optimization (BSO) algorithm using reference vectors to divide the population into subpopulations and guide the individuals i) in each subpopulation to search in promising directions and 2) between subpopulations to maintain diversity. We also design a new mutation operator. On a widely used LSMOPs test suites with 1000 decision variables, 2 objectives, and 3 objectives, we evaluate LMOBSO-DV's effectiveness in comparison to other several state-of-the-art algorithms. The results of the experiment show that our proposed approach, LMOBSO-DV, outperforms the other studied algorithms.

Keywords: Brain storm optimization · Multi-objective optimization · Large-scale optimization

1 Introduction

A multi-objective optimization problem (MOP) is one that has numerous conflicting objective functions. [14,21–23]. A MOP is referred to as a large-scale multi-objective optimization problem (LSMOPs) when the decision vector's dimensionality surpasses 1000 [12,17,20,32]. However, traditional multi-objective evolutionary algorithms (MOEAs) are severely affected by the exponential growth of the search space, which is known as dimensional disaster.

Various large-scale MOEAs (LMOEAs) have been designed to address the dimensional catastrophe, which are generally classified into three categories, namely, methods based on grouping of decision variables, methods based on decision space reduction and methods based on novel search strategies. Methods based on grouping of decision variables employ some sort of grouping technique to divide large-scale issues into smaller ones, which are then solved one by one with the existing optimizer [2,13,25,29]. Methods based on decision space reduction try to find a latent low-dimensional search space for large-scale problems by some strategies, and there are two main strategies, one is a problem transformation-based approach that optimizes a low-dimensional space of weight variables instead of optimizing a large-scale search space [8,15,38,39], and the other depends on decreasing dimensionality, which lowers the number of decision variables by machine learning dimensionality reduction strategies [16]. Methods based on novel search strategies aim to investigate novel reproduction operators to enhance search efficiency in original large-scale search spaces [1,9,10,31,37].

LMOEAs based on decision variable grouping and LMOEAs based on decision space reduction increase the efficiency of solving LSMOPs by decreasing the number of the decision variables. However, searching in a low-dimensional space causes the population to gravitate towards local optimums because the entire search space is not explored completely. The existing LMOEAs based on novel search strategies can enhance the search ability of solving LSMOPs to some extent, but they are still plagued by computation slowness because of the potentially slow convergence rate.

The brain storm optimization (BSO) algorithm [18,26–28] is oriented to clustering centers and learns probabilistically toward other individuals, which can improve the right convergence-diversity balance. To alleviate the shortcomings of existing LMOEAs, we improve the BSO algorithm and propose the large-scale multi-objective brain storm optimization algorithm based on direction vectors and variance analysis (LMOBSO-DV). Specifically, we use BSO algorithm that uses reference vectors to split the current population into distinct subpopulations, and then in each subpopulation, let the poorly performing individuals learn from the well performing individuals as a way to improve population convergence, and in different subpopulations, learn from each other as a way to improve population diversity, and we also consider the effect of different decision variables on the search efficiency and propose a new mutation operator. The following are our main contributions.

1. Using BSO algorithm, clustering is performed using reference vectors to split the current population into distinct subpopulations and to generate new individuals by defining direction vectors within or between subpopulations to improve search efficiency.
2. A new mutation operator is proposed, by which the search space can be fully explored.

2   Related Work

2.1   BSO

Motivated by the process of brainstorming, Shi et al. [18, 26–28] proposed BSO algorithm, which imitates the activity of the human mind and uses the idea of continuous convergence and divergence operations to find the optimum. It employs clustering for locating the local optimum and then generates offsprings by mutation, balancing convergence and diversity. The basic process of BSO algorithm is as follows.

1. Population initialization
2. Clustering
3. Generate new individuals
4. Updating populations
5. If the limit amount of evaluations is achieved, then output the optimum individual; otherwise, return to step 2

BSO has three major operations: clustering, new solution generation, and mutation. For clustering, BSO employs KMeans [34], and the best-performing individual in each cluster serves as the cluster center. When generating offsprings, individuals are selected with specific probability from four ways of generating individuals, i.e., one cluster center, individuals in one cluster, two cluster centers, or two individuals belong to different clusters, and then weighted to generate offsprings. On this premise, offsprings are produced through Gaussian mutation, which is defined as follows.

$$X^i_{new} = X^i + N(\mu, \sigma^2) \times m \tag{1}$$

where X^i and X^i_{new} are the i-th dimension of the current individual and the offspring, respectively. Having μ as the mean and σ as the standard deviation, $N(\mu, \sigma^2)$ is a Gaussian random function. m is a variable that quantifies the input of Gaussian random values, which is defined as follows.

$$m = rand() \times \log sig(\frac{\frac{t_{max}}{2} - t}{k}) \tag{2}$$

where $rand()$ returns a random number between 0 and 1. $\log sig()$ is a logarithmic sigmoid transfer function. t_{max} and t are the maximum iterations and the current iteration, respectively. The slope of the $\log sig()$ function is controlled by the user-defined parameter k.

2.2   LMOEA

To handle with LSMOPs, a variety of LMOEAs have been developed. These algorithms are classified into three categories: LMOEAs based on grouping of decision variables, LMOEAs based on decision space reduction, and LMOEAs based on novel search strategies.

LMOEAs Based on Grouping of Decision Variables. The work [2] proposed CCGDE3, which subdivided all decision variables into various groups and subsequently optimized them using co-evolutionary framework. Ma et al. [24] proposed MOEA/DVA, which analyzes decision variables and categorizes them into three subgroups: convergence-related, diversity-related and mixed variables, respectively, before optimizing them using different strategies. Song et al. [29] proposed MOEA/D-RDG by combining MOEA/D and random dynamic grouping. LMEA [36] analyzed the attributes of decision variables using angle-based cluster analysis and optimized the two kinds of variables with convergence-related and diversity-related optimization methods, respectively.

LMOEAs Based on Decision Space Reduction. Zille et al. [39] developed a weighted optimization framework (WOF) to achieve search space dimensionality reduction by grouping decision variables into numerous groups and assigning a weight to each group, transforming the original high-dimensional search space into the low-dimensional weight space. Two weight variables are established in the LSMOF framework [11] to sample the points along two search directions to find a better individual to decrease the number of the decision variables and improve the efficiency of solving LSMOPs. Liu et al. [16] developed the PCA-MOEA algorithm, which apply PCA to LSMOPs.

LMOEAs Based on Novel Search Strategies. Cheng et al. [4] proposed a probabilistic model-based method, namely IM-MOEA, in which a Gaussian process-based inverse model is employed to transfer individuals from objective vectors to decision vectors, and the inverse model was utilized to sample points from the objective space to generate offspring. To solve LSMOPs, LMOCSO [33] employs a competitive grouping optimizer. It obtains good outcomes by utilizing a competitive method that enables losers to learn from victors. An adaptive offspring generation strategy for solving LSMOP is proposed in DGEA [9]. It partitions the solution into dominated and non-dominated solutions using non-dominated sort, and generates new individuals along the direction vector between the pair.

3   The Proposed Method

We introduce our proposed algorithm LMOBSO-DV in detail in this section. We extend the BSO method to solve LSMOPs effectively. As shown in Fig. 1, we begin with a random initialization operation, then we initialize some reference vectors, and subdivide the population into multiple subpopulations based on the reference vectors, which differs from the original BSO clustering in the decision space. After clustering, we look for promising directions from within or among subpopulations in order to generate promising solutions. On this basis, decision variables with low exploitation are fully explored using our new mutation operator. Afterwards, utilizing the environmental selection to keep the better individuals.

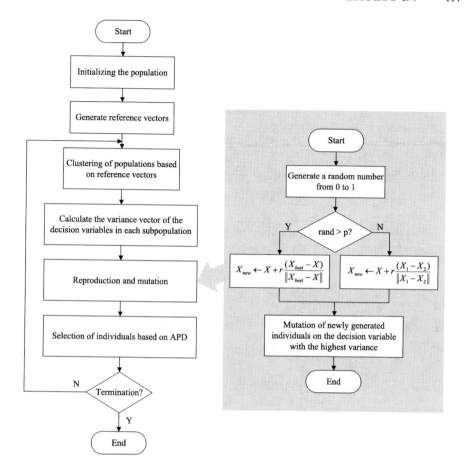

Fig. 1. The LMOBSO-DV algorithm flowchart

3.1 Reproduction

The proposed offspring generation strategys aim to generate optimal solutions by finding promising directions within and among populations, as shown in lines 12-17 of Algorithm 1. To measure the solutions and also to increase the pressure for convergence and maintain diversity, we choose to use the angle-penalized distance (APD) proposed in RVEA [5].

$$\text{APD}(\theta_{i,j}, \mathbf{f}'_i) = \left(1 + \left(\frac{t}{t_{\max}}\right)^\alpha \times \frac{M \cdot \theta_{i,j}}{\min_{i \in \{1, \cdots, |V|\}, i \neq j} \langle \mathbf{v}_i, \mathbf{v}_j \rangle}\right) \times \|\mathbf{f}'_i\| \quad (3)$$

where t and $t_{\max}$ is the current iterations and the maximum iterations, respectively. α is a parameter, which is defined by user. M is the number of objective functions. $\mathbf{f}'_i$ is the objective vector of the current individual. RV is the reference

---

Algorithm 1: LMOBSO-DV

Input: N (Population size), N_{ref} (Reference vectors's number), $t_{\max}$ (Maximum iterations)

Output: P (Population at the end)

1 $t \leftarrow 0$;
2 $P \leftarrow$ Pop_Init(N);
3 $RV \leftarrow$ Uniform_Reference_Vector(N_{ref});
4 **while** $t < t_{\max}$ **do**
5 $\quad$ $Clusters, APDs \leftarrow$ Clustering(P, RV);
6 $\quad$ $p \leftarrow \frac{t}{t_{\max}}$;
7 $\quad$ **foreach** $pop \in Clusters$ and $apd \in APDs$ **do**
8 $\quad\quad$ $X_{best} \leftarrow \arg\min(apd)$;
9 $\quad\quad$ $Vars \leftarrow$ GetVariances(pop);
10 $\quad\quad$ $d \leftarrow \arg\max(Vars)$;
11 $\quad\quad$ **foreach** $X \in pop$ **do**
12 $\quad\quad\quad$ $r \leftarrow N(0,1)$;
13 $\quad\quad\quad$ **if** $rand() > p$ **then**
14 $\quad\quad\quad\quad$ $X_{new} \leftarrow X + r\frac{X_{best}-X}{\|X_{best}-X\|}$;
15 $\quad\quad\quad$ **else**
16 $\quad\quad\quad\quad$ $X_1, X_2 \leftarrow$ From other two subpopulations, choose two individuals at random;
17 $\quad\quad\quad\quad$ $X_{new} \leftarrow X + r\frac{X_1-X_2}{\|X_1-X_2\|}$;
18 $\quad\quad\quad$ **end**
19 $\quad\quad\quad$ $X_{new}^d \leftarrow rand() \times (UB(X^d) - LB(X^d)) + LB(X^d)$;
20 $\quad\quad\quad$ $APD_{X_{new}} \leftarrow$ GetAPD(X_{new});
21 $\quad\quad\quad$ $APD_X \leftarrow$ GetAPD(X);
22 $\quad\quad\quad$ **if** $APD_{X_{new}} < APD_X$ **then**
23 $\quad\quad\quad\quad$ $X \leftarrow X_{new}$;
24 $\quad\quad\quad$ **end**
25 $\quad\quad$ **end**
26 $\quad$ **end**
27 **end**

---

vectors set. $\theta_{i,j}$ denotes the acute angle formed by the objective vector $\mathbf{f}'_i$ and the reference vector $\mathbf{v}_j$.

We propose two types of offspring generation strategies, one for offspring generation in populations and the other for offspring generation between populations. For the former, we first use APD to measure the superiority of the solutions in the subpopulation and find the optimal individual X_{best} in the current subpopulation according to the APD value, and then for the current solution X, we construct the direction vector $X_{best} - X$ of the current individual to the optimal individual, and generate the convergent individual X_{new} on the direction vector, as shown in Eq. (4); for the latter, we select any two subpopulations from other subpopulations besides the current one, and then randomly select two individuals X_1 and X_2 from them respectively, and then construct a direc-

---

Algorithm 2: GetVariances

Input: *pop* (population)
Output: *Variances* (variances vector)
1 $X \leftarrow$ GetDecisionMatrix(*pop*);
2 $n, d \leftarrow$ size(X);
3 $Vars \leftarrow$ zeros(d);
4 **for** $i \leftarrow 1 : d$ **do**
5 $\quad | \quad Vars(i) \leftarrow$ var($X(:, i)$);
6 **end**

---

tion vector $X_1 - X_2$ based on this, and generate the solution of diversity on this direction vector, as shown in Eq. (5).

$$X_{new} \leftarrow r \frac{X_{best} - X}{\|X_{best} - X\|} + X \tag{4}$$

$$X_{new} \leftarrow r \frac{X_1 - X_2}{\|X_1 - X_2\|} + X \tag{5}$$

We choose the offspring generation strategy based on a probability $p = t/t_{max}$, utilizing the first strategy if the random decimal number is greater than p and the second strategy otherwise. Based on this, our algorithm will tend to be convergent in the early phase and diverse in the later phase.

3.2 Mutation

Gaussian mutation operator [7,35] is not suitable for LSMOPs, so we developed a new mutation operator. Firstly, based on the current subpopulation, we analyze the variance for each decision variable, as shown in Algorithm 2. The larger the variance, the more we think the current decision variable is not explored sufficiently, so we find the decision variable with the largest variance and perform mutation on it, so that multiple populations are purposefully explored for different decision variables, which improves the search efficiency for large-scale search space. The mutation operator is defined as follows.

$$X_{new}^i = \left(UB\left(X^i\right) - LB\left(X^i\right)\right) \times rand() + LB\left(X^i\right) \tag{6}$$

where X^i is the i-th dimension of the current individual X, $UB(X^i)$ and $LB(X^i)$ is the upper and lower bound of X^i, respectively, and $rand()$ returns a random decimal number from 0 to 1.

4 Experiments

4.1 Algorithms for Comparison and Test Suites

We evaluate the efficacy of our proposed LMOBSO-DV algorithm to three other state-of-the-art LMOEAs, namely MOEA/DVA [24], LSMOF [11] and DGEA

[9], and the parameters settings of the chosen algorithms follow their papers. We choose to experiment on LSMOP1-9 test problems [6,17] with 1000 decision variables, 2 objectives, and 3 objectives. All experiments are run on PlatEMO [30].

Each algorithm run independently on LSMOPs test suites 20 times. As stated in Table 1, the results of the experiment are statistically assessed with the Wilcoxon rank sum test with a significance level of 0.05. "+" is used if the current algorithm is considerably better than LMOBSO-DV; "−" is used if the current method is considerably worse than LMOBSO-DV; and "≈" is used if the current algorithm is similar to LMOBSO-DV.

4.2 Parameter Settings and Performance Indicator

We set the number of reference vectors to 10, the population size to 300, and the maximum number of evaluations to 80000 for all test instances.

It is known that the inverse generation distance (IGD) [3,19,40] can indicate both the algorithm's convergence and diversity. The lower the IGD number, the more effective the search algorithm. So we choose IGD as the performance indicator of the algorithm. Assuming that $\mathcal{S}$ is the sampling point on the true pareto front and $\mathcal{P}$ is the non-dominated solutions set of the final population, IGD is defined as follows.

$$\text{IGD}(\mathcal{P}, \mathcal{S}) = \frac{\sum_{\mathbf{y} \in \mathcal{S}} [\min_{\mathbf{x} \in \mathcal{P}} \|\mathcal{F}(\mathbf{x}) - \mathbf{y}\|]}{|\mathcal{S}|} \tag{7}$$

where $|\mathcal{S}|$ is the size of $\mathcal{S}$, and we set it to 10000 in this paper.

4.3 Results and Discussion

Table 1 gives the IGD statistical results for all algorithms on the LSMOP1-9 test problems with 1000 decision variables, 2 objectives, and 3 objectives. As can be seen in Table 1, LMOBSO-DV achieved 11 of the 18 best results for IGD, outperforming MOEA/DVA, LSMOF and DGEA overall. Compared to LSMOF, LMOBSO-DV performed poorly on two 2-objective test problems and three 3-objective test problems, but outperformed LSMOF on six 2-objective test problems and five 3-objective test problems. Compared to DGEA, LMOBSO-DV performs poorly on the two 2-objective test problems, comparable to DGEA on the three 3-objective test problems, and better than DGEA on the other test problems.

The statistical results show that the IGD values of LMOBSO-DV outperform the three algorithms compared, which proves that our proposed algorithm is capable of achieving both good convergence and diversity, and performs well in solving LSMOPs, demonstrating that our proposed algorithm LMOBSO-DV may improve the efficiency of large-scale space search.

Table 1. Comparison results of IGD indicators for different algorithms on 1000-D, 2-M and 3-M LSMOPs test suites. The greatest performing algorithm is highlighted in each row.

Problem	M	D	MOEA/DVA	LSMOF	DGEA	LMOBSO-DV
LSMOP1	2	1000	1.5026e+1 (1.75e-1) −	6.3254e-1 (2.42e-2) −	4.7553e-1 (2.20e-1) −	**4.5421e-1 (2.31e-1)**
	3		1.3257e+1 (2.11e-1) −	**6.1213e-1 (7.32e-3) +**	6.8631e-1 (1.53e-1) −	6.3673e-1 (2.13e-2)
LSMOP2	2	1000	3.7631e-2 (1.45e-4) −	1.1280e-2 (3.02e-4) −	**8.4123e-3 (5.34e-4) +**	8.9831e-3 (2.74e-4)
	3		5.2349e-2 (1.36e-3) −	5.9278e-2 (2.21e-3) −	3.9753e-2 (1.13e-3) ≈	**3.937e-2 (4.23e-4)**
LSMOP3	2	1000	5.4255e+2 (4.24e+2) −	**1.3711e+0 (7.12e-4) +**	1.4053e+0 (8.23e-2) +	1.4731e+0 (4.23e-3)
	3		2.0114e+2 (1.32e+2) −	8.4067e-1 (7.06e-5) −	9.5380e-1 (5.18e-1) ≈	**8.3068e-1 (1.26e-4)**
LSMOP4	2	1000	7.4323e-2 (1.82e-4) −	2.3303e-2 (4.21e-4) −	2.1423e-2 (1.82e-3) −	**2.0340e-2 (6.32e-4)**
	3		1.1422e-1 (2.13e-3) −	1.2310e-1 (2.28e-3) −	7.2198e-2 (6.07e-3) −	**6.9317e-2 (1.01e-3)**
LSMOP5	2	1000	2.0017e+1 (4.01e-1) −	**7.1349e-1 (2.12e-16) ≈**	5.2359e+0 (1.53e+0) −	7.1360e-1 (2.23e-16)
	3		1.3443e+1 (6.24e-1) −	5.2237e-1 (1.25e-1) −	6.8201e-1 (5.37e-1) −	**5.097e-1 (3.04e-3)**
LSMOP6	2	1000	1.4413e+3 (2.02e+3) −	3.1136e-1 (3.32e-4) −	6.1131e-1 (2.01e-1) −	**2.0076e-1 (1.13e-1)**
	3		3.5267e+4 (5.18e+3) −	**7.2313e-1 (1.10e-2) +**	4.1345e+1 (8.23e+1) −	1.2143e+0 (1.38e-3)
LSMOP7	2	1000	6.5239e+4 (3.11e+3) −	1.1027e+0 (1.42e-3) −	3.2312e+3 (1.11e+3) −	**1.0013e+0 (2.30e-3)**
	3		9.1241e+2 (6.21e+2) −	8.6238e-1 (2.20e-2) ≈	8.9429e-1 (6.10e-2) ≈	**8.5098e-1 (3.81e-2)**
LSMOP8	2	1000	1.8123e+1 (4.09e-1) −	**7.1309e-1 (2.28e-16) +**	2.3243e+0 (1.19e+0) −	7.1509e-1 (2.24e-16)
	3		6.3123e-1 (1.23e-1) −	3.1038e-1 (5.01e-2) −	2.2324e-1 (2.05e-1) −	**1.4919e-1 (3.42e-2)**
LSMOP9	2	1000	4.9236e+1 (1.26e-0) −	7.9823e-1 (1.43e-3) −	2.4234e+0 (9.03e+0) −	**7.3621e-01(4.17e-02)**
	3		1.2469e+2 (2.61e+0) −	**1.2314e+0 (2.12e-1) +**	3.2174e+1 (7.93e+0) −	1.4396e+0 (6.12e-16)
+/ − / ≈			0/18/0	5/11/2	2/13/3	

5 Conclusion

We propose LMOBSO-DV to improve the efficiency of solving LSMOPs. Using reference vectors clustering, multiple subpopulations are divided and the proposed two offspring generation strategies are used in each subpopulation. In the early phase, the algorithm tends to be convergent, where all individuals in the subpopulation learn from the best ones and generate new individuals on the direction vector; in the later phase, the algorithm tends to be diverse, where the generation of individuals is performed in multiple subpopulations. Also, we believe that the exploration of decision variables is crucial to solve LSMOPs, so we propose to use the variance of each decision variable in the population to decide which decision variable to explore. Finally, environmental selection is performed to update the population.

On the LSMOPs test suites, the proposed algorithm and other several state-of-the-art LMOEAs, namely MOEA/DVA, LSMOF, and DGEA, are compared. The experimental results show that LMOBSO-DV outperforms the other algorithms compared, and that it can effectively solve LSMOPs.

Acknowledgments. This work was supported in part by the Fundamental Re-search Funds for the Central Universities No. N2117005, the Joint Funds of the Natural Science Foundation of Liaoning Province und Grant 2021-KF-11-01 and the Fundamental Research Funds for the Central Universities.

References

1. Abdi, Y., Feizi-Derakhshi, M.R.: Hybrid multi-objective evolutionary algorithm based on search manager framework for big data optimization problems. Appl. Soft Comput. **87**, 105991 (2020)
2. Antonio, L.M., Coello, C.A.C.: Use of cooperative coevolution for solving large scale multiobjective optimization problems. In: 2013 IEEE Congress on Evolutionary Computation, pp. 2758–2765. IEEE (2013)
3. Chen, H., Zhu, Y., Hu, K., Ma, L.: Bacterial colony foraging algorithm: combining chemotaxis, cell-to-cell communication, and self-adaptive strategy. Inf. Sci. **273**, 73–100 (2014)
4. Cheng, R., Jin, Y., Narukawa, K., Sendhoff, B.: A multiobjective evolutionary algorithm using gaussian process-based inverse modeling. IEEE Trans. Evol. Comput. **19**(6), 838–856 (2015)
5. Cheng, R., Jin, Y., Olhofer, M., Sendhoff, B.: A reference vector guided evolutionary algorithm for many-objective optimization. IEEE Trans. Evol. Comput. **20**(5), 773–791 (2016)
6. Cheng, R., Jin, Y., Olhofer, M., et al.: Test problems for large-scale multiobjective and many-objective optimization. IEEE Trans. Cybern. **47**(12), 4108–4121 (2016)
7. Cheng, S., Qin, Q., Chen, J., Shi, Y.: Brain storm optimization algorithm: a review. Artif. Intell. Rev. **46**(4), 445–458 (2016). https://doi.org/10.1007/s10462-016-9471-0
8. He, C., Cheng, R., Tian, Y., Zhang, X.: Iterated problem reformulation for evolutionary large-scale multiobjective optimization. In: 2020 IEEE Congress on Evolutionary Computation (CEC), pp. 1–8. IEEE (2020)
9. He, C., Cheng, R., Yazdani, D.: Adaptive offspring generation for evolutionary large-scale multiobjective optimization. IEEE Trans. Syst. Man Cybern. Syst. **52**(2), 786–798 (2020)
10. He, C., Huang, S., Cheng, R., Tan, K.C., Jin, Y.: Evolutionary multiobjective optimization driven by generative adversarial networks (GANs). IEEE Trans. Cybern. **51**(6), 3129–3142 (2020)
11. He, C., et al.: Accelerating large-scale multiobjective optimization via problem reformulation. IEEE Trans. Evol. Comput. **23**(6), 949–961 (2019)
12. Hong, W.J., Yang, P., Tang, K.: Evolutionary computation for large-scale multi-objective optimization: a decade of progresses. Int. J. Autom. Comput. **18**(2), 155–169 (2021)
13. Li, M., Wei, J.: A cooperative co-evolutionary algorithm for large-scale multi-objective optimization problems. In: Proceedings of the Genetic and Evolutionary Computation Conference Companion, pp. 1716–1721 (2018)
14. Li, N., Ma, L., Yu, G., Xue, B., Zhang, M., Jin, Y.: Survey on evolutionary deep learning: Principles, algorithms, applications and open issues. arXiv preprint arXiv:2208.10658 (2022)
15. Liu, R., Liu, J., Li, Y., Liu, J.: A random dynamic grouping based weight optimization framework for large-scale multi-objective optimization problems. Swarm Evol. Comput. **55**, 100684 (2020)
16. Liu, R., Ren, R., Liu, J., Liu, J.: A clustering and dimensionality reduction based evolutionary algorithm for large-scale multi-objective problems. Appl. Soft Comput. **89**, 106120 (2020)
17. Liu, S., Lin, Q., Wong, K.C., Li, Q., Tan, K.C.: Evolutionary large-scale multi-objective optimization: Benchmarks and algorithms. IEEE Trans. Evol. Comput. (2021)

18. Ma, L., Cheng, S., Shi, Y.: Enhancing learning efficiency of brain storm optimization via orthogonal learning design. IEEE Trans. Syst. Man Cybern. Syst. **51**(11), 6723–6742 (2020)

19. Ma, L., Hu, K., Zhu, Y., Niu, B., Chen, H., He, M.: Discrete and continuous optimization based on hierarchical artificial bee colony optimizer. J. Appl. Math. **2014** (2014)

20. Ma, L., Huang, M., Yang, S., Wang, R., Wang, X.: An adaptive localized decision variable analysis approach to large-scale multiobjective and many-objective optimization. IEEE Trans. Cybern. **52**(7), 6684–6696 (2021)

21. Ma, L., et al.: Learning to optimize: reference vector reinforcement learning adaption to constrained many-objective optimization of industrial copper burdening system. IEEE Trans. Cybern. (2021)

22. Ma, L., Wang, X., Huang, M., Lin, Z., Tian, L., Chen, H.: Two-level master-slave RFID networks planning via hybrid multiobjective artificial bee colony optimizer. IEEE Trans. Syst. Man Cybern. Syst. **49**(5), 861–880 (2017)

23. Ma, L., Wang, X., Huang, M., Zhang, H., Chen, H.: A novel evolutionary root system growth algorithm for solving multi-objective optimization problems. Appl. Soft Comput. **57**, 379–398 (2017)

24. Ma, X., et al.: A multiobjective evolutionary algorithm based on decision variable analyses for multiobjective optimization problems with large-scale variables. IEEE Trans. Evol. Comput. **20**(2), 275–298 (2015)

25. Miguel Antonio, L., Coello Coello, C.A.: Decomposition-based approach for solving large scale multi-objective problems. In: Handl, J., Hart, E., Lewis, P.R., López-Ibáñez, M., Ochoa, G., Paechter, B. (eds.) PPSN 2016. LNCS, vol. 9921, pp. 525–534. Springer, Cham (2016). https://doi.org/10.1007/978-3-319-45823-6_49

26. Shi, Y.: Brain storm optimization algorithm. In: Tan, Y., Shi, Y., Chai, Y., Wang, G. (eds.) ICSI 2011. LNCS, vol. 6728, pp. 303–309. Springer, Heidelberg (2011). https://doi.org/10.1007/978-3-642-21515-5_36

27. Shi, Y.: Brain storm optimization algorithm in objective space. In: 2015 IEEE Congress on Evolutionary Computation (CEC), pp. 1227–1234. IEEE (2015)

28. Shi, Y., Xue, J., Wu, Y.: Multi-objective optimization based on brain storm optimization algorithm. Int. J. Swarm Intell. Res. (IJSIR) **4**(3), 1–21 (2013)

29. Song, A., Yang, Q., Chen, W.N., Zhang, J.: A random-based dynamic grouping strategy for large scale multi-objective optimization. In: 2016 IEEE Congress on Evolutionary Computation (CEC), pp. 468–475. IEEE (2016)

30. Tian, Y., Cheng, R., Zhang, X., Jin, Y.: Platemo: a matlab platform for evolutionary multi-objective optimization [educational forum]. IEEE Comput. Intell. Mag. **12**(4), 73–87 (2017)

31. Tian, Y., Lu, C., Zhang, X., Tan, K.C., Jin, Y.: Solving large-scale multiobjective optimization problems with sparse optimal solutions via unsupervised neural networks. IEEE Trans. Cybern. **51**(6), 3115–3128 (2020)

32. Tian, Y., et al.: Evolutionary large-scale multi-objective optimization: a survey. ACM Comput. Surv. (CSUR) **54**(8), 1–34 (2021)

33. Tian, Y., Zheng, X., Zhang, X., Jin, Y.: Efficient large-scale multiobjective optimization based on a competitive swarm optimizer. IEEE Trans. Cybern. **50**(8), 3696–3708 (2019)

34. Zeng, R., Su, M., Yu, R., Wang, X.: CD^2: fine-grained 3D mesh reconstruction with twice chamfer distance. ACM Trans. Multimedia Comput. Commun. Appl. (2023). https://doi.org/10.1145/3582694

35. Zhang, B., Wang, X., Ma, L., Huang, M.: Optimal controller placement problem in internet-oriented software defined network. In: 2016 International Conference on Cyber-Enabled Distributed Computing and Knowledge Discovery (CyberC), pp. 481–488. IEEE (2016)
36. Zhang, X., Tian, Y., Cheng, R., Jin, Y.: A decision variable clustering-based evolutionary algorithm for large-scale many-objective optimization. IEEE Trans. Evol. Comput. 22(1), 97–112 (2016)
37. Zhang, Y., Wang, G.G., Li, K., Yeh, W.C., Jian, M., Dong, J.: Enhancing MOEA/D with information feedback models for large-scale many-objective optimization. Inf. Sci. 522, 1–16 (2020)
38. Zille, H., Ishibuchi, H., Mostaghim, S., Nojima, Y.: Weighted optimization framework for large-scale multi-objective optimization. In: Proceedings of the 2016 on Genetic and Evolutionary Computation Conference Companion, pp. 83–84 (2016)
39. Zille, H., Ishibuchi, H., Mostaghim, S., Nojima, Y.: A framework for large-scale multiobjective optimization based on problem transformation. IEEE Trans. Evol. Comput. 22(2), 260–275 (2017)
40. Zitzler, E., Thiele, L., Laumanns, M., Fonseca, C.M., Da Fonseca, V.G.: Performance assessment of multiobjective optimizers: an analysis and review. IEEE Trans. Evol. Comput. 7(2), 117–132 (2003)

Modified Brain Storm Optimization Algorithm for Solving Multimodal Multiobjective Optimization Problems

Yue Liu[1], Shi Cheng[1]([✉]) [iD], Xueping Wang[1], Yuyuan Shan[1], and Hui Lu[2]

[1] School of Computer Science, Shaanxi Normal University, Shaanxi 710119, China
{liuyuemelody,cheng,wxp,shanyuyuan}@snnu.edu.cn
[2] School of Electronic and Information Engineering, Beihang University,
Beijing 100191, China
mluhui@buaa.edu.cn

Abstract. Finding more solutions from different Pareto-optimal sets (PSs) in the decision space and maintaining the diversity of solutions at the Pareto front (PF) in the objective space are the two objectives of multimodal multiobjective optimization (MMO). This paper proposes a brain storm optimization algorithm with a non-dominated special crowding distance sorting strategy (BSO-SCD) to ensure a sufficient number of good enough solutions and diversity in both the decision and objective spaces. Based on the brain storm optimization (BSO) algorithm, the BSO-SCD algorithm enhances the ability of diversity maintenance and has three steps. Firstly, k-means clustering divides the population into multiple subpopulations in the decision space to help locate various optimal solutions. Secondly, the non-dominated special crowding distance sorting (SCD) strategy, considering the diversity of decision space and objective space simultaneously, is used to find each subpopulation cluster center and stored in an archive. Finally, the solutions in the archive are selected after the iteration. The performance of the BSO-SCD algorithm and the other five algorithms is verified on 12 MMO benchmark functions. Experimental results show that the BSO-SCD algorithm could find as many equivalent PSs as possible in the decision space and guarantee a good PF distribution in the objective space.

Keywords: Brain storm optimization · multimodal multiobjective optimization · multiobjective optimization · multimodal optimization

1 Introduction

In the real world, many problems have two or more conflicting objectives to be optimized, which are called multiobjective optimization problems. The mul-

This work is partially supported by National Natural Science Foundation of China (Grant No. 61806119), Fundamental Research Funds for the Central Universities (No. GK202201014).

Y. Tan et al. (Eds.): ICSI 2023, LNCS 13968, pp. 425–437, 2023.
https://doi.org/10.1007/978-3-031-36622-2_35

timodal multiobjective optimization problem (MMOP) [4] is an extended case of the multiobjective optimization problem in which, for the same Pareto front (PF) in the objective space, at least two equivalent Pareto-optimal sets (PSs) exist in the decision space. MMO requires finding as many equivalent PSs in the decision space and ensuring a good PF distribution in the objective space. MMO provides multiple optimal solutions after one run, which can offer more choices for decision-makers.

Various improved swarm intelligence algorithms have been proposed to solve MMOPs. Liang [4] et al. proposed a decision space-based niching NSGA-II (DN-NSGAII) algorithm. The experimental results show that the algorithm can find as many PSs as possible. Liang [5] et al. proposed a self-organizing multiobjective PSO (SMPSO-MM) algorithm. The self-organizing mapping network in this algorithm can find the distribution structure of the population and establish the neighborhood in the decision space. The elite learning strategy avoids premature convergence and enhances the global search capability. Li [3] et al. proposed a DE based on reinforcement learning with a fitness ranking (DE-RLFR) algorithm. The hierarchical state distribution of the population determined by fitness ranking is used to determine the evolutionary direction and mutation strategy. DE algorithm is combined with reinforcement learning to improve the efficiency of population search in the solution space.

Niching technique [9], such as the clustering strategy, is a common approach to handling multimodal optimization problems in swarm intelligence algorithms. Brain storm optimization (BSO) algorithm performs well on multimodal optimization problems due to the grouping strategy that divides the solutions into multiple clusters during the search process. Cheng [1] et al. used several BSO variants to solve multimodal optimization problems. Experiments were conducted on test functions and nonlinear systems of equations, and the experimental results showed that BSO is effective in solving multimodal problems. The traditional BSO algorithm is unable to solve MMOPs because it does not combine the properties of multiobjective optimization problems for the environment selection of the obtained solutions. Therefore, we propose a BSO based on non-dominated special crowding distance sorting (BSO-SCD) algorithm for solving MMOPs.

The rest paper is organized as follows: Section 2 introduces the knowledge of MMOPs, including problem definition, algorithm evaluation metrics, and the basics of BSO algorithms; Sect. 3 gives the details of the BSO-SCD algorithm; Sect. 4 analyzes the experimental results of the improved algorithm in 12 benchmark problems; Sect. 5 concludes the paper and future work.

2   Background Knowledge

2.1   Multiobjective Optimization

In general, the mathematical representation of the minimization multiobjective optimization problem [6] is shown in Eq. (1).

$$\min \boldsymbol{f}(\boldsymbol{x}) = \{f_1(\boldsymbol{x}), f_2(\boldsymbol{x}), \cdots, f_m(\boldsymbol{x})\} \tag{1}$$
$$\text{s.t.} g_i(\boldsymbol{x}) \leq 0, i = 1, 2, \cdots, k;$$
$$h_j(\boldsymbol{x}) = 0, j = 1, 2, \cdots, p.$$

where $\boldsymbol{x} = \{x_1, x_2, \cdots, x_n\}$ is an n-dimensional decision vector, $\boldsymbol{f}$ is an m-dimensional objective vector, $g_i(\boldsymbol{x}) \leq 0$ $(i = 1, 2, \cdots, k)$ are inequality constraints, $h_j(\boldsymbol{x}) = 0$ $(j = 1, 2, \cdots, p)$ are equality constraints. The n-dimensional space consisting of all possible values of the decision vector $\boldsymbol{x}$ is the decision space, and the m-dimensional space composed of all possible values of the objective vector is the objective space.

- Pareto dominate: suppose that $\boldsymbol{x}$ and $\boldsymbol{y}$ are two feasible solutions of a minimization multiobjective optimization problem, for $\forall i = 1, 2, \cdots, m$, $f_i(\boldsymbol{x}) \leq f_i(\boldsymbol{y})$ and $\exists j \in \{1, 2, \cdots, m\}$, $f_j(\boldsymbol{x}) < f_j(\boldsymbol{y})$, then $\boldsymbol{x}$ is said to Pareto dominate $\boldsymbol{y}$. If any other solution does not dominate a feasible solution, then this solution is called a non-dominated solution.
- Pareto-optimal set (PS): the set of all non-dominated solutions in the decision space is called the Pareto-optimal set.
- Pareto front (PF): the set of all the objective values in the objective space corresponding to the PS is called the Pareto front.

2.2   Multimodal Multiobjective Optimization Problems

MMOPs [4] are a class of problems in which at least two equivalent Pareto optimal solutions exist in the decision space for the same point on PF in the objective space. That is, for the same PF in the objective space in an MMOP, at least two equivalent PSs exist in the decision space. As shown in Fig. 1, the left coordinate system represents the decision space, and the right coordinate system represents the objective space. PS_1 and PS_2 are two equivalent PSs corresponding to the same PF in the objective space. The MMOPs require finding as many equivalent PSs in the decision space as possible and having a good PF distribution in the objective space.

2.3   MMO Evaluation Metrics

In this paper, Pareto sets proximity (PSP) [8] and inverted generational distance in the objective space (IGDF) [10] are selected as performance evaluation metrics for MMOPs. PSP is an evaluation metric used to measure the performance of algorithms in decision space, while IGDF is used to evaluate the performance of

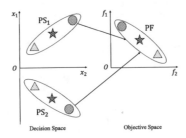

Fig. 1. Illustration of a multimodal multiobjective optimization problem.

the algorithm in objective space. A larger PSP and a smaller IGDF imply that the algorithm performs well in both decision space and objective space.

The definition of PSP is

$$PSP = \frac{CR}{IGDX} \tag{2}$$

$$IGDX(PS, PS^*) = \frac{\sum_{v \in PS^*} d(v, PS)}{|PS^*|} \tag{3}$$

$$CR = (\prod_{l=1}^{n} \delta_l)^{\frac{1}{2n}} \tag{4}$$

$$\delta_l = \begin{cases} 1, & \text{if } V_l^{\max} = V_l^{\min} \\ 0, & \text{if } v_l^{\min} \geq V_l^{\max} || v_l^{\max} \leq V_l^{\min} \\ (\frac{\min(v_l^{\max}, V_l^{\max}) - \max(v_l^{\min}, V_l^{\min})}{V_l^{\max} - V_l^{\min}})^2, & \text{otherwise.} \end{cases} \tag{5}$$

where CR (cover rate) indicates the overlap ratio between the obtained PS in the decision space and the true PS; $IGDX$ indicates the inverted generational distance in the decision space; PS is obtained Pareto-optimal set; PS^* is true Pareto-optimal set; $d(v, PS)$ is the minimum Euclidean distance between v and all points in PS; $|PS^*|$ is the number of solutions in PS^*; n is the dimension of the decision space; $V_l^{\max}$ and $V_l^{\min}$ are the maximum and minimum values of the lth variable in PS^*; $v_l^{\max}$ and $v_l^{\max}$ are the maximum and minimum values of the lth variable in PS. The PSP reflects not only the degree of convergence of the obtained PS but also the degree of similarity between the obtained PS and the true PS. The larger the PSP, the better performance of the algorithm.

The definition of IGDF is

$$IGDF(PF, PF^*) = \frac{\sum_{v \in PF^*} d(v, PF)}{|PF^*|} \tag{6}$$

where PF is obtained PF; PF^* is true PF; $d(v, PF)$ is the minimum Euclidean distance between v and all points in PF; $|PF^*|$ is the number of solutions in PF^*. IGDF indicates the distribution of PF in the objective space. The smaller the IGDF, the better performance of the algorithm.

2.4   Brain Storm Optimization Algorithm

The BSO algorithm is an algorithm that simulates the brainstorming process of human problem solving [7]. It is assumed that in a D-dimensional search space, there exists a population of m individuals. The value of the ith individual in the dth dimension is denoted as x_i^d. The new individuals are generated from one or two previous individuals. The BSO algorithm uses Eq. (7) and Eq. (8) to generate new individuals.

$$x_{new}^d = x_{old}^d + \xi(t) \times N(\mu, \sigma) \tag{7}$$

$$\xi(t) = \text{logsig}(\frac{0.5 \times T - t}{k}) \times \text{rand}(0,1) \tag{8}$$

where x_{old}^d is the value of x_{old} that was selected to generate the new individual in the dth dimension; x_{new}^d is the value of x_{new} that was selected to generate the new individual in the dth dimension; $N(\mu, \sigma)$ is a Gaussian random function with mean μ and variance σ; $\xi(t)$ is a step function used to balance the convergence speed of the algorithm; logsig() is an S-type logarithmic transfer function for nonlinearly reducing the search step size; T is the maximum number of iterations; t is the current number of iterations; k is the factor that changes the slope of the logsig() function; rand(0, 1) is a random number between (0, 1).

3   Modified Brain Storm Optimization

3.1   Algorithm Framework

The main framework of BSO-SCD proposed in this paper is shown in Algorithm 1. In BSO-SCD, firstly, the archive is initialized to preserve the optimal solutions. Next, k-means clustering is performed on the population in the decision space to divide the population into multiple sub-populations. In swarm intelligence algorithms, clustering is a common niching method to deal with multimodal optimization problems. The k-means clustering comes with the BSO algorithm to help locate multiple Pareto optimal solutions. The k-means clustering performed in the decision space can guarantee the diversity of solutions in the decision space.

Then, the subpopulations are sorted by non-dominated special crowding distance to find the cluster centers of each cluster and save each cluster center into the archive. Then new individuals are generated according to the BSO algorithm. The Pareto dominance relationship between the new and original individuals is compared for each newly generated individual. One randomly selected solution will be kept for the next iteration of the population when the two solutions are not dominated by each other. Otherwise, the dominant solution will be kept, and the dominated solution will be abandoned between the original and new solutions, i.e., the better solution will be kept for the next iteration. Finally, the archive is sorted by non-dominated special crowding distance, and the result is saved into SortedArchive. The MAXSIZE-Archive individuals with a dominant performance in SortedArchive are output as results.

Algorithm 1: The basic flow of BSO-SCD algorithm

Input: Initial population Popu, population size PopuSize, the number of clusters
ClusterNum, maximum of iteration MaxIteration, the maximum capacity of
archive MAXSIZE-Archive

1 Evaluate initial population;
2 Initialize archive set Archive;
3 **while** *iteration < MaxIteration* **do**
4   cluster=kmeans(Popu,ClusterNum);
5   **for** *cluIndex=1:ClusterNum* **do**
6   cluster[cluIndex]=non-dominated-scd-sort(cluster[cluIndex]);   // Non-dominated
        special crowding distance sorting for each cluster
7   centers[cluIndex]=cluster[cluIndex][1];   // Each cluster center is the first
        individual after the clusters are sorted by non-dominated special crowding
        distance
8   Archive=Archive ∪ centers[cluIndex];   // Store each cluster center into Archive
9   **for** *i=1: PopuSize* **do**
10   newIndi=generate new individuals according to Eq. (7) and Eq. (8);
11   Popu[i]= The non-dominated solution in both the newly generated individual
        newIndi and the original value Popu[i]. One randomly selected solution will be
        kept for the next iteration of the population when the two solutions are not
        dominated by each other;
12 SortedArchive=non-dominated-scd-sort(Archive);
Output: The MAXSIZE-Archive individuals in SortedArchive with superior performance.

The k-means clustering of BSO is a common niching mechanism for solving multimodal problems, which helps to locate multiple optimal solutions. Considering the multiobjective characteristics, the non-dominated special crowding distance sorting in BSO-SCD performs environment selection of the obtained solutions to filter out the eligible optimal solutions. BSO-SCD algorithm takes into account the multimodal and multiobjective characteristics of MMOPs.

3.2 Special Crowding Distance

The special crowding distance [8] can consider the diversity of decision space and objective space and help maintain solutions that are far away from the decision space but close to the objective space. The special crowding distance dis is calculated as shown in Eq. (9).

$$
dis_i = \begin{cases} \max(CD_i^x, CD_i^f), & \text{if } CD_i^x > CD_{\text{avg}}^x \text{ or } CD_i^f > CD_{\text{avg}}^f \\ \min(CD_i^x, CD_i^f), & \text{otherwise.} \end{cases} \tag{9}
$$

where dis_i is the special crowding distance of the ith individual, CD_i^x is crowding distance of the ith individual in the decision space, CD_i^f is crowding distance of the ith individual in the objective space, CD_{avg}^x is average crowding distance in the decision space, CD_{avg}^f is average crowding distance in the objective space. When the crowding distance of the ith individual in the decision space or objective space is larger than the average crowding distance, dis takes the larger one. Otherwise, it takes the smaller one. Therefore, dis could maintain the diversity of decision space and objective space.

3.3 Non-dominated Special Crowding Distance Sorting

The non-dominated special crowding distance sorting (SCD) strategy combines the Pareto dominance relationship and crowding distance for environment selection to select the better individuals considering the multiobjective characteristics of MMOPs. The SCD strategy is used to find the center of each cluster and output the final Pareto optimal solutions found. The process of SCD strategy is shown in Algorithm 2, which is mainly divided into two steps: (1) according to the non-dominated sorting algorithm, all individuals in the population are divided into dominance ranks; (2) the special crowding distance of all individuals in the same dominance rank is calculated. Based on the magnitude of the calculated distance values, all individuals in the same dominance rank are sorted in descending order of distance. For the individuals in the same dominance rank, the larger the distance value, the higher the ranking. After the non-dominated special crowding distance sorting, the first-ranked individual is the non-dominated individual with the largest distance value. The SCD strategy can take into account the diversity of both decision space and objective space.

Algorithm 2: Non-dominated special crowding distance sorting

Input: Population P
1 **for** p in P **do**
2 $\quad$ Calculate the number of solutions that dominate it n_p and the set of individuals it dominates S_p ;
3 $\quad$ **if** $n_p = 0$ **then**
4 $\quad\quad$ $F_1 = F_1 \cup p$; $//$ F_1 is PF with Pareto dominance rank of 1

5 $i = 1$; $//$ i indicates different PF ranks
6 **while** $F_i \neq \phi$ **do**
7 $\quad$ $Q = \phi$; $//$ Q save the next Pareto front set
8 $\quad$ **for** p in F_i **do**
9 $\quad\quad$ **for** q in S_p **do**
10 $\quad\quad\quad$ $n_q = n_q - 1$;
11 $\quad\quad\quad$ **if** $n_q = 0$ **then**
12 $\quad\quad\quad\quad$ $Q = Q \cup \{q\}$;
13 $\quad$ $i = i + 1; F_i = Q$;
14 **for** $i = 1 : length(F)$ **do**
15 $\quad$ SCD was calculated for each individual in according to Eq. (9);
16 $\quad$ F_i is sorted in descending order of special crowding distance;
17 SortedP$= F_1 \cup F_2 \cup \cdots \cup F_{length(F)}$;
Output: Sorted population SortedP.

---

4 Experimental Results and Analysis

4.1 Algorithm Parameter Settings

In this paper, 12 benchmark functions [8] are selected to verify the performance of BSO-SCD. As shown in Table 1, the first column shows the 12 benchmark functions' names, and the second and third columns are the dimensions of the objective vector and the decision vector, respectively. The fourth column indicates the number of equivalent PSs of the same PF, and the fifth column indicates whether multiple equivalent PSs overlap. The more the number of equivalent PSs of the same PF, the more difficult the MMOPs are. The more equivalent PSs overlap, the more complex the problem is.

Table 1. Selected ten multimodal multiobjective test functions.

	Function name	m	n	number of PS	Overlap
f_1	MMF1	2	2	2	No
f_2	MMF2	2	2	2	No
f_3	MMF3	2	2	2	Yes
f_4	MMF4	2	2	4	No
f_5	MMF5	2	2	4	No
f_6	SYM-PART-simple	2	2	9	Yes
f_7	SYM-PART-rotated	2	2	9	Yes
f_8	Omni-test ($n = 3$)	2	3	27	Yes
f_9	Omni-test ($n = 4$)	2	4	72	Yes
f_{10}	Omni-test ($n = 5$)	2	5	360	Yes
f_{11}	Omni-test ($n = 10$)	2	10	$3 * 10!$	Yes
f_{12}	Omni-test ($n = 15$)	2	15	$3 * 15!$	Yes

In order to verify the performance of the proposed BSO-SCD algorithm, five MMO algorithms are selected for comparison. These five algorithms are DN-NSGAII [4], Omni-optimizer [2], SPEA2 [11], MO-RING-PSO-SCD [8], and SMPSO-MM [5]. The experimental parameters of all comparison algorithms were set in accordance with the original paper. In the experiments, the parameters were set as follows.

- the population size: $PopuSize = 800$,
- the number of clusters: $ClusterNum = 20$,
- the maximum of iteration: $MaxIteration = 200$,
- the capacity of archive: $MAXSIZE - Archive = 800$,

In addition, all algorithms were run 20 times independently on 12 test functions.

4.2   Experimental Results

The mean and standard variance of PSP obtained from the experiments of BSO-SCD and the other five algorithms are shown in Table 2. Compared with the other five MMO algorithms, BSO-SCD has the best performance on the eight test functions MMF2, MMF3, SYM-PART-rotated, Omni-test ($n = 3$), Omni-test ($n = 4$), Omni-test ($n = 5$), Omni-test ($n = 10$), and Omni-test ($n = 15$).

The k-means clustering in BSO-SCD helps to locate multiple optimal solutions, and the SCD strategy simultaneously takes into account the diversity of the objective space and the decision space, thus finding more high-quality solutions. BSO-SCD outperforms DN-NSGAII on all test functions and outperforms Omni-optimizer and SPEA2 on all test functions except function MMF1. SMPSO-MM outperforms several other algorithms on the other four test functions, such as MMF1 and MMF4. Compared with other MMO algorithms, BSO-SCD performs better on MMOPs with more dimensions (Omni-test $(n=4)$ $\sim$ Omni-test $(n=15)$). In addition, the results of the BSO-SCD algorithm on the standard variance of PSP are the minimum for most benchmark functions, which shows the proposed algorithm's stability.

Table 2. PSP average and standard deviation values obtained by different algorithms.

Function	BSO-SCD	DN-NSGAII	Omni-optimizer	SPEA2	MO-RING-PSO-SCD	SMPSO-MM
MMF1	46.50	46.20	47.92	54.68	67.12	**81.64**
	(2.87)	(5.34)	(5.40)	(1.41)	(3.53)	(3.13)
MMF2	**218.39**	67.810	85.86	71.71	108.35	132.96
	(18.95)	(30.93)	(60.78)	(4.29)	(11.49)	(20.31)
MMF3	**261.98**	78.77	72.72	97.94	140.92	158.19
	(18.60)	(35.63)	(33.39)	(21.56)	(12.85)	(16.04)
MMF4	80.35	38.84	37.25	72.05	113.95	**135.39**
	(3.07)	(9.88)	(11.75)	(10.41)	(5.25)	(5.19)
MMF5	28.96	14.77	14.50	26.19	33.21	**39.13**
	(1.29)	(1.81)	(1.35)	(3.24)	(1.10)	(1.19)
SYM-PART-simple	27.40	0.39	0.35	0.17	21.95	**37.47**
	(1.23)	(0.15)	(0.14)	(0.01)	(1.67)	(3.54)
SYM-PART-rotated	**30.39**	4.48	0.93	0.52	18.72	14.42
	(1.15)	(4.77)	(1.13)	(0.03)	(0.93)	(1.28)
Omni-test $(n=3)$	**15.87**	1.04	0.89	0.23	7.70	4.88
	(0.65)	(0.16)	(0.15)	(0.10)	(1.58)	(1.32)
Omni-test $(n=4)$	**2.67**	0.39	0.40	0.13	1.14	0.86
	(0.37)	(5.03e-02)	(7.91e-02)	(6.75e-02)	(6.52e-02)	(6.59e-02)
Omni-test $(n=5)$	**0.81**	0.29	0.34	8.45e-02	0.56	0.45
	(5.56e-03)	(5.11e-02)	(1.34e-02)	(2.57e-02)	(3.15e-02)	(1.73e-02)
Omni-test $(n=10)$	**0.25**	0.11	0.10	1.67e-02	0.20	0.19
	(1.09e-02)	(4.17e-02)	(4.30e-02)	(4.17e-03)	(1.58e-02)	(9.31e-03)
Omni-test $(n=15)$	**0.16**	8.33e-02	7.27e-02	1.14e-02	0.14	0.14
	(7.17e-03)	(2.08e-02)	(2.74e-02)	(2.00e-03)	(9.20e-03)	(4.58e-03)

The mean and standard variance of IGDF obtained from the experiments of BSO-SCD and the other five algorithms are shown in Table 3. The lowest IGDF values are obtained by Omni-optimizer on MMF1, MMF2, MMF3, MMF5, SYM-PART-simple, SYM-PART-rotated and Omni-test $(n=3)$ test functions compared to the other five algorithms. SMPSO-MM obtains the lowest IGDF values on MMF4, Omni-test $(n=10)$ and Omni-test $(n=15)$ functions. BSO-SCD obtains the lowest IGDF values on Omni-test $(n=4)$ and Omni-test$(n=5)$ functions. The differences between the IGDF values obtained by BSO-SCD on the other ten benchmark functions and those obtained by the other algorithms are not very significant. The fact that the IGDF values of all algorithms are very close is due to the fact that all algorithms consider the distribution in the objective space.

Table 3. IGDF average and standard deviation values obtained by different algorithms.

Function	BSO-SCD	DN-NSGAII	Omni-optimizer	SPEA2	MO-RING-PSO-SCD	SMPSO-MM
MMF1	2.81e-03	8.39e-04	**6.80e-04**	8.07e-04	9.83e-04	7.17e-04
	(3.98e-04)	(7.23e-05)	**(3.21e-05)**	(7.49e-05)	(6.04e-05)	(3.86e-05)
MMF2	4.00e-03	1.48e-03	**1.09e-03**	5.11e-03	5.13e-03	3.79e-03
	(2.68e-04)	(3.33e-03)	**(2.07e-03)**	(1.05e-03)	(5.53e-04)	(2.70e-04)
MMF3	3.24e-03	7.41e-04	**6.38e-04**	3.91e-03	3.73e-03	3.13e-03
	(2.58e-04)	(1.39e-03)	**(4.08e-05)**	(3.37e-04)	(3.56e-04)	(2.98e-03)
MMF4	1.58e-03	7.92e-04	6.95e-04	8.39e-04	9.05e-04	**6.65e-04**
	(1.30e-04)	(6.18e-05)	(4.15e-05)	(1.99e-05)	(4.61e-05)	**(3.35e-05)**
MMF5	1.78e-03	8.06e-04	**6.64e-04**	8.07e-04	9.50e-04	7.26e-04
	(8.22e-05)	(3.80e-05)	**(2.66e-05)**	(2.54e-05)	(3.69e-05)	(2.20e-05)
SYM-PART-simple	6.64e-03	3.23e-03	**3.00e-03**	3.31e-03	9.42e-03	5.12e-03
	(8.40e-04)	(4.42e-04)	**(3.01e-04)**	(4.72e-04)	(1.16e-03)	(7.36e-04)
SYM-PART-rotated	5.79e-03	3.93e-03	**3.29e-03**	3.96e-03	1.14e-02	1.38e-02
	(5.02e-04)	(5.37e-04)	**(3.51e-04)**	(6.28e-04)	(1.89e-03)	(2.19e-03)
Omni-test	1.34e-02	2.99e-03	**2.48e-03**	7.37e-03	1.71e-02	9.78e-01
($n = 3$)	(6.71e-04)	(1.31e-04)	**(1.13e-04)**	(1.79e-03)	(1.16e-03)	(1.76e-03)
Omni-test	**1.85**	1.99	1.99	1.99	1.94	1.92
($n = 4$)	**(3.11e-03)**	(7.48e-04)	(2.85e-04)	(2.48e-04)	(2.68e-03)	(2.35e-02)
Omni-test	**2.72**	2.99	2.99	2.98	2.86	2.78
($n = 5$)	**(2.27e-02)**	(1.23e-03)	(5.79e-04)	(2.42e-04)	(1.32e-02)	(2.53e-02)
Omni-test	6.83	7.96	7.97	7.90	6.52	**6.01**
($n = 10$)	(8.03e-02)	(7.63e-03)	(6.28e-03)	(7.92e-02)	(0.15)	**(0.18)**
Omni-test	10.49	12.84	12.89	12.76	9.14	**8.36**
($n = 15$)	(0.14)	(3.05e-02)	(1.76e-02)	(9.25e-02)	(0.33)	**(0.48)**

4.3   Experimental Analyses

The PSs and PF obtained by BSO-SCD and the other five algorithms on the Omni-test ($n = 3$) test function are shown in Fig. 2 and Fig. 3, respectively. From Fig. 3, it can be seen that there is no significant difference between the PF results found by BSO-SCD and the other five algorithms, which is because all algorithms consider the distribution of the objective space. However, the number of equivalent PSs obtained by BSO-SCD is greater than the number of PSs obtained by the other algorithms. PSs obtained by BSO-SCD also outperform the PSs obtained by the other algorithms. In addition, DN-NSGAII, Omni-optimizer, and SPEA2 obtain fewer equivalent PSs. On the Omni-test ($n = 3$) test function, BSO-SCD finds more complete PSs and as many equivalent PSs as possible compared to the other five algorithms. The k-means clustering in BSO-SCD helps to locate multiple optimal solutions at once, and the special crowding distance takes into account the diversity of decision space and objective space, thus finding more high-quality multiple equivalent optimal solutions.

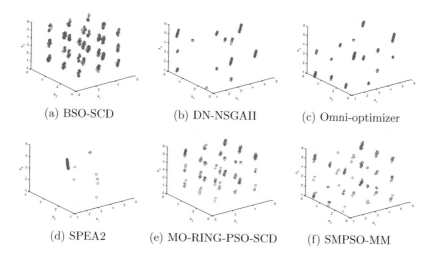

Fig. 2. PSs of Omni-test (n = 3) function obtained by different algorithms.

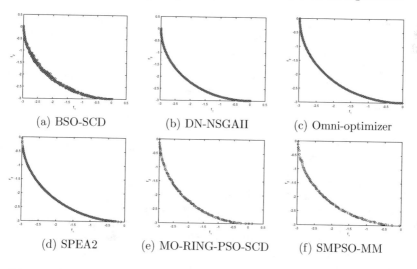

Fig. 3. PF of Omni-test ($n = 3$) function obtained by different algorithms.

5   Conclusion and Future Work

This paper proposes a modified BSO algorithm with an SCD strategy (BSO-SCD) for solving MMOPs. The BSO-SCD algorithm divides the population into multiple subpopulations in the decision space using k-means clustering. Then the cluster centers of each subpopulation are found using the SCD strategy, and each cluster center is stored in an archive. Finally, the SCD strategy is performed on the archive after the iteration. The k-means clustering in BSO-SCD can help locate multiple optimal solutions, while the SCD strategy could simultaneously maintain the diversity of decision space and objective space.

The BSO-SCD could effectively solve MMOPs, and in some test problems, its performance is superior to the other compared algorithms, especially in solving high-dimensional MMOPs. However, the performance of the BSO-SCD algorithm could be better in some test functions. Designing more effective strategies is necessary to improve the BSO-SCD algorithm's performance to solve MMOPs. At the same time, with the continuous progress of MMOPs, MMOPs with high-dimensional characteristics and irregular PF will increase the problem's difficulty. It is the next research step to improve the solution performance of MMOPs from both algorithm and problem aspects.

References

1. Cheng, S., Chen, J., Lei, X., Shi, Y.: Locating multiple optima via brain storm optimization algorithms. IEEE Access **6**, 17039–17049 (2018). https://doi.org/10.1109/ACCESS.2018.2811542
2. Deb, K., Tiwari, S.: Omni-optimizer: a procedure for single and multi-objective optimization. In: Coello Coello, C.A., Hernández Aguirre, A., Zitzler, E. (eds.) EMO 2005. LNCS, vol. 3410, pp. 47–61. Springer, Heidelberg (2005). https://doi.org/10.1007/978-3-540-31880-4_4
3. Li, Z., Shi, L., Yue, C., Shang, Z., Qu, B.: Differential evolution based on reinforcement learning with fitness ranking for solving multimodal multiobjective problems. Swarm Evol. Comput. **49**, 234–244 (2019). https://doi.org/10.1016/j.swevo.2019.06.010
4. Liang, J.J., Yue, C.T., Qu, B.Y.: Multimodal multi-objective optimization: a preliminary study. In: 2016 IEEE Congress on Evolutionary Computation (CEC), pp. 2454–2461 (2016). https://doi.org/10.1109/CEC.2016.7744093
5. Liang, J., Guo, Q., Yue, C., Qu, B., Yu, K.: A self-organizing multi-objective particle swarm optimization algorithm for multimodal multi-objective problems. In: Tan, Y., Shi, Y., Tang, Q. (eds.) ICSI 2018. LNCS, vol. 10941, pp. 550–560. Springer, Cham (2018). https://doi.org/10.1007/978-3-319-93815-8_52
6. Liu, Y., Yen, G.G., Gong, D.: A multimodal multiobjective evolutionary algorithm using two-archive and recombination strategies. IEEE Trans. Evol. Comput. **23**(4), 660–674 (2019). https://doi.org/10.1109/TEVC.2018.2879406
7. Shi, Y.: An optimization algorithm based on brainstorming process. Int. J. Swarm Intell. Res. (IJSIR) **2**(4), 35–62 (2011). https://doi.org/10.4018/jsir.2011100103
8. Yue, C., Qu, B., Liang, J.: A multiobjective particle swarm optimizer using ring topology for solving multimodal multiobjective problems. IEEE Trans. Evol. Comput. **22**(5), 805–817 (2018). https://doi.org/10.1109/TEVC.2017.2754271
9. Zeng, C.: Improved swarm intelligence algorithm and application for multimodal optimization problems. Master's thesis, Beijing University of Chemical Technology, Beijing (2020). https://doi.org/10.26939/d.cnki.gbhgu.2020.001010
10. Zhou, A., Zhang, Q., Jin, Y.: Approximating the set of pareto-optimal solutions in both the decision and objective spaces by an estimation of distribution algorithm.

IEEE Trans. Evol. Comput. **13**(5), 1167–1189 (2009). https://doi.org/10.1109/TEVC.2009.2021467

11. Zitzler, E., Laumanns, M., Thiele, L.: SPEA2: improving the strength pareto evolutionary algorithm. Technical report, TIK-Report 103, Swiss Federal Institute of Technology (ETH), Zurich, Switzerland (2001)

Balancing Measurement Efficiency, Test Diversity and Security for Item Selection in On-the-Fly Assembled Multistage Adaptive Testing via a Multi-objective Evolutionary Algorithm

Xiaoshu Xiang[1], Ling Wu[1], Haiping Ma[1(✉)], and Xingyi Zhang[2]

[1] Institutes of Physical Science and Information Technology, Anhui University,
Hefei 230039, China
hpma@ahu.edu.cn
[2] School of Artificial Intelligence, Anhui University,
Hefei 230039, China

Abstract. On-the-fly assembled multistage adaptive testing (OMST) is a recently emerging computer-administered test form and has been employed in a variety of large-scale examinations. Item selection is a key challenge in OMST, which adaptively assembles real-time question modules by selecting a set of questions tailored for examinees from item bank. Existing question module assembly recognizes the measurement efficiency, test diversity and security as crucial evaluation criteria for item selection. Nevertheless, most of the studies on OMST concentrate on improving the measurement efficiency in item selection, while few of them devote to making a trade-off between the three criteria. In this paper, we propose to use a dynamic indicator-based multi-objective evolutionary algorithm with reference point adaptation, termed AR-DMOEA, for striking a balance between the three criteria in the item selection of OMST. A diversity-strengthened population initialization strategy is suggested to generate diverse individuals in the initial iteration of AR-DMOEA. A set of knowledge-guided offspring generation operators are designed to reproduce high-quality offspring individuals during the optimization of AR-DMOEA. Empirical results on five OMST datasets demonstrate that AR-DMOEA effectively balances the measurement efficiency, test diversity and security in the item selection of OMST. The effectiveness of the suggested initialization strategy and offspring generation operators are also validated by comparison between AR-DMOEA with the two strategies and that without the two strategies.

Keywords: Multi-objective optimization algorithm · On-the-fly assembled multistage adaptive testing · Item selection

ⓒ The Author(s), under exclusive license to Springer Nature Switzerland AG 2023
Y. Tan et al. (Eds.): ICSI 2023, LNCS 13968, pp. 438–451, 2023.
https://doi.org/10.1007/978-3-031-36622-2_36

1   Introduction

On-the-fly assembled multistage adaptive testing (OMST) is a novel and promising computerized adaptive testing (CAT) form [15,19,27], which properly combines the typical CAT [5] and multi-stage testing (MST) [25]. The OMST is equipped with the advantages of both CAT and MST, which allows examinees to modify answers over stages and comprehensively manages non-statistical constraints [26]. Different from typical CAT and MST, OMST assembles question modules in real time for examinees' online adaptive testing, which is inspired by the adaptive presentation of CAT [4]. Therefore, the OMST is able to quickly and accurately evaluate potential ability of an examinee in a specific field by adaptively assembling question module that tests the examinee's learning ability.

For effective question module assembly, it is necessary to improve the item selection method to provide examinees with questions that are capable of discriminating their ability, which thus attracts much attention from researchers [8]. A significant criteria for item selection of OMST is the measurement efficiency for examinees' ability [10,23]. To measure an examinee's ability, the indicator "validity" was suggested for representing that whether the testing content aligns with the testing purpose [17]. Furthermore, some non-statistical constraints such as test diversity [12,13,20] and test security [1,3,11,18] are also considered for practical applications [12,14,21]. The test diversity mainly depends on the knowledge coverage of selected questions, which is regarded as an important factor for comprehensively estimating an examinee's ability [20]. The test security mainly refers to the question exposure control, i.e., the selection times of questions for examinations [1]. If a question is frequently answered by different examinees, the communication between the examinees may cause a leak of question bank [11].

Regardless of the great significance of the measurement efficiency, test diversity and test security, few studies devote to simultaneously optimizing the three criteria in item selection of OMST [6]. Therefore, in this study we propose to use a dynamic indicator-based multi-objective evolutionary algorithm with reference point adaptation, termed AR-DMOEA, to achieve a balance between the three criteria in the item selection of OMST. The proposed AR-DMOEA intends to produce high-quality question sets in terms of the three criteria, and maintain a good balance between the three criteria. To summarize, the main contributions of this study are as follows.

1. We propose to use a dynamic indicator-based multi-objective evolutionary algorithm AR-DMOEA for item selection optimization of OMST. The proposed AR-DMOEA is able to achieve a good balance between the measurement efficiency, test diversity and test security.
2. In the proposed AR-DMOEA, a diversity-strengthened population initialization strategy is suggested to generate diverse high-quality individuals in the initial iteration of AR-DMOEA. A set of knowledge-guided offspring generation operators are tailored to accelerate the convergence of AR-DMOEA optimization.

3. The proposed AR-DMOEA is applied to five typical OMST datasets and compared with three state-of-the-art item selection methods. Empirical results demonstrate that AR-DMOEA effectively balances the measurement efficiency, test diversity and security in the item selection of OMST. The effectiveness of the suggested initialization strategy and offspring generation operators are also validated by comparison between AR-DMOEA with the two strategies and that without the two strategies.

The remainder of this paper is organized as follows. Section 2 introduces preliminaries of item selection of OMST. Section 3 elaborates the proposed AR-DMOEA, and Sect. 4 presents empirical results of the proposed AR-DMOEA on several typical item selection datasets. Finally, the conclusions and outlines of our future work are drawn in Sect. 5.

2   Preliminaries

This section first presents the procedure and mathematical formulation of OMST and then introduces existing work on approaches to item selection of OMST.

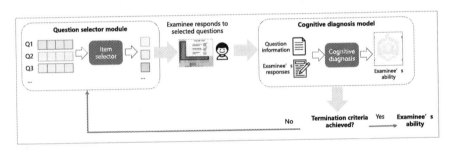

Fig. 1. The procedure of OMST.

The procedure of OMST is shown in Fig. 1, which contains a certain number of stages according to specific examination requirements. Each stage of OMST consists of two components: the cognitive diagnosis model and the question module. The cognitive diagnosis model estimates the probability that an examinee correctly answers a set of questions selected in last stage. The question module selects a set of questions for the testing of next stage. The two components cooperatively work in each stage so that the testing for the examinee is updated over stages according to the examinee's ability. In each stage, an item selector is first used to obtain a set of questions, which are then displayed to the examinee for responding. Next, the properties of the selected questions and the examinee's responses are input to a cognitive diagnosis model for estimating the examinee's ability. The examinee's ability is updated and will be used in the item selection of next stage. The above two steps repeat until the termination criteria are achieved, and finally the examinee's ability is obtained.

The item selection in OMST is formulated as a tri-objective optimization problem that simultaneously optimizes the measurement efficiency, test diversity and security of selected items. Assume $S = \{s_1, s_2, ..., s_N\}$ is a set of N students, $Q = \{q_1, q_2, ..., q_M\}$ is a set of M questions, and $\{c_1, c_2, ..., c_K\}$ is a set of K knowledge points. A binary matrix G denotes the relationship between questions and knowledge points, where $G_{jk} = 1$ means that the question q_j is related to the knowledge point c_k, and otherwise is irrelevant to the point. The examinee exercising response logs $\mathcal{R}$ are denoted as a set of triple $<s_i, q_j, r_{ij}>$, where $s_i \in S, q_j \in Q$ and $r_{ij} \in \{0, 1\}$ represent the response score of examinee s_i on question q_j. The vector $\mathbf{x} = (x_1, x_j, ..., x_M)$ represents an individual in the item selection optimization, where the vector dimension is the size of the question bank Q, and each element x_j represents whether the question q_j is selected, all the dimensions of $x_j = 1$ form the set of selected questions Q^T. Formally, the three objectives are defined as follows.

Measurement Efficiency (f_1). The exact measurement efficiency is obtained only after an examinee answers a set of selected questions, thus it is difficult to compute exact measurement efficiency during the item selection of OMST. In this case, the expected model change (EMC) is typically considered as an indicator that is able to reflect the measurement efficiency, which indicates the quantity of information contained in a set of selected questions. The questions containing a large amount of information generally facilitates more accurate estimation of an examinee's ability. For an examinee s_i, the EMC of the question q_j is calculated as:

$$EMC_j = \mathbb{E}_{r_{ij} \sim p} \Delta \mathcal{M}(<s_i, q_j, r_{ij}>) \tag{1}$$

$$p = \mathcal{M}(s_i, q_j | \theta_i) \tag{2}$$

The first objective is to maximize the EMC value of selected questions:

$$\text{Max} \quad f_1(\mathbf{x}) = \sum_{j=1}^{M} EMC_j \times x_j \tag{3}$$

Test Diversity (f_2). The test diversity of a question set is represented as the knowledge coverage of the questions. The more knowledge points a question set contains, the more comprehensive assessment the question set makes for examinees' ability. The knowledge coverage of a question set is formulated as:

$$\text{Max} \quad f_2(\mathbf{x}) = \frac{\sum_{k=1}^{K} 1[\exists q_j \in Q^T, G_{jk} = 1]}{K} \tag{4}$$

Test Security (f_3). The test diversity of a question set is computed as the chi-square value of exposure frequency of the questions, which is calculated as:

$$\text{Min} \quad f_3(\mathbf{x}) = \frac{\sum_{j=1}^{M} (fr_j/N - TL/M)^2}{TL/M} \tag{5}$$

where fr_j is the exposed frequency of question q_j, and TL is the test length of the OMST process.

In terms of optimization objectives or considered criteria, item selection algorithms of OMST can be roughly categorized into single-objective (or single-criterion) algorithms and double-objective ones. The maximum Fisher information (MFI) is a commonly used criteria in item selection, targeting the questions that contain much information for high measurement efficiency [16]. Regardless of the effectiveness of MFI in improving measurement efficiency, it causes that the questions with high differentiation are overexposed to a variety of examinees, since MFI is proportional to question differentiation. Thereafter, the model-agnostic adaptive testing (MAAT) was proposed to consider both the measurement efficiency and the knowledge coverage in two phases [2]. The question sets containing much information will be selected in the first phase, and then the question set with the maximum knowledge coverage will be determined from these sets. This two-phase item selection improves the test diversity of selected questions, which is, however, under the premise of the question sets with much information.

Few studies devote to simultaneously improving the measurement efficiency, test diversity and security in item selection of OMST. Therefore, in this study we integrating the three-criteria-based item selection into a multi-objective dynamic optimization framework, aiming at striking a good balance between the three criteria. For constraint control in item selection, we adopt the maximum priority index method (MPI) [7] and the Sympson-Hetter exposure control method (SH) [27] in the proposed AR-DMOEA.

3   The Proposed AR-DMOEA

This section first presents the main procedure of the proposed AR-DMOEA, then the details of the proposed initialization and offspring strategies are explained.

3.1   Procedure of the Proposed AR-DMOEA

Due to the fact that three objectives need to be simultaneously optimized in item selection, we propose to use a many-objective evolutionary optimization framework, namely, AR-MOEA, to address the item selection problem. The AR-MOEA uses a new indicator to distinguish non-dominated individuals that do not contribute to quality of population, and develops a reference point adaptive method for calculation of the indicator in each generation. The AR-MOEA is used as a dynamic optimization algorithm since the OMST is a dynamic multi-stage testing process. The flowchart of the proposed AR-DMOEA at each stage is presented in Fig. 2. First, an external archive is used to store the optimal individuals obtained from previous stages, which are regarded as initial population at current stage. A diversity-strengthened population initialization strategy will be used to create initial population at the beginning of the first stage. Then binary

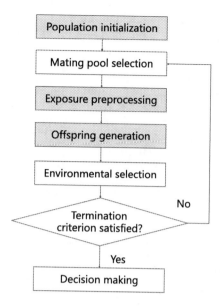

Fig. 2. The flowchart of the proposed AR-DMOEA.

tournament selection is used for individual mating, and an exposure preprocessing method is utilized to control the uniform distribution of question exposure. Next, a set of knowledge-guided operators are tailored for the offspring generation in item selection of OMST. The parent population and offspring population are united and the environmental selection is implemented to select population in next generation. The second to the fifth steps repeat until the termination criterion is satisfied, and a decision is finally made to obtain a set of questions for examination at next stage.

3.2 Details of the Proposed AR-DMOEA

This section elaborates three key steps in the procedure of the proposed AR-DMOEA: exposure preprocessing, population initialization and offspring generation.

Exposure Preprocessing. An exposure preprocessing method is applied before item selection to guarantee uniform distribution of question exposure. The exposure preprocessing method first chooses $log_2 M$ questions with high exposure, and then removes these questions from the question bank with probability p. In this way, the usage of low-exposure questions can be enhanced, and thus the computational cost is reduced by avoiding search for questions with high exposure in item selection. Specifically, the probability p is calculated as follows:

$$p = \frac{std(EX_Q)}{std(EX_Q) + 1} \tag{6}$$

Algorithm 1: Population Initialization

Input: Q (question bank), P (population from previous stage), Non (Non-dominated questions from the first stage), m (expected number of selected questions)

Output: P (initial population)

1 **if** *at the first stage of item selection* **then**
2 | $P \leftarrow$ Null
3 | $Non \leftarrow$ Null
4 | $EMC, COV, EXP \leftarrow$ calculate the EMC, knowledge coverage and exposure values for each question in the bank Q
5 | $C \leftarrow$ KMeans(EMC, COV, EXP, m)
6 | **for** $i=1$:ClusterSizeof(C) **do**
7 | | $Non_i \leftarrow$ identify non-dominated questions in cluster C_i
8 | **for** $i=1$:Sizeof(P) **do**
9 | | $S \leftarrow$ Null
10 | | $S \leftarrow$ randomly select a question from each cluster in Non and add the question to S
11 | | $P_i \leftarrow$ set the decision values of the questions in S to 1 to obtain an individual, and add the individual to P

12 **else**
13 | **for** $i=1$:Sizeof(P) **do**
14 | | $n \leftarrow$ Sum(P_i)
15 | | **if** $n < m$ **then**
16 | | | $P_i \leftarrow$ randomly select $m - n$ questions from clusters in Non, and set the corresponding decision values to 1 in P_i
17 | | | /*at most one question for each cluster in Non*/

18 Return P

---

where EX_Q denotes the exposure set of all questions in question bank Q, $std(.)$ refers to the variance, p reflects the exposure dispersion degree of question bank Q, and the greater the dispersion, the greater the p value.

Population Initialization. Due to the fact that three objectives are optimized in the proposed AR-DMOEA, it is difficult to obtain high-quality population in initial generation of item selection. As a result, the optimization algorithm may slowly converge in searching for high-quality question sets. Hence, we propose a diversity-strengthened population initialization strategy to create high-quality initial population in item selection, which is presented in Algorithm 1. The proposed initialization strategy main consists of four steps: First, the EMC, knowledge coverage and exposure values are computed for each question in the question bank Q. Next, all the questions in Q are clustered into m clusters (denote as a cluster set C) based on the EMC, knowledge coverage and exposure of each question, where m is an expected number of selected questions according to expertise. Then for each cluster in C, the non-dominated questions are identified using non-dominated sorting method based on the EMC, knowledge coverage and exposure of each question. Finally, an individual can be created by randomly selecting one question from non-dominated questions in each cluster. An initial population will be created by repeating the final step, which is not only of high quality in terms of the three objectives but holds good diversity and stochastic property due to the random selection.

The above steps are used for the first stage of item selection. For later stages, the population from previous stage maintains and is modified by using some parameters computed in the first stage. First, the number of selected questions n is obtained for each individual in the population from previous stage. If n is smaller than m, we randomly select $m - n$ questions from the clusters in the same way as described above, so that the individual is improved and achieved the expected number of selected questions.

Offspring Generation. To further speed up the convergence of the optimization, a knowledge-guided offspring generation strategy is proposed. Basically, classical binary genetic operators are adopted to generate offspring individuals, including one point crossover and bitwise mutation. Then knowledge-guided offspring generation operators are designed to guide the population to higher EMC value and lower exposure. Specifically, for each knowledge point that is not selected, the questions containing the knowledge point are identified from the question bank. The identified questions are sorted using non-dominated sorting method based on the EMC and exposure values of these questions. Finally, the decision variable corresponding to a random question in the non-dominated questions will be set to 1 for some individuals. This not only guarantees high knowledge coverage, but quickly leads the population to high EMC and low exposure in searching for question sets.

Question Maintenance. To achieve the expected number of selected questions, a question maintenance strategy is designed at the last generation of evolution. Similar to population initialization strategy, if the number of selected questions is less than m, then the $m - n$ decision variables with larger EMC value and lower exposure are selected from question bank and the corresponding decisions are set to one. If the number of selected questions is larger than m, the decision variable with the largest EMC value under each knowledge point is set to zero for the nonzero decision variables, and the $m - n$ decision variables with larger EMC value and the lower exposure are selected from the remainder and set to 0.

Decision Making. At the last generation of each stage, the optimizer of item selection produces a Pareto set with many individuals. A decision making method should be considered to select one individual from the Pareto set whose selected questions can be provided to an examinee. In this study, the individuals in the Pareto set are sorted using non-dominated sorting based on

EMC and knowledge coverage values of the individuals. Combining the EMC and knowledge coverage, the individual with the best comprehensive indicator will be determined as the final individual, since the EMC and knowledge coverage values play an important part in examinees' ability estimation.

4   Empirical Study

This section first introduces the experimental setting in empirical study. Next, the proposed AR-DMOEA is compared to three state-of-the-art item selection algorithms, and the corresponding empirical results are presented in tables and figures. Finally, the effectiveness of the strategies tailored for AR-DMOEA is validated.

4.1   Experimental Setting

Datasets: Five real-world educational datasets are used for evaluating the performance of the proposed AR-DMOEA, which are presented in Table 1. The five datasets are collected from online tutoring systems or educational platforms, which record examinees' practice logs and answers to the questions presented in these systems or platforms. For reliability of empirical results, we delete the knowledge points that have less than 10 related questions, the questions that are answered by less than 50 students, and the students that answer less than 10 questions. Each dataset is divided into 80% training set and 20% testing set. The training set is imported the CDMs to initially learn some question parameters before testing, while the testing set is used to simulate an adaptive testing environment.

Table 1. Dataset description

Dataset	ASSIST09	ASSIST17	NIPS-EDU	JUNYI	EXAM
Num.students	1,473	1,704	4,918	55,775	1,443
Num.questions	903	1,884	900	668	329
Num.concepts	22	96	57	39	19
Num.records	58,427	376,586	1,382,362	1,781,668 5	96,553
Records/student	39.7	221	281.1	31.9	66.9
Records/question	64.7	199.9	1535.9	2667.2	293.40

Comparison Algorithms and Parameter Setting: Three typical item selection algorithms are used for comparison with the proposed AR-DMOEA, namely, maximum Fisher information (MFI) [16], model-agnostic adaptive testing (MAAT) [2] and traditional on-the-fly assembled multistage adaptive testing (OMST_tra) [27]. The parameters in MAAT and MFI are set to the same values as those in their original references. For the other two algorithms, the number of stages and the module size are typically set to 3 and 15, respectively. The threshold for OMST_tra is set to 0.15 and the number of iterations is set to 10 for the calibration phase in OMST_tra.

Evaluation Metrics: The measurement efficiency, test diversity and security are used to evaluate the performance of each compared algorithm, which are denoted as test accuracy, knowledge coverage and exposure, respectively. The overall performance of the three criteria is represented with a comprehensive indicator in the field of multi-objective optimization, namely, hypervolume (HV), which also indicates the balance of the three criteria [24].

4.2 Comparison with State-of-the-Art Item Selection Algorithms

Table 2 presents the measurement efficiency, test diversity and security achieved by MFI, MAAT, OMST_tra and the proposed AR-DMOEA, where two typical cognitive diagnosis models are used for examinees' ability estimation: the item response theory model (IRT) [9] and the neural cognitive diagnosis model (NCD) [22]. It is observed from Table 2 that the proposed AR-DMOEA achieves the highest knowledge coverage on all of the five datasets under either of the two diagnosis models. On three of the five datasets, the proposed AR-DMOEA achieves over 95% knowledge coverage. For exposure of selected questions, the proposed AR-DMOEA achieves the best result on most of the datasets under either of the two diagnosis models. Despite that the proposed AR-DMOEA does not obtain the best EMC value in most cases, it achieves the second best and gets very close to the best value in most cases. It can be drawn that the proposed AR-DMOEA overall holds a competitive performance to state-of-the-art item selection algorithms in item selection of OMST.

Table 2. The measurement efficiency, test diversity and security achieved by MFI, MAAT, OMST_tra and the proposed AR-DMOEA

CDM	Method	ASSIST09		ASSIST17		NIPS-EDU		JUNYI		EXAM	
		AUC		AUC		AUC		AUC		AUC	
		COV	CHI	COV	CHI	COV	CHI	COV	CHI	COV	CHI
IRT	MFI	**0.6885**		0.7175		0.7117		0.7537		**0.7047**	
		0.7301	174	0.5559	402	0.7355	313	0.646	209	0.8681	93
	MAAT	**0.6885**		0.718		**0.7126**		0.7543		0.7046	
		0.7513	167	0.6079	395	0.7673	290	0.675	267	0.9081	83
	OMST_tra	0.6874		0.7171		0.7116		**0.7547**		0.7030	
		0.8578	86	0.6872	280	0.7362	**113**	0.8511	208	0.9319	**33**
	AR-DMOEA	0.6881		**0.7184**		0.7122		0.7544		0.7040	
		0.9707	85	**0.738**	151	**0.8916**	140	**0.869**	110	**0.9829**	42
NCD	MAAT	0.7234		0.7427		0.7765		0.8217		0.7498	
		0.6797	94	0.6234	252	0.9244	137	0.6985	179	0.8326	47
	OMST_tra	0.7159		0.7209		0.7618		0.8019		0.7437	
		0.8572	53	0.65	178	0.7729	84	0.7734	86	0.9019	35
	AR-DMOEA	0.7216		0.7352		0.7703		0.8082		0.7474	
		0.9418	53	**0.7573**	107	**0.9526**	70	**0.8891**	72	**0.9769**	30

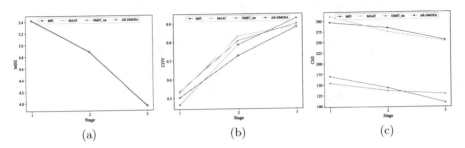

Fig. 3. Change of the three criteria obtained by the proposed AR-DMOEA over item selection stages on the "Exam" dataset.

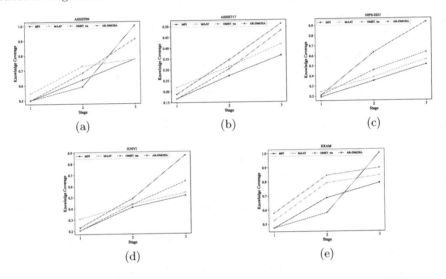

Fig. 4. Coverage obtained by MFI, MAAT, OMST_tra and AR-DMOEA.

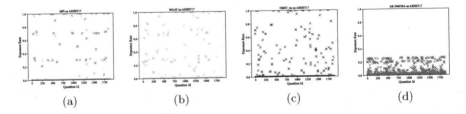

Fig. 5. Exposure obtained by MFI, MAAT, OMST_tra and AR-DMOEA on the "Assist17" dataset.

More specifically, Fig. 3 presents the change of the three criteria obtained by the proposed AR-DMOEA over item selection stages on the "Exam" dataset. Here we use "MSE" to measure the difference between the examinees' ability estimated during item selection and that computed based on the real answer cases of the testing dataset. It is seen from Fig. 3 that the proposed AR-DMOEA holds similar estimation error to state-of-the-art item selection algorithms over stages, and exhibits

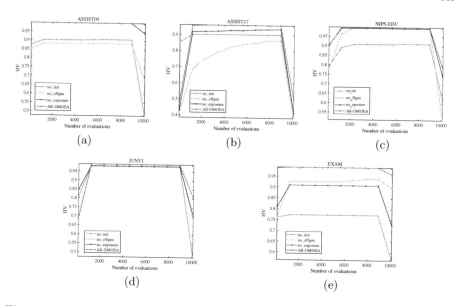

Fig. 6. Change of HV values achieved by the proposed AR-DMOEA and the AR-MOEAs without the proposed initialization, offspring generation and exposure control strategies.

higher performance in seeking for item sets of high knowledge coverage and low exposure in comparison with state-of-the-art item selection algorithms. According to Fig. 3, the proposed AR-DMOEA starts with the lowest knowledge coverage, but ends with the highest one. Similar performance can be found in Fig. 4, where the proposed AR-DMOEA significantly improves the knowledge coverage on all of the five datasets. Furthermore, Fig. 5 intuitively depicts the exposure obtained by MFI, MAAT, OMST_tra and AR-DMOEA on the "Assist17" dataset. As can be observed from Fig. 5, the questions found by the proposed AR-DMOEA are mostly of low exposure level, while the questions found by state-of-the-art item selection algorithms have a wide range of exposure levels.

4.3   Effectiveness of the Proposed Strategies

To validate the effectiveness of the proposed initialization, offspring generation and exposure control strategies, Fig. 6 depicts the change of HV values achieved by the proposed AR-DMOEA, the AR-MOEA without the proposed initialization strategy (no_init), the AR-MOEA without the proposed offspring generation strategy (no_offgen), and the AR-MOEA without the proposed exposure control strategy (no_exposure) during the item selection. As shown in Fig. 6, the initialization strategy is effective in enhancing the quality of initial individuals on three of the five datasets, while the offspring generation strategy leads to higher HV values on all of the five datasets. The exposure control strategy also contributes to higher HV values on the five datasets. To summarize, the proposed strategies collaboratively assists the proposed AR-DMOEA in striking a good balance between the three crucial criteria of item selection.

5   Conclusion

In this paper, we have proposed a dynamic tri-objective evolutionary algorithm AR-DMOEA to address the item selection problem of OMST, which has incorporated a diversity-strengthened population initialization strategy and a set of knowledge-guided offspring generation operators. Based on the empirical results, we have demonstrated the competitiveness of the proposed AR-DMOEA and the effectiveness of the proposed strategies.

Despite that the proposed AR-DMOEA has achieved a good balance between three item selection criteria, the measurement efficiency still needs to be further improved since the testing accuracy achieved by state-of-the-art item selection algorithms get quite close to each other. Therefore, developing or enhancing cognitive diagnosis models for item selection is an interesting research topic due to the fact that the models significantly influence the testing accuracy. On the other hand, we may also propose a new indicator to approximate testing accuracy so that the measurement efficiency can be optimized during the item selection.

Acknowledgment. This work was supported in part by the National Key Research and Development Project, Ministry of Science and Technology, China under Grant 2018AAA0101302, the National Natural Science Foundation of China under Grant 62202006, 62107001, Anhui Provincial Natural Science Foundation under Grant 2108085QF272, 2208085QF194, and the University Synergy Innovation Program of Anhui Province (GXXT-2021-004).

References

1. Barrada, J.R., Veldkamp, B.P., Olea, J.: Multiple maximum exposure rates in computerized adaptive testing. Appl. Psychol. Meas. **33**(1), 58–73 (2009)
2. Bi, H., et al.: Quality meets diversity: a model-agnostic framework for computerized adaptive testing. In: Proceedings of the IEEE International Conference on Data Mining, pp. 42–51. IEEE (2020)
3. Carroll, I.A.: Precision-based item selection for exposure control in computerized adaptive testing. Ph.D. thesis, University of Kansas (2017)
4. Chang, H.H.: Psychometrics behind computerized adaptive testing. Psychometrika **80**(1), 1–20 (2015)
5. Chang, H.H., Ying, Z.: A global information approach to computerized adaptive testing. Appl. Psychol. Meas. **20**(3), 213–229 (1996)
6. Chen, S.Y., Ankenmann, R.D., Spray, J.A.: The relationship between item exposure and test overlap in computerized adaptive testing. J. Educ. Meas. **40**(2), 129–145 (2003)
7. Cheng, Y., Chang, H.H.: The maximum priority index method for severely constrained item selection in computerized adaptive testing. Br. J. Math. Stat. Psychol. **62**(2), 369–383 (2009)

8. Du, Y., Li, A., Chang, H.-H.: Utilizing response time in on-the-fly multistage adaptive testing. In: Wiberg, M., Culpepper, S., Janssen, R., González, J., Molenaar, D. (eds.) IMPS 2017. SPMS, vol. 265, pp. 107–117. Springer, Cham (2019). https://doi.org/10.1007/978-3-030-01310-3_10

9. Embretson, S.E., Reise, S.P.: Item Response Theory. Psychology Press (2013)

10. Ghosh, A., Lan, A.: Bobcat: bilevel optimization-based computerized adaptive testing. arXiv preprint arXiv:2108.07386 (2021)

11. Gür, R., Hamide, D.G.: The effect of item exposure control methods on measurement precision and test security under different measurement conditions in computerized adaptive testing. Egitim ve Bilim **45**(202) (2020)

12. Han, K.C.T.: Components of the item selection algorithm in computerized adaptive testing. J. Educ. Eval. Health Prof. **15** (2018)

13. Lin, C.J., Chang, H.H.: Item selection criteria with practical constraints in cognitive diagnostic computerized adaptive testing. Educ. Psychol. Measur. **79**(2), 335–357 (2019)

14. Linden, W.J.: Constrained adaptive testing with shadow tests. In: Elements of Adaptive Testing, pp. 31–55 (2009)

15. Liu, S., Cai, Y., Tu, D.: On-the-fly constraint-controlled assembly methods for multistage adaptive testing for cognitive diagnosis. J. Educ. Meas. **55**(4), 595–613 (2018)

16. Lord, F.M.: Applications of Item Response Theory to Practical Testing Problems. Routledge (1980)

17. Luecht, R.M., De Champlain, A., Nungester, R.J.: Maintaining content validity in computerized adaptive testing. Adv. Health Sci. Educ. **3**(1), 29–41 (1998)

18. Luecht, R., Nungester, R., Hadadi, A.: Heuristic-based CAT: balancing item information, content and exposure. In: Proceedings of the Annual Meeting of the National Council on Measurement in Education (NCME), New York (1996)

19. Luo, X., Wang, X.: Dynamic multistage testing: a highly efficient and regulated adaptive testing method. Int. J. Test. **19**(3), 227–247 (2019)

20. Sari, H.İ., Huggins-Manley, A.C.: Examining content control in adaptive tests: computerized adaptive testing vs. computerized adaptive multistage testing. Educ. Sci. Theory Pract. **17**(5) (2017)

21. Wainer, H.: Rescuing computerized testing by breaking Zipf's law. J. Educ. Behav. Stat. **25**(2), 203–224 (2000)

22. Wang, F., et al.: Neural cognitive diagnosis for intelligent education systems. In: Proceedings of the AAAI Conference on Artificial Intelligence, vol. 34, pp. 6153–6161 (2020)

23. Wang, S., Lin, H., Chang, H.H., Douglas, J.: Hybrid computerized adaptive testing: from group sequential design to fully sequential design. J. Educ. Meas. **53**(1), 45–62 (2016)

24. While, L., Hingston, P., Barone, L., Huband, S.: A faster algorithm for calculating hypervolume. IEEE Trans. Evol. Comput. **10**(1), 29–38 (2006)

25. Zenisky, A., Hambleton, R.K., Luecht, R.M.: Multistage testing: issues, designs, and research. In: Elements of Adaptive Testing, pp. 355–372 (2009)

26. Zheng, Y., Chang, H.H.: Multistage testing, on-the-fly multistage testing, and beyond. In: Advancing Methodologies to Support both Summative and Formative Assessments, pp. 21–39 (2014)

27. Zheng, Y., Chang, H.H.: On-the-fly assembled multistage adaptive testing. Appl. Psychol. Meas. **39**(2), 104–118 (2015)

CCGSK-PSO: Constrained Multi-objective Particle Swarm Optimization Algorithm Based on Co-evolution and Gaining-Sharing Knowledge

Liyan Qiao and Sibo Hou[✉]

Department of Automatic Test and Control, Harbin Institute of Technology, Harbin, China
hsssee@163.com

Abstract. Constrained multi-objective optimization is facing huge challenges due to the complex constraints and the restraint of multiple optimization objectives. The most difficult problem need to solve is to achieve the balance between optimization objective convergence and feasibility under constraints as far as possible. To solve this problem, a constrained multi-objective particle swarm optimization algorithm based on co-evolution and gaining-sharing knowledge mechanism is proposed, which is called CCGSK-PSO. Based on the evolutionary strategy of co-evolution, the evolutionary process conducts two types of co-evolutionary in each iteration: the co-evolution of unconstrained evolutionary population and constrained evolutionary population, and the co-evolution of objective convergence subpopulation and constrained subpopulation in constrained evolutionary population. The objective convergence subpopulation is mainly responsible for exploring the objective convergence boundary information, while learning from the individuals in constrained subpopulation in the neighborhood to guide the population to effectively cross the infeasible regions. The constrained subpopulation is mainly responsible for exploring the constraint boundary information, while guiding the population to converge to the elite individuals. Finally, the obtained solution set is widely and uniformly distributed along the constrained Pareto front. The experimental results on 37 test problems in three benchmark suites show the effectiveness of CCGSK-PSO.

Keywords: Constrained multi-objective optimization · Co-evolution · Gaining-sharing Knowledge mechanism

1 Introduction

Optimization problem generally exists in scientific theoretical research and practical engineering applications. From the perspective of global optimization, single objective optimization often has one-sided limitation of the optimization

© The Author(s), under exclusive license to Springer Nature Switzerland AG 2023
Y. Tan et al. (Eds.): ICSI 2023, LNCS 13968, pp. 452–463, 2023.
https://doi.org/10.1007/978-3-031-36622-2_37

perspective, it is difficult to provide decision-makers with an overall optimal solution. Then the multi-objective optimization problem (MOP) is proposed.

According to whether it is the problem with constraints, MOP can be divided into unconstrained multi-objective optimization problem and constrained multi-objective optimization problem (CMOP). Because there are often complex constraints when solving practical application problems, it is of practical significance to study CMOP. The solution of CMOP should consider the constraints while optimizing multiple objectives. The existence of constraints makes the optimal solution under unconstrained conditions may not meet the constraint conditions. It is necessary to design constrained multi-objective optimization evolutionary algorithm (CMOEA) to effectively solve the solution set that meets the constraint conditions and makes each objective reach the optimal. The CMOEA faces two challenges, one is to balance the convergence and the diversity of solutions, and the other is to balance the optimization of objectives and the constraint conditions. In the iterative process of CMOEA, the optimal solutions of the following three cases need to be retained: (1) The objective is optimal and the constraint is satisfied; (2) The objective is suboptimal and meets the constraints; (3) The objective is optimal, the constraints are satisfied within the acceptable range. CMOEA can effectively use the individual information of the evolutionary population to quickly solve the problem after balancing the optimization of objectives and the constraint conditions. The existence of constraints increases the complexity of the problem, especially when the objective space is relatively complex, there may be multiple infeasible regions that are difficult to cross. These infeasible regions are continuous or discrete, and the area of infeasible regions is large or small, which is very easy to fall into local optimum. Therefore, it is relatively difficult to solve.

In the recent years, researchers have proposed multiple different kinds of CMOEA for solving CMOP. Professor Kalyanmoy Deb's research team proposed CMOEA/D [1], which uses the probability of constraint domination and combines MOEA/D to deal with CMOP. Professor Xin Yao's research team proposed C-TAEA [2], which is the constrained version of the multi-objective Two-Archive evolutionary algorithm. The constraint violation value is added to the Two-Archive technology to deal with the CMOP. The research team of Professor Zhun Fan and Professor Xinye Cai proposed the push and pull search framework. Based on this framework, PPS [3] was proposed, which can effectively guide the population to the feasible region of the Pareto front through push and pull search methods. Although many scientific research teams have carried out long-term research on CMOEA, the proposed algorithms still have the shortcomings and fault of low efficiency in finding constraint boundaries, the inability of individuals to effectively cross infeasible regions and easy to fall into local optimum during evolutionary iteration when dealing with multi-objective optimization problems under complex constraints. Especially when dealing with the small and discrete feasible regions and the situation that the unconstrained Pareto front is far away from the constrained Pareto front, it faces serious challenges in exploring and finding the Pareto optimal solution in the feasible region. Moreover, the existing research theories and achievements have not carried out special research on the constrained multi-objective optimization algorithm based on swarm intelligence.

The advantage of swarm intelligence algorithm to solve such problems is that it can effectively use the interaction between group information and individual information to balance the exploration and acquisition of objective convergence boundary information and constraint boundary information.

Inspired by the gaining-sharing knowledge mechanism of GSK [4], based on the swarm intelligence optimization algorithm, this paper proposes a new method CCGSK-PSO for CMOP. Based on the idea of co-evolution, CCGSK-PSO integrates the particle swarm optimization algorithm in swarm intelligence optimization algorithm into the gaining-sharing knowledge mechanism, so that the particle individual information and the particle swarm information can fully interact under the gaining-sharing knowledge mechanism. CCGSK-PSO can effectively transfer the objective convergence boundary and constraint boundary information. At the same time, CCGSK-PSO divides the whole population into convergence subpopulation and constrained subpopulation. The two subpopulations effectively balance the exploration and optimization of objectives and constraints through co-evolution, so as to achieve the distribution and convergence of the population and realize the feasibility of the population.

2   The Proposed CCGSK-PSO

2.1   General Framework

The basic structure diagram of CCGSK-PSO is shown in Fig. 1. Firstly, like other CMOEA, randomly initialize a particle swarm with N individual solutions in the entire decision space. This initial population is the parent population of the first iteration. CCGSk-PSO adopts two-level co-evolution strategy and hybrid gaining-sharing knowledge mechanism with PSO to accelerate the solving of CMOP.

2.2   Hybrid Gaining-Sharing Knowledge Mechanism

Primary Gaining-Sharing Knowledge Stage. In the primary gaining-sharing knowledge stage, each individual tries to obtain knowledge from the closest and mostly trusted individuals who belonging to his or her neighbor small group. At the same time, because of his curiosity and desire to explore others, he also tries to share knowledge with some individuals who do not belong to this small group or any group. Therefore, the following primary scheme can be used for information exchange among individuals in the population.

1. All individuals are arranged in ascending order according to the aggregation function value of the objective and constraint violation degree: x_{best}, ..., x_{i-1}, x_i, x_{i+1}, ..., x_{worst}.

2. Then, two different the nearest individuals are selected for each x_i, namely the best individual closest to the current individual x_{i-1} and the worst individual closest to the current individual x_{i+1}, constitute the source of knowledge acquisition. In addition, another individual x_r is randomly selected as the source

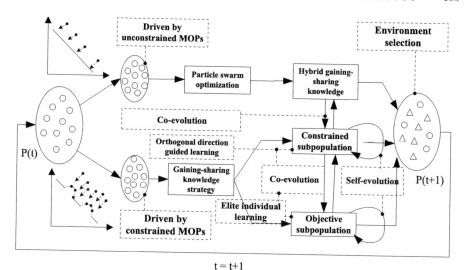

t = t+1

Fig. 1. Framework of the Proposed CCGSK-PSO

of shared knowledge.Note: at this stage, the best two individuals and the worst two individuals are used to update the best and worst individuals. If x_i is the global best, select the two nearest positive best individuals, as shown below: x_{bset}, x_{bset+1}, x_{bset+2}. If x_i is the worst in the whole population, select the two nearest individuals with the worst in reverse, as shown below:$x_{worst-2}$, $x_{worst-1}$, x_{worst}. Among them, it will also be mentioned below, k_f is a real number greater than 0 and means knowledge factor, which controls the total amount of knowledge obtained and shared by others in each generation of individuals. $k_r \in [0,1]$ is the knowledge ratio, which controls the total amount of acquired and shared knowledge transferred, also inherited, between individuals of each generation. k_r is also the ratio between current experience and acquired experience.

Advanced Gaining-Sharing Knowledge Stage. This stage involves using the available information and appropriate knowledge of different categories of individuals, also the best, better and worst individuals in a specific population. Utilization refers to the influence of others, including good and bad on an individual. Therefore, the following scheme can be used to update each individual.

1. After ranking all individuals in ascending order according to the objective function, they will be divided into three categories: the best individual, the better or medium individual, and the worst individual.
2. Then, for each individual x_i, $100p\%$ individuals in the current N size population to form the gain part, while the third vector is from the middle $N - (2 \times 100p\%)$ randomly selected individuals form a shared part. Where $p \in [0,1]$, $p = 0.1$, $10\%N$ is appropriate.

Algorithm 1: Hybrid gaining-sharing knowledge mechanism with PSO

Input: P(current population), V(current velocities), N(swarm size),D(total dimensions)

Output: Offspring (Offspring individuals)

1   i ← According to aggregate functions, which comprehensively consider convergence, diversity and constraints, arrange particles in ascending order,then get the position i;

2   /* primary gaining-sharing knowledge stage */;

3   **for** *each particle a_i in P* **do**

4   **for** *j = 1 : D* **do**

5   **if** *rand ≤ k_r (Knowledge ratio)* **then**

6   **if** $f(a_i) > f(a_r)$ **then**

7   Select corresponding learning individual b;

8           $V_{p,k}(t+1) = R_1(k,t)V_{p,k} + R_2(k,t)\left(b_{p,k}(t) - a_{p,k}(t)\right);$

9           $a_{ij}^{new} = a_i + k_f * [(a_{i-1} - a_{i+1}) + (a_r - a_i)] + V_{p,k}(t+1);$

10   **else**

11   Select corresponding learning individual b;

12           $V_{p,k}(t+1) = R_1(k,t)V_{p,k} + R_2(k,t)\left(b_{p,k}(t) - a_{p,k}(t)\right);$

13           $a_{ij}^{new} = a_i + k_f * [(a_{i-1} - a_{i+1}) + (a_i - a_r)] + V_{p,k}(t+1);$

14   **end**

15   **else**

16   Select corresponding learning individual b;

17         $V_{p,k}(t+1) = R_1(k,t)V_{p,k} + R_2(k,t)\left(b_{p,k}(t) - a_{p,k}(t)\right);$

18         $a_{ij}^{new} = a_{ij}^{old} + V_{p,k}(t+1);$

19   **end**

20   **end**

21   **end**

22   /* advanced gaining-sharing knowledge stage */;

23   **for** *each particle a_i in P* **do**

24   **for** *j = 1 : D* **do**

25   **if** *rand ≤ k_r (Knowledge ratio)* **then**

26   **if** $f(a_i) > f(a_m)$ **then**

27   Select corresponding learning individual b;

28           $V_{p,k}(t+1) = R_1(k,t)V_{p,k} + R_2(k,t)\left(b_{p,k}(t) - a_{p,k}(t)\right);$

29           $a_{ij}^{new} = a_i + k_f * [(a_{best} - a_{worst}) + (a_m - a_i)] + V_{p,k}(t+1);$

30   **else**

31   Select corresponding learning individual b;

32           $V_{p,k}(t+1) = R_1(k,t)V_{p,k} + R_2(k,t)\left(b_{p,k}(t) - a_{p,k}(t)\right);$

33           $a_{ij}^{new} = a_i + k_f * [(a_{best} - a_{worst}) + (a_i - a_m)] + V_{p,k}(t+1);$

34   **end**

35   **else**

36   Select corresponding learning individual b;

37         $V_{p,k}(t+1) = R_1(k,t)V_{p,k} + R_2(k,t)\left(b_{p,k}(t) - a_{p,k}(t)\right);$

38         $a_{ij}^{new} = a_{ij}^{old} + V_{p,k}(t+1);$

39   **end**

40   **end**

41   **end**

42   Offspring ← Individual offspring;

43   return Offspring

Hybrid Gaining-Sharing Knowledge Mechanism with PSO. Combined with the co-evolution strategy, the two auxiliary subpopulations in the algorithm need to complete the process of self-evolution and information interaction at the same time during each evolutionary iteration. The information interaction process adopts the hybrid gaining-sharing knowledge mechanism with PSO, which is shown in Algorithm 1. CCGSK-PSO adopts different information interaction strategies according to different stages of evolution.

2.3   Two-Level Co-evolution Strategy

In order to better explore the constraint boundary information, CCGSK-PSO adopts the objective convergence subpopulation and constraint subpopulation co-evolution strategy, and the two subpopulations interact with the objective convergence boundary information and constraint boundary information during the evolutionary process. If the population size is N individuals, the two subpopulations each contain $N/2$ individuals. The division standard is determined according to the proportion of individual carrying constraint boundary information. In order to explore the constraint boundary information, the algorithm should retain the individuals at the constraint boundary as much as possible. Because the individual at the constraint boundary is extremely sensitive to disturbance, small disturbance may change the feasibility of the individual, so the algorithm uses Gaussian perturbation method to explore the constraint boundary information. Each individual in the population is subject to k times of Gaussian random perturbation in its small range of fields, and the feasibility of the perturbed individual is judged according to the CV value. If the number of feasible individuals in the perturbed k individuals is m, hen the number of infeasible individuals is $k - m$, then the proportion of carrying constraint boundary information is measured by $|k - 2m|/k$. Generally, k is a fixed parameter predetermined by the algorithm, so comparing the $|k - 2m|$ value of each individual after Gaussian perturbation can measure the amount of constraint boundary information carried. The smaller the $|k - 2m|$ value, the greater the constraint boundary information is carried. The population individuals are sorted from small to large according to the $|k - 2m|$ value. The first $N/2$ individuals constitute constrained subpopulation, and the remaining $N/2$ individuals constitute objective convergence subpopulation.

After the main population with size N is classified into two auxiliary subpopulations with size $N/2$, the two auxiliary subpopulations enter the stage of information interaction of self-evolution and coevolution respectively. The constrained auxiliary subpopulation uses the orthogonal direction guided learning method, which is shown in Fig. 2, to explore more information about the constraint boundary. The auxiliary subpopulation obtained after its evolution contains more information about the constraint boundary. At the same time, it extracts the convergence boundary information of the objective convergence auxiliary subpopulation and learns towards to the direction of convergence. The objective convergence auxiliary subpopulation uses the elite individual learning method, which is shown in Fig. 3, to explore more information about the convergence Pareto front boundary. The auxiliary subpopulation obtained after its self-evolution contains more objective convergence boundary information. At the same time, it extracts the constraint boundary information of the constrained auxiliary subpopulation and learns from the constraint boundary direction. Note that the coevolutionary strategies here are all weak coevolutionary strategies similar to CCMO [5] to avoid the problem of premature population convergence caused by too strong coevolutionary strategies.

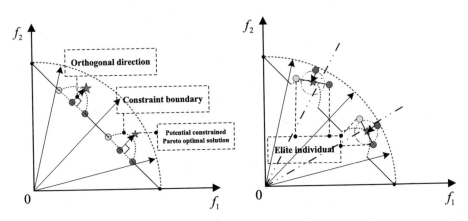

Fig. 2. Orthogonal Direction Guided Learning Strategy

Fig. 3. Elite Individual Guided Learning Strategy

3   Numerical Experiment and Result Analysis

3.1   Experimental Parameter Settings and Performance Index

The experiment is set to explore and verify the optimization performance of CCGSK-PSO algorithm. This paper compares the performance of CCGSK-PSO algorithm and other four advanced constrained multi-objective optimization algorithms on 37 test problems of MW [6], LIR-COMP [7] and DAS-COMP [8] benchmark suites. MW, LIR-COMP and DAS-COMP benchmark suites have different types of constraint feasible regions, it poses a serious challenge for constrained multi-objective optimization algorithm to balance objective optimization and constraint satisfaction. The population size of each tested algorithm is set to 100. Due to the complexity of the test problems of LIR-COMP and DAS-COMP standard benchmark suites, the maximum number of iterations on the MW standard benchmark suite is set to 100, and the maximum number of iterations on the LIR-COMP and DAS-COMP standard benchmark suites is set to 500. The comparison algorithm selects four advanced constrained multi-objective optimization algorithms: CMOEA/D, NSGA-II [9], Ship [10] and ToP [11].

Without losing fairness, algorithm comparison experiments are conducted in the same experimental environment. In this case, each comparison algorithm runs independently for 30 times in each standard benchmark suites, and takes the mean and IQR [12] of IGD [13] indicator to measure the average test performance and algorithm stability.

3.2 Experimental Results and Analysis

Table 1, Table 2 and Table 3 are the performance comparison results of IGD indicators on MW, LIR-CMOP and DAS-CMOP standard benchmark suites between CCGSK-PSO algorithm and other four constrained multi-objective optimization algorithms with good performance. Where+,-,=respectively represent the number of problems that the performance of other comparison algorithms is superior to, inferior to, or approximate to the CCGSK-PSO proposed in this paper.

Table 1 is the data table of algorithm comparison experiments on the 14 test problems of MW1-14. From the data in Table 1, we can see that on the constrained multi-objective optimization standard benchmark suite MW, CCGSK-PSO proposed in this paper achieves the best performance on the 11 test problems of MW1-7, MW9, and MW11-13, and outperforms the other four excellent comparison algorithms on 78.57% of the test problems. Therefore, it can be seen that the proposed CCGSK-PSO is not only competent for complex solution scenarios with discrete feasible regions, but also can effectively solve and obtain the best performance in different distribution types of solution scenarios.Figure 4 shows the convergence and distribution comparison of CMOEA on the MW3 problem.

Table 2 is the data table of algorithm comparison experiments on the 14 test problems of LIR-CMOP1-14. From the data in Table 2, we can see that CCGSK-PSO proposed in this paper achieves the best performance on all 14 test problems and outperforms the other four excellent comparison algorithms on all test problems. Whether it is for the test scenarios where the unconstrained Pareto Front is included in the feasible region, on such test problems LIR-CMOP5-6, LIR-CMOP13, or for the test scenarios where the unconstrained Pareto Front intersects with the feasible region, on such test problems LIR-CMOP9-12, or for the test scenarios where the unconstrained Pareto Front is far from the feasible region, on such test problems LIR-CMOP1-4, LIR-CMOP7-8, LIR-CMOP14, proposed CCGSK-PSO is obviously superior to the other four comparison algorithms. Therefore, it can be seen that the proposed CCGSK-PSO is also competent in complex optimization scenarios such as narrow feasible region, Pareto Front is blocked by a large block of infeasible region, Pareto Front is composed of disjoint discrete segments or sparse points, and feasible region is limited to a curve. CCGSK-PSO can effectively solve and obtain the best performance in different distribution types of solution scenarios.

It can be seen from Table 3 that the proposed CCGSK-PSO algorithm achieves the best performance on the seven test problems DAS-CMOP1-6 and DAS-CMOP9, and outperforms the other four excellent comparison algorithms on 77.77% of the test problems. Figure 5 shows the simulation results of the CCGSK-PSO algorithm on the DAS-CMOP benchmark suite.

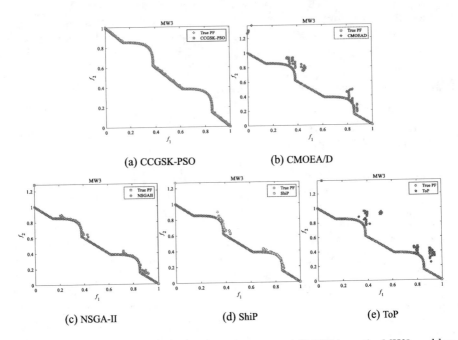

(a) CCGSK-PSO   (b) CMOEA/D

(c) NSGA-II   (d) ShiP   (e) ToP

Fig. 4. the convergence and distribution comparison of CMOEA on the MW3 problem

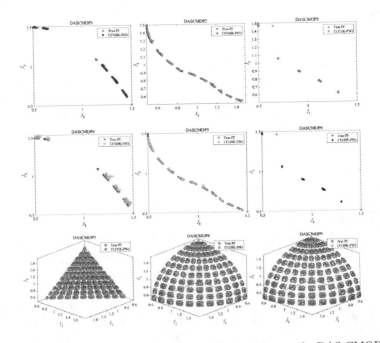

Fig. 5. the simulation results of the CCGSK-PSO algorithm on the DAS-CMOP benchmark suite

Table 1. Median and IQR (in brackets) of IGD metric on the MW benchmark suite

Problem	C-MOEA/D	NSGA-II	ShiP	ToP	CCGSK-PSO
MW1	2.7548e-1 (1.43e-1) -	1.1479e-1 (1.35e-1) -	1.3801e-1 (1.16e-1) -	8.6693e-1 (1.96e-1) -	**4.8111e-3 (9.38e-4)**
MW2	6.1336e-2 (1.95e-2) -	3.6364e-2 (4.61e-2) -	4.1948e-2 (9.07e-3) -	9.8276e-2 (3.52e-2) -	**4.1883e-2 (8.24e-3)**
MW3	9.0745e-2 (8.60e-2) -	3.5933e-2 (4.56e-2) -	1.1861e-1 (1.16e-1) -	4.2931e-1 (3.13e-1) -	**7.8369e-3 (3.20e-4)**
MW4	1.8809e-1 (1.09e-1) -	8.9693e-2 (5.01e-2) -	1.0690e-1 (1.96e-2) -	3.7558e-1 (2.87e-1) -	**8.6939e-2 (5.46e-3)**
MW5	2.1940e-1 (2.70e-1) -	9.1251e-2 (4.05e-2) -	2.7028e-1 (3.03e-1) -	5.6082e-1 (4.25e-1) -	**1.2591e-2 (2.79e-3)**
MW6	3.3721e-1 (1.85e-1) =	3.5809e-1 (1.82e-1) =	5.3467e-1 (1.03e-1) -	6.4629e-1 (3.98e-1) -	**6.2918e-2 (2.90e-2)**
MW7	1.0475e-2 (4.73e-4) -	1.9016e-1 (2.42e-1) -	2.5572e-2 (1.76e-2) -	3.5357e-1 (2.82e-1) -	**9.2007e-3 (2.08e-3)**
MW8	9.3695e-2 (4.81e-2) =	**9.2765e-2 (4.36e-2) =**	1.2076e-1 (5.79e-2) -	1.1274e+0 (7.18e-3) -	1.0988e-1 (1.86e-2)
MW9	9.8386e-2 (2.19e-2) -	6.1315e-1 (1.59e-1) -	6.0749e-1 (3.56e-1) -	8.1315e-1 (5.56e-1) -	**9.9029e-3 (1.44e-3)**
MW10	**1.1742e-1 (3.78e-2) =**	2.4793e-1 (2.11e-1) -	1.6703e-1 (9.54e-2) -	9.2331e-1 (3.64e-1) -	1.3966e-1 (1.75e-2)
MW11	7.0819e-1 (7.86e-3) -	7.3321e-1 (2.37e-2) -	7.6781e-1 (5.93e-2) -	8.9662e-1 (9.38e-2) -	**1.0409e-2 (8.29e-4)**
MW12	2.3960e-1 (7.41e-2) -	5.3556e-1 (1.80e-1) -	4.7281e-1 (2.00e-1) -	7.9558e-1 (2.47e-1) -	**7.5355e-3 (5.21e-4)**
MW13	3.6052e-1 (2.12e-1) =	3.6831e-1 (2.68e-1) -	1.0256e+0 (5.43e-1) -	1.6910e+0 (3.39e-1) -	**1.6126e-1 (3.09e-2)**
MW14	9.4761e-1 (2.55e-1) =	**8.8778e-1 (6.95e-2) =**	1.0085e+0 (1.50e-1) -	1.0407e+0 (2.84e-1) =	9.1109e-1 (1.09e-1)
±/=	0/8/6	0/7/7	0/9/5	0/1/13	

Table 2. Median and IQR (in brackets) of IGD metric on the LIR-CMOP benchmark suite

Problem	C-MOEA/D	NSGA-II	ShiP	ToP	CCGSK-PSO
LIR-CMOP1	2.7395e-1 (2.56e-2) =	3.1393e-1 (2.90e-2) =	9.3752e-1 (7.56e-2) -	3.3451e-1 (1.85e-2) =	**2.6447e-1 (5.70e-2)**
LIR-CMOP2	2.5829e-1 (2.23e-2) -	2.6846e-1 (1.88e-2) -	8.7873e-1 (3.93e-2) -	2.9036e-1 (1.20e-2) -	**1.6453e-1 (8.39e-2)**
LIR-CMOP3	2.9338e-1 (3.14e-2) =	2.8088e-1 (4.28e-2) =	9.5315e-1 (1.67e-2) -	3.4215e-1 (3.31e-1) -	**1.9530e-1 (9.96e-2)**
LIR-CMOP4	2.6391e-1 (5.67e-2) =	2.5810e-1 (2.21e-2) =	9.3664e-1 (2.85e-2) -	3.1953e-1 (1.10e-2) -	**2.3300e-1 (7.51e-2)**
LIR-CMOP5	1.2280e+0 (6.73e-3) -	1.2196e+0 (4.21e-3) -	1.2338e+0 (5.48e-3) -	1.2422e+0 (8.29e-3) -	**2.3854e-1 (2.34e-1)**
LIR-CMOP6	1.3465e+0 (5.02e-4) -	1.3456e+0 (1.28e-4) -	1.3454e+0 (7.74e-5) -	1.3478e+0 (3.48e-3) -	**1.5892e-1 (9.82e-2)**
LIR-CMOP7	1.4702e+0 (4.79e-1) -	1.1573e+0 (7.42e-1) -	1.3861e+0 (6.61e-1) -	1.6827e+0 (7.07e-4) -	**7.2256e-2 (4.01e-2)**
LIR-CMOP8	1.6830e+0 (7.19e-4) -	1.6819e+0 (2.56e-4) -	8.3573e-1 (7.73e-1) -	1.6832e+0 (1.21e-3) -	**3.8010e-2 (2.75e-2)**
LIR-CMOP9	8.6640e-1 (9.37e-2) -	9.2002e-1 (2.13e-1) -	1.0322e+0 (6.17e-2) -	5.6783e-1 (1.35e-1) -	**4.0131e-1 (6.39e-2)**
LIR-CMOP10	7.8035e-1 (1.84e-1) -	9.1836e-1 (3.32e-2) -	1.0135e+0 (4.50e-2) -	4.7310e-1 (1.38e-1) -	**4.3544e-1 (9.25e-2)**
LIR-CMOP11	8.1851e-1 (1.68e-1) -	8.4383e-1 (7.30e-2) -	9.3388e-1 (1.32e-1) -	5.6912e-1 (1.27e-1) -	**1.9893e-1 (9.00e-2)**
LIR-CMOP12	7.0738e-1 (1.83e-1) -	7.9949e-1 (2.11e-1) -	1.0584e+0 (1.19e-1) -	3.3263e-1 (5.63e-2) -	**8.7356e-2 (2.22e-2)**
LIR-CMOP13	1.3057e+0 (5.76e-2) -	1.3245e+0 (4.34e-3) -	1.3273e+0 (2.41e-3) -	1.3509e+0 (5.66e-2) -	**2.1910e-1 (2.21e-2)**
LIR-CMOP14	1.2616e+0 (5.88e-4) -	1.2809e+0 (1.61e-3) -	1.2816e+0 (1.96e-3) -	1.3097e+0 (5.42e-2) -	**1.7114e-1 (3.66e-2)**
±/=	0/11/3	0/11/3	0/14/0	0/11/3	

Table 3. Median and IQR (in brackets) of IGD metric on the DAS-CMOP benchmark suite

Problem	C-MOEA/D	NSGA-II	ShiP	ToP	CCGSK-PSO
DAS-CMOP1	7.3677e-1 (3.22e-2) -	7.3746e-1 (2.23e-2) -	7.4801e-1 (2.64e-2) -	7.5232e-1 (6.78e-3) -	**8.3578e-3 (2.05e-3)**
DAS-CMOP2	2.8674e-1 (3.37e-2) -	2.9578e-1 (5.74e-2) -	3.0172e-1 (4.12e-2) -	7.8095e-1 (3.43e-2) -	**7.5558e-3 (1.02e-3)**
DAS-CMOP3	3.6355e-1 (4.56e-2) -	3.4580e-1 (2.18e-3) -	3.4698e-1 (3.41e-3) -	7.3885e-1 (5.13e-2) -	**1.1797e-1 (1.33e-1)**
DAS-CMOP4	3.7691e-1 (1.01e-1) =	4.4752e-1 (6.18e-2) -	5.4190e-1 (7.01e-2) -	9.2186e-1 (1.58e-1) -	**1.2923e-1 (1.60e-1)**
DAS-CMOP5	4.0321e-1 (2.55e-1) =	5.8026e-1 (3.26e-2) =	6.5603e-1 (2.30e-2) -	7.9889e-1 (4.79e-2) -	**1.1458e-1 (1.85e-1)**
DAS-CMOP6	6.0319e-1 (1.94e-1) -	6.4491e-1 (1.07e-1) -	7.1599e-1 (1.45e-1) -	9.3754e-1 (2.25e-1) -	**3.2721e-1 (7.47e-2)**
DAS-CMOP7	**5.6793e-2 (1.42e-2) =**	6.6239e-2 (3.74e-2) =	7.5044e-2 (2.85e-2) -	2.7728e-1 (3.92e-2) -	5.6982e-2 (1.42e-2)
DAS-CMOP8	1.9161e-1 (1.99e-1) =	**7.0611e-2 (1.68e-2) =**	1.0600e-1 (5.07e-2) =	4.3639e-1 (4.61e-2) -	7.3539e-2 (1.09e-2)
DAS-CMOP9	4.9927e-1 (9.62e-2) =	4.6973e-1 (1.10e-1) -	4.6489e-1 (6.60e-2) -	6.8577e-1 (1.36e-1) -	**2.0884e-1 (2.28e-1)**
±/=	0/4/5	0/5/4	0/6/3	0/9/0	

Based on the comparison experiments of the algorithms on the three standard benchmark suites of MW, LIR-CMOP and DAS-CMOP, it can be seen that the proposed CCGSK-PSO can well balance the constraint feasibility and objective convergence, effectively solve constrained multi-objective optimization problems, and be competent for multiple types of constrained multi-objective optimization scenarios. CCGSK-PSO can obtain the solution set with better convergence and distribution while satisfying the constraints. From the comparative experimental data, we can see the comprehensive ability of CCGSK-PSO to deal with multi-objective optimization problems with multiple types of constraints. At the same time, by comparing the standard deviation data, it can be seen that the performance of CCGSK-PSO is relatively stable, and it can play a stable role in solving constrained multi-objective optimization problems.

4   Conclusion and Remark

This paper studies the existing problems of constrained multi-objective optimization algorithm, such as poor exploration ability of constrained feasible space, low efficiency of solving feasible solutions that meet the optimal conditions, and difficulty in balancing objective convergence and constrained feasibility. Through three representative and challenging constrained multi-objective optimization standard benchmark suites MW, LIR-CMOP, DAS-CMOP, the performance test covers all kinds of optimization scenarios with different distribution relations between the constrained feasible region and the unconstrained Pareto Front. At the same time, the three standard benchmark suites have conducted performance tests on complex constraint scenarios such as the narrow feasible region separated by the large constraint infeasible region and the discrete feasible region separated by the complex constraint. The proposed CCGSK-PSO and the other four excellent constrained multi-objective optimization algorithms are tested on a total of 37 classic test problems. After comparing and analyzing the experimental results, the following conclusions are obtained:

1) A new coevolutionary strategy is proposed, which aims at the poor ability of exploring the constrained feasible space and the low efficiency of solving the feasible solution meeting the optimal conditions of the CMOEA.
2) A new hybrid mechanism of gaining-sharing knowledge with PSO is proposed. The evolutionary mechanism realizes the information learning between the objective convergence subpopulation and the constrained subpopulation.
3) A CMOEA based on co-evolution and gaining-sharing knowledge mechanism is proposed. The experimental results show that the performance of CCGSK-PSO is significantly better than other excellent comparison algorithms, and the obtained population has good convergence, diversity and constraint feasibility. CCGSK-PSO has achieved excellent solution results.

This paper mainly studies the defects and problems of constrained multi-objective optimization algorithm. The experimental results fully show the feasibility and effectiveness of CCGSK-PSO algorithm. However, the algorithm proposed in this paper does not improve the distribution of the population and the environment selection strategy, and the follow-up work will continue to improve in the above aspects.

References

1. Jain, H., Deb, K.: An evolutionary many-objective optimization algorithm using reference-point based non-dominated sorting approach, part II: handling constraints and extending to an adaptive approach. IEEE Trans. Evol. Comput. **18**(4), 602–622 (2014)

2. Li, K., Chen, R., Fu, G., Yao, X.: Two-archive evolutionary algorithm for constrained multi-objective optimization. IEEE Trans. Evol. Comput. **23**(2), 303–315 (2018)

3. Fan, Z., et al.: Push and pull search for solving constrained multi-objective optimization problems. Swarm Evol. Comput. **44**(2), 665–679 (2019)

4. Mohamed, A.W., Hadi, A.A., Mohamed, A.K.: Gaining-sharing knowledge based algorithm for solving optimization problems: a novel nature-inspired algorithm. Int. J. Mach. Learn. Cybern. **11**(7), 1501–1529 (2020)

5. Tian, Y., Zhang, T., Xiao, J., Zhang, X., Jin, Y.: A coevolutionary framework for constrained multi-objective optimization problems. IEEE Trans. Evol. Comput. **25**(1), 102–116 (2021)

6. Ma, Z., Wang, Y.: Evolutionary constrained multiobjective optimization: test suite construction and performance comparisons. IEEE Trans. Evol. Comput. **23**(6), 972–986 (2019)

7. Fan, Z., et al.: An improved epsilon constraint-handling method in MOEA/D for CMOPs with large infeasible regions. Soft. Comput. **23**(23), 12491–12510 (2019). https://doi.org/10.1007/s00500-019-03794-x

8. Fan, Z., et al.: Difficulty adjustable and scalable constrained multi-objective test problem toolkit. Evol. Comput. **28**(3), 339–378 (2020)

9. Deb, K., Pratap, A., Agarwal, S., Meyarivan, T.: A fast and elitist multiobjective genetic algorithm: NSGA-II. IEEE Trans. Evol. Comput. **6**(2), 182–197 (2002)

10. Ma, Z., Wang, Y.: Shift-based penalty for evolutionary constrained multiobjective optimization and its application. IEEE Trans. Cybern. **53**(1), 18–30 (2023)

11. Liu, Z., Wang, Y.: Handling constrained multiobjective optimization problems with constraints in both the decision and objective spaces. IEEE Trans. Evol. Comput. **23**(5), 870–884 (2019)

12. Liu, Y., Yen, G.G., Gong, D.: A multi-modal multi-objective evolutionary algorithm using two-archive and recombination strategies. IEEE Trans. Evol. Comput. **23**(4), 660–674 (2019)

13. Coello, C.A.C., Cortes, N.C.: Solving multiobjective optimization problems using an artificial immune system. Genet. Program. Evolvable Mach. **6**(2), 163–190 (2005)

A Diversified Multi-objective Particle Swarm Optimization Algorithm for Unsupervised Band Selection of Hyperspectral Images

Yuze Zhang⬡, Lingjie Li$^{(\boxtimes)}$⬡, Zhijiao Xiao$^{(\boxtimes)}$⬡, Qiuzhen Lin⬡, and Zhong Ming⬡

Shenzhen University, Nanshan District, Shenzhen 518060, China
{zhangyuze2021,lilingjie2017}@email.szu.edu.cn,
{cindyxzj,qiuzhlin,mingz}@szu.edu.cn

Abstract. Evolutionary algorithm (EA) with good search capability has been successfully extended to a mainstream band selection (BS) technique for hyperspectral images (HSIs). However, most of the existing methods still face two challenges: 1) falling into local optimum due to the single search strategy; 2) ignoring the problem of potential duplicate bands. To address these issues, this paper proposes an effective unsupervised BS method by using a diversified multi-objective particle swarm optimization (PSO) algorithm, called DPSO-BS. First, a new unsupervised BS model is designed, which applies the information entropy and structural similarity measure as two optimization objectives. Then, two complementary PSO search strategies are proposed to solve the above constructed BS model. In addition, a self-repair mechanism is designed to correct the offending solutions with duplicate bands. Experimental results on three HSI datasets demonstrate that DPSO-BS outperforms several state-of-the-art BS methods.

Keywords: Hyperspectral image · Band selection · Particle swarm optimization · Multiobjective optimization

1 Introduction

Hyperspectral image (HSI) remote sensing technology applies hyperspectral sensors to simultaneously image the target area in tens to hundreds of consecutive bands, where the HSI is obtained from these spatial and spectral information by combining images and spectral [2,3]. With the development of HSI remote sensing technology, the application of HSI tends to be extensive, and the amount of information contained in HSI will also be more enormous. Therefore, HSI classification techniques face a series of challenges when handling high-dimensional HSIs. First, the highly redundant data makes the process of training and convergence of the model difficult. Second, as the HSI dimension increases, it may

Y. Tan et al. (Eds.): ICSI 2023, LNCS 13968, pp. 464–475, 2023.
https://doi.org/10.1007/978-3-031-36622-2_38

lead to the *Hughes phenomenon* [9], i.e., the classification accuracy decreases significantly when the data dimension is high with the limited training data.

To avoid these problems, band selection (BS) has become an efficient technique in HSI because it not only helps to reduce the dimensionality of the data, but also improves the accuracy of HSI classification. According to the availability of label information, BS methods can be classified into supervised methods [13], semi-supervised methods [1] and unsupervised methods [5]. Particularly, unsupervised BS methods are preferred over the first two types of BS methods because they do not require labels when processing HSI.

Recently, evolutionary algorithms (EAs) with good search capability have been successfully extended as promising unsupervised BS techniques [16,21], called EA-based BS methods. Specifically, the BS problems are first converted into multi-objective optimization problems (MOPs), and then some effective natural-inspired EAs, such as genetic algorithm [15], artificial bee colony algorithm [21], particle swarm optimization algorithm [12], etc., are employed to explore the optimal subset of bands. Nevertheless, most existing EA-based BS methods still face two challenges: 1) most of them are based on some single search strategy that tend to fall into the local optimum, 2) most of them ignore the problem of potential duplicate bands in the solution vectors.

To address the above issues, this paper proposes an effective unsupervised BS method that applies a diversified multi-objective particle swarm optimization algorithm, called DPSO-BS. First, a new robust unsupervised multi-objective BS model is designed, which adopts the information entropy [14] and the structural similarity measure (SSIM) [20] as two metrics to evaluate the quality of the subset of bands, simultaneously. Specifically, the information entropy is used to measure the amount of information contained in the bands, while the SSIM is adopted to measure the correlation between the bands. Second, to effectively solve the above designed BS model, two complementary particle swarm optimization (PSO) search strategies are designed in DPSO-BS, where the traditional PSO search strategy is used to accelerate the search process, and the SA-PSO search strategy based on the idea of simulated annealing [17] is used to improve the diversity of population to avoid the local optimum. In addition, a self-repair mechanism is designed in DPSO-BS to repair the offending solutions with the duplicate bands.

The structure of this paper is organized as follows. Section 2 describes the designed unsupervised multi-objective BS model. Then, Sect. 3 introduces the details of the proposed DPSO-BS. Section 4 provides the experimental comparison results and discussion. Finally, Sect. 5 gives our conclusions and some potential paths for future research.

2   Proposed Band Selection Model

There are two main aspects to be considered when designing a robust BS model, i.e., the selected bands contain as much information as possible and the correlation between the selected bands is as weak as possible. To this end, this paper proposes a new robust unsupervised multi-objective BS model, which adopts the information entropy to measure the amount of information contained in the bands and the SSIM to measure the correlation between the bands. Mathematically, the designed BS model containing two minimization objective functions is formulated as:

$$
\min F(\mathbf{x}) = \begin{cases} f_1(\mathbf{x}) = 1/\sum\limits_{i=1}^{k} H(x_i) \\ \\ f_2(\mathbf{x}) = \frac{2}{k(k-1)} \sum\limits_{i=1}^{k} \sum\limits_{j=i+1}^{k} S(x_i, x_j) \end{cases}, \qquad (1)
$$

where $\mathbf{x}$ means a solution vector, k is the number of selected bands, and x_i and x_j denote the i-th and j-th bands in $\mathbf{x}$, respectively. Two optimization objective functions contained in the designed BS model are introduced as below.

1) *Information entropy*: $f_1(\mathbf{x})$ describes the amount of information in the selected bands, where $H(x_i)$ denotes the information entropy, computed by,

$$
H(x_i) = -\sum p(x_i^g) \log p(x_i^g)
$$
$$
\text{s.t.} \sum p(x_i^g) = 1 \qquad (2)
$$

where x_i^g denotes the grayscale level in band x_i, $p(x_i^g)$ denotes the proportion of x_i^g in all grayscale levels. For a HSI, each band is considered as the set of outputs of a random variable, and the grayscale histogram of the band as its corresponding probability distribution.

2) *Structural similarity measure (SSIM)*: $f_2(\mathbf{x})$ represents the correlation between the selected bands, where $S(x_i, x_j)$ denotes the SSIM, calculated by:

$$
S(x_i, x_j) = \frac{\left(2\mu_{x_i}\mu_{x_j} + \varepsilon_1\right)\left(2\sigma_{x_i x_j} + \varepsilon_2\right)}{\left(\mu_{x_i}^2 + \mu_{x_j}^2 + \varepsilon_1\right)\left(\sigma_{x_i}^2 + \sigma_{x_j}^2 + \varepsilon_2\right)}, \qquad (3)
$$

where x_i and $x_j \in \Omega$ are two different image signals with size $M \times N$. μ_{x_i} and μ_{x_j} indicate the luminance of image signals, calculated by,

$$
\mu_{x_i} = (1/(M \times N)) \sum\limits_{p=1}^{M \times N} x_{ip}, \qquad (4)
$$

where x_{ip} denotes the grayscale value of the pixel. σ_{x_i} and σ_{x_j} represent the contrast of the image signal, computed by

$$
\sigma_{x_i} = \sqrt{\frac{1}{M \times N - 1} \sum\limits_{p=1}^{M \times N} (x_{ip} - \mu_{x_i})^2}. \qquad (5)
$$

Moreover, as suggested in [20], ε_1 and ε_2 are set as $(0.01 \times L)^2$ and $(0.03 \times L)^2$, respectively, where L is the dynamic range of the pixel values.

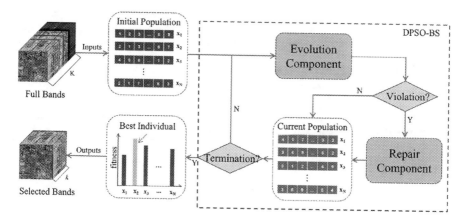

Fig. 1. Schematic of DPSO-BS

3 Methodology

To effectively solve BS model designed in (1), this paper proposes a diversified PSO algorithm, called DPSO-BS, which adopts two different and complementary PSO search strategies to explore a promising subset of bands. The general framework of DPSO-BS is provided in Fig. 1. First, an original HSI dataset consisting of K bands is input and then the initialization process is executed. Specifically, for each individual ($\mathbf{x}_i$) in the initial population, they are encoded with an equal-length integer code, and each gene position of each individual takes values in the integer ranging from 0 to K. For example, $\mathbf{x}(12345)$ indicates that the subset of selected bands is $[1, 2, 3, 4, 5]$. After that, DPSO-BS enters the main evolutionary loop, consisting of two core components (i.e., evolution component and repair component). Particularly, the evolution component aims to evolve the population and generate more individuals of high qualities, and the repair component is used to repair the offending solutions with duplicate bands. Finally, when the termination condition is reached, the individual with the best fitness value is selected from the population, and the bands it contains are output as the final optimal subset of bands.

3.1 Proposed Optimization Algorithm

Algorithm 1 gives the pseudo-code of the proposed DPSO-BS. First, the initialization process is performed in lines 1–2, where the initial population P and some relevant parameters are initialized. Second, DPSO-BS enters the main evolutionary loop, which consists of two different search strategies, a self-repair mechanism, and an alternating operator selection mechanism. To be specific, where two different PSO search strategies (i.e., the traditional PSO search strategy and the simulated annealing-inspired PSO search strategy) are alternatively performed in line 6 and line 8, respectively, introduced as follows:

Algorithm 1. DPSO-BS

Input: Population size N, maximum number of iteration GX, number of selected
  bands k, threshold Q.
Output: The final selected band subset with k bands.
1: Randomly initialize a population: P;
2: Initialize $A = P$, $flag = 1$, $UR_t = 0$;
3: **for** $t = 1 : GX$ **do**
4:    **for** $i = 1 : N$ **do**
5:        **if** $flag == 1$ **then**
6:            Evolve solution $\mathbf{x}_i$ by using **PSO**;
7:        **else**
8:            Evolve solution $\mathbf{x}_i$ by using **SA-PSO**;
9:        **end if**
10:        **if** $\mathbf{x}_i$ is violation **then**
11:            Repair $\mathbf{x}_i$ by **Algorithm 2**;
12:        **end if**
13:        $A = A \cup \mathbf{x}_i$;
14:    **end for**
15:    Obtain NP by the fast non-dominated sorting;
16:    Calculate the current update rate UR_t by (9);
17:    **if** $UR_t < Q$ **then**
18:        $flag = -flag$;
19:    **end if**
20: **end for**
21: Select the solution with best fitness value in NP;

---

1) Traditional PSO search strategy: According to the principle of traditional PSO search strategy [11], the velocity and position of each particle (i.e., $V_i(t)$ and $X_i(t)$) are updated based on its corresponding personal-best and global-best particles (i.e., X_{pbest_i} and X_{gbest_i}), defined as follows:

$$V_i(t + 1) = \omega \times V_i(t) + c_1 \times r_1 \times (X_{pbest_i} - X_i(t))$$
$$+ c_2 \times r_2 \times (X_{gbest} - X_i(t)), \tag{6}$$

$$X_i(t + 1) = X_i(t) + V_i(t + 1), \tag{7}$$

where ω is the inertia weight and is set to 0.5, t is the number of iterations, c_1 and c_2 are two learning factors and both are set to 0.5, and $r_1, r_2 \in [0, 1]$.

2) SA-PSO search strategy: In order to avoid falling into local optimum, this paper proposes a new PSO search strategy, namely the simulated annealing-inspired PSO search strategy (SA-PSO), which aims to improve the diversity by reusing historical solutions from previous iterations. In the proposed SA-PSO, the velocity of each particle is updated by,

$$V_i(t + 1) = \omega \times V_i(t) + F \times (X_{r_1} - X_{r_2}) + F \times (X_{r_3} - X_{r_4}), \tag{8}$$

where F means the scaling factor (we set $F = 0.5$ in this paper). X_{r_1} is randomly selected from the non-dominated population NP at current iteration, X_{r_3} is

---

Algorithm 2. Self-Repair mechanism

Input: Offending solution $\mathbf{x}_i$, non-dominated set NP, number of selected bands k.
Output: Non-offending new solution $\mathbf{x}_{new}$.

1: Randomly select a solution $\mathbf{x}_j$ from NP;
2: $S_1 = \mathbf{x}_i \cap \mathbf{x}_j$;
3: $S_2 = \mathbf{x}_i \cup \mathbf{x}_j$;
4: **if** rand $\leq \sigma$ **then**
5:     $\mathbf{x}_{new} \leftarrow$ Select $|S_1|$ bands from S_1 and $k - |S_1|$ bands from $S_2 \backslash S_1$;
6: **else**
7:     $\mathbf{x}_{new} \leftarrow$ Randomly select k bands from S_2;
8: **end if**

---

randomly selected from the population P, X_{r_2} and X_{r_4} are randomly selected from both the population P and the historical archive A. Then, the position of particles are also updated by using (7).

Furthermore, in order to enable the above two search operators to cooperate better, this paper designs an effective alternating operator selection mechanism, which can select the most suitable search operators according to the current update rate UR_t of the population, calculated by,

$$UR_t = count/N, \tag{9}$$

where *count* represents the number of solutions that are updated by the offspring generated by using the search operator (i.e., PSO or SA-PSO) in the last iteration, and N and t mean the population size and the current iteration, respectively. That is, a smaller value of UR_t implies that the current population is stagnate, as few well promising offspring are generated in the last iteration. Therefore, it is necessary to switch the current search strategy. Specifically, UR_t is compared with a predefined threshold Q (we set $Q = 0.8$ in this paper) and the current search operator is switched if and only if UR_t is less than Q.

After that, as shown in lines 15–16, the non-dominated population NP is obtained by the fast non-dominated sorting method [8] and UR_t are recomputed. Finally, the solution with best fitness function value in NP is selected in line 21 and its corresponding bands are output as the final selected optimal subset of bands. Particularly, the sigmod function is used in the fitness function to balance the impact of two optimization objectives in (1), formulated as follows,

$$fitness(\mathbf{x}) = 1/(1 + e^{f_1(\mathbf{x})}) * 1/(1 + e^{f_2(\mathbf{x})}). \tag{10}$$

3.2 Self-repair Mechanism

Due to the existence of offending solutions that contain duplicate bands, this paper proposes a self-repair mechanism (SRM) to perform an extra repair operation on the offending solutions. Algorithm 2 presents the pseudo-code of the proposed SRM. For an offending solution $\mathbf{x}_i$ to be repaired, the detailed steps

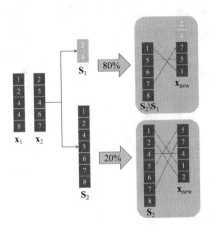

Fig. 2. A simple example of self-repair mechanism.

of the SRM mechanism are as follow. First, a solution $\mathbf{x}_j$ is randomly selected from NP, and then the dominant gene set $S_1 = \mathbf{x}_i \cap \mathbf{x}_j$ and the union gene set $S_2 = \mathbf{x}_i \cup \mathbf{x}_j$ are obtained. After that, as shown in lines 5–9, according to the control probability σ (we set $\sigma = 0.8$ in this paper), two different repair operations are performed alternatively to repair the offending solution and then a new non-offending solution $\mathbf{x}_{new}$ is obtained. Figure 2 gives a simple example of the repair process, where $\mathbf{x}_1$ is the offending individual with five gene positions including two duplicate bands and a individual $\mathbf{x}_2$ is randomly selected from the non-dominant population. Finally, a new and non-offending individual $\mathbf{x}_{new}$ is obtained by using the proposed repair mechanism. By this way, the problem of potential duplicate bands can be addressed.

4   Experimental Studies

4.1   Experimental Setup

Datasets: To verify the performance of the proposed DPSO-BS, three common HSI datasets are adopted in the experimental studies. To be specific, the Indian Pines (IP) dataset contains 16 land covers in 200 bands, with each band consisting of 145 × 145 pixels, the Pavia University (PU) dataset contains 9 land covers, described by 610 × 340 pixels in 103 bands, and the Botswana (BW) dataset has 1476 × 256 pixels in 145 bands to describe 14 land covers. Figure 3 gives the reference maps of these three datasets and the feature categories they contain.

Comparison Methods: Four state-of-the-art methods are adopted for performance comparisons. MVPCA [4] is a joint band prioritization and decorrelation method to select promising band subset. E-FDPC [10] prioritizes each band by maximizing the local density and inter-cluster distance to obtain a subset of bands. ASPS-MN [18] partitions the HSI cube into multiple subcubes and then

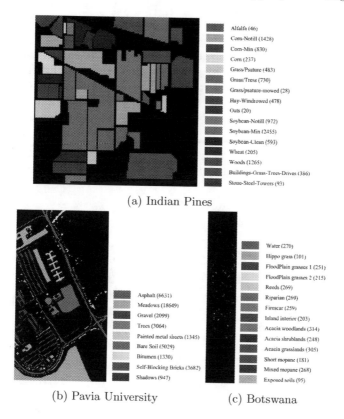

(a) Indian Pines

Alfalfa (46)
Corn-Notill (1428)
Corn-Min (830)
Corn (237)
Grass/Pasture (483)
Grass/Tress (730)
Grass/psature-mowed (28)
Hay-Windrowed (478)
Oats (20)
Soybean-Notill (972)
Soybean-Min (2455)
Soybean-Clean (593)
Wheat (205)
Woods (1265)
Buildings-Grass-Trees-Drives (386)
Stone-Steel-Towers (93)

(b) Pavia University

Asphalt (6631)
Meadows (18649)
Gravel (2099)
Trees (3064)
Painted metal sheets (1345)
Bare Soil (5029)
Bitumen (1330)
Self-Blocking Bricks (3682)
Shadows (947)

(c) Botswana

Water (270)
Hippo grass (101)
FloodPlain grasses 1 (251)
FloodPlain grasses 2 (215)
Reeds (269)
Riparian (269)
Firescar (259)
Island interior (203)
Acacia woodlands (314)
Acacia shrublands (248)
Acacia grasslands (305)
Short mopane (181)
Mixed mopane (268)
Exposed soils (95)

Fig. 3. The reference maps of Indian pines, Pavia University and Botswana

selects the band with the least noise in each subcube. FNGBS [19] selects the promising bands based on the local density and information entropy.

Parameter Settings: k-nearest neighborhood (KNN) [6] and support vector machine (SVM) [7] are used as two classifiers to evaluate the quality of the selected bands, where k is set to 5 for KNN, and the RBF kernel is used and the penalties C and gamma are set to 1×10^4 and 0.5, respectively, for SVM. To be fair, the population size N and the maximum number of iteration GX both are set to 100 for all compared algorithms, and other unique parameters are set according to their corresponding references. In addition, 20% of the samples are randomly selected for training test, and the remaining 80% of the samples are used for testing. Two widely used metrics, i.e., overall accuracy (OA) and Kappa coefficient, are adopted for result comparisons. Note that all the results are collected after twenty independently runs.

4.2   Results and Discussions

Figure 4 provides the classification results of each compared algorithm based on different numbers of selected bands on two classifiers and three HSI datasets. Some conclusions can be drawn from these figures.

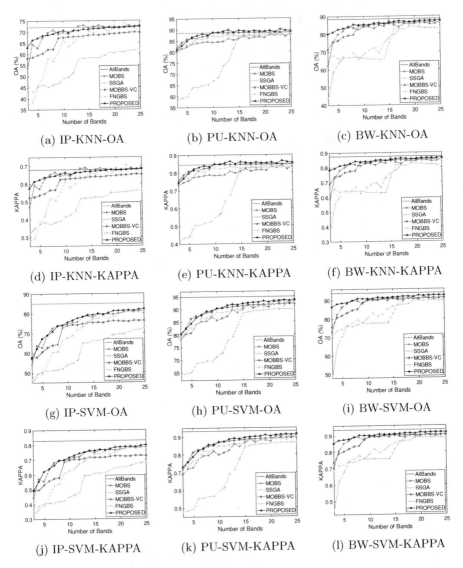

(a) IP-KNN-OA   (b) PU-KNN-OA   (c) BW-KNN-OA

(d) IP-KNN-KAPPA   (e) PU-KNN-KAPPA   (f) BW-KNN-KAPPA

(g) IP-SVM-OA   (h) PU-SVM-OA   (i) BW-SVM-OA

(j) IP-SVM-KAPPA   (k) PU-SVM-KAPPA   (l) BW-SVM-KAPPA

Fig. 4. Results of compared methods with different numbers of selected bands on two classifiers and three HSI datasets.

Table 1. Average classification results (i.e., OA and KAPPA) on two classifiers and three HSI datasets.

Dataset	Classifier	Measure	DPSO-BS	FNGBS	MVPCA	EFDPC	ASPS-MN
IP	KNN	OA	**70.26**	69.83	54.02	66.41	63.74
		KAPPA	**0.659**	0.654	0.474	0.616	0.585
	SVM	OA	**76.20**	75.41	61.27	71.64	70.19
		KAPPA	**0.725**	0.714	0.543	0.670	0.650
PU	KNN	OA	**88.13**	85.26	76.57	87.66	86.56
		KAPPA	**0.840**	0.798	0.679	0.834	0.819
	SVM	OA	**90.43**	88.17	80.00	89.76	88.69
		KAPPA	**0.871**	0.836	0.720	0.862	0.847
BW	KNN	OA	**85.61**	83.19	72.67	83.52	76.48
		KAPPA	**0.844**	0.817	0.704	0.822	0.745
	SVM	OA	**91.08**	89.06	80.60	88.45	84.44
		KAPPA	**0.903**	0.881	0.790	0.875	0.831

1) Results and analysis on IP dataset: DPSO-BS achieves better results overall than the other algorithms under two different classifiers. Specifically, under the KNN classifier, the classification result of FNGBS exceeds the all bands baseline at 13 bands, which is predictable according to the Hughes phenomenon, but decreases when the number of bands increases. DPSO-BS also exceeds the baseline of all bands at 24 bands and achieves superior classification results compared to other algorithms at a smaller number of bands. Considering the results under the SVM classifier, the classification results of all algorithms fail to reach that of all bands baseline, but our proposed DPSO-BS achieves significantly better classification results than other algorithms for both smaller and larger numbers of bands.

2) Results and analysis on PU dataset: First, under the KNN classifier, we can clearly see that only the classification result of DPSO-BS is higher than that of the baseline of all bands after 11 bands, while other compared algorithms show worse classification results than the baseline of all bands in all number of bands. Second, under the SVM classifier, DPSO-BS also outperforms other compared algorithms in all bands, and is only slightly inferior to E-FPDC in bands 4–6.

3) Results and analysis on BW dataset: At any of the two classifiers, the classification results of DPSO-BS are significantly higher than other compared algorithms based on different numbers of selected bands. Specifically, under the KNN classifier, the classification results of DPSO-BS are mostly better than other algorithms, especially achieving significantly better results at a smaller number of bands than other algorithms. Second, under the SVM classifier, it can be seen that the classification results of DPSO-BS are better than other algorithms in all bands.

In addition, the average classification results, including OA and KAPPA, of these algorithms are summarized in Table 1. Obviously, DPSO-BS achieves the best classification results on three different datasets and two different classifiers.

Summarily, experimental results on three HSI datasets (i.e., IP, PU and BW) and two classifiers (i.e., KNN and SVM) demonstrate that the proposed DPSO-BS is superior to other algorithms in terms of the classification accuracy both in different numbers of the selected bands and the overall average accuracy. Moreover, it is worth pointing out that DPSO-BS only needs to select fewer bands to obtain a similar classification accuracy compared to other methods.

5   Conclusion

In this paper, we proposed a new robust unsupervised multi-objective BS model, which adopts the information entropy and SSIM to evaluate the quality of bands. Then, we designed DPSO-BS to effectively address the constructed BS model, which applies two complementary PSO search strategies (i.e., traditional PSO and SA-PSO) to guarantee the exploitation and exploration during the search process, simultaneously. Finally, experimental results on three common HSI datasets (i.e., IP, PU and BW) and two classifiers (i.e., KNN and SVM) demonstrated that our proposed DPSO-BS is superior to several state-of-the-art methods. Regarding the future work, we will investigate the performance of evolutionary transfer optimization for heyperspectral band selection.

Acknowledgment. This work was supported by the National Natural Science Foundation of China (NSFC) under Grants 62002230, 61836005 and 62272315; in part by the Natural Science Foundation of Guangdong Province under grant 2023A1515011296; in part by the Guangdong Regional Joint Foundation Key Project under Grant 2022B1515120076, and in part by the Shenzhen Science and Technology Program under Grants JCYJ20220531101411027 and JCYJ20190808164211203.

References

1. Aghaee, R., Momeni, M., Moallem, P.: Semisupervised band selection from hyperspectral images using levy flight-based genetic algorithm. IEEE Trans. Geosci. Remote Sens. Lett. **19**, 1–5 (2022)
2. Benediktsson, J.A., Chanussot, J., Moon, W.M.: Very high-resolution remote sensing: challenges and opportunities [point of view]. Proc. IEEE **100**(6), 1907–1910 (2012)
3. Bioucas-Dias, J.M., Plaza, A., Camps-Valls, G., Scheunders, P., Nasrabadi, N., Chanussot, J.: Hyperspectral remote sensing data analysis and future challenges. IEEE Geosci. Remote Sens. Mag. **1**(2), 6–36 (2013)
4. Chang, C.I., Du, Q., Sun, T.L., Althouse, M.L.: A joint band prioritization and band-decorrelation approach to band selection for hyperspectral image classification. IEEE Trans. Geosci. Remote Sens. **37**(6), 2631–2641 (1999)
5. Chen, C., Wan, Y., Ma, A., Zhang, L., Zhong, Y.: A decomposition-based multiobjective clonal selection algorithm for hyperspectral image feature selection. IEEE Trans. Geosci. Remote Sens. **60**, 5541516 (2022)

6. Cover, T., Hart, P.: Nearest neighbor pattern classification. IEEE Trans. Inf. Theory **13**(1), 21–27 (1967)
7. Cristianini, N., Shawe-Taylor, J., et al.: An Introduction to Support Vector Machines and Other Kernel-Based Learning Methods. Cambridge University Press, Cambridge (2000)
8. Deb, K., Pratap, A., Agarwal, S., Meyarivan, T.: A fast and elitist multiobjective genetic algorithm: NSGA-II. IEEE Trans. Evol. Comput. **6**(2), 182–197 (2002)
9. Hughes, G.: On the mean accuracy of statistical pattern recognizers. IEEE Trans. Inf. Theory **14**(1), 55–63 (1968)
10. Jia, S., Tang, G., Zhu, J., Li, Q.: A novel ranking-based clustering approach for hyperspectral band selection. IEEE Trans. Geosci. Remote Sens. **54**(1), 88–102 (2015)
11. Kennedy, J., Eberhart, R.: Particle swarm optimization. In: Proceedings of IEEE International Conference on Neural Networks, vol. 4, pp. 1942–1948. IEEE (1995)
12. Paul, A., Chaki, N.: Band selection using spectral and spatial information in particle swarm optimization for hyperspectral image classification. Soft. Comput. **26**(6), 2819–2834 (2022). https://doi.org/10.1007/s00500-022-06821-6
13. Paul, A., Chaki, N.: Supervised data-driven approach for hyperspectral band selection using quantization. Geocarto Int. **37**(8), 2312–2322 (2022)
14. Shannon, C.E.: A mathematical theory of communication. Bell Syst. Tech. J. **27**(3), 379–423 (1948)
15. Singh, P.S., Karthikeyan, S.: Enhanced classification of remotely sensed hyperspectral images through efficient band selection using autoencoders and genetic algorithm. Neural Comput. Appl. **34**(24), 21539–21550 (2022)
16. Song, M., Liu, S., Xu, D., Yu, H.: Multiobjective optimization-based hyperspectral band selection for target detection. IEEE Trans. Geosci. Remote Sens. (2022)
17. Wang, J., Ye, M., Xiong, F., Qian, Y.: Cross-scene hyperspectral feature selection via hybrid whale optimization algorithm with simulated annealing. IEEE J. Sel. Top. Appl. Earth Obs. Remote Sens. **14**, 2473–2483 (2021)
18. Wang, Q., Li, Q., Li, X.: Hyperspectral band selection via adaptive subspace partition strategy. IEEE J. Sel. Top. Appl. Earth Obs. Remote Sens. **12**(12), 4940–4950 (2019)
19. Wang, Q., Li, Q., Li, X.: A fast neighborhood grouping method for hyperspectral band selection. IEEE Trans. Geosci. Remote Sens. **59**(6), 5028–5039 (2020)
20. Wang, Z., Bovik, A.C., Sheikh, H.R., Simoncelli, E.P.: Image quality assessment: from error visibility to structural similarity. IEEE Trans. Image Process. **13**(4), 600–612 (2004)
21. Yong, Z., Chun-lin, H., Xian-fang, S., Xiao-yan, S.: A multi-strategy integrated multi-objective artificial bee colony for unsupervised band selection of hyperspectral images. Swarm Evol. Comput. **60**, 100806 (2021)

Differential Evolution Using Special Sorting for Multimodal Multi-objective Optimization with Local Pareto Sets

Li Yan, Yiran Li, Boyang Qu[✉], Baihao Qiao, Hongxin Duan, and Shunge Guo

School of Electronic and Information, Zhongyuan University of Technology, Zhengzhou, China
quboyang@zut.edu.cn

Abstract. Multimodal multi-objective optimization problems (MMOPs) can be categorized into two types based on the nature of their Pareto sets (PSs): those with global PSs only and those with both global and local PSs. However, most existing multimodal multi-objective evolutionary algorithms (MMEAs) are designed for the former type, and few works focus on the latter type of MMOPs. Thus, this paper proposes an improved differential evolution algorithm that uses a new-designed special sorting to solve the MMOP with local PSs. Specifically, the special sorting is realized based on the global and local non-dominated sorting. By measuring the ranks of individuals in both the global and local regions, the potential local optimal solutions can be preserved while ensuring the global convergence of population. In addition, an elite-based differential mutation strategy is introduced to generate offspring with good convergence and diversity that can improve the search capability of the population. The experimental results on CEC2019 test suite demonstrate the effectiveness and superiority of the proposed algorithm on finding and retaining both global and local optimal solutions.

Keywords: Multimodal multi-objective optimization · Local Pareto set · Special sorting · Differential mutation strategy

1 Introduction

There are many optimization problems in practice, such as logistics distribution [1], feature selection [2], recommendation systems [3], which always involve multiple objectives to be optimized. These problems are called multi-objective optimization problems (MOPs) [4, 5]. In general, these objectives are in conflict with each other, and improving one objective may lead to the deterioration of other objectives. Therefore, finding a single optimal solution for MOPs is not feasible. Instead, the result of multi-objective optimization is a set of mutually non-dominated optimal solutions, known as the Pareto set (PS). The set of objective values corresponding to this PS is referred to as the Pareto front (PF).

A MOP can be considered as a multimodal multi-objective optimization problem (MMOP) [6] when it satisfies one of the following two conditions: 1) The problem has at least two equivalent global PSs. 2) The problem has at least one local PS. Multimodal multi-objective optimization strives to find all Pareto solution sets as completely as possible, aiming to provide more options for decision-makers with different preferences. However, most of the current multimodal multi-objective optimization algorithms (MMEAs) [7–10] focus on the MMOPs only with multiple global PSs. Liang et al. [11] proposed a decision space based niching NSGAII named DN_NSGAII which is able to find almost all global PSs. A ring-topology-based multimodal multi-objective particle swarm algorithm (MO_Ring_PSO_SCD) proposed by Yue et al. [12] employs a non-parameter ring topology niching to prevent premature convergence while proposing a special crowding distance (SCD) to take into account both decision and objective space diversity. A multimodal multi-objective optimization differential evolution algorithm, MMODE, is proposed in [13], which designed a solution pre-selection mechanism and a boundary processing method to improve the diversity of generated offspring. Liu et al. [14] proposed a novel multimodal multi-objective evolutionary algorithm using two-archive and recombination strategies (TriMOEA-TA&R), in which the independence of decision variables is analyzed from a mathematical point of view and two archives of convergence and diversity are introduced to synergistically obtain solutions with excellent performance.

Although the ability to locate all global Pareto solutions is crucial for MMEAs, it is also important to identify both global and local solutions. Decision-makers may tend to sacrifice a slight objective value to make a more desirable choice. To address this, Yue et al. [15] proposed a multimodal multi-objective genetic algorithm (MMOGA), in which the individuals are limited to mate and compete only within the same niching, thereby preserving both global and local solutions. However, since MMOGA relies solely on local information to guide population evolution and environmental selection, it may not always retain the complete Pareto front without the global information.

This paper proposes a differential evolution algorithm using a novel special sorting mechanism (abbreviated to SSDE-MM) for solving MMOPs with local PS. The main contributions of this paper are as follows:

1) A special sorting mechanism is designed to identify both global and local Pareto solutions. The sorting information of each individual within both the global and local is integrated into a new convergence criterion, enabling both global and local PSs to be preserved in environmental selection.

2) An elite-based differential mutation strategy is introduced, which employs elite individuals with superior convergence and diversity located near the current individuals to guide the generation of offspring. This strategy facilitates faster convergence and the improvement of the population diversity.

The rest of this paper is organized as follows: Section 2 presents the relevant definition of MMOP and the basic DE. Section 3 elaborates on the proposed SSDE-MM. Experimental results and related discussions are given in Sect. 4. Section 5 summarizes the whole paper.

2 Preliminary Work

2.1 Related Definitions of MMOP

A minimization MOP can be represented as follows [16]:

$$\min F(x) = \{f_1(x), f_2(x), ..., f_m(x)\} \tag{1}$$
$$s.t. \; x = (x_1, x_2, ..., x_n) \in \Omega$$

where m and n represent the dimensions of objective space and decision space, respectively. For two solutions x_a and x_b, if for $\forall i = 1, 2, ..., m$, there are $f_i(x_a) \leq f_i(x_b)$, and $\exists j \in [1, m]$, let $f_j(x_a) \leq f_j(x_b)$, then x_a is said to dominate x_b. A solution is a global optimal solution if it is not dominated by any other solutions. Moreover, a solution can be regarded as a local optimal solution if it is not dominated by any other solution within its neighborhood [17]. The set of global optimal solutions is the global PS, and the set of local optimal solutions is the local PS. And their mappings in objective space are the global PF and local PF, respectively. A MOP can be defined as a MMOP if it has more than one global PS or at least one local PS in decision space. The two cases of MMOP are shown in Fig. 1.

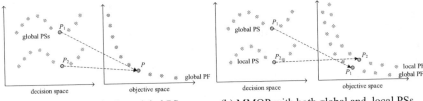

(a) MMOP with equivalent global PSs   (b) MMOP with both global and local PSs

Fig. 1. Two types of MMOPs

2.2 Differential Evolution

Differential evolution (DE) [18, 19] is a classic evolutionary algorithm that is on par with other algorithms such as genetic algorithms (GA) [20], particle swarm optimization (PSO) [21], and ant colony optimization [22], and is known for its simplicity and robustness. Differential evolution differs from other evolutionary algorithms mainly in its unique evolution approach of generating mutant individuals by using the differences between individuals in the current population. In differential evolution, the differential mutation strategy plays a crucial role in generating offspring. Selecting or designing appropriately differential mutation strategies would improve the performance of differential evolution. Specifically, selecting differential individuals to generate differential vectors is crucial for the effectiveness of the differential mutation strategy. Here taking DE/rand/2 and DE/best/2 as the examples, which can be formulated as follows:

DE/rand/2:

$$v_i = x_{r1} + F \cdot (x_{r2} - x_{r3} + x_{r4} - x_{r5}) \tag{2}$$

DE/best/2:

$$v_i = x_{best} + F \cdot (x_{r1} - x_{r2} + x_{r3} - x_{r4}) \tag{3}$$

where v_i is the difference vector, F is the scale factor, x_{r1}, x_{r2}, x_{r3}, x_{r4}, x_{r5} are five different individuals randomly selected from current population. Designing or adopting suitable differential mutation strategies could improve search ability and convergence speed of DE.

3 Proposed Method

The algorithm framework of SSDE-MM is similar to our previous work MMODE. What makes SSDE-MM special is the sorting method for individuals in environmental selection and the proposed differential mutation strategy, which is detailed in the following.

3.1 Global and Local Rank Based Special Sorting

In traditional non-dominated based MMEAs, the global non-dominated rank of individuals is the first criterion for selecting individuals in environmental selection. During the process of population evolution, individuals with lower non-dominated ranks have a greater chance of survival. Even in the final stage of the evolution, it is highly probable that only individuals with the non-dominated rank of 1 can survive, so local PS are difficult to be preserved. As shown in Fig. 2, P1 and P2 are a global optimal solution and a local optimal solution, respectively. Following their distribution in objective space, it can be seen that P_2 is dominated by P_1, so the non-dominated rank of P_2 is higher than that of P_1, and P_2 may be eliminated in environmental selection as the population evolves. However, P_2 is a local optimal solution, namely, it can be regarded as a non-dominated solution in the local region. Seen from the decision space in Fig. 2, it is dominant in its local range, that is, the local non-dominated rank of P_2 is 1. Therefore, if this local rank can be adopted, P_2 can survive through environmental selection.

According to the definition of local optimal solutions given in Sect. 2.1, local optimal solutions are always dominant in their local range. However, the individuals that prevail in local range during the population evolution are not always local optimal solutions, but may be individuals that have not yet converged. If such individuals can also survive through environmental selection, it will slow down the convergence speed and even affect the final convergence of the algorithm. However, since these individuals are not optimal solutions, they often have a poor global non-dominated rank. If the global non-dominated rank of such poor individuals is adopted, they will be eliminated in the process of environmental selection, which will avoid the waste of computing resources and accelerate the convergence speed and convergence of the algorithm.

To maintain both global and local Pareto sets and prevent relying solely on local information, a special sorting criterion based on global and local non-dominated ranks of individuals is proposed, and a new rank value, denoted as GLR, can be calculated and assigned to each individual resultantly:

$$GLR_i = \begin{cases} LR_i, & \text{if } GR_i \leq mean\_GR \\ GR_i, & \text{otherwise} \end{cases} \tag{4}$$

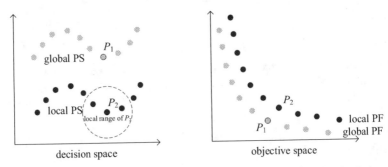

Fig. 2. Relationship between global optimal solution and local optimal solution

where LR_i and GR_i are the local non-dominated rank and global non-dominated rank of the ith individual, respectively. *mean_GR* is the average of the global non-dominated rank of all individuals in population. The new *GLR* not only considers the local information of an individual but also integrates its global information, trying to preserve the local solutions and simultaneously maintain the convergence and completeness of the obtained PSs. In this study, the K-means clustering algorithm [24] is used to partition the local range. The reason for choosing K-means is because it is popular and simple and effective. Of course, other clustering methods or neighborhood partitioning methods can theoretically be integrated into the proposed strategy.

Algorithm 1 *global and local rank based special sorting*

Input: population P, cluster number k
Output: sorted population P

```
1:      [F₁, ..., Fₛ] = non-dominated sorting (P)
2:      [C₁, ..., Cₖ] = K-means clustering algorithm (P)
3:      for i = 1 to k
4:          [R₁, ..., Rₜ] = non-dominated sorting (Cᵢ)
5:      end
6:      Calculate GLR of each individual by Eq. (4)
7:      merge C₁, ..., Cₖ into P
8:      return P
```

The process of special sorting is shown in Algorithm 1. First, the global non-dominated sorting is performed on the entire population to obtain the global non-dominated rank of each individual. Next, the population is divided into k clusters using the K-means algorithm. Then, the local non-dominated sorting is carried out in each cluster to obtain the local non-dominated ranks of individuals. In this way, the GLR of each individual can be calculated according to Eq. 4. Finally, the sorted population P is output.

3.2 Elite-Based Differential Mutation Strategy

In differential evolution, the differential mutation strategy plays a crucial role in generating offspring. By generating offspring with better convergence and diversity, the performance of the algorithm can be improved. The selection of differential individuals to generate differential vectors is critical to the efficiency of the differential mutation strategy. To address this issue, an elite-based differential mutation strategy is proposed:

$$v_i = x_{elite} + F \cdot (x_{r1} - x_{r2} + x_{r3} - x_{r4}) \tag{5}$$

The selection of x_{elite} is the most significant part, and its detailed process is shown in Algorithm 2. First, for the current individual x_i, all individuals in the population with better GLR than GLR_i form the candidate $Pool$ for x_{elite}. Then, individuals with top 50% crowding in $Pool$ are further filtered as the candidate $Pool'$ of x_{elite}. Further, individuals in $Pool'$ belonging to the same cluster as x_i form the candidate $Pool''$. Finally, the x_{elite} for x_i is obtained by roulette selection in $Pool''$ based on their crowding. In addition, x_{r1}, x_{r2}, x_{r3}, and x_{r4} are randomly selected in current population.

Algorithm 2 *selection of elite individuals*

Input: population P; population size NP
Output: *elite* of each individual
1.   **for** $i = 1$ to NP
2. $Pool$ = all individuals in P with $GLR => GLR_i$
3. $Pool'$ = individuals with top 50% crowding in $Pool$
4. $Pool''$ = individuals in $Pool'$ which are in the same cluster as x_i
5.   *elite* = individuals in $Pool''$ perform roulette selection based on crowding
6.   **end**
7.   **return** *elite*

Following this approach, the elite individuals obtained are always in the vicinity of the current individual and exhibit better convergence and diversity. Furthermore, roulette selection ensures that the elite individuals for different individuals are not always the same, thereby ensuring diversity in the generated offspring. The selected elite individuals guide the population to converge quickly and distribute evenly. Moreover, the random selection of the other four differential individuals provides additional information and diversity the differential individuals.

4 Experiments

4.1 Experimental Settings and Comparison Algorithms

CEC2019 test suite was adopted to verify the performance of the proposed SSDE-MM algorithm. CEC2019 test suite includes 22 test functions, among which MMF10, MMF11, MMF12, MMF13, MMF15, and MMF15_a are 6 functions with local PSs. In the experiments, the experimental settings are the same as the initial settings of the test suite. The population size is set to 100*$Nvar$ and the maximum number of evaluations is set to 5000*$Nvar$, $Nvar$ denotes the dimensionality of decision space. The performance of the algorithm was evaluated using the $IGDx$ and $IGDf$ performance metrics in decision space and objective space. Smaller values of them indicate better performance.

The proposed SSDE-MM was compared with five state-of-the-art MMEAs, i.e., DN-NSGAII, MO_Ring_PSO_SCD, MMODE, TriMOEA-TA&R, and MMOGA. The parameter settings of these algorithms are consistent with the original literature. For SSDE-MM, the scale factor F of difference mutations was set to 0.5, and the crossover probability Cr was also set to 0.5. Each algorithm was independently run 31 times on each test function of CEC2019. To observe the difference between SSDE-MM and the compared algorithms, the Wilcoxon rank-sum test with a significance level of $\alpha = 0.05$ was employed.

4.2 Experimental Results and Analysis

Table 1 and Table 2 show the performance of SSDE-MM and five other comparison algorithms on $IGDx$ and $IGDf$. From Table 1, it can be seen that in terms of the $IGDx$ indicator, SSDE-MM ranked first in 14 out of the 22 test problems, and its performance on MMF7 and Omni_test problems was comparable to that of the first-ranked algorithm. It is noteworthy that on problems with local PS, the performance of SSDE-MM is significantly better than the other four algorithms designed to search for global PS, with a difference of more than one order of magnitude. Although MMOGA performs better than the other four algorithms on problems with local PS, it is still inferior to SSDE-MM, and its performance is also inferior to SSDE-MM on problems with only global PS. Overall, SSDE-MM has a significant advantage over other comparison algorithms in the decision space. From Table 2, it can be seen that the performance advantage of SSDE-MM in objective space is not as obvious as in decision space, which may be due to the special sorting mechanism that relaxes the requirements for convergence compared to the non-dominated rank-based sorting. However, it still ranked first in 11 out of the 22 test problems in terms of the $IGDf$ indicator. Especially on test problems with local Pareto solutions, SSDE-MM still has a huge advantage in objective space compared to other comparison algorithms, even compared to MMOGA. Its performance on problems with only global PSs is also not inferior. In summary, SSDE-MM performs well in both decision space and objective space, and has better performance than other comparison algorithms. The global and local optimal solutions found by SSDE-MM algorithm are more comprehensive, and its performance is not inferior to those algorithms specifically designed to search for global PSs.

To provide a more intuitive comparison, Fig. 3 shows the PSs and PFs obtained by SSDE-MM and five other algorithms on MMF11. Comparing the PSs and PFs obtained by SSDE-MM and MMOGA in Fig. 2(a) and Fig. 2(b), it can be seen that the PS and PF obtained by SSDE-MM almost completely cover the true PS and PF, and the diversity is better distributed and more uniform. The PS and PF obtained by MMOGA are relatively sparse, and the convergence is not as good as that of SSDE-MM. From the objective space, MMOGA did not find part of PF. From Fig. 2(c)–(f), it can be seen that the other four algorithms can only find global PS and PF of MMF11, they cannot retain the local PS and PF of the problem. The PS and PF obtained by TriMOEA-TA&R have better convergence, but the upper half of the global PF is not covered. The convergence and diversity of the individuals obtained by the other three algorithms that only search for global PS are comparable to those of SSDE-MM.

Table 1. The *IGDx* values obtained by all algorithms

Problems	MO_Ring_PSO_SCD	DN-NSGAII	MMODE	TriMOEA-TA&R	MMOGA	SSDE-MM
MMF1	0.04843±0.00185(+)	0.0046±0.00099(+)	0.04663±0.0023(+)	0.0738±0.00796(+)	0.08984±0.00979(+)	**0.04201±0.00111**
MMF2	0.03784±0.0072(+)	0.12756±0.08469(+)	0.02478±0.00588(+)	0.07246±0.02826(+)	0.11264±0.04671(+)	**0.01544±0.0032**
MMF3	0.02875±0.00695(+)	0.08566±0.02901(+)	0.02248±0.00727(+)	0.08059±0.03224(+)	0.09288±0.02646(+)	**0.01431±0.00888**
MMF4	**0.0274±0.00152(+)**	0.07919±0.01432(+)	0.029±0.00238(-)	0.14185±0.17802(≈)	0.05382±0.00388(+)	0.03211±0.002
MMF5	0.0848±0.00439(+)	0.17561±0.02008(+)	0.08456±0.00578(+)	0.11285±0.00964(+)	0.15823±0.0146(+)	**0.0742±0.00288**
MMF6	**0.07233±0.00314(+)**	0.14485±0.01972(+)	0.07536±0.00612(+)	0.09208±0.00825(+)	0.12641±0.00902(+)	**0.06529±0.00265**
MMF7	**0.02673±0.00193(≈)**	0.05158±0.01337(+)	0.03074±0.00366(+)	0.04586±0.0185(+)	0.04589±0.00529(+)	0.028±0.00238
MMF8	0.06719±0.00606(≈)	0.23921±0.07603(+)	**0.06624±0.00584(-)**	0.35943±0.14065(+)	0.33796±0.11839(+)	0.07042±0.00464
MMF9	0.00831±0.00058(+)	0.02116±0.00767(+)	0.00652±0.00041(≈)	**0.00313±0.00008(-)**	0.01252±0.00092(+)	0.00662±0.00065
MMF10	0.16898±0.01021(+)	0.17198±0.04741(+)	0.1919±0.01346(+)	0.20138±0.00006(+)	0.06355±0.02219(+)	**0.03353±0.06522**
MMF11	0.21502±0.02395(+)	0.25041±0.00035(+)	0.24909±0.00035(+)	0.25236±0.00006(+)	0.01568±0.00304(+)	**0.00791±0.00075**
MMF12	0.1874±0.04397(+)	0.24669±0.00056(+)	0.24474±0.0005(+)	0.24781±0.00069(+)	0.05045±0.02636(+)	**0.00387±0.0005**
MMF13	0.23796±0.01376(+)	0.28145±0.01803(+)	0.25735±0.00181(+)	0.27256±0.00906(+)	0.15548±0.01521(+)	**0.06911±0.00697**
MMF14	0.0535±0.00111(+)	0.09623±0.00862(+)	0.07829±0.00585(+)	**0.03661±0.00034(-)**	0.07223±0.0025(+)	0.05262±0.00234
MMF15	0.15441±0.01634(+)	0.22639±0.01999(+)	0.21603±0.01339(+)	0.27104±0.00032(+)	0.14486±0.02374(+)	**0.05643±0.00272**
MMF1_z	0.03535±0.00145(+)	0.08045±0.01791(+)	0.03613±0.0038(+)	0.07444±0.01309(+)	0.08729±0.01468(+)	**0.03108±0.00148**
MMF1_e	0.46711±0.12049(+)	0.98398±0.40185(+)	1.73936±0.75(+)	1.71588±0.62889(+)	1.11316±0.24731(+)	**0.26646±0.03569**
MMF14_a	0.06095±0.00206(-)	0.11987±0.00754(+)	0.09798±0.00829(+)	**0.05692±0.00137(-)**	0.08104±0.00432(+)	0.0659±0.00233
MMF15_a	0.16365±0.00898(+)	0.21152±0.01527(+)	0.18499±0.01211(+)	0.2203±0.00161(+)	0.19398±0.01517(+)	**0.07065±0.00494**
SYM_PART_simple	0.19004±0.03234(+)	4.71764±2.16578(+)	0.06792±0.00722(+)	**0.02524±0.02416(-)**	2.80646±1.21577(+)	0.06438±0.00337
SYM_PART_rotated	0.24564±0.22692(+)	4.29166±1.54918(+)	0.07892±0.01093(≈)	2.13494±1.27611(+)	3.63964±1.18036(+)	**0.07863±0.00379**
Omni_test	0.384±0.10028(+)	1.42752±0.20264(+)	**0.67492±0.18434(≈)**	0.26432±0.13265(+)	1.40821±0.21176(+)	0.07268±0.00448
+/≈/-	18/2/2	22/0/0	17/3/2	17/1/4	22/0/0	

"+", "-" and "=" represent that SSDE-MM performs significantly better than, worse than, and similar to the compared algorithms

Table 2. The *IGDf* values obtained by all algorithms

Problems	MO_Ring_PSO_SCD	DN-NSGAII	MMODE	TriMOEA-TA&R	MMOGA	SSDE-MM
MMF1	0.00378±0.00018(+)	0.0046±0.00099(+)	0.00329±0.00014(+)	0.00495±0.00118(+)	0.01093±0.0017(+)	**0.00305±0.00014**
MMF2	0.02082±0.00352(+)	0.02678±0.01838(+)	0.01404±0.00534(≈)	0.02914±0.01932(+)	0.08222±0.04562(+)	**0.01191±0.00278**
MMF3	0.01518±0.0024(+)	0.01567±0.00982(+)	0.01113±0.00212(+)	0.02215±0.0072(+)	0.06262±0.02529(+)	**0.01015±0.00638**
MMF4	0.00361±0.00031(-)	0.00314±0.00026(-)	**0.00305±0.00021(-)**	0.05448±0.08325(-)	0.00974±0.00126(+)	0.00457±0.00022
MMF5	0.00365±0.00015(-)	0.00388±0.00038(-)	**0.00315±0.00011(-)**	0.00457±0.00114(≈)	0.01319±0.00251(+)	0.00412±0.00016
MMF6	0.00355±0.00017(-)	0.00374±0.00033(-)	**0.00303±0.0001(-)**	0.00421±0.00185(-)	0.01047±0.00183(+)	0.00399±0.00024
MMF7	0.00381±0.00027(+)	0.00396±0.00036(+)	**0.00303±0.0001(-)**	0.00424±0.00126(+)	0.00746±0.00102(+)	0.00361±0.00023
MMF8	0.00491±0.00056(-)	**0.00405±0.00084(-)**	0.00425±0.00028(-)	0.00344±0.00015(+)	0.01159±0.00174(+)	0.00528±0.00039
MMF9	0.01588±0.0016(+)	0.01411±0.00206(≈)	0.01318±0.00118(≈)	0.06927±0.00441(+)	0.02615±0.00338(+)	**0.01316±0.00143**
MMF10	0.20539±0.01949(+)	0.20093±0.04161(+)	0.17879±0.01794(+)	0.22907±0.00474(+)	0.11451±0.03477(+)	**0.0414±0.04827**
MMF11	0.08698±0.00646(+)	0.09807±0.00187(+)	0.09168±0.00552(+)	0.16092±0.00946(+)	0.05025±0.01141(+)	**0.02803±0.00285**
MMF12	0.06028±0.00997(+)	0.08332±0.00024(+)	0.08279±0.00217(+)	0.08611±0.00127(+)	0.10582±0.04553(+)	**0.0086±0.00179**
MMF13	0.09749±0.01663(+)	0.14778±0.01625(+)	0.14062±0.01459(+)	0.24448±0.00596(+)	0.07074±0.01356(+)	**0.02534±0.00396**
MMF14	0.07994±0.00196(≈)	0.11024±0.00663(+)	**0.07834±0.00462(-)**	0.08649±0.00105(+)	0.12267±0.00468(+)	0.08018±0.00278
MMF15	0.17328±0.00315(+)	0.21443±0.00724(+)	0.18152±0.00338(+)	0.2065±0.0087(+)	0.20017±0.00718(+)	**0.1194±0.00353**
MMF1_z	0.00361±0.00014(+)	0.00374±0.00053(+)	0.00323±0.00014(≈)	0.00429±0.00068(+)	0.01432±0.00432(+)	**0.00318±0.00015**
MMF1_e	0.01207±0.00177(-)	0.02609±0.0219(≈)	0.00466±0.00071(-)	**0.01095±0.01232(-)**	0.0913±0.0462(+)	0.01716±0.00633
MMF14_a	**0.07763±0.00193(-)**	0.1229±0.01013(+)	0.07977±0.00217(-)	0.07857±0.00127(-)	0.10905±0.00456(+)	0.08354±0.0024
MMF15_a	0.17368±0.00204(+)	0.22659±0.00933(+)	0.17047±0.00335(+)	0.19425±0.00372(+)	0.20747±0.00299(+)	**0.13039±0.00698**
SYM_PART_simple	0.04033±0.00554(+)	0.01284±0.00162(-)	0.01575±0.00143(≈)	0.0318±0.00368(+)	0.13149±0.02913(+)	**0.01574±0.0012**
SYM_PART_rotated	0.04657±0.00717(+)	0.01509±0.0018(-)	**0.01514±0.0014(-)**	0.02915±0.00421(+)	0.17331±0.06313(+)	0.0174±0.00166
Omni_test	0.04233±0.00684(+)	**0.00779±0.00036(-)**	0.01058±0.00094(-)	0.01874±0.0036(+)	0.05822±0.0173(+)	0.0114±0.00078
+/≈/-	15/1/6	13/2/7	8/4/10	17/1/4	22/0/0	

"+", "-" and "=" represent that SSDE-MM performs significantly better than, worse than, and similar to the compared algorithms

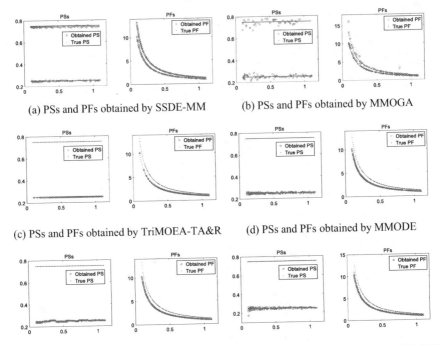

(a) PSs and PFs obtained by SSDE-MM   (b) PSs and PFs obtained by MMOGA

(c) PSs and PFs obtained by TriMOEA-TA&R   (d) PSs and PFs obtained by MMODE

(e) PSs and PFs obtained by DN_NSGAII (f) PSs and PFs obtained by MO_Ring_PSO_SCD

Fig. 3. The PSs and PFs obtained by all the compared algorithms on MMF11

4.3 Parameter Sensitivity Analysis

In this section, we analyze the sensitivity of SSDE-MM on parameter k which determines the number of clusters for the K-means clustering algorithm used in calculating the *GLR*. To investigate the impact of different k values on algorithm performance, we conduct four different experiments with k values of 5, 10, 15, and 20, respectively. Each experiment is run 31 times on all the 22 test functions of the CEC2019 test suite, and the mean values of *IGDx* and *IGDf* are shown in Fig. 4. We observe that different k values do not have a significant effect on algorithm performance for most of the tested problems. However, on a few more challenging problems, the performance of SSDE-MM is affected when k is too small. This may be because when the number of clusters is too small, some local optimal solutions are assigned to the same cluster as the global optimal solutions, so they cannot be retained. Additionally, excessive number of clusters will waste computational resources. Therefore, in this paper, a value of 10 is recommended for the parameter k in SSDE-MM.

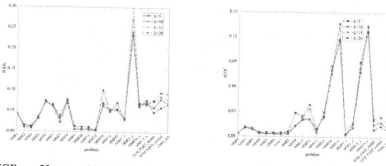

(a) *IGDx* on 22 test problems with different k (b) *IGDf* on 22 test problems with different k

Fig. 4. The *IGDx* and *IGDf* performance of SSDE-MM with different k

4.4 Analysis of the Proposed Elite-Based Differential Mutation Strategy

The two strategies proposed in the SSDE-MM algorithm, i.e., global and local rank based special sorting and elite-based differential mutation strategy. The effectiveness of global and local rank based special sorting in retaining global and local PSs has been validated in Sect. 4.2. In this section, two variants of SSDE-MM were designed to verify the effectiveness of elite-based differential mutation strategy. SSDE-MM-rand refers to SSDE-MM that uses DE/rand/2 differential mutation strategy, while SSDE-MM-best refers to SSDE-MM that uses DE/best/2 differential mutation strategy. Figure 5 shows the distribution of PSs obtained by three algorithms on MMF3. As can be seen from Fig. 5, the PSs obtained by SSDE-MM has better and more complete diversity compared to those obtained by SSDE-MM-rand and SSDE-MM-best, which are relatively sparse and have lower coverage than SSDE-MM. The comparative experimental results of SSDE-MM, SSDE-MM-rand, and SSDE-MM-best fully demonstrate the effectiveness of the elite-based differential mutation strategy.

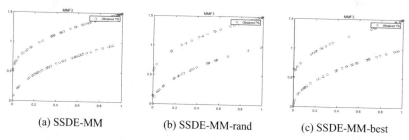

(a) SSDE-MM   (b) SSDE-MM-rand   (c) SSDE-MM-best

Fig. 5. The PSs obtained by SSDE-MM and two variants on MMF3

5   Conclusion

In this paper, we propose a novel differential evolution using special sorting for multimodal multi-objective optimization, which can simultaneously find global and local PSs. The algorithm utilizes a special sorting method that combines global sorting and local sorting to the preserve global and local optimal solutions. Among which, the K-means clustering algorithm is used for local range partitioning. In future studies, we will explore additional local range partitioning methods to assess their impact on algorithm performance. In addition, an elite-based differential mutation strategy is introduced to improve the convergence and diversity of the generated offspring. Finally, the effectiveness of SSDE-MM is experimentally verified. However, there is still room for further performance improvement of SSDE-MM, which will be explored in future studies.

Acknowledgements. This work was supported by the National Natural Science Foundation of China (62103456, 61976237, 61922072, and 61876169), Science and Technology Innovation Team of Colleges and Universities in Henan Province (22IRTSTHN015), "Central Plains Thousand Talents Plan"-Top Talents in Central Plains (ZYQR201810162), Natural Science Foundation of Henan Province (212300410321, 202300410511), Research Award Fund for Outstanding Young Teachers in Henan Provincial Institutions of Higher Education (2021GGJS111, 2020GGJS141).

References

1. Konstantakopoulos, G.D., Gayialis, S.P., Kechagias, E.P.: Vehicle routing problem and related algorithms for logistics distribution: a literature review and classification. Oper. Res. 1–30 (2020)
2. Jha, K., Saha, S.: Incorporation of multimodal multiobjective optimization in designing a filter based feature selection technique. Appl. Soft Comput. **98**, 106823 (2021)
3. Ko, H., Lee, S., Park, Y., Choi, A.: A survey of recommendation systems: recommendation models, techniques, and application fields. Electronics **11**(1), 141 (2022)
4. Deb, K.: Multi-objective optimisation using evolutionary algorithms: an introduction. In: Wang, L., Ng, A., Deb, K. (eds.) Multi-objective Evolutionary Optimisation for Product Design and Manufacturing, pp. 3–34. Springer, London (2011). https://doi.org/10.1007/978-0-85729-652-8_1
5. Giagkiozis, I., Purshouse, R.C., Fleming, P.J.: An overview of population-based algorithms for multi-objective optimisation. Int. J. Syst. Sci. **46**(9), 1572–1599 (2015)
6. Tanabe, R., Ishibuchi, H.: A review of evolutionary multimodal multiobjective optimization. IEEE Trans. Evol. Comput. **24**(1), 193–200 (2019)
7. Liang, J., Guo, Q., Yue, C., Qu, B., Yu, K.: A self-organizing multi-objective particle swarm optimization algorithm for multimodal multi-objective problems. In: Tan, Y., Shi, Y., Tang, Q. (eds.) ICSI 2018. LNCS, vol. 10941, pp. 550–560. Springer, Cham (2018). https://doi.org/10.1007/978-3-319-93815-8_52
8. Yan, L., Li, G.S., Jiao, Y.C., Qu, B.Y., Yue, C.T., Qu, S.K.: A performance enhanced niching multi-objective bat algorithm for multimodal multi-objective problems. In: 2019 IEEE Congress on Evolutionary Computation (CEC), pp. 1275–1282. IEEE (2019)
9. Liang, J., et al.: A clustering-based differential evolution algorithm for solving multimodal multi-objective optimization problems. Swarm Evol. Comput. **60**, 100788 (2021)

10. Qu, B., Li, C., Liang, J., Yan, L., Yu, K., Zhu, Y.: A self-organized speciation based multi-objective particle swarm optimizer for multimodal multi-objective problems. Appl. Soft Comput. **86**, 105886 (2020)

11. Liang, J.J., Yue, C.T., Qu, B.Y.: Multimodal multi-objective optimization: a preliminary study. In: 2016 IEEE Congress on Evolutionary Computation (CEC), pp. 2454–2461. IEEE (2016)

12. Yue, C., Qu, B., Liang, J.: A multiobjective particle swarm optimizer using ring topology for solving multimodal multiobjective problems. IEEE Trans. Evol. Comput. **22**(5), 805–881 (2017)

13. Liang, J., et al.: Multimodal multiobjective optimization with differential evolution. Swarm Evol. Comput. **44**, 1028–1059 (2019)

14. Liu, Y., Yen, G.G., Gong, D.: A multimodal multiobjective evolutionary algorithm using two-archive and recombination strategies. IEEE Trans. Evol. Comput. **23**(4), 660–674 (2018)

15. Yue, C.T., Liang, J.J., Suganthan, P.N., Qu, B.Y., Yu, K.J., Liu, S.: MMOGA for solving multimodal multiobjective optimization problems with local pareto sets. In: 2020 IEEE Congress on Evolutionary Computation (CEC), pp. 1–8. IEEE (2020)

16. Deb, K.: Multi-objective optimization. Search Methodol. **2014**, 403–449 (2014)

17. Deb, K., Pratap, A., Agarwal, S., Meyarivan, T.A.M.T.: A fast and elitist multiobjective genetic algorithm: NSGA-II. IEEE Trans. Evol. Comput. **6**(2), 182–197 (2002)

18. Zelinka, I., Snasael, V., Abraham, A. (eds.): Handbook of Optimization: From Classical to Modern Approach, vol. 38. Springer, Heidelberg (2012). https://doi.org/10.1007/978-3-642-30504-7

19. Pant, M., Zaheer, H., Garcia-Hernandez, L., Abraham, A.: Differential evolution: a review of more than two decades of research. Eng. Appl. Artif. Intell. **90**, 103479 (2020)

20. Katoch, S., Chauhan, S.S., Kumar, V.: A review on genetic algorithm: past, present, and future. Multimedia Tools Appl. **80**(5), 8091–8126 (2020). https://doi.org/10.1007/s11042-020-10139-6

21. Poli, R., Kennedy, J., Blackwell, T.: Particle swarm optimization: an overview. Swarm Intell. **1**, 33–57 (2007)

22. Dorigo, M., Birattari, M., Stutzle, T.: Ant colony optimization. IEEE Comput. Intell. Mag. **1**(4), 28–39 (2006)

23. Li, X., Epitropakis, M.G., Deb, K., Engelbrecht, A.: Seeking multiple solutions: an updated survey on niching methods and their applications. IEEE Trans. Evol. Comput. **21**(4), 518–538 (2016)

24. Hartigan, J.A., Wong, M.A.: Algorithm AS 136: a k-means clustering algorithm. J. Roy. Stat. Soc. Ser. C (Appl. Stat.) **28**(1), 100–108 (1979)

Author Index

© The Editor(s) (if applicable) and The Author(s), under exclusive license
to Springer Nature Switzerland AG 2023
Y. Tan et al. (Eds.): ICSI 2023, LNCS 13968, pp. 489–492, 2023.
https://doi.org/10.1007/978-3-031-36622-2

Printed in the United States
by Baker & Taylor Publisher Services